JACARANDA MATHS QUEST

JACARANDA MATHS QUEST
GENERAL MATHEMATICS 12

VCE UNITS 3 AND 4 | SEVENTH EDITION

JACARANDA MATHS QUEST
GENERAL MATHEMATICS 12

VCE UNITS 3 AND 4 | SEVENTH EDITION

PAULINE HOLLAND

MARK BARNES

JENNIFER NOLAN

GEOFF PHILLIPS

jacaranda
A Wiley Brand

Seventh edition published 2023 by
John Wiley & Sons Australia, Ltd
155 Cremorne Street, Cremorne, Vic 3121

First edition published 2016
Second edition published 2019

Typeset in 10.5/13 pt TimesLTStd

ISBN: 978-1-119-87638-0

The covers of the *Jacaranda Maths Quest VCE Mathematics* series are the work of Victorian artist Lydia Bachimova.

Lydia is an experienced, innovative and creative artist with over 10 years of professional experience, including five years of animation work with Walt Disney Studio in Sydney. She has a passion for hand drawing, painting and graphic design.

Illustrated by diacriTech and Wiley Composition Services

Typeset in India by diacriTech

A catalogue record for this book is available from the National Library of Australia

Printed in Singapore
M121037_110822

Contents

About this resource

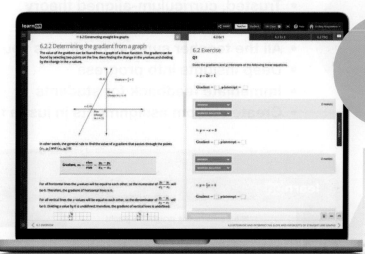

Everything you need
for your students
to succeed

JACARANDA MATHS QUEST
GENERAL MATHEMATICS 12 VCE UNITS 3 AND 4 | SEVENTH EDITION

Developed by expert Victorian teachers for VCE students

Tried, tested and trusted. The NEW Jacaranda VCE Mathematics series continues to deliver curriculum-aligned material that caters to students of all abilities.

Completely aligned to the VCE Mathematics Study Design

Our expert author team of practising teachers and assessors ensures 100 per cent coverage of the new VCE Mathematics Study Design (2023–2027).

Everything you need for your students to succeed, including:

- **NEW!** Access targeted question sets including exam-style questions and all relevant past VCAA exam questions since 2013. Ensure assessment preparedness with practice School-assessed coursework.

- **NEW!** Be confident your students can get unstuck and progress, in class or at home. For every question online they receive immediate feedback and fully worked solutions.

- **NEW!** Teacher-led videos to unpack concepts, plus VCAA exam questions, exam-style questions and worked examples to fill learning gaps after COVID-19 disruptions.

Learn online with Australia's most

- Trusted, curriculum-aligned theory
- Engaging, rich multimedia
- All the teacher support resources you need
- Deep insights into progress
- Immediate feedback for students
- Create custom assignments in just a few clicks.

Practical teaching advice and ideas for each lesson provided in teachON

Each lesson linked to the Key Knowledge (and Key Skills) from the VCE Mathematics Study Design

Reading content and rich media including embedded videos and interactivities

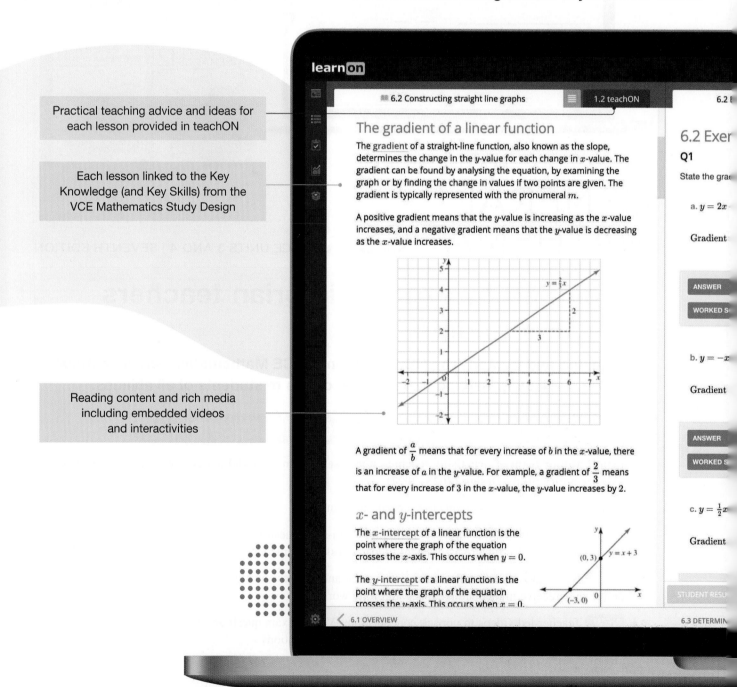

powerful learning tool, learnON

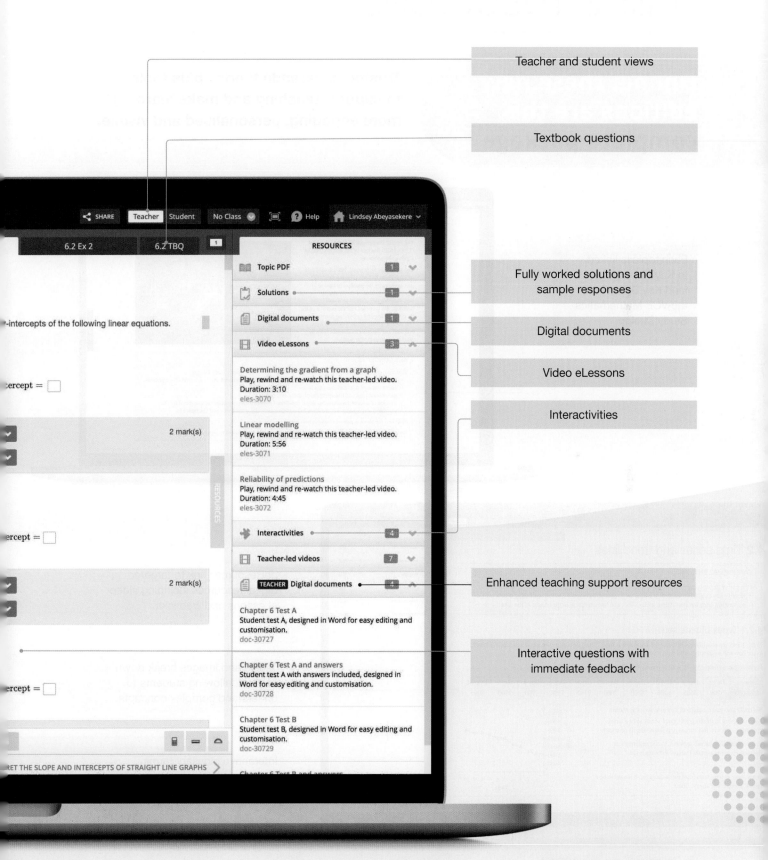

Teacher and student views

Textbook questions

Fully worked solutions and sample responses

Digital documents

Video eLessons

Interactivities

Enhanced teaching support resources

Interactive questions with immediate feedback

Get the most from your online resources

Online, these new editions are the complete package

Trusted Jacaranda theory, plus tools to support teaching and make learning more engaging, personalised and visible.

Each topic is linked to Key Knowledge (and Key Skills) from the VCE Mathematics Study Design.

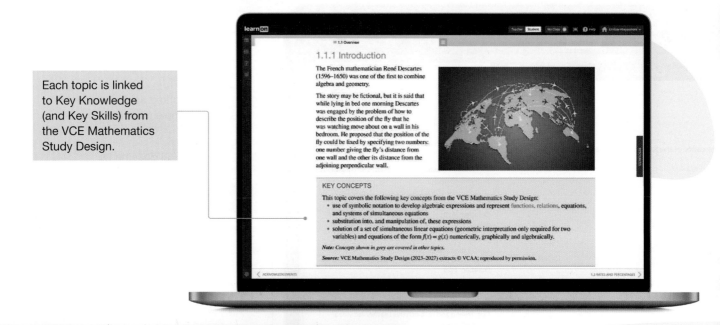

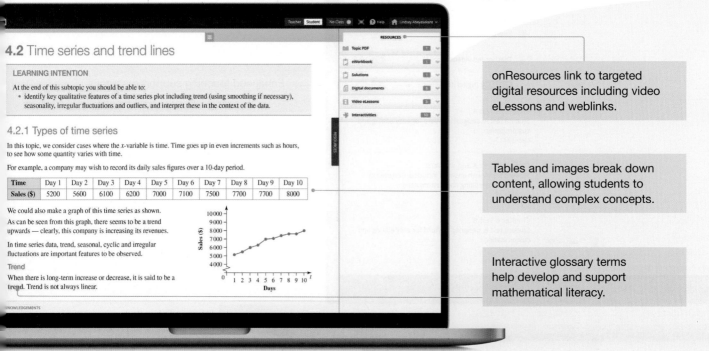

onResources link to targeted digital resources including video eLessons and weblinks.

Tables and images break down content, allowing students to understand complex concepts.

Interactive glossary terms help develop and support mathematical literacy.

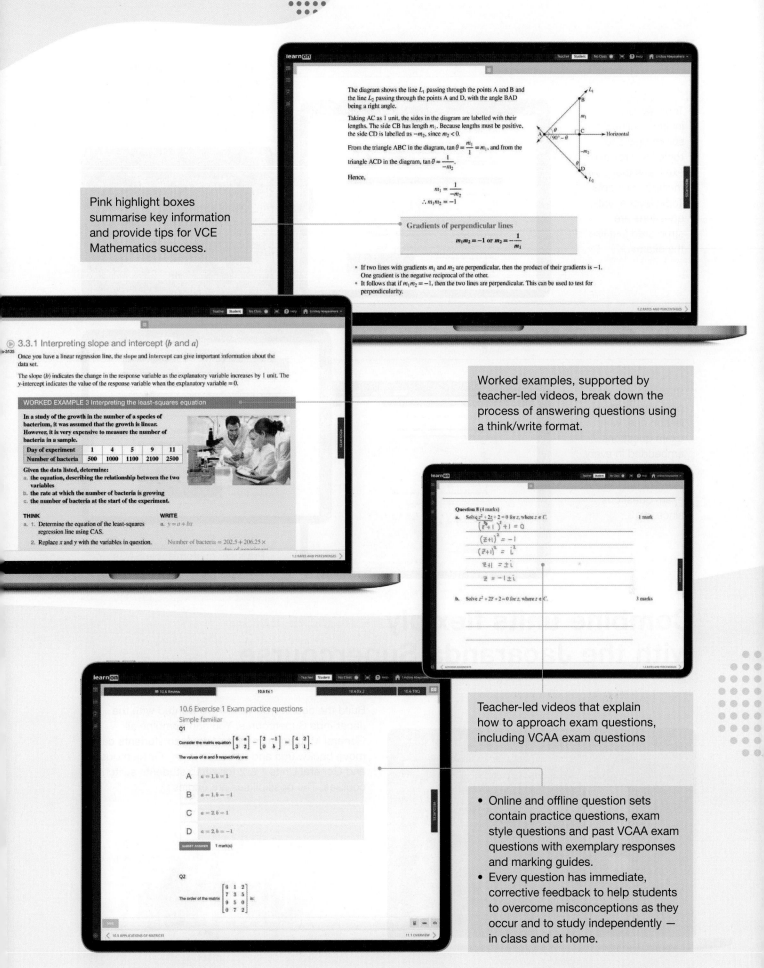

Pink highlight boxes summarise key information and provide tips for VCE Mathematics success.

The diagram shows the line L_1 passing through the points A and B and the line L_2 passing through the points A and D, with the angle BAD being a right angle.

Taking AC as 1 unit, the sides in the diagram are labelled with their lengths. The side CB has length m_1. Because lengths must be positive, the side CD is labelled as $-m_2$, since $m_2 < 0$.

From the triangle ABC in the diagram, $\tan \theta = \frac{m_1}{1} = m_1$, and from the triangle ACD in the diagram, $\tan \theta = \frac{1}{-m_2}$.

Hence,

$$m_1 = \frac{1}{-m_2}$$
$$\therefore m_1 m_2 = -1$$

Gradients of perpendicular lines

$$m_1 m_2 = -1 \text{ or } m_2 = -\frac{1}{m_1}$$

- If two lines with gradients m_1 and m_2 are perpendicular, then the product of their gradients is -1. One gradient is the negative reciprocal of the other.
- It follows that if $m_1 m_2 = -1$, then the two lines are perpendicular. This can be used to test for perpendicularity.

3.3.1 Interpreting slope and intercept (b and a)

Once you have a linear regression line, the slope and intercept can give important information about the data set.

The slope (b) indicates the change in the response variable as the explanatory variable increases by 1 unit. The y-intercept indicates the value of the response variable when the explanatory variable $= 0$.

WORKED EXAMPLE 3 Interpreting the least-squares equation

In a study of the growth in the number of a species of bacterium, it was assumed that the growth is linear. However, it is very expensive to measure the number of bacteria in a sample.

Day of experiment	1	4	5	9	11
Number of bacteria	500	1000	1100	2100	2500

Given the data listed, determine:
a. the equation, describing the relationship between the two variables
b. the rate at which the number of bacteria is growing
c. the number of bacteria at the start of the experiment.

THINK

a. 1. Determine the equation of the least-squares regression line using CAS.

2. Replace x and y with the variables in question.

WRITE

a. $y = a + bx$

Number of bacteria $= 202.5 + 206.25 \times$ day of experiment

Worked examples, supported by teacher-led videos, break down the process of answering questions using a think/write format.

Question 8 (4 marks)

a. Solve $z^2 + 2z + 2 = 0$ for z, where $z \in C$.   1 mark

$$\left(z^2 + 1\right)^2 + 1 = 0$$
$$\left(z + 1\right)^2 = -1$$
$$\left(z + 1\right)^2 = i^2$$
$$z + 1 = \pm i$$
$$z = -1 \pm i$$

b. Solve $z^2 + 2\bar{z} + 2 = 0$ for z, where $z \in C$.   3 marks

10.6 Exercise 1 Exam practice questions
Simple familiar

Q1

Consider the matrix equation $\begin{bmatrix} 6 & a \\ 3 & 2 \end{bmatrix} - \begin{bmatrix} 2 & -1 \\ 0 & b \end{bmatrix} = \begin{bmatrix} 4 & 2 \\ 3 & 1 \end{bmatrix}$.

The values of a and b respectively are:

A $a = 1, b = 1$

B $a = 1, b = -1$

C $a = 2, b = 1$

D $a = 2, b = -1$

SUBMIT ANSWER   1 mark(s)

Q2

The order of the matrix $\begin{bmatrix} 6 & 1 & 2 \\ 7 & 3 & 5 \\ 9 & 5 & 0 \\ 0 & 7 & 2 \end{bmatrix}$ is:

Teacher-led videos that explain how to approach exam questions, including VCAA exam questions

- Online and offline question sets contain practice questions, exam style questions and past VCAA exam questions with exemplary responses and marking guides.
- Every question has immediate, corrective feedback to help students to overcome misconceptions as they occur and to study independently — in class and at home.

Topic reviews

Topic reviews include online summaries and topic level review exercises that cover multiple concepts. Topic level exam questions are structured just like the exams.

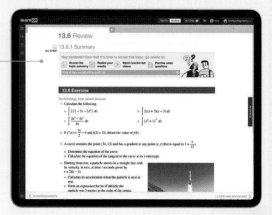

End-of-topic exam questions include relevant past VCE exam questions and are supported by teacher-led videos.

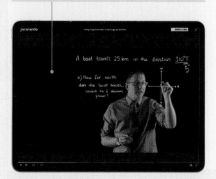

Get exam-ready!

Students can start preparing from lesson one, with exam questions embedded in every lesson — with relevant past VCAA exam questions since 2013.

Practice, customisable SACs available to build student competence and confidence.

Combine units flexibly with the Jacaranda Supercourse

Build the course you've always wanted with the Jacaranda Supercourse. You can combine all General Mathematics Units 1 to 4, so students can move backwards and forwards freely. Or Methods and General Units 1 & 2 for when students switch courses. The possibilities are endless!

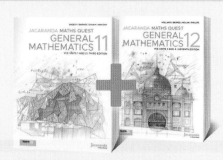

A wealth of teacher resources

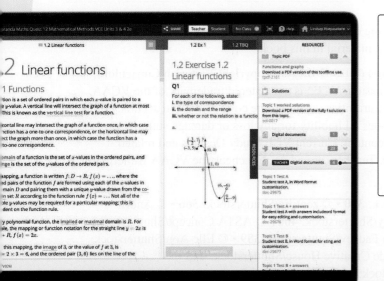

Enhanced teacher support resources, including:
- work programs and curriculum grids
- teaching advice and additional activities
- quarantined topic tests (with solutions)
- quarantined SACs (with worked solutions and marking rubrics)

Customise and assign

A testmaker enables you to create custom tests from the complete bank of thousands of questions (including past VCAA exam questions).

Reports and results

Data analytics and instant reports provide data-driven insights into performance across the entire course.

Show students (and their parents or carers) their own assessment data in fine detail. You can filter their results to identify areas of strength and weakness.

Acknowledgements

The authors and publisher would like to thank the following copyright holders, organisations and individuals for their assistance and for permission to reproduce copyright material in this book.

Selected extracts from the VCE Mathematics Study Design (2023–2027) are copyright Victorian Curriculum and Assessment Authority (VCAA), reproduced by permission. VCE* is a registered trademark of the VCAA. The VCAA does not endorse this product and makes no warranties regarding the correctness and accuracy of its content. To the extent permitted by law, the VCAA excludes all liability for any loss or damage suffered or incurred as a result of accessing, using or relying on the content. Current VCE Study Designs and related content can be accessed directly at www.vcaa.vic.edu.au. Teachers are advised to check the VCAA Bulletin for updates.

• © Maridav/Shutterstock: **260** • © sylv1rob1/Shutterstock: **350** • © zentilia/Shutterstock: **366** • © Andrii Yalanskyi/Shutterstock: **397** • © Andy Dean Photography/Shutterstock: **400** • © ASTA Concept/Shutterstock: **360** • © Daxiao Productions/Shutterstock: **381** • © Dima Fadeev/Shutterstock: **389** • © Elle Aon/Shutterstock: **396** • © Hryshchyshen Serhii/Shutterstock: **403** • © Inside Creative House/Shutterstock: **448** • © insta_photos/Shutterstock: **395** • © Jacob Lund/Shutterstock: **411** • © LightField Studios/Shutterstock: **407** • © NDAB Creativity/Shutterstock: **408** • © Teerawat Anothaistaporn/Shutterstock: **401** • © Yuriy K/Shutterstock: **423** • Monkey Business Images/Shutterstock: **153** • NickJulia/Shutterstock: **2** • Studio Romantic/Shutterstock: **430** • wavebreakmedia/Shutterstock: **283** • © 2xSamara.com/Shutterstock: **112** • © adriaticfoto/Shutterstock: **530** • © aelitta/Shutterstock: **556** • © Alex Cimbal/Shutterstock: **204** • © Alex Mit/Shutterstock: **337** • © alexandar.design/Shutterstock: **386** • © Alexander Lukatskiy/ Shutterstock: **27** • © Alexander Raths/Shutterstock: **92** • © Alexander Tolstykh/Shutterstock: **461** • © ALPA POD/Shutterstock: **661** • © altafulla/Shutterstock: **429** • © Anatoliy Kosolapov/Shutterstock: **451** • © Andrii Yalanskyi/Shutterstock: **325** • © Angelo Giampiccolo/Shutterstock: **63** • © Anna Jedynak/Shutterstock: **337** • © Anton Balazh/Shutterstock: **617** • © Anton_AV/Shutterstock: **365** • © anythings/Shutterstock: **89** • © ASTA Concept/Shutterstock: **576** • © Atstock Productions/Shutterstock: **369** • © BearFotos/Shutterstock: **493, 494** • © Blaj Gabriel/Shutterstock: **60, 662** • © bogdanhoda/Shutterstock: **349** • © bonchan/Shutterstock: **6** • © Catarina Belova/Shutterstock: **413** • © ChameleonsEye/Shutterstock: **344** • © Charles Smith/Corbis/VCG/Getty Images: **599** • © Cherry-Merry/Shutterstock: **14** • © Daisy Daisy/Shutterstock: **532** • © Dar1930/Shutterstock: **155** • © David Tadevosian/Shutterstock: **17** • © Dean Drobot/Shutterstock: **413** • © DenPhotos/Shutterstock: **494** • © Digital Storm/Shutterstock: **340** • © Digital Vision: **422** • © DimaBerlin/Shutterstock: **428** • © Dmitry Kalinovsky/Shutterstock: **346** • © Dmytro Zinkevych/Shutterstock: **662** • © Don Mammoser/Shutterstock: **23** • © dotsock/Shutterstock: **10** • © DOUGBERRY/iStockphoto: **33** • © Dragon Images/Shutterstock: **121** • © Duet PandG/Shutterstock: **531** • © Elenamiv/Shutterstock: **237** • © Elizaveta Galitckaia/Shutterstock: **194** • © EQRoy/Shutterstock: **660** • © Eric Isselee/Shutterstock: **194** • © Eric Isselee/Shutterstock: **37** • © Erica Lorimer Images/Shutterstock: **427** • © Ermolaev Alexander/Shutterstock: **481** • © ESB Professional/Shutterstock: **36, 110, 246** • © ESB Professional/Shutterstock: **91** • © Evgeny Atamanenko/Shutterstock: **26** • © f11photo/Shutterstock: **160** • © fizkes/Shutterstock: **113, 433** • © Flashon Studio/Shutterstock: **165, 402** • © FotoDuets/Shutterstock: **328** • © Fotokostic/Shutterstock: **435** • © GraphEGO/Shutterstock: **121** • © Gurgen Bakhshetyan/Shutterstock: **443** • © hedgehog94/Shutterstock: **382, 489** • © HI_Pictures/Shutterstock: **462** • © Hong Vo/Shutterstock: **59** • © Ian 2010/Shutterstock: **218** • © ifong/Shutterstock: **336** • © Image 100: **262** • © Inside Creative House/Shutterstock: **448** • © insta_photos/Shutterstock: **425** • © Jackson Stock Photography/Shutterstock: **493** • © James Steidl/Shutterstock: **359** • © Jan Kratochvila/Shutterstock: **132** • © Jandrie Lombard/Shutterstock: **67** • © jazz3311/Shutterstock: **619, 624** • © Jeffrey Schmieg/Shutterstock: **338** • © Jiri Hera/Shutterstock: **80** • © Kanitta Kuha/Shutterstock: **555** • © katatonia82/Shutterstock: **676** • © Katty2016/Shutterstock: **22** • © Kevin Wells Photography/Shutterstock: **30** • © kurhan/Shutterstock: **444** • © Kzenon/Shutterstock: **669** • © Lee Torrens/Shutterstock: **84** • © LightField Studios/Shutterstock: **185, 186** • © Ljupco Smokovski/Shutterstock: **437** • © lucadp/Shutterstock: **618** • © M. Unal Ozmen/Shutterstock: **286** • © Makistock/Shutterstock: **370** • © Maks Narodenko/Shutterstock: **289** • © marijaf/Shutterstock: **328** • ©

mariva2017/Shutterstock: **88** • © Matej Kastelic/Shutterstock: **668** • © Max kegfire/Shutterstock: **34** • © Maxim Tupikov/Shutterstock: **600** • © Maxx-Studio/Shutterstock: **343, 495** • © Mejini Neskah/Shutterstock: **119** • © metriognome/Shutterstock: **127** • © Michael Leslie/Shutterstock: **68** • © michaeljung/Shutterstock: **667** • © michelaubryphoto/Shutterstock: **46** • © Mikbiz/Shutterstock: **481** • © Milleflore Images/Shutterstock: **126** • © Milos Vucicevic/Shutterstock: **56** • © mokjc/Shutterstock: **119** • © Monkey Business Images/Shutterstock: **120, 417, 533** • © Monkey Business Images/Shutterstock: **43** • © My Ocean Production/Shutterstock: **9** • © Neale Cousland/Shutterstock: **151, 439, 482** • © Nejron Photo/Shutterstock: **664** • © NeoLeo/Shutterstock: **524** • © New Africa/Shutterstock: **128** • © No formal credit line required.: **470, 558, 575** • © oblong1/Shutterstock: **534** • © Oksana Shufrych/Shutterstock: **13** • © Oleksandr Lysenko/Shutterstock: **416** • © Oleksiy Mark/Shutterstock: **529** • © Olga Kashubin/Shutterstock: **420** • © oliveromg/Shutterstock: **284** • © OSTILL is Franck Camhi/Shutterstock: **134** • © Patryk Kosmider/Shutterstock: **246** • © PHOTOCREO Michal Bednarek/Shutterstock: **238** • © Phovoir/Shutterstock: **187, 196** • © Pressmaster/Shutterstock: **341** • © Protasov AN/Shutterstock; cowboy54/Shutterstock; Peter Waters/Shutterstock: **678** • © puhhha/Shutterstock: **666** • © Quad Design/Shutterstock: **333** • © Rachel Brunette/Shutterstock: **182** • © ranjith ravindran/Shutterstock: **109** • © Rawpixel.com/Shutterstock: **327** • © Rawpixel.com/Shutterstock: **69, 452** • © RIA Novosti/Alamy: **516** • © Rido/Shutterstock: **4, 451, 522** • © Ronnachai Palas/Shutterstock: **4** • © Samuel Borges Photography/Shutterstock: **268** • © Sashkin/Shutterstock: **222** • © sasimoto/Shutterstock: **189** • © Sergey Nivens/Shutterstock: **1** • © Sergey Novikov/Shutterstock: **71** • © Sergiy Bykhunenko/Shutterstock: **191** • © Sergiy Nigeruk/Shutterstock: **415** • © simona pilolla 2/Shutterstock: **320** • © Singkham/Shutterstock: **330** • © SofiaV/Shutterstock: **557** • © Sonate/Shutterstock: **377** • © SpeedKingz/Shutterstock: **426** • © Standret/Shutterstock: **142** • © steamroller_blues/Shutterstock: **336** • © studiovin/Shutterstock: **81** • © Supertrooper/Shutterstock: **345** • © Svetlana Foote/Shutterstock: **6** • © Svetography/Shutterstock: **90** • © Syda Productions/Shutterstock: **45, 56, 66, 425** • © syhun/Shutterstock: **35** • © sylv1rob1/Shutterstock: **344** • © Take Photo/Shutterstock: **11** • © Tatyana Vyc/Shutterstock: **5** • © The_Molostock/Shutterstock: **427** • © tommaso79/Shutterstock: **415, 628** • © Tony Bowler/Shutterstock: **15** • © TonyNg/Shutterstock: **647** • © Tsekhmister/Shutterstock: **188** • © TsipiLevin/Shutterstock: **319** • © Tyler Olson/Shutterstock: **521** • © UfaBizPhoto/Shutterstock: **195** • © Unkas Photo/Shutterstock: **634** • © Urbanscape/Shutterstock: **256** • © v.gi/Shutterstock: **414** • © Valentin Valkov/Shutterstock: **255** • © Valentyn Volkov/Shutterstock: **255** • © Vdovichenko Denis/Shutterstock: **540** • © Veronica Louro/Shutterstock: **453** • © Vladi333/Shutterstock: **181** • © wavebreakmedia/Shutterstock: **222, 267** • © Wittybear/Shutterstock: **428** • © woe/Shutterstock: **512** • © Yellow duck/Shutterstock; pattang/Shutterstock; Denis Tabler/Shutterstock; Nattika/Shutterstock; rangizzz/Shutterstock: **3** • © Yuganov Konstantin/Shutterstock: **294** • © zlikovec/Shutterstock: **30** • ©Tyler Olson/Shutterstock: **70**

Every effort has been made to trace the ownership of copyright material. Information that will enable the publisher to rectify any error or omission in subsequent reprints will be welcome. In such cases, please contact the Permissions Section of John Wiley & Sons Australia, Ltd.

1 Investigating data distributions

LEARNING SEQUENCE

Fully worked solutions for this topic are available online.

1.1 Overview

1.1.1 Introduction

Due to advances in technology, the collection, analysis and interpretation of data is at an unprecedented level in the world today. These advances ensure that after the collection of large datasets, the analysis and results can be distributed very quickly. It is very important that future planning and decisions are based on current empirical data as change is fast-paced and costly in the modern world. In 1880, a census conducted in the USA took 7 years to publish the information collected. The very detailed dataset collected by the Australian census in August 2016 was released in June 2017.

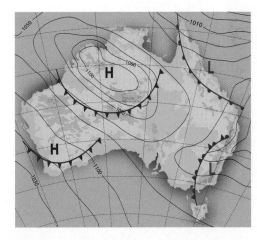

Data has the power to initiate change, and businesses and governments ignore it at their peril. Meteorologists, urban planners, health scientists, traffic engineers, pharmaceutical companies, environmental scientists, educators and agriculturalists use data to monitor change and make future predictions. These changes may be as small as choosing to block off a suburban street to stop too many people using it as a short cut and creating traffic problems, or as big as ensuring there is enough food production to feed an ever-increasing world population. Decisions that require major societal change or large investment need to be justified. Empirical evidence can provide this justification; therefore, it is important that everyone has some knowledge and understanding of the collection and interpretation of data.

KEY CONCEPTS

This topic covers the following key concepts from the VCE Mathematics Study Design:
- types of data
- representation, display and description of the distributions of categorical variables: data tables, two-way frequency tables and their associated segmented bar charts
- representation, display and description of the distributions of numerical variables: dot plots, stem plots, histograms; the use of a logarithmic (base 10) scale to display data ranging over several orders of magnitude and their interpretation in terms of powers of ten
- use of the distribution(s) of one or more categorical or numerical variables to answer statistical questions
- summary of the distributions of numerical variables; the five-number summary and boxplots (including the use of the lower fence ($Q1 - 1.5 \times IQR$) and upper fence ($Q3 + 1.5 \times IQR$) to identify and display possible outliers); the sample mean and standard deviation and their use in comparing data distributions in terms of centre and spread
- the normal model for bell-shaped distributions and the use of the 68–95–99.7% rule to estimate percentages and to give meaning to the standard deviation; standardised values (z-scores) and their use in comparing data values across distributions.

Source: VCE Mathematics Study Design (2023–2027) extracts © VCAA; reproduced by permission.

1.2 Types of data

1.2.1 Types of data

Data can be classified as either **numerical** or **categorical**. The methods we use to display data depend on the type of information we are dealing with.

There are two types of categorical data:
- **nominal** data
- **ordinal** data.

There are two types of numerical data:
- **discrete** data
- **continuous** data.

Determining types of data

The information in the flow chart shown can be used to determine the type of data being considered.

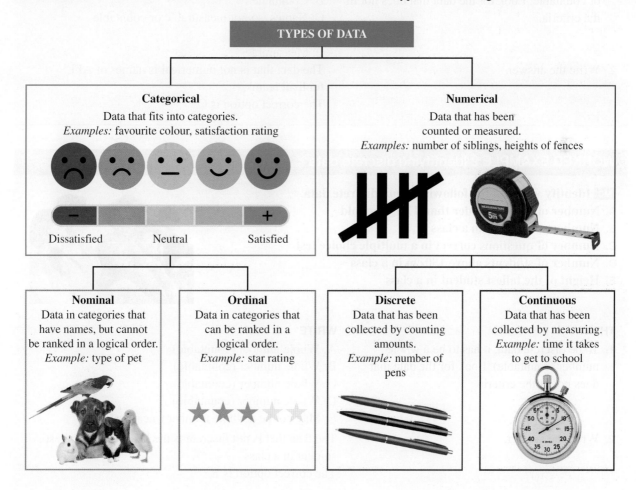

1.2.2 Discrete and continuous data

Data are said to be *discrete* when a variable can take only certain fixed values. For example, if we counted the number of children per household in a particular suburb, the data obtained would always be whole numbers starting from 0. A value in between, such as 2.5, would clearly not be possible. If objects can be counted, then the data are discrete.

Continuous data are obtained when a variable takes any value between two values. If the heights of students in a school were obtained, then the data could consist of any values between the smallest and largest heights. The values recorded would be restricted only by the precision of the measuring instrument. If variables can be measured, then the data are continuous.

tlvd-3530

WORKED EXAMPLE 1 Identifying numerical data

MC Identify which of the following is *not* numerical data.

A. Maths test results
B. Ages (in years)
C. Names of AFL football teams
D. Heights of students in a class
E. Lengths of bacteria

THINK	WRITE
1. To be numerical data, it has to be measurable or countable. Look for the data that does not fit the criteria.	A: Measurable B: Countable C: Names, so not measurable or countable D: Measurable E: Measurable
2. Write the answer.	The data that is not numerical is names of AFL football teams. The correct option is C.

WORKED EXAMPLE 2 Identifying discrete data

MC Identify which of the following is *not* discrete data.

A. Number of students older than 17.5 years old
B. Number of children in a class
C. Number of questions correct in a multiple choice test
D. Number of students above 180 cm in a class
E. Height of the tallest student in a class

THINK	WRITE
1. To be discrete data, it has to be a whole number (countable). Look for the data that does not fit the criteria.	A: Whole number (countable) B: Whole number (countable) C: Whole number (countable) D: Whole number (countable) E: May not be a whole number (measurable)
2. Write the answer.	The data that is not discrete is the height of the tallest student in a class. The correct option is **E**.

1.2 Exercise

1. **MC** **WE1** Identify which of the following is *not* numerical data.
   - A. Number of students in a class
   - B. The number of supporters at an AFL match
   - C. The amount of rainfall in a day
   - D. Finishing positions in the Melbourne Cup
   - E. The number of coconuts on a palm tree

2. **MC** Identify which of the following is *not* categorical data.
   - A. Preferred political party
   - B. Gender
   - C. Hair colour
   - D. Salaries
   - E. Religion

3. Write whether each of the following represents numerical or categorical data.
   - a. The heights, in centimetres, of a group of children
   - b. The diameters, in millimetres, of a collection of ball bearings
   - c. The numbers of visitors to an exhibit each day
   - d. The modes of transport that students in Year 12 take to school
   - e. The 10 most-watched television programs in a week
   - f. The occupations of a group of 30-year-olds

4. Identify which of the following represent categorical data.
   - a. The number of subjects offered to VCE students at various schools
   - b. Life expectancies
   - c. Species of fish
   - d. Blood groups
   - e. Years of birth
   - f. Countries of birth
   - g. Tax brackets

5. **MC** **WE2** Identify which of the following is *not* discrete data.
   - A. Number of players in a netball team
   - B. Number of goals scored in a football match
   - C. The average temperature in March
   - D. The number of Melbourne Storm members
   - E. The number of twins in Year 12

6. **MC** Identify which of the following is *not* continuous data.
   - A. The weight of a person
   - B. The number of shots missed in a basketball game
   - C. The height of a sunflower in a garden
   - D. The length of a cricket pitch
   - E. The time taken to run 100 m

7. For each set of numerical data identified in Question **3** above, state whether the data are discrete or continuous.

8. **MC** An example of a numerical variable is:

   A. attitude to 4-yearly elections (for or against).
   B. year level of students.
   C. the total attendance at Carlton football matches.
   D. position in a queue at the pie stall.
   E. television channel numbers shown on a dial.

9. **MC** The weight of each truck-load of woodchips delivered to the wharf during a one-month period was recorded. This is an example of:

   A. categorical and discrete data.
   B. discrete data.
   C. continuous and numerical data.
   D. continuous and categorical data.
   E. numerical and discrete data.

10. When reading the menu at a local Chinese restaurant, you notice that the dishes are divided into sections. The sections are labelled chicken, beef, duck, vegetarian and seafood. State what type of data this is.

11. **MC** NASA collects data on the distance to other stars in the universe. The distance is measured in light years. State what type of data is being collected.

   A. Discrete
   B. Continuous
   C. Nominal
   D. Ordinal
   E. Bivariate

12. **MC** The number of blue, red, yellow and purple flowers in an award-winning display is counted. State what type of data is being collected.

   A. Nominal
   B. Ordinal
   C. Discrete
   D. Continuous
   E. Bivariate

13. Students in a performing arts class watch a piece of modern dance and are then asked to rate the quality of the dance as poor, average, above average or excellent. State what type of data is being collected.

14. Given the set of data: 12, 6, 21, 15, 8, 2, describe what type of numerical data this data set is.

15. If a tennis tournament seeds the players to organise the draw, state what type of categorical data this is.

1.2 Exam questions

Question 1 (1 mark)

Source: VCE 2021, Further Mathematics Exam 1, Section A Core, Q1; © VCAA

MC The percentaged segmented bar chart below shows the *age* (under 55 years, 55 years and over) of visitors at a travel convention, segmented by *preferred travel destination* (domestic, international).

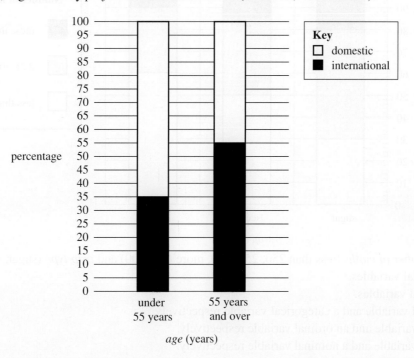

The variables *age* (under 55 years, 55 years and over) and *preferred travel destination* (domestic, international) are

A. both categorical variables.
B. both numerical variables.
C. a numerical variable and a categorical variable respectively.
D. a categorical variable and a numerical variable respectively.
E. a discrete variable and a continuous variable respectively.

Question 2 (1 mark)

Source: VCE 2020, Further Mathematics Exam 1, Section A, Q7; © VCAA

MC Data relating to the following five variables was collected from insects that were caught overnight in a trap:

- *colour*
- *name of species*
- *number of wings*
- *body length* (in millimetres)
- *body weight* (in milligrams)

The number of these variables that are discrete numerical variables is

A. 1
B. 2
C. 3
D. 4
E. 5

▷ **Question 3 (1 mark)**

Source: VCE 2017, Further Mathematics Exam 1, Section A, Core, Q7; © VCAA

MC A study was conducted to investigate the association between the *number of moths* caught in a moth trap (less than 250, 250–500, more than 500) and the *trap type* (sugar, scent, light). The results are summarised in the percentaged segment bar chart below.

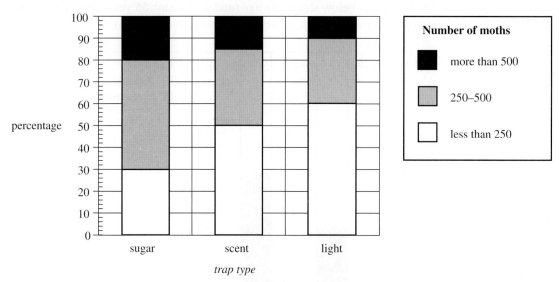

The variables *number of moths* (less than 250, 250–500, more than 500) and *trap type* (sugar, scent, light) are
- **A.** both nominal variables.
- **B.** both ordinal variables.
- **C.** a numerical variable and a categorical variable respectively.
- **D.** a nominal variable and an ordinal variable respectively.
- **E.** an ordinal variable and a nominal variable respectively.

More exam questions are available online.

1.3 Stem plots

LEARNING INTENTION

At the end of this subtopic you should be able to:
- construct 2-digit, 3-digit and split stem plots.

1.3.1 Constructing stem plots

A *stem-and-leaf plot*, or **stem plot** for short, is a way of ordering and displaying a set of data, with the advantage that all of the raw data is kept. Since all individual values are listed, it is only suitable for smaller data sets (up to about 50 observations).

The stem plot shows the ages of people attending an advanced computer class.

Stem	Leaf
1	6
2	2 2 3
3	0 2 4 6
4	2 3 6 7
5	3 7
6	1

Key: 2 | 2 = 22 years old

The ages of the members of the class are 16, 22, 22, 23, 30, 32, 34, 36, 42, 43, 46, 47, 53, 57 and 61.

Stem plots

A stem plot is constructed by splitting the numerals of a record into two parts — the stem, which in the above case is the first digit, and the leaf, which is always the last digit.

With every stem-and-leaf plot it is important to include a key so it is clear what the data values represent.

In cases where numerous leaves are attached to one stem (meaning that the data is heavily concentrated in one area), the stem can be subdivided. Stems are commonly subdivided into halves or fifths. By splitting the stems, we get a clearer picture about the data variation.

The following stem plots show the splitting of the stems into halves and fifths.

Stems subdivided into halves

Stem	Leaf
4	3
4*	7 7 8 8 9 9 9
5	0 0 0 0 1 2 2 3
5*	

Key: 4 | 3 = 43 cm

Stems subdivided into fifths

Stem	Leaf
2	0
2	2 2
2	5
2	6
2	8 8
3	
3	3 3 3
3	
3	7 7
3	8 9 9 9

Key: 2 | 0 = 20 cm

tlvd-3531

WORKED EXAMPLE 3 Constructing a 2-digit stem plot

The number of cars sold in a week at a large car dealership over a 20-week period is shown.

16, 12, 8, 7, 26, 32, 15, 51, 29, 45,
19, 11, 6, 15, 32, 18, 43, 31, 23, 23

Construct a stem plot to display the number of cars sold in a week at the dealership.

THINK

1. In this example the observations are one- or two-digit numbers and so the stems will be the digits referring to the 'tens', and the leaves will be the digits referring to the units.

   Work out the lowest and highest numbers in the data to determine what the stems will be.

WRITE

Lowest number = 6
Highest number = 51
Use stems from 0 to 5.

2. Before we construct an ordered stem plot, construct an unordered stem plot by listing the leaf digits in the order they appear in the data.

Stem	Leaf
0	876
1	6259158
2	6933
3	221
4	53
5	1

3. Now rearrange the leaf digits in numerical order to create an ordered stem plot.

Include a key so that the data can be understood by anyone viewing the stem plot.

Stem	Leaf
0	678
1	1255689
2	3369
3	122
4	35
5	1

Key: 2 | 3 = 23 cars

WORKED EXAMPLE 4 Constructing a 3-digit stem plot

The masses (in kilograms) of the members of an under-17 football squad are shown.

70.3, 65.1, 72.9, 66.9, 68.6, 69.6, 70.8,
72.4, 74.1, 75.3, 75.6, 69.7, 66.2, 71.2,
68.3, 69.7, 71.3, 68.3, 70.5, 72.4, 71.8

Display the data in a stem plot.

THINK

1. In this case the observations contain 3 digits. The last digit is always the leaf and so in this case the digit referring to the tenths becomes the leaf and the two preceding digits become the stem.

Work out the lowest and highest numbers in the data to determine what the stems will be.

WRITE

Lowest number = 65.1
Highest number = 75.6
Use stems from 65 to 75.

2. Construct an unordered stem plot. Note that the decimal points are omitted since we are aiming to present a quick visual summary of data.

Stem	Leaf
65	1
66	9 2
67	
68	6 3 3
69	6 7 7
70	3 8 5
71	2 3 8
72	9 4 7
73	
74	1
75	3 6

3. Construct an ordered stem plot. Provide a key.

Stem	Leaf
65	1
66	2 9
67	
68	3 3 6
69	6 7 7
70	3 5 8
71	2 3 8
72	4 4 9
73	
74	1
75	3 6

Key: 74 | 1 = 74.1 kg

WORKED EXAMPLE 5 Constructing a split stem plot

A set of golf scores for a group of professional golfers trialling a new 18-hole golf course is shown on the following stem plot.

Stem	Leaf
6	1 6 6 7 8 9 9 9
7	0 1 1 2 2 3 7

Key: 6 | 1 = 61

Produce another stem plot for these data by splitting the stems into:

a. halves
b. fifths.

THINK

a. By splitting the stem 6 into halves, any leaf digits in the range 0–4 appear next to the 6, and any leaf digits in the range 5–9 appear next to the 6*.
Likewise for the stem 7.

WRITE

a.
Stem	Leaf
6	1
6*	6 6 7 8 9 9 9
7	0 1 1 2 2 3
7*	7

Key: 6 | 1 = 61

b. Alternatively, to split the stems into fifths, each stem would appear five times. Any 0s or 1s are recorded next to the first 6. Any 2s or 3s are recorded next to the second 6. Any 4s or 5s are recorded next to the third 6. Any 6s or 7s are recorded next to the fourth 6 and, finally, any 8s or 9s are recorded next to the fifth 6.

This process would be repeated for those observations with a stem of 7.

b. Stem	Leaf
6	1
6	
6	
6	6 6 7
6	8 9 9 9
7	0 1 1
7	2 2 3
7	
7	7
7	

Key: 6 | 1 = 61

 Resources

Interactivities Stem plots (int-6242)
Create stem plots (int-6495)

1.3 Exercise

Students, these questions are even better in jacPLUS

 Receive immediate feedback and access sample responses

 Access additional questions

 Track your results and progress

Find all this and MORE in jacPLUS

1. **WE3** The number of iPads sold in a month from a department store over 16 weeks is shown.

   28, 31, 18, 48, 38, 25, 21, 16,
   33, 42, 35, 39, 49, 30, 29, 28

   Construct a stem plot to display the number of iPads sold over the 16 weeks.

2. The money (correct to the nearest dollar) earned each week by a busker over an 18-week period is shown.

   5, 19, 11, 27, 23, 35, 18, 42, 29,
   31, 52, 43, 37, 41, 39, 45, 32, 36

   Construct a stem plot for the busker's weekly earnings. Comment on the busker's earnings.

3. **WE4** The test scores (as percentages) of a student in a Year 12 General Maths class are shown.

   88.0, 86.8, 92.1, 89.8, 92.6, 90.4, 98.3, 94.3, 87.7,
   94.9, 98.9, 92.0, 90.2, 97.0, 90.9, 98.5, 92.2, 90.8

   Display the data in a stem plot.

4. The heights of members of a squad of basketballers are given in metres.

$$1.96, \quad 1.85, \quad 2.03, \quad 2.21, \quad 2.17, \quad 1.89, \quad 1.99, \quad 1.87,$$
$$1.95, \quad 2.03, \quad 2.09, \quad 2.05, \quad 2.01, \quad 1.96, \quad 1.97, \quad 1.91$$

Construct a stem plot for these data.

5. The ages of those attending an embroidery class are given.

$$39, \quad 68, \quad 51, \quad 57, \quad 63, \quad 51, \quad 37, \quad 42,$$
$$63, \quad 49, \quad 52, \quad 61, \quad 58, \quad 59, \quad 49, \quad 53$$

Construct a stem plot for these data and draw a conclusion.

6. **MC** The observations shown on the stem plot are:
   A. 4, 10, 27, 28, 29, 31, 34, 36, 41
   B. 14, 10, 27, 28, 29, 29, 31, 34, 36, 41, 41
   C. 4, 22, 27, 28, 29, 29, 30, 31, 34, 36, 41, 41
   D. 14, 22, 27, 28, 29, 30, 30, 31, 34, 36, 41, 41
   E. 4, 2, 27, 28, 29, 29, 30, 31, 34, 36, 41

Stem	Leaf
0	4
1	
2	2 7 8 9 9
3	0 1 4 6
4	1 1

Key: 2 | 5 = 25

7. The ages of the parents of a class of children attending an inner-city kindergarten are given.

$$32, \quad 30, \quad 19, \quad 28, \quad 25, \quad 29, \quad 32, \quad 28, \quad 29, \quad 34,$$
$$32, \quad 35, \quad 39, \quad 30, \quad 37, \quad 33, \quad 29, \quad 35, \quad 38, \quad 33$$

Construct a stem plot for these data.
Based on your display, comment on the statement 'Parents of kindergarten children are young' (less than 30 years old).

8. The number of hit outs made by the ruckmen in each AFL team in Round 6, 2022, is recorded below:

Club	Hit outs
Adelaide Crows	52
Brisbane Lions	41
Carlton Blues	18
Collingwood Magpies	36
Essendon Bombers	36
Fremantle Dockers	50
Geelong Cats	46
Gold Coast Suns	55
GWS Giants	77

Club	Hit outs
Hawthorn Hawks	27
Melbourne Demons	39
North Melbourne Kangaroos	33
Port Adelaide Power	54
Richmond Tigers	27
St Kilda Saints	19
Sydney Swans	40
West Coast Eagles	28
Western Bulldogs	29

Construct a stem plot to display these data.
Identify which three teams had the highest performing ruckmen.

9. The weekly median rental price for a 2-bedroom unit in a number of Melbourne suburbs is given in the following table.

Suburb	Weekly rental ($)
Alphington	400
Box Hill	365
Brunswick	410
Burwood	390
Clayton	350
Essendon	350
Hampton	430
Ivanhoe	395
Kensington	406
Malvern	415

Suburb	Weekly rental ($)
Moonee Ponds	373
Newport	380
North Melbourne	421
Northcote	430
Preston	351
St Kilda	450
Surrey Hills	380
Williamstown	330
Windsor	423
Yarraville	390

Construct a stem plot for these data and comment on it.

10. **WE5** LeBron James's scores for his last 16 games are shown in the following stem plot.

Stem	Leaf
2	0 2 2 5 6 8 8
3	3 3 3 7 7 8 9 9 9

Key: 2 | 0 = 20

Produce another stem plot for these data by splitting the stems into:

a. halves
b. fifths.

11. The data shown give the head circumference (correct to the nearest cm) of 16 four-year-old children.

48,　49,　47,　52,　51,　50,　49,　48,
50,　50,　53,　52,　43,　47,　49,　50

Construct a stem plot for head circumference, using:

a. the stems 4 and 5
b. the stems 4 and 5 split into halves
c. the stems 4 and 5 split into fifths.

12. For each of the following, write down all the pieces of data shown on the stem plot. The key used for each stem plot is 3 | 2 = 32.

a.
Stem	Leaf
0	1 2
0*	5 8
1	2 3 3
1*	6 6 7
2	1 3 4
2*	5 5 6 7
3	0 2

b.
Stem	Leaf
1	0 1
2	3 3
3	0 5 9
4	1 2 7
5	5
6	2

c.
Stem	Leaf
10	1 2
11	5 8
12	2 3 3
13	6 6 7
14	1 3 4
15	5 5 6 7

d.
Stem	Leaf
5	0 1
5	3 3
5	4 5 5
5	6 6 7
5	9

e.
Stem	Leaf
0	1 4
0*	5 8
1	0 2
1*	6 9 9
2	1 1
2*	5 9

13. A random sample of 20 screws is taken and the length of each is recorded to the nearest millimetre.

$$23, \quad 15, \quad 18, \quad 17, \quad 17, \quad 19, \quad 22, \quad 19, \quad 20, \quad 16,$$
$$20, \quad 21, \quad 19, \quad 23, \quad 17, \quad 19, \quad 21, \quad 23, \quad 20, \quad 21$$

Construct a stem plot for screw length using:

a. the stems 1 and 2
b. the stems 1 and 2 split into halves
c. the stems 1 and 2 split into fifths.

Use your plots to help you comment on the screw lengths.

14. The first 20 scores that came into the clubhouse in a local golf tournament are shown.

$$102, \quad 98, \quad 83, \quad 92, \quad 85, \quad 99, \quad 104, \quad 112, \quad 88, \quad 91,$$
$$78, \quad 87, \quad 90, \quad 94, \quad 83, \quad 93, \quad 72, \quad 100, \quad 92, \quad 88$$

Construct a stem plot for these data.

15. Golf handicaps are designed to even up golfers on their abilities. Their handicap is subtracted from their score to create a net score.

$$76, \quad 76, \quad 73, \quad 74, \quad 69, \quad 72, \quad 73, \quad 86, \quad 73, \quad 72,$$
$$75, \quad 74, \quad 77, \quad 73, \quad 75, \quad 75, \quad 71, \quad 71, \quad 68, \quad 67$$

Construct a stem plot on the golfers' net scores above and comment on how well the golfers are handicapped.

16. The following data represents percentages for a recent General Maths test.

$$63, \quad 71, \quad 70, \quad 89, \quad 88, \quad 69, \quad 76, \quad 83, \quad 93, \quad 80, \quad 73,$$
$$77, \quad 91, \quad 75, \quad 81, \quad 84, \quad 87, \quad 78, \quad 97, \quad 89, \quad 98, \quad 60$$

Construct a stem plot for the test percentages, using:

a. the stems 6, 7, 8 and 9
b. the stems 6, 7, 8 and 9 split into halves.

17. The following data was collected from a company that compared the battery life (measured in minutes) of two different Ultrabook computers. To complete the test, they ran a series of programs on the two computers and measured how long it took for the batteries to go from 100% to 0%.

| Computer 1 | 358 | 376 | 392 | 345 | 381 | 405 | 363 | 380 | 352 | 391 | 410 | 366 |
| Computer 2 | 348 | 355 | 361 | 342 | 355 | 362 | 353 | 358 | 340 | 346 | 357 | 352 |

a. Draw a back-to-back stem plot (using the same stem) of the battery life of the two Ultrabook computers.
b. Use the stem plot to compare and comment on the battery life of the two Ultrabook computers.

18. The heights of 20 Year 8 and Year 10 students, chosen at random, are measured to the nearest centimetre. The data collected is shown in the table.

| Year 8 | 151 | 162 | 148 | 153 | 165 | 157 | 172 | 168 | 155 | 164 | 175 | 161 | 155 | 160 | 149 | 155 | 163 | 171 | 166 | 150 |
| Year 10 | 167 | 164 | 172 | 158 | 169 | 159 | 174 | 177 | 165 | 156 | 154 | 160 | 178 | 176 | 182 | 152 | 167 | 185 | 173 | 178 |

a. Draw a back-to-back stem plot of the data.
b. Comment on what the stem plot tells you about the heights of Year 8 and Year 10 students.

Question 1 (1 mark)

Source: VCE 2016, Further Mathematics Exam 1, Section A, Q3; © VCAA

`MC` The stem plot below displays 30 temperatures recorded at a weather station.

	Temperature	Key: 2|2 = 2°C
2	2 2 4 4	
2	5 7 8 8 8 8 8 8 9 9 9 9	
3	1 2 3 3 4 4 4	
3	5 6 7 7 7 7	
4	1	

The modal temperature is

A. 2.8°C　　　**B.** 2.9°C　　　**C.** 3.7°C　　　**D.** 8.0°C　　　**E.** 9.0°C

Question 2 (1 mark)

Source: VCE 2013, Further Mathematics, Exam 1, Section A Core, Q1; © VCAA

`MC` The following ordered stem plot shows the percentage of homes connected to broadband internet for 24 countries in 2007.

Key 1\|6 = 16%

1	
1	6 7
2	0 1 1 3 4 4
2	5 7 8 9
3	0 0 1 1 1 2 2 3
3	5 7 8 8
4	

The number of these countries with more than 22% of homes connected to broadband internet in 2007 is

A. 4　　　**B.** 5　　　**C.** 19　　　**D.** 20　　　**E.** 22

Question 3 (1 mark)

`MC` The marks students received in a recent test are displayed in the stem plot. Determine how many outliers the data contains.

Stem	Leaf
0	7 8
1	8
2	7 8 8 9
3	2 4 5 6 8 9 9
4	0 2 8 9 9
5	0

Key: 2\|7 = 27

A. 0　　　**B.** 1　　　**C.** 2　　　**D.** 3　　　**E.** 4

More exam questions are available online.

1.4 Dot plots, frequency tables, histograms, bar charts and logarithmic scales

LEARNING INTENTION

At the end of this subtopic you should be able to:
- construct dot plots
- construct histograms from lists of data and frequency tables
- construct bar charts and segmented bar charts
- apply logarithmic (base 10) scales to display data ranging over several orders of magnitude.

1.4.1 Dot plots

In **dot plots**, a single dot represents each data value. Dot plots are used to display discrete data where values are not spread out very much. They are also used to display categorical data.

When representing discrete data, dot plots have a scaled horizontal axis and each data value is indicated by a dot above this scale. The end result is a set of vertical 'lines' of evenly spaced dots.

For example, the following dot plot shows scores for a soccer team. The lowest score is 1 and the highest score is 5.

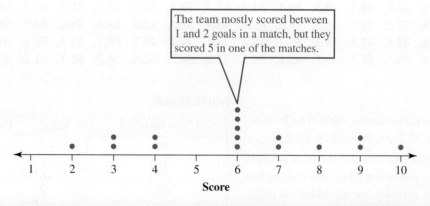

The team mostly scored between 1 and 2 goals in a match, but they scored 5 in one of the matches.

Score

tlvd-3532

WORKED EXAMPLE 6 Constructing a dot plot

The number of hours per week spent on art by 18 students is shown.

$$4, \quad 0, \quad 3, \quad 1, \quad 3, \quad 4, \quad 2, \quad 2, \quad 3,$$
$$4, \quad 1, \quad 3, \quad 2, \quad 5, \quad 3, \quad 2, \quad 1, \quad 0$$

Display the data as a dot plot.

THINK

1. Determine the lowest and highest scores and then draw a suitable scale.

2. Represent each score by a dot on the scale.

DRAW

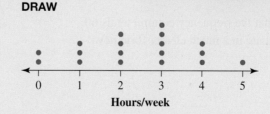

Hours/week

1.4.2 Frequency tables and histograms

A **histogram** is a useful way of displaying large data sets (say, over 50 observations). The vertical axis on the histogram displays the frequency and the horizontal axis displays class intervals of the variable (e.g. height or income).

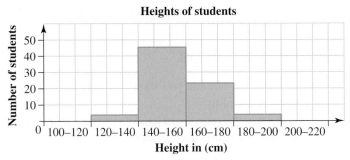

Heights of students

The vertical bars in a histogram are adjacent with no gaps between them, as we generally consider the numerical data scale along the horizontal axis as continuous. Note, however, that histograms can also represent discrete data. It is common practice to leave a small gap before the first bar of a histogram.

When data are given in raw form — that is, as a list of figures in no particular order — it is helpful to first construct a **frequency table** before constructing a histogram.

tlvd-3533

WORKED EXAMPLE 7 Constructing a histogram from a list of data

The data below show the distribution of masses (in kilograms) of 60 students in Year 7 at Northwood Secondary College. Construct a histogram to display the data more clearly.

45.7, 45.8, 45.9, 48.2, 48.3, 48.4, 34.2, 52.4, 52.3, 51.8, 45.7, 56.8, 56.3, 60.2, 44.2,
53.8, 43.5, 57.2, 38.7, 48.5, 49.6, 56.9, 43.8, 58.3, 52.4, 54.3, 48.6, 53.7, 58.7, 57.6,
45.7, 39.8, 42.5, 42.9, 59.2, 53.2, 48.2, 36.2, 47.2, 46.7, 58.7, 53.1, 52.1, 54.3, 51.3,
51.9, 54.6, 58.7, 58.7, 39.7, 43.1, 56.2, 43.0, 56.3, 62.3, 46.3, 52.4, 61.2, 48.2, 58.3

THINK

1. First construct a frequency table. The lowest data value is 34.2 and the highest is 62.3.

   Divide the data into class intervals. If we start the first class interval at 30 kg and end the last class interval at 65 kg, we would have a range of 35.

   If each interval was 5 kg, we would then have 7 intervals, which is a reasonable number of class intervals.

   While there are no set rules about how many intervals there should be, somewhere between 5 and 15 class intervals is usual.

   Complete a tally column using one mark for each value in the appropriate interval. Add up the tally marks and write them in the frequency column.

2. Check that the frequency column totals 60. The data are in a much clearer form now.

WRITE/DRAW

Class interval	Tally	Frequency
30–	|	1
35–	||||	4
40–	JHT ||	7
45–	JHT JHT JHT |	16
50–	JHT JHT JHT	15
55–	JHT JHT ||||	14
60–	|||	3
Total		60

3. Use the information given in the first and third columns of the frequency table to construct a histogram.

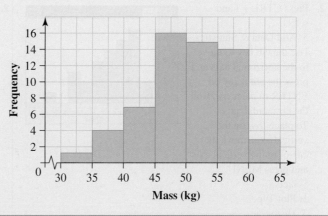

When constructing a histogram to represent continuous data, as in Worked example 7, the bars will sit between two values on the horizontal axis, which represent the class intervals. When dealing with discrete data, the bars should appear above the middle of the value they represent.

WORKED EXAMPLE 8 Constructing a histogram from a frequency table

The marks out of 20 received by 30 students for a book-review assignment are given in the frequency table.

Mark	12	13	14	15	16	17	18	19	20
Frequency	2	7	6	5	4	2	3	0	1

Display this data on a histogram.

THINK

In this case we are dealing with integer values (discrete data). Since the horizontal axis should show a class interval, we extend the base of each of the columns on the histogram halfway either side of each score.

DRAW

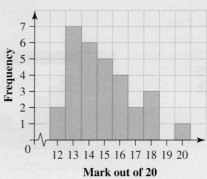

TI	THINK	DISPLAY/WRITE

1. On a Lists & Spreadsheet page, enter the data into the lists named Mark and Freq.

CASIO	THINK	DISPLAY/WRITE

1. On a Statistics screen, enter the data into the lists named Mark and Freq.

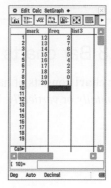

2. Press CTRL + I and select:
5: Add Data & Statistics Use the tab key to move to the horizontal axis and press the click key or ENTER to place *mark* on the horizontal axis.
To place *freq* on the vertical axis, move the cursor to the vertical axis and press MENU.
Select:
1: Plot Type
3: Histogram
Press MENU:
2: Plot Properties
9: Add Y Summary List
Select freq.

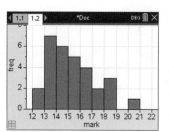

3. As the data in this example is discrete change the column width by pressing MENU and selecting:
2: Plot Properties
2: Histogram Properties
2: Bin Settings
1: Equal Bin Width
Complete the fields as shown.
Press ENTER.
To Zoom out, press MENU and select:
5: Window/Zoom
2: Zoom – Data

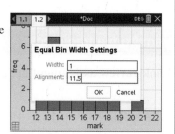

4. The histogram is shown on the screen.

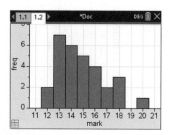

2. Tap:
SetGraph
Setting
Complete the fields under Set
StatGraphs as:
Draw: On
Type: Histogram
XList: main\mark
Freq: main\freq
Tap Set to lock your selection.
Tap graph icon.

3. Complete the Set Interval fields as:
HStart: 11.5
HStep: 1
Tap OK.

4. The histogram is shown on the screen.

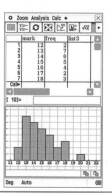

1.4.3 Bar charts

A **bar chart** consists of bars of equal width *separated* by small, equal spaces that may be arranged either *horizontally* or *vertically*. Bar charts are often used to display categorical data.

In bar charts, the frequency is graphed against a variable as shown in both of the following figures.

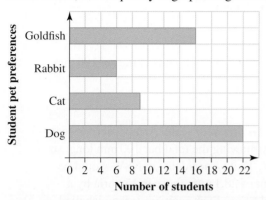

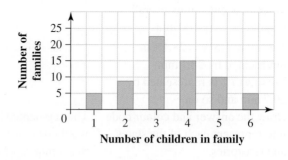

The variable may or may not be numerical. However, if it is, the variable should represent discrete data because the scale is broken by the gaps between the bars.

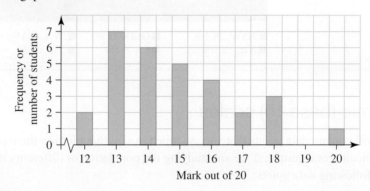

Segmented bar charts

A **segmented bar chart** is a single bar that is used to represent all the data being studied. It is divided into segments, with each segment representing a particular group of the data. Generally, the information is presented as percentages and so the total bar length represents 100% of the data.

WORKED EXAMPLE 9 Constructing a segmented bar chart

The table shown represents fatal road accidents for one year in Australia. Construct a segmented bar chart to represent this data.

| Year | Accidents involving fatalities | | | | | | | | |
	NSW	Vic.	Qld	SA	WA	Tas.	NT	ACT	Aust.
2008	376	278	293	87	189	38	67	12	1340

Source: Australian Bureau of Statistics 2010, *Year Book Australia 2009–10*, cat. no. **1301.0**. ABS. Canberra, table 24.20, p. 638.

THINK

1. To draw a segmented bar chart, the data need to be converted to percentages.

WRITE/DRAW

State	Number of accidents	Percentage
NSW	376	$376 \div 1340 \times 100\% = 28.1\%$
Vic.	278	$278 \div 1340 \times 100\% = 20.7\%$
Qld	293	$293 \div 1340 \times 100\% = 21.9\%$
SA	87	$87 \div 1340 \times 100\% = 6.5\%$
WA	189	$189 \div 1340 \times 100\% = 14.1\%$
Tas.	38	$38 \div 1340 \times 100\% = 2.8\%$
NT	67	$67 \div 1340 \times 100\% = 5.0\%$
ACT	14	$12 \div 1340 \times 100\% = 0.9\%$

2. To draw the segmented bar chart to scale, decide on its overall length, let's say 100 mm.

Measure a line 100 mm in length.

3. Therefore NSW = 28.1% is represented by 28.1 mm.
Vic = 20.7% is represented by 20.7 mm and so on.

Measure off each segment and check it adds to the set 100 mm.

4. Draw the answer and colour code it to represent each of the states and territories.

The segmented bar chart is drawn to scale. An appropriate scale would be constructed by drawing the total bar 100 mm long, so that 1 mm represents 1%. That is, accidents in NSW would be represented by a segment of 28.1 mm, those in Victoria by a segment of 20.7 mm and so on. Each segment is then labelled directly, or a key may be used.

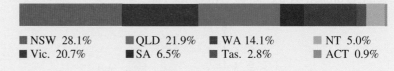

- NSW 28.1%
- Vic. 20.7%
- QLD 21.9%
- SA 6.5%
- WA 14.1%
- Tas. 2.8%
- NT 5.0%
- ACT 0.9%

1.4.4 Logarithmic (base 10) scales

Sometimes a data set will contain data points that vary so much in size that plotting them using a traditional scale becomes very difficult. For example, if we are studying the population of different cities in Australia, we might end up with the following data points:

City	Population
Adelaide	1 304 631
Ballarat	98 543
Brisbane	2 274 460
Cairns	146 778
Darwin	140 400
Geelong	184 182
Launceston	86 393
Melbourne	4 440 328
Newcastle	430 755
Shepparton	49 079
Wagga Wagga	55 364

A histogram splitting the data into class intervals of 100 000 would then appear as follows:

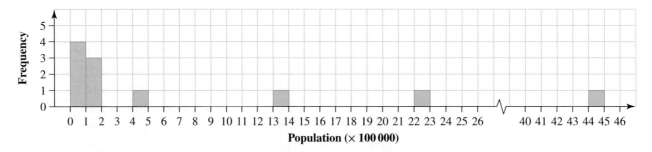

A way to overcome this is to write the numbers in **logarithmic (log) form**.

Logarithmic (log) form

The log of a number is the power of 10 that creates this number.

$$\log(10) = \log\left(10^1\right) = 1$$
$$\log(100) = \log\left(10^2\right) = 2$$
$$\log(1000) = \log\left(10^3\right) = 3$$
$$\vdots$$
$$\log(10^n) = n$$

Not all logarithmic values are integers, so use the log key on CAS to determine exact logarithmic values.

For example, from our previous example showing the population of different Australian cities:

$$\log(4\,440\,328) = 6.65 \text{ (correct to 2 decimal places)}$$
$$\log(184\,182) = 5.27 \text{ (correct to 2 decimal places)}$$
$$\log(49\,079) = 4.69 \text{ (correct to 2 decimal places)}$$

We can then group our data using class intervals based on log values (from 4 to 7) to come up with the following frequency table and histogram.

Population	Log (population)	Frequency
10 000–	4–5	4
100 000–	5–6	4
1 000 000–	6–7	3

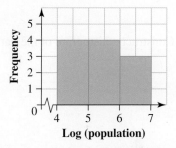

WORKED EXAMPLE 10 Constructing a histogram using a log (base 10) scale

The following table shows the average weights of 10 different adult mammals.

Mammal	Weight (kg)
African elephant	4800
Black rhinoceros	1100
Blue whale	136 000
Giraffe	800
Gorilla	140
Humpback whale	30 000
Lynx	23
Orang-utan	64
Polar bear	475
Tasmanian devil	7

Display the data in a histogram using a log (base 10) scale.

THINK

1. Use CAS to calculate the logarithmic values of all of the weights, for example:
   log (4800) = 3.68 (correct to 2 decimal places)

WRITE/DRAW

Weight	Log (weight (kg))
4800	3.68
1100	3.04
136 000	5.13
800	2.90
140	2.15
30 000	4.48
23	1.36
64	1.81
475	2.68
7	0.85

2. Group the logarithmic weights into class intervals and create a frequency table for the groupings.

Log (weight (kg))	Frequency
0–1	1
1–2	2
2–3	3
3–4	2
4–5	1
5–6	1

3. Construct a histogram of the data set.

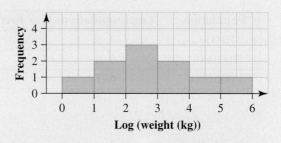

TI | THINK

1. On a Lists & Spreadsheet page, enter the data into the list named 'weight'.
   Label Column B as 'lgwght' and in the formula cell for Column B, complete the entry line as:
   = log (*weight*)
   Press ENTER.

DISPLAY/WRITE

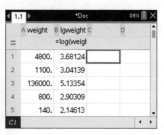

CASIO | THINK

1. On a Statistics screen, enter the data into List1 and rename the list 'weight'.
   Label List2 as 'lgwght'.
   Place the cursor in the Cal cell at the base of the List2 column and complete the entry line as:
   = log (*weight*)
   Tap EXE.

DISPLAY/WRITE

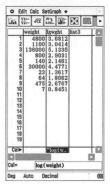

2. Press CTRL + I and select:
   5: Add Data & Statistics
   Use the tab key to move to the
   horizontal axis and press the
   click key or ENTER to place
   '*lgwght*' on the horizontal
   axis.
   Press MENU and select:
   1: Plot Type
   3: Histogram
   Press ENTER.

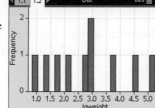

2. Tap:
   SetGraph
   Setting
   Complete the fields under
   Set StatGraphs as:
   Draw: On
   Type: Histogram
   XList: main\lgwght
   Freq: 1
   Tap Set to lock your
   selection.
   Tap graph icon.

3. To change the column width,
   press MENU and select:
   2: Plot Properties
   2: Histogram Properties
   2: Bin Settings
   1: Equal Bin Width
   Complete the fields as shown.
   Press ENTER.

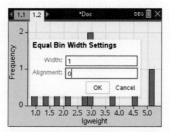

3. Complete the Set Interval
   fields as:
   HStart: 11.5
   HStep: 1
   Tap OK.

4. To zoom out, press MENU
   and select:
   5: Window/Zoom
   2: Zoom-Data

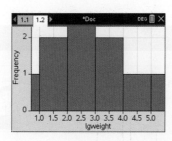

4. The histogram is shown
   on the screen.

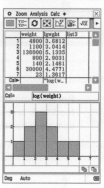

5. The histogram is shown on
   the screen.

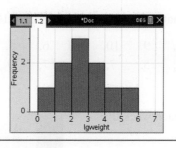

Interpreting log (base 10) values

If we are given values in logarithmic form, by raising 10 to the power of the logarithmic number we can determine the conventional number.

For example, the number 3467 in log (base 10) form is 3.54, and $10^{3.54} = 3467$.

We can use this fact to compare values in log (base 10) form, as shown in Worked example 11.

WORKED EXAMPLE 11 Application of a log (base 10) scale

The Richter Scale measures the magnitude of earthquakes using a log (base 10) scale.
Calculate how many times stronger an earthquake of magnitude 7.4 is than one of magnitude 5.2.
Give your answer correct to the nearest whole number.

THINK	WRITE
1. Calculate the difference between the magnitude of the two earthquakes.	$7.4 - 5.2 = 2.2$
2. Raise 10 to the power of the difference in magnitudes.	$10^{2.2} = 158.49$ (correct to 2 decimal places)
3. Write the answer in a sentence.	The earthquake of magnitude 7.4 is 158 times stronger than the earthquake of magnitude 5.2.

 Resources

 Interactivities Create a histogram (int-6494)
Dot plots, frequency tables and histograms, and bar charts (int-6243)
Create a bar chart (int-6493)

1.4 Exercise

Students, these questions are even better in jacPLUS

Receive immediate feedback and access sample responses

Access additional questions

Track your results and progress

Find all this and MORE in jacPLUS

1. **WE6** The number of questions completed for maths homework each night by 16 students is shown.

$$5, \quad 6, \quad 5, \quad 9, \quad 10, \quad 10, \quad 6, \quad 8,$$
$$10, \quad 9, \quad 8, \quad 5, \quad 7, \quad 8, \quad 7, \quad 9$$

Display the data as a dot plot.

2. The data shown represent the number of hours each week that 40 people spent on household chores.

$$2, \quad 5, \quad 2, \quad 0, \quad 8, \quad 7, \quad 8, \quad 5,$$
$$1, \quad 0, \quad 2, \quad 1, \quad 8, \quad 0, \quad 4, \quad 2,$$
$$2, \quad 9, \quad 8, \quad 5, \quad 7, \quad 5, \quad 4, \quad 2,$$
$$1, \quad 2, \quad 9, \quad 8, \quad 1, \quad 2, \quad 8, \quad 5,$$
$$8, \quad 10, \quad 0, \quad 3, \quad 4, \quad 5, \quad 2, \quad 8$$

Display these data on a bar chart and a dot plot.

3. The number of hours spent on homework for a group of 20 Year 12 students each week is shown.

$$15, \quad 15, \quad 18, \quad 21, \quad 20, \quad 21, \quad 24, \quad 24, \quad 20, \quad 18,$$
$$18, \quad 21, \quad 22, \quad 20, \quad 24, \quad 20, \quad 24, \quad 21, \quad 18, \quad 15$$

Display the data as a dot plot.

4. **WE7** The data shows the distribution of heights (in cm) of 40 students in Year 12.

167,	172,	184,	180,	178,	166,	154,	150,	164,	161,
187,	159,	182,	177,	172,	163,	179,	181,	170,	176,
177,	162,	172,	184,	188,	179,	189,	192,	164,	160,
166,	169,	163,	185,	178,	183,	190,	170,	168,	159

Construct a histogram to display the data more clearly.

5. Construct a frequency table for each of the following sets of data.

   a. 4.3, 4.5, 4.7, 4.9, 5.1, 5.3, 5.5, 5.6, 5.2, 3.6, 2.5, 4.3, 2.5, 3.7, 4.5, 6.3, 1.3

   b. 11, 13, 15, 15, 16, 18, 20, 21, 22, 21, 18, 19, 20, 16, 18,
      20, 16, 10, 23, 24, 25, 27, 28, 30, 35, 28, 27, 26, 29, 30,
      31, 24, 28, 29, 20, 30, 32, 33, 29, 30, 31, 33, 34

   c. 0.4, 0.5, 0.7, 0.8, 0.8, 0.9, 1.0, 1.1, 1.2, 1.0, 1.3, 0.4, 0.3, 0.9, 0.6

   Using the frequency tables above, construct a histogram by hand for each set of data.

6. The number of dogs at an RSPCA kennel each week is shown.

7,	6,	2,	12,	7,	9,	12,	10,	5,	7,	9,	4,	5,	9,	3,
2,	10,	8,	9,	7,	9,	10,	9,	4,	3,	8,	9,	3,	7,	9

   a. Construct a frequency table for these data.
   b. Construct a histogram by hand.
   c. Using CAS, construct a histogram from the data and compare it to the histogram in part **b**.

7. **WE8** The number of fish caught by 30 anglers in a fishing competition is given in the frequency table shown.

Fish	0	1	2	3	4	5	7
Frequency	4	7	4	6	5	3	1

Display these data on a histogram.

8. The number of fatal car accidents in Victoria each week is given in the frequency table for a year.

Fatalities	0	1	2	4	6	7	9	10	13
Frequency	12	3	6	10	7	6	5	2	1

Display these data on a histogram.

9. **WE9** The following table shows the number of goals that the leading goal kicker for each of the 18 AFL teams scored in total in the 2021 season plus finals:

Club	Goals
Adelaide Crows	48
Brisbane Lions	55
Carlton Blues	58
Collingwood Magpies	34
Essendon Bombers	41
Fremantle Dockers	37
Geelong Cats	62
Gold Coast Suns	47
GWS Giants	45

Club	Goals
Hawthorn Hawks	33
Melbourne Demons	59
North Melbourne Kangaroos	42
Port Adelaide Power	48
Richmond Tigers	51
St Kilda Saints	38
Sydney Swans	51
West Coast Eagles	42
Western Bulldogs	48

Construct a segmented bar chart to represent this data.

10. Information about adult participation in sport and physical activities in 2021–22 is shown in the following table.

Participation in sport and physical activities, 2021–22

Age group (years)	Number (×1000)
18–24	1406.4
25–34	2088.3
35–44	2011.2
45–54	1795.2
55–64	1386.5
65 and over	1243.9
Total	**9931.5**

Draw a segmented bar graph to compare the participation of all persons from various age groups. Comment on the statement, 'Only young people participate in sport and physical activities'.

11. A group of students was surveyed, asking how many children were in their family. The data is shown in the table.

Number of children	1	2	3	4	5	6	9
Number of families	12	18	24	10	8	3	1

Construct a bar chart that displays the data.

12. **WE10** The following table shows the average weights of 10 different adult mammals.

Mammal	Weight (kg)
Black wallaroo	18
Capybara	55
Cougar	63
Fin whale	70 000
Lion	175
Ocelot	9
Pygmy rabbit	0.4
Red deer	200
Quokka	4
Water buffalo	725

Display the data in a histogram using a log (base 10) scale, using class intervals of width 1.

13. **MC** The following graph shows the weights of animals.
If a gorilla has a weight of 207 kilograms, then its weight is between that of:

A. monkey and jaguar.
B. horse and triceratops.
C. guinea pig and monkey.
D. jaguar and horse.
E. none of the above.

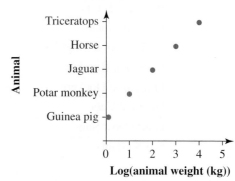

14. **MC** The following table shows a variety of top speeds.

F1 racing car	370 000 m/h
V8 supercar	300 000 m/h
Cheetah	64 000 m/h
Space shuttle	28 000 000 m/h
Usain Bolt	34 000 m/h

The correct value, to 2 decimal places, for a cheetah's top speed using a log (base 10) scale would be:

A. 5.57   **B.** 5.47   **C.** 7.45   **D.** 4.81   **E.** 11.07

15. **MC** The following graph represents the capacity of five Victorian dams.

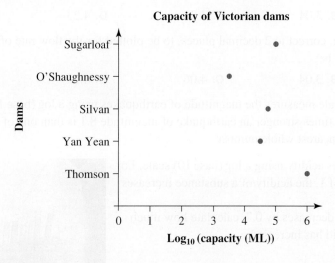

Capacity of Victorian dams

a. The capacity of Thomson Dam is closest to:

    A. 100 000 ML

    B. 1 000 000 ML

    C. 500 000 ML

    D. 5000 ML

    E. 40 000 ML

b. The capacity of Sugarloaf Dam is closest to:

    A. 100 000 ML

    B. 1 000 000 ML

    C. 500 000 ML

    D. 5000 ML

    E. 40 000 ML

c. The capacity of Silvan Dam is between the range:

    A. 1 and 10 ML

    B. 10 and 100 ML

    C. 1000 and 10 000 ML

    D. 10 000 and 100 000 ML

    E. 100 000 and 1 000 000 ML

The following information about the flow rates of world-famous waterfalls refers to Questions 16 and 17.

Waterfalls	Flow rate
Victoria Falls	$1088 \, \text{m}^3/\text{s}$
Niagara Falls	$2407 \, \text{m}^3/\text{s}$
Celilo Falls	$5415 \, \text{m}^3/\text{s}$
Khane Phapheng Falls	$11610 \, \text{m}^3/\text{s}$
Boyoma Falls	$17000 \, \text{m}^3/\text{s}$

16. **MC** The correct value, correct to 2 decimal places, to be plotted for the flow rate of Niagara Falls using a log (base 10) scale would be:

   A. 3.38   **B.** 3.04   **C.** 4.06   **D.** 4.23   **E.** 5.10

17. **MC** The correct value, correct to 2 decimal places, to be plotted for the flow rate of Victoria Falls using a log (base 10) scale would be:

   A. 3.38   **B.** 3.04   **C.** 4.06   **D.** 4.23   **E.** 5.10

18. **WE11** The Richter scale measures the magnitude of earthquakes using a log (base 10) scale. Determine how many times stronger an earthquake of magnitude 8.1 is than one of magnitude 6.9. Give your answer correct to the nearest whole number.

19. The pH scale measures acidity using a log (base 10) scale. For each decrease in pH of 1, the acidity of a substance increases by a factor of 10.
    If a liquid's pH value decreases by 0.7, calculate how much the acidity of the liquid has increased.

20. Using the information provided in the table:

Short-term Australian resident departures by major destination					
	2015 ($\times$ 1000)	2016 ($\times$ 1000)	2017 ($\times$ 1000)	2018 ($\times$ 1000)	2019 ($\times$ 1000)
New Zealand	815.8	835.4	864.7	902.1	921.1
United States of America	376.1	426.3	440.3	479.1	492.3
United Kingdom	375.1	404.2	412.8	428.5	420.3
Indonesia	335.1	319.7	194.9	282.6	380.7
China (excluding Special Administrative Regions (SARs))	182.0	235.1	251.0	284.3	277.3
Thailand	188.2	202.7	288.0	374.4	404.1
Fiji	175.4	196.9	202.4	200.3	236.2
Singapore	159.0	188.5	210.9	221.5	217.8
Hong Kong (SAR of China)	152.6	185.7	196.3	206.5	213.1
Malaysia	144.4	159.8	168.0	181.3	191.0

   a. calculate the proportion of residents who travelled in 2019 to each of the countries listed
   b. draw a segmented bar graph showing the major destinations of Australians travelling abroad in 2016.

Question 1 (1 mark)

Source: VCE 2021, Further Mathematics Exam 1, Section A Core, Q2; © VCAA

MC The percentaged segmented bar chart below shows the *age* (under 55 years, 55 years and over) of visitors at a travel convention, segmented by *preferred travel destination* (domestic, international).

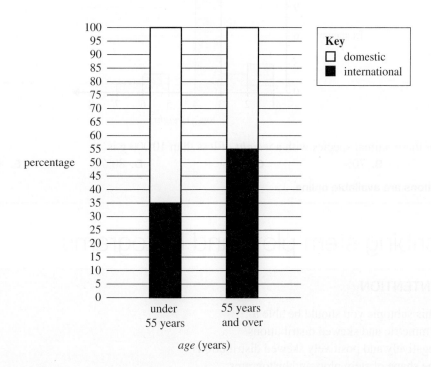

The data displayed in the percentaged segmented bar chart supports the contention that there is an association between *preferred travel destination* and *age* because

- **A.** more visitors favour international travel.
- **B.** 35% of visitors under 55 years favour international travel.
- **C.** 45% of visitors 55 years and over favour domestic travel.
- **D.** 65% of visitors under 55 years favour domestic travel while 45% of visitors 55 years and over favour domestic travel.
- **E.** the percentage of visitors who prefer domestic travel is greater than the percentage of visitors who prefer international travel.

Question 2 (1 mark)

Source: VCE 2020, Further Mathematics Exam 1, Section A, Q6; © VCAA

MC A percentaged segmented bar chart would be an appropriate graphical tool to display the association between *month of the year* (January, February, March, etc.) and the

- **A.** *monthly average rainfall* (in millimetres).
- **B.** *monthly mean temperature* (in degrees Celsius).
- **C.** *annual median wind speed* (in kilometres per hour).
- **D.** *monthly average rainfall* (below average, average, above average).
- **E.** *annual average temperature* (in degrees Celsius).

Question 3 (1 mark)

Source: VCE 2020, Further Mathematics Exam 1, Section A, Q5; © VCAA

MC The histogram below shows the distribution of *weight*, in grams, for a sample of 20 animal species. The histogram has been plotted on a $\log_{10}$ scale.

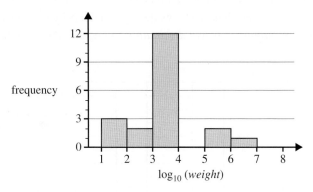

The percentage of these animal species with a *weight* of less than 10 000 g is

A. 17% **B.** 70% **C.** 75% **D.** 80% **E.** 85%

More exam questions are available online.

1.5 Describing stem plots and histograms

LEARNING INTENTION

At the end of this subtopic you should be able to:
- identify symmetric and skewed distributions
- identify negatively and positively skewed distributions
- describe the shape of stem plots and histograms.

1.5.1 Types of distributions

The type of distribution describes the shape of the data. There are two types of distribution: **symmetric** and **skewed** distributions.

Symmetric distributions

The data shown in the histogram can be described as symmetric.

The stem plot could also describe a symmetric distribution.

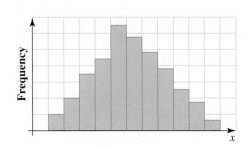

Stem	Leaf
0	7
1	2 3
2	2 4 5 7 9
3	0 2 3 6 8 8
4	7 8 9 9
5	2 7 8
6	1 3

Key: $0|7 = 7$

There is a single peak and the data trail off on both sides of this peak in roughly the same fashion.

The single peak for these data occurs at 3 on the stem. On either side of the peak, the number of observations reduces in approximately matching fashion.

Skewed distributions

Each of the histograms shown are examples of skewed distributions.

The figure shows data that are **negatively skewed**.

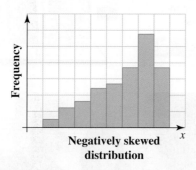

Negatively skewed
distribution

The data in this case peaks to the right and trails off
to the left.

The figure shows data that are **positively skewed**.

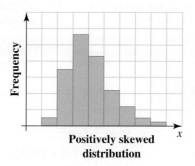

Positively skewed
distribution

The data in this case peaks to the left and trails off to
the right.

tlvd-3534

WORKED EXAMPLE 12 Describing the shape of a stem plot

**The ages of a group of people who were taking out their
first home loan is shown.**

Stem	Leaf
1	9 9
2	1 2 4 6 7 8 8 9
3	1 1 1 2 3 4 7
4	1 3 5 6
5	2 3
6	7

Key: 1 | 9 = 19 years old

Describe the shape of the distribution of these data.

THINK

Check whether the distribution is symmetric or skewed. The peak of
the data occurs at the stem 2. The data trails off as the stems increase
in value. This seems reasonable since most people would take out a
home loan early in life to give themselves time to pay it off.

WRITE

The data are positively skewed.

1. **WE12** The ages of a group of people when they bought their first car are shown.

Stem	Leaf
1	7 7 8 8 8 8 9 9
2	0 0 1 2 3 6 7 8 9
3	1 4 7 9
4	4 8
5	3

Key: $1 | 7 = 17$ years old

Describe the shape of the distribution of the data.

2. The ages of women when they gave birth to their first child are shown.

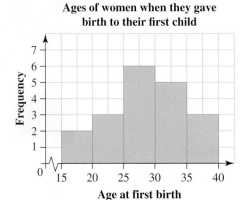

Ages of women when they gave birth to their first child

Describe the shape of the distribution of the data.

3. **MC** The distribution of the data shown in this stem plot could be described as:

A. negatively skewed.
B. negatively skewed and symmetric.
C. positively skewed.
D. positively skewed and symmetric.
E. symmetric.

Stem	Leaf
0	1
0	2
0	4 4 5
0	6 6 6 7
0	8 8 8 8 9 9
1	0 0 0 1 1 1 1
1	2 2 2 3 3 3
1	4 4 5 5
1	6 7 7
1	8 9

Key: $1 | 8 = 18$

4. **MC** The distribution of the data shown in the histogram could be described as:

A. negatively skewed.
B. negatively skewed and symmetric.
C. positively skewed.
D. positively skewed and symmetric.
E. symmetric.

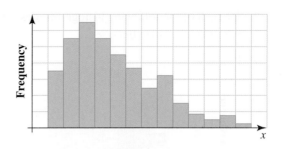

5. For each of the following stem plots, describe the shape of the distribution of the data.

a.
Stem	Leaf
0	1 3
1	2 4 7
2	3 4 4 7 8
3	2 5 7 9 9 9 9
4	1 3 6 7
5	0 4
6	4 7
7	1

Key: 1 | 2 = 12

b.
Stem	Leaf
1	3
2	6
3	3 8
4	2 6 8 8 9
5	4 7 7 7 8 9 9
6	0 2 2 4 5

Key: 2 | 6 = 2.6

c.
Stem	Leaf
2	3 5 5 6 7 8 9 9
3	0 2 2 3 4 6 6 7 8 8
4	2 2 4 5 6 6 6 7 9
5	0 3 3 5 6
6	2 4
7	5 9
8	2
9	7
10	

Key: 10 | 4 = 04

d.
Stem	Leaf
1	
1*	5
2	1 4
2*	5 7 8 8 9
3	1 2 2 3 3 3 4 4
3*	5 5 5 6
4	3 4
4*	

Key: 2 | 4 = 24

e.
Stem	Leaf
3	
3	8 9
4	0 0 1 1 1
4	2 3 3 3 3 3 4
4	5 5 5
4	6 7
4	8

Key: 4 | 3 = 0.43

f.
Stem	Leaf
60	2 5 8
61	1 3 3 6 7 8 9
62	0 1 2 4 6 7 8 8 9
63	2 2 4 5 7 8
64	3 6 7
65	4 5 8
66	3 5
67	4

Key: 62 | 3 = 623

6. The average number of product enquiries per day received by a group of small businesses who advertised on a website is given below.

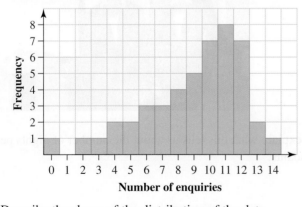

Describe the shape of the distribution of the data.

7. For each of the following histograms, describe the shape of the distribution of the data and comment on the existence of any outliers.

a.
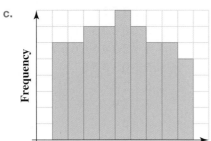

b.
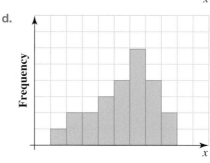

c.

d.

e.

f.

8. The number of nights per month spent interstate by a group of flight attendants is shown in the stem plot.

Stem	Leaf
0	0 0 1 1
0	2 2 3 3 3 3 3 3 3 3
0	4 4 5 5 5 5 5
0	6 6 6 6 7
0	8 8 8 9
1	0 0 1
1	4 4
1	5 5
1	7

Key: 1 | 4 = 14 nights

Describe the shape of distribution of the data and explain what this tells us about the number of nights per month spent interstate by this group of flight attendants.

9. The mass (correct to the nearest kilogram) of each dog at a dog obedience school is shown in the stem plot.

Stem	Leaf
0	4
0*	5 7 9
1	1 2 4 4
1*	5 6 6 7 8 9
2	1 2 2 3
2*	6 7

Key: 0 | 4 = 4 kg

a. Describe the shape of the distribution of the data.
b. Comment on what this information tells us about this group of dogs.

10. The amount of pocket money (correct to the nearest 50 cents) received each week by students in a Grade 6 class is illustrated in the histogram.

a. Describe the shape of the distribution of the data.
b. Draw conclusions about the amount of pocket money received weekly by this group of students.

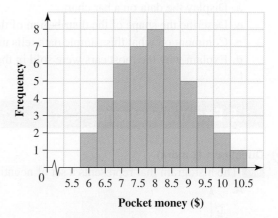

11. Statistics were collected on the number of goals kicked by forwards over 3 weeks in the AFL. This is displayed in the histogram.

a. Describe the shape of the histogram.
b. Use the histogram to determine:

i. the number of players who kicked 3 or more goals over the 3 weeks
ii. the percentage of players who kicked between 2 and 6 goals inclusively over the 3 weeks.

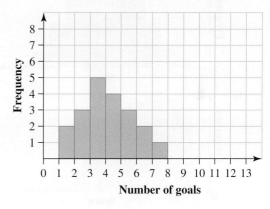

12. The number of hours a group of students exercise each week is shown in the stem plot.

Stem	Leaf
0	0 0 0 0 1 1 1
0	2 2 2 3
0	4 4
0	6
0	8 8 9
1	0 0 1
1	2 2 2 3

Key: 0 | 1 = 1

a. Describe the shape of the distribution of the data.
b. Comment on what this sample data tells us about this group of students.

13. The stem plot shows the age of players in two bowling teams.

Club A	Stem	Club B
1	4	
	5	5 7 8 9
	6	0 2 3 5 7
6 5 4 3	7	0 1 2
8 6 5 4 3 2 1	8	8
	9	0

    Key: 5 | 5 = 55

    a. Describe the shape of the distribution of Club A and Club B.
    b. Comment on the make-up of Club A compared to Club B.
    c. State how many players are over the age of 70 from:

      i. Club A   ii. Club B.

14. The following table shows the number of cars sold at a dealership over eight months.

Month	April	May	June	July	August	September	October	November
Cars sold	9	14	27	21	12	14	10	18

   a. Display the data on a bar chart.
   b. Describe the shape of the distribution of the data.
   c. Comment on what this sample data tells us about car sales over these months.
   d. Explain why the most cars were sold in the month of June.

1.5 Exam questions

Question 1 (1 mark)

MC The histogram that represents a set of negatively skewed data is

A.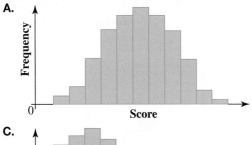

B.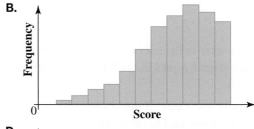

C.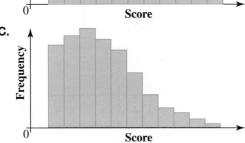

D.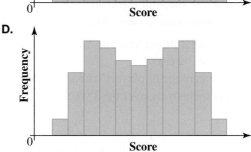

E.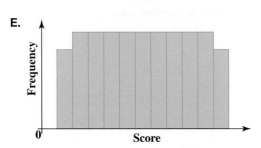

Question 2 (1 mark)

MC The shape of the distribution of the data in the graph is best described as

A. symmetrical.
B. positively skewed.
C. negatively skewed.
D. bimodal.
E. positively skewed with an outlier.

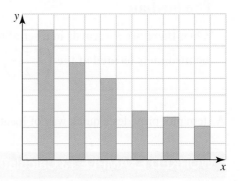

Question 3 (1 mark)

MC How is the distribution of the data in the following stem plot best described?

A. negatively skewed
B. negatively skewed and symmetric
C. positively skewed
D. positively skewed and symmetric
E. symmetric

stem	leaf
0	1
0	2
0	4 4 5
0	5 6 5 7
0	8 8 8 8 9 9
1	0 0 0 1 1 1 1
1	2 2 2 3 3 3
1	4 4 5 5
1	6 7 7
1	8 9

Key 1 | 8 = 18

More exam questions are available online.

1.6 Summary statistics

> **LEARNING INTENTION**
>
> At the end of this subtopic you should be able to:
> - calculate the summary statistics — the median and mode
> - calculate the summary statistics — the interquartile range and range.

1.6.1 Summary statistics — the median and mode

Summary statistics are calculated because they can provide information about:
- where the centre of the distribution lies
- how the distribution is spread.

Median

The **median** is the midpoint of an ordered set of data. Half the data are less than or equal to the median and half the data are greater than or equal to the median.

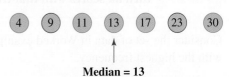

The median

When there are *n* records in a set of ordered data, the median is located at the $\left(\dfrac{n+1}{2}\right)$ th position.

Median = 11

A stem plot provides a quick way of locating a median since the data in a stem plot are already ordered.

WORKED EXAMPLE 13 Calculating the median from a stem plot

Consider the stem plot below, which contains 22 observations. Calculate the median.

Stem	Leaf
2	3 3
2*	5 7 9
3	1 3 3 4 4
3*	5 8 9 9
4	0 2 2
4*	6 8 8 8 9

Key: 3 | 4 = 34

THINK

1. Calculate the median position, where $n = 22$.

2. Determine the 11th and 12th terms.

3. The median is halfway between the 11th and 12th terms.

WRITE

$$\text{Median} = \left(\dfrac{n+1}{2}\right) \text{th position}$$
$$= \left(\dfrac{22+1}{2}\right) \text{th position}$$
$$= 11.5\text{th position}$$

11 th term $= 35$
12 th term $= 38$

Median $= 36.5$

Mode

The mode is the data value that occurs most often. It is a weak measure of the centre of data because it may be a value that is close to the extremes of the data.

> ### The mode
>
> **The mode is the score that occurs most often; that is, it is the score with the highest frequency. If there is more than one score with the highest frequency, then all scores with that frequency are the modes.**

Consider the set of data in Worked example 13: the mode is 48 since it occurs three times and hence is the score with the highest frequency.

1.6.2 Summary statistics — the range and interquartile range

Range

The range of a set of data is the difference between the highest and lowest values in that set. It is one way of measuring the spread of the data.

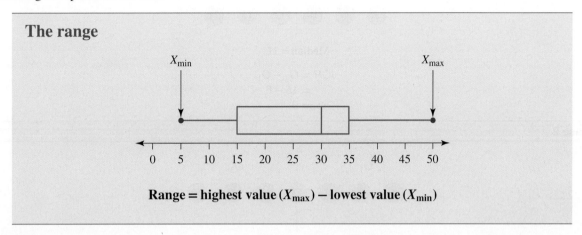

The range

Range = highest value (X_{max}) − lowest value (X_{min})

Interquartile range

The **quartiles** divide a set of data in quarters. The symbols used to refer to quartiles are Q_1, Q_2 and Q_3.

The middle quartile, Q_2, is the median.

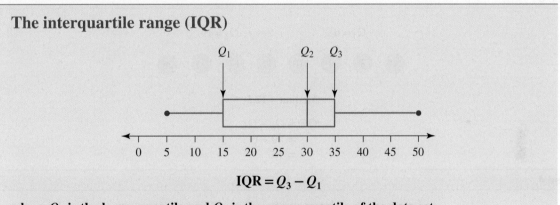

The interquartile range (IQR)

IQR = $Q_3 - Q_1$

where Q_1 is the lower quartile and Q_3 is the upper quartile of the data set.

The interquartile range gives us the range of the middle 50% of values in a set of data.

There are four steps to locating Q_1 and Q_3.

Step 1. Write down the data in ordered form from lowest to highest.

Step 2. Locate the median; that is, locate Q_2.

Step 3. Now consider just the lower half of the set of data. Determine the middle score. This score is Q_1.

Step 4. Now consider just the upper half of the set of data. Determine the middle score. This score is Q_3.

The four cases given below illustrate this method.

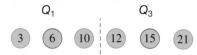

Case 1

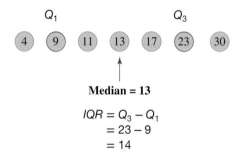

Q_1 Q_3

3 6 10 | 12 15 21

Median = 11

$IQR = Q_3 - Q_1$
$= 15 - 6$
$= 9$

Case 2

Q_1 Q_3

4 9 11 13 17 23 30

↑
Median = 13

$IQR = Q_3 - Q_1$
$= 23 - 9$
$= 14$

Case 3

$Q_1 = 6$ $Q_3 = 19$

1 3 | 9 10 | 15 | 17 | 21 26

Median = 12.5

$IQR = Q_3 - Q_1$
$= 19 - 6$
$= 13$

Case 4

$Q_1 = 10$ $Q_3 = 23$

2 7 | 13 14 17 | 19 21 | 25 29

↑
Median = 17

$IQR = Q_3 - Q_1$
$= 23 - 10$
$= 13$

tlvd-3535

WORKED EXAMPLE 14 Calculating the IQR

The ages of the patients who attended the casualty department of an inner-suburban hospital on one particular afternoon are shown below.

$$14, \quad 3, \quad 27, \quad 42, \quad 19, \quad 17, \quad 73,$$
$$60, \quad 62, \quad 21, \quad 23, \quad 2, \quad 5, \quad 58,$$
$$33, \quad 19, \quad 81, \quad 59, \quad 25, \quad 17, \quad 69$$

Calculate the interquartile range of these data.

THINK	WRITE
1. Order the data.	2, 3, 5, 14, 17, 17, 19, 19, 21, 23, 25, 27, 33, 42, 58, 59, 60, 62, 69, 73, 81
2. Determine the median.	The median is 25, since ten scores lie below it and ten lie above it.
3. Determine the middle score of the lower half of the data.	For the scores: 2, 3, 5, 14, 17, 17, 19, 19, 21, 23, the middle score is 17. So, $Q_1 = 17$.
4. Determine the middle score of the upper half of the data.	For the scores: 27, 33, 42, 58, 59, 60, 62, 69, 73, 81, the middle score is halfway between 59 and 60, 59.5. So, $Q_3 = 59.5$.
5. Calculate the interquartile range.	$\begin{aligned} IQR &= Q_3 - Q_1 \\ &= 59.5 - 17 \\ &= 42.5 \end{aligned}$
6. Write the answer in a sentence.	The interquartile range of the given data is 42.5.

TI | THINK	DISPLAY/WRITE	CASIO | THINK	DISPLAY/WRITE
1. On a Lists & Spreadsheet page, enter the data into Column A and rename the column 'age'.	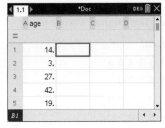	1. On a Statistics screen, enter the data into List 1 and rename the list 'age'.	
2. Press MENU and select: 4: Statistics 1: Stat Calculations 1: One-Variable Statistics Complete the field: Num of Lists: 1 Press ENTER. Complete the fields under One-Variable Statistics as shown. Press ENTER or click OK.		2. Tap: Calc One-variable Complete the fields under Set Calculation as: XList: main\age Freq: 1 Tap OK.	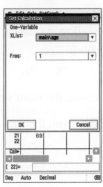

3. The One-Variable Statistics are displayed in Column D.

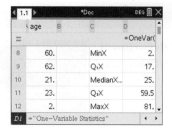

4. The answer is displayed on the screen.

$$Q_1 = 17$$
$$Q_3 = 59.5$$
$$\text{IQR} = 59.5 - 17$$
$$= 42.5$$

3. The One-Variable Statistics are displayed.

4. The answer is displayed on the screen.

$$Q_1 = 17$$
$$Q_3 = 59.5$$
$$\text{IQR} = 59.5 - 17$$
$$= 42.5$$

 Resources

Interactivities The median, the interquartile range, the range and the mode (int-6244)

Mean, median, mode and quartiles (int-6496)

1.6 Exercise

Students, these questions are even better in jacPLUS

 Receive immediate feedback and access sample responses

 Access additional questions

 Track your results and progress

Find all this and MORE in jacPLUS

1. **WE13** The stem plot shows 30 observations. Calculate the median value.

Stem	Leaf
2	1 1 3 4 4 4
2*	5 5 7 8 9
3	0 0 1 3 3 3 3
3*	6 6 7 9
4	0 1 1
4*	6 7 9 9 9

Key: $2 | 1 = 21$

2. The following data represents the number of goals scored by a netball team over the course of a 16-game season.

28, 36, 24, 46, 37, 21, 49, 32, 33, 41, 47, 29, 45, 52, 37, 24

Calculate the team's median number of goals for the season.

3. **WE14** From the following data, calculate the interquartile range.

33, 21, 39, 45, 31, 28, 15, 13, 16,
21, 49, 26, 29, 30, 21, 37, 27, 19,
12, 15, 24, 33, 37, 10, 23, 28, 39

4. The ages of a sample of people surveyed at a concert are shown.

21, 25, 24, 18, 19, 16, 19, 27, 32, 24,
15, 20, 31, 24, 29, 33, 27, 18, 19, 21

Calculate the interquartile range of the data.

5. The data shows the amount of money spent (to the nearest dollar) at the school canteen by a group of students in a week.

3, 5, 7, 12, 15, 10, 8, 9, 21, 5,
7, 9, 13, 15, 7, 3, 4, 2, 11, 8

Calculate the interquartile range for the data set.

6. The amount of money, in millions, changing hands through a large stock company, in one-minute intervals, was recorded as follows.

45.8, 48.9, 46.4, 45.7, 43.8, 49.1, 42.7, 43.1, 45.3, 48.6,
41.9, 40.0, 45.9, 44.7, 43.9, 45.1, 47.1, 49.7, 42.9, 45.1

Calculate the interquartile range for the data.

7. Write the median, range and mode of the sets of data shown in the following stem plots.
The key for each stem plot is $3 \mid 4 = 34$.

a.
Stem	Leaf
0	7
1	2 3
2	2 4 5 7 9
3	0 2 3 6 8 8
4	4 7 8 9 9
5	2 7 8
6	1 3

b.
Stem	Leaf
0	0 0 1 1
0	2 2 3 3
0	4 4 5 5 5 5 5 5 5 5
0	6 6 6 6 7
0	8 8 8 9
1	0 0 1
1	3 3
1	5 5
1	7
1	

c.
Stem	Leaf
0	1
0	2
0	4 4 5
0	6 6 6 7
0	8 8 8 8 9 9
1	0 0 0 1 1 1 1
1	2 2 2 3 3 3
1	4 4 5 5
1	6 7 7
1	8 9

d.
Stem	Leaf
3	1
3	
3	
3	6
3	8 9
4	0 0 1 1 1
4	2 2 3 3 3 3
4	4 5 5 5
4	6 7
4	9

8. For each of the following sets of data, determine the median and the range.

   a. 2, 4, 6, 7, 9
   b. 12, 15, 17, 19, 21
   c. 3, 4, 5, 6, 7, 8, 9
   d. 3, 5, 7, 8, 12, 13, 15, 16
   e. 12, 13, 15, 16, 18, 19, 21, 23, 24, 26
   f. 3, 8, 4, 2, 1, 6, 5

9. a. The number of cars that used the drive-in at a burger restaurant during each
   hour, from 7.00 am until 10.00 pm on a particular day, is shown below.

   $$14, \quad 18, \quad 8, \quad 9, \quad 12,$$
   $$24, \quad 25, \quad 15, \quad 18, \quad 25,$$
   $$24, \quad 21, \quad 25, \quad 24, \quad 14$$

   Calculate the interquartile range of this set of data.

   b. On the same day, the number of cars stopping during each hour when a nearby fried chicken restaurant
   was open is shown.

   $$7, \quad 9, \quad 13, \quad 16, \quad 19, \quad 12, \quad 11, \quad 18, \quad 20, \quad 19, \quad 21, \quad 20, \quad 18, \quad 10, \quad 14$$

   Calculate the interquartile range of these data.

   c. Comment on what these values suggest about the two restaurants.

10. Write down a set of data for which $n = 5$, the median is 6 and the range is 7. Determine if this is the only set
    of data with these parameters.

11. Give an example of a data set where:

    a. the lower quartile equals the lowest score
    b. the IQR is zero.

12. **MC** The quartiles for a set of data are calculated and found to be $Q_1 = 13$, $Q_2 = 18$ and $Q_3 = 25$. Select the
    true statement from the following options.

    A. The interquartile range is 5.
    B. The interquartile range is 7.
    C. The interquartile range is 12.
    D. The median is 12.
    E. The median is 19.

13. For each of the following sets of data, calculate the median, interquartile range, range and mode.

    a. 16, 12, 8,   7, 26, 32, 15, 51, 29, 45,
       19, 11, 6, 15, 32, 18, 43, 31, 23, 23

    b. 22, 25, 27, 36, 31, 32, 39, 29, 20, 30,
       23, 25, 21, 19, 29, 28, 31, 27, 22, 29

    c. 1.2, 2.3, 4.1, 2.4, 1.5, 3.7, 6.1, 2.4, 3.6, 1.2,
       6.1, 3.7, 5.4, 3.7, 5.2, 3.8, 6.3, 7.1, 4.9

14. For each set of data shown in the stem plots, calculate the median, interquartile range, range and mode.

a.
Stem	Leaf
2	3 5 5 6 7 8 9 9
3	0 2 2 3 4 6 6 7 8 8
4	2 2 4 5 6 6 6 7 9
5	0 3 3 5 6
6	2 4
7	5 9
8	2
9	7
10	
11	4

Key: 4 | 2 = 42

b.
Stem	Leaf
1	4
1*	
2	1 4
2*	5 7 8 8 9
3	1 2 2 2 4 4 4 4
3*	5 5 5 6
4	3 4
4*	

Key: 2 | 1 = 21

Compare these values for both data sets.

15. For the data in the stem plots shown, calculate the range, median, mode and interquartile range.

a.
Stem	Leaf
0	1
1	1 4 7
2	3 4 6 7
3	4 6 7 8 9
4	2 3 6 6
5	2 3 5
6	7
7	3

Key: 0 | 1 = 1

b.
Stem	Leaf
40	3 5 7
41	1 1 1 3 4 6 7 8 9
42	0 2 3 6 7 8 9
43	2 3 3 6 8
44	1 2
45	0

Key: 40 | 3 = 403

16. From the following set of data, calculate the median and mode.

$$4, \quad 7, \quad 9, \quad 12, \quad 15, \quad 2, \quad 3, \quad 7, \quad 9, \quad 4, \quad 7,$$
$$9, \quad 2, \quad 8, \quad 13, \quad 5, \quad 3, \quad 7, \quad 5, \quad 9, \quad 7, \quad 10$$

17. The following data represents distances (in metres) thrown during a javelin competition.

$$40.3, \quad 42.8, \quad 41.0, \quad 50.3, \quad 52.2, \quad 46.1, \quad 44.5,$$
$$41.6, \quad 44.3, \quad 47.4, \quad 45.1, \quad 48.8, \quad 46.1, \quad 44.5,$$
$$45.3, \quad 42.9, \quad 41.1, \quad 49.0, \quad 47.5, \quad 40.8, \quad 51.1$$

Use CAS to calculate the interquartile range and median.

18. The data below shows the distribution of golf scores for one day of an amateur tournament.
    a. Use CAS to calculate the median, interquartile range, mode and range.
    b. A golf handicap is a numerical measure that can be used to level the playing field when players of varying abilities compete against each other on a golf course. A par score for a hole is the expected number of shots it should take to complete that hole.
    If par for the course is 72, comment on what the average handicap should be of the players listed.

$$111, \quad 93, \quad 103, \quad 85, \quad 81, \quad 90, \quad 101, \quad 95, \quad 84,$$
$$93, \quad 101, \quad 85, \quad 87, \quad 85, \quad 93, \quad 100, \quad 86, \quad 91,$$
$$93, \quad 95, \quad 93, \quad 99, \quad 95, \quad 93, \quad 92, \quad 96, \quad 93,$$
$$97, \quad 93, \quad 93, \quad 97, \quad 96, \quad 92 \quad 100, \quad 95, \quad 104$$

1.6 Exam questions

▶ **Question 1 (1 mark)**
Source: VCE 2020, Further Mathematics Exam 1, Section A, Q4; © VCAA

MC The histogram below shows the distribution of the *forearm circumference*, in centimetres, of 252 men. Assume that the *forearm circumference* values were all rounded to one decimal place.

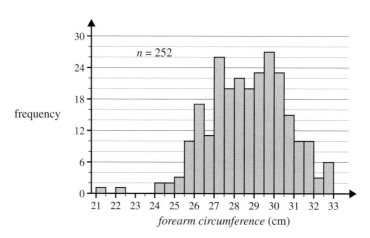

The third quartile (Q_3) for this distribution could be
A. 29.3   **B.** 29.8   **C.** 30.3   **D.** 30.8   **E.** 31.3

▶ **Question 2 (1 mark)**
Source: VCE 2019, Further Mathematics Exam 1, Section A, Q4; © VCAA

MC The stem plot below shows the distribution of mathematics *test scores* for a class of 23 students.

Key: $4\,|\,2 = 42$ $n = 23$

```
4 | 0 1 4 4
5 | 2 7 9 9 9
6 | 5 6 8 8 9 9
7 | 0 0 5 6 7 8
8 | 5 9
```

For this class, the range of *test scores* is
A. 22   **B.** 40   **C.** 45   **D.** 49   **E.** 89

▶ **Question 3 (1 mark)**
Source: VCE 2019, Further Mathematics Exam 1, Section A, Q5; © VCAA

MC The stem plot below shows the distribution of mathematics *test scores* for a class of 23 students.

Key: $4\,|\,2 = 42$ $n = 23$

```
4 | 0 1 4 4
5 | 2 7 9 9 9
6 | 5 6 8 8 9 9
7 | 0 0 5 6 7 8
8 | 5 9
```

For this class, the interquartile range (IQR) of *test scores* is
A. 14.5   **B.** 17.5   **C.** 18   **D.** 24   **E.** 49

More exam questions are available online.

1.7 The five-number summary and boxplots

1.7.1 Boxplots

The five-number summary statistics that we studied in the previous subtopic (min_x, Q_1, Q_2, Q_3, max_x) can be illustrated very neatly in a special diagram known as a **boxplot** (or box-and-whisker diagram). The diagram is made up of a box with straight lines (whiskers) extending from opposite sides of the box.

A boxplot displays the minimum and maximum values of the data together with the quartiles and is drawn with a labelled scale. The length of the box is given by the interquartile range. A boxplot gives us a very clear visual display of how the data are spread out.

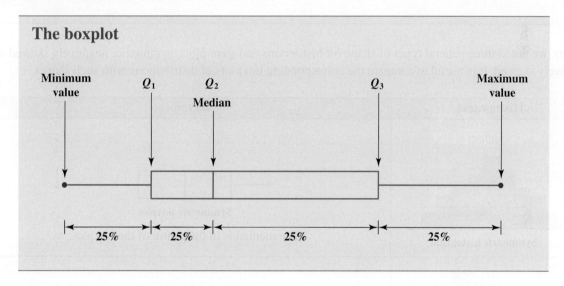

Boxplots can be drawn horizontally or vertically.

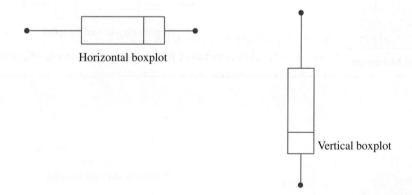

The boxplot below shows the distribution of the part-time weekly earnings of a group of Year 12 students.

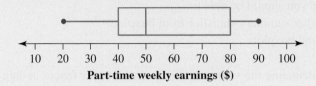

Part-time weekly earnings ($)

Calculate the range, median and interquartile range for the data.

THINK

1. Range = maximum value − minimum value
   The minimum value is 20 and the maximum value is 90.

2. The median is located at the bar inside the box.

3. The ends of the box are at 40 and 80. Substitute the value of $Q_1 = 40$ and $Q_3 = 80$ in the formula, IQR $= Q_3 - Q_1$.

WRITE

Range $= 90 - 20$
$\quad\quad = 70$

Median $= 50$

$Q_1 = 40$ and $Q_3 = 40$
IQR $= 80 - 40$
$\quad\quad = 40$

Earlier, we noted three general types of shape for histograms and stem plots: symmetric, negatively skewed and positively skewed. It is useful to compare the corresponding boxplots of distributions with such shapes.

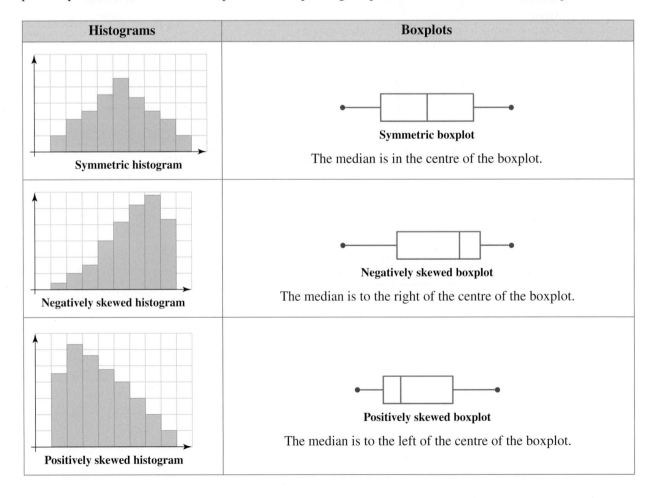

Histograms	Boxplots
Symmetric histogram	**Symmetric boxplot** The median is in the centre of the boxplot.
Negatively skewed histogram	**Negatively skewed boxplot** The median is to the right of the centre of the boxplot.
Positively skewed histogram	**Positively skewed boxplot** The median is to the left of the centre of the boxplot.

Explain whether or not the histogram and the boxplot shown below could represent the same data.

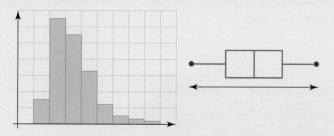

THINK	WRITE
The histogram shows a distribution that is positively skewed. The boxplot shows a distribution that is approximately symmetric.	The histogram and the boxplot could not represent the same data since the histogram shows a distribution that is positively skewed and the boxplot shows a distribution that is approximately symmetric.

The results (out of 20) of oral tests in a Year 12 Indonesian class are:

$$15, \quad 12, \quad 17, \quad 8, \quad 13, \quad 18, \quad 14, \quad 16, \quad 17, \quad 13, \quad 11, \quad 12$$

Display these data using a boxplot and discuss the shape obtained.

THINK	WRITE/DRAW
1. Find the lowest and highest scores, Q_1, the median (Q_2) and Q_3 by first ordering the data.	8, 11, 12, 12, 13, 13, 14, 15, 16, 17, 17, 18 The median score is 13.5. The lower half of the scores are: 8, 11, 12, 12, 13, 13. So, $Q_1 = 12$ The upper half of the scores are: 14, 15, 16, 17, 17, 18. So, $Q_3 = 16.5$ The lowest score is 8. The highest score is 18.
2. Using these five-number summary statistics, draw the boxplot.	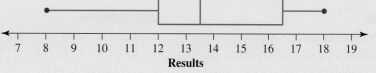
3. Consider the spread of each quarter of the data.	The scores are grouped around 12 and 13, as well as around 17 and 18 with 25% of the data in each section. The scores are more spread elsewhere.

1.7.2 Outliers

When one observation lies well away from other observations in a set, we call it an outlier. Sometimes an outlier occurs because data have been incorrectly obtained or misread. For example, here we see a histogram showing the weights of a group of 5-year-old children.

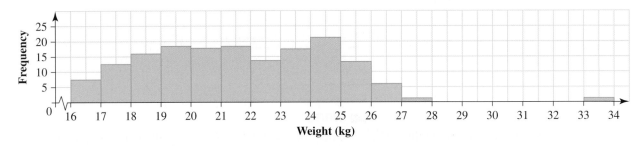

The outlier, 33, may have occurred because a weight was incorrectly recorded as 33 rather than 23 or perhaps there was a child in this group who, for some medical reason, weighed a lot more than their counterparts. When an outlier occurs, the reasons for its occurrence should be checked.

The lower and upper fences

We can identify outliers by calculating the values of the lower and upper fences in a data set. Values that lie either below the lower fence or above the upper fence are outliers.

> ### Calculating outliers
>
> The lower fence $= Q_1 - 1.5 \times \text{IQR}$
>
> The upper fence $= Q_3 + 1.5 \times \text{IQR}$

An outlier is not included in the boxplot, but should instead be plotted as a point beyond the end of the whisker.

tlvd-3536

WORKED EXAMPLE 18 Constructing a boxplot with outliers

The times (in seconds) achieved by the 12 fastest runners in the 100 m sprint at a school athletics meeting are listed below.

$$11.2, \quad 12.3, \quad 11.5, \quad 11.0, \quad 11.6, \quad 11.4,$$
$$11.9, \quad 11.2, \quad 12.7, \quad 11.3, \quad 11.2, \quad 11.3$$

Draw a boxplot to represent the data. Describe the shape of the distribution and comment on the existence of any outliers.

THINK	WRITE/DRAW
1. Determine the five-number summary statistics by first ordering the data and obtaining the interquartile range.	11.0, 11.2, 11.2, 11.2, 11.3, 11.3, 11.4, 11.5, 11.6, 11.9, 12.3, 12.7 Lowest score $= 11.0$ Highest score $= 12.7$ Median $= Q_2 = 11.35$, $Q_1 = 11.2$, $Q_3 = 11.75$ IQR $= 11.75 - 11.2$ $ = 0.55$

2. Identify any outliers by calculating the values of the lower and upper fences.

$$Q_1 - 1.5 \times \text{IQR} = 11.2 - 1.5 \times 0.55$$
$$= 10.375$$

The lowest score lies above the lower fence of 10.375, so there is no outlier below.

$$Q_3 + 1.5 \times \text{IQR} = 11.75 + 1.5 \times 0.55$$
$$= 12.575$$

The score 12.7 lies above the upper fence of 12.575, so it is an outlier and 12.3 becomes the end of the upper whisker.

3. Draw the boxplot with the outlier.

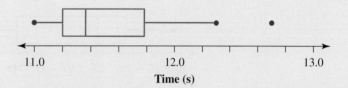

4. Describe the shape of the distribution. Data peak to the left and trail off to the right with one outlier.

The data are positively skewed with 12.7 seconds being an outlier. This may be due to incorrect timing or recording but more likely the top 11 runners were significantly faster than the other competitors in the event.

TI	THINK	DISPLAY/WRITE	CASIO	THINK	DISPLAY/WRITE

TI | THINK

1. On a Lists & Spreadsheet page, enter the data into Column A and rename the column '*time*'.

DISPLAY/WRITE

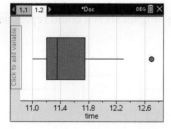

2. Press Ctrl + I and select:
5: Add Data and Statistics
Move the cursor to select '*time*' on the horizontal axis.
Press MENU and select:
1: Plot Type
2: Box Plot

3. The answer is displayed on the screen.

The data is positively skewed with 12.7 seconds being an outlier.

CASIO | THINK

1. On a Statistics screen, enter the data into List 1 and rename the list '*time*'.

DISPLAY/WRITE

2. Tap:
SetGraph
Settings
Complete the fields under Set StatGraphs as:
Draw: On
Type: MedBox
XList: main\time
Freq: 1
Tick Show Outliers.
Tap Set.

3. The answer is displayed on the screen.

The data is positively skewed with 12.7 seconds being an outlier.

1. **WE15** Write down the range, median and interquartile range for the data in the boxplot shown.

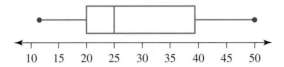

2. Determine the median, range and interquartile range of the data displayed in the boxplot shown.

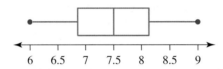

3. For the boxplots shown, write down the range, interquartile range and median of the distributions that each one represents.

a.
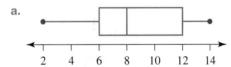

b.
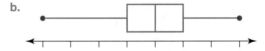

c.
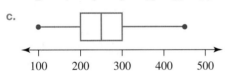

d.
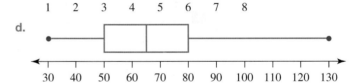

4. **WE16** Explain whether or not the histogram and the boxplot shown could represent the same data.

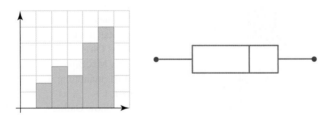

5. Explain whether the histogram and boxplot could represent the same data.

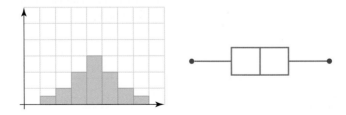

6. Match each histogram below with the boxplot that could show the same distribution.

a.
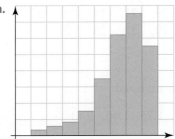

b.
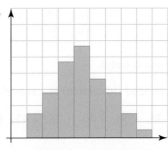

c.
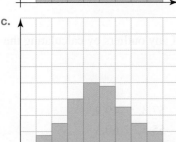

d.
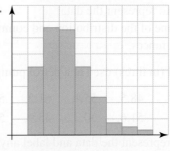

i.

ii.

iii.

iv.

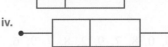

7. **WE17** The results for a Physics test (out of 50) are shown.

$$32, \quad 38, \quad 42, \quad 40, \quad 37, \quad 26, \quad 46, \quad 36, \quad 50,$$
$$41, \quad 48, \quad 50, \quad 40, \quad 38, \quad 32, \quad 35, \quad 28, \quad 30$$

Display the data using a boxplot and discuss the shape obtained.

8. The number of hours Year 12 students spend at their part-time job per week is shown.

$$4, \quad 8, \quad 6, \quad 5, \quad 12, \quad 8, \quad 16, \quad 4, \quad 7, \quad 10, \quad 8, \quad 20, \quad 12, \quad 7, \quad 6, \quad 4, \quad 8$$

Display the data using a boxplot and discuss what the boxplot shows in relation to part-time work of Year 12 students.

9. For each of the following sets of data, construct a boxplot.
   a. 3, 5, 6, 8, 8, 9, 12, 14, 17, 18
   b. 3, 4, 4, 5, 5, 6, 7, 7, 7, 8, 8, 9, 9, 10, 10, 12
   c. 4.3, 4.5, 4.7, 4.9, 5.1, 5.3, 5.5, 5.6
   d. 11, 13, 15, 15, 16, 18, 20, 21, 22, 21, 18, 19, 20, 16, 18, 20

10. **MC** For the distribution shown in the boxplot, it is true to say that:
    A. the median is 30.
    B. the median is 45.
    C. the interquartile range is 10.
    D. the interquartile range is 30.
    E. the interquartile range is 60.

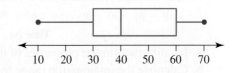

11. The number of clients seen each day over a 15-day period by a tax consultant is:

$$3, 5, 2, 7, 5, 6, 4, 3, 4, 5, 6, 6, 4, 3, 4$$

Represent the data on a boxplot.

12. The maximum daily temperatures (in °C) for the month of October in Melbourne are:

18, 26, 28, 23, 16, 19, 21, 27, 31, 23, 24, 26, 21, 18, 26, 27
23, 21, 24, 20, 19, 25, 27, 32, 29, 21, 16, 19, 23, 25, 27

Represent the data on a boxplot.

13. **WE18** The heights jumped (in metres) at a school high jump competition are listed:

1.35, 1.30, 1.40, 1.38, 1.45, 1.48, 1.30, 1.36, 1.45, 1.75, 1.46, 1.40

a. Draw a boxplot to represent the data.
b. Describe the shape of the distribution and comment on the existence of any outliers by calculating the lower and upper fences.

14. The amount of fuel (in litres) used at a petrol pump for 16 cars is listed:

48.5, 55.1, 61.2, 58.5, 46.9, 49.2, 57.3, 49.9,
51.6, 30.3, 45.9, 50.2, 52.6, 47.0, 55.5, 60.3

Draw a boxplot to represent the data and label any outliers.

15. The number of rides that 16 children had at the annual show are listed below.

8, 5, 9, 4, 9, 0, 8, 7, 9, 2, 8, 7, 9, 6, 7, 8

a. Draw a boxplot to represent the data. Describe the shape of the distribution and comment on the existence of any outliers.
b. Use CAS to draw a boxplot for the data.

16. A concentration test was carried out on 40 students in VCE. The test involved the use of a computer mouse and the ability to recognise multiple images. The less time required to complete the activity, the better the student's ability to concentrate.
The data are shown by the parallel boxplots.

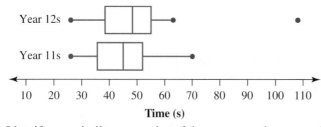

a. Identify two similar properties of the concentration spans for Year 12s and Year 11s.
b. Calculate the interquartile range for Year 12s and Year 11s.
c. Comment on the existence of an outlier in the Year 12s' data by calculating the lower and upper fences.

17. You work in the marketing department of a perfume company. You surveyed people who purchased your perfume, asking them how many times a week they used it.

7,   2,   5,   4,   7,   5,   7,   2,   5,   4,   3,   5,
7,   7,   9,   8,   5,   6,   5,   3,   15,   8,   7,   5

Analyse the data by drawing a boxplot and comment on the existence of any outliers by calculating the lower and upper fences.

18. For the data set shown:

11,   11,   14,   16,   19,   22,   24,   25,   25,   27,   28,   28,   36,   38,   38,   39

a. construct a boxplot by hand
b. comment on the presence of outliers by calculating the lower and upper fences
c. construct a boxplot using CAS and compare the two boxplots.

1.7 Exam questions

Question 1 (1 mark)

Source: VCE 2017, Further Mathematics Exam 1, Section A, Core, Q1; © VCAA

MC The boxplot shows the distribution of the forearm *circumference*, in centimetres, of 252 people.

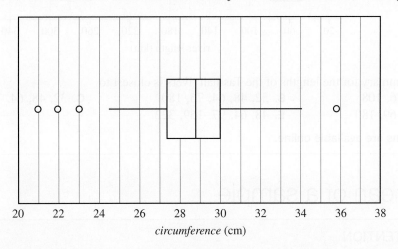

circumference (cm)

The percentage of these 252 people with a forearm *circumference* of less than 30 cm is closest to

   A. 15%   **B.** 25%   **C.** 50%   **D.** 75%   **E.** 100%

Question 2 (1 mark)

Source: VCE 2017, Further Mathematics Exam 1, Section A, Core, Q2; © VCAA

MC The boxplot shows the distribution of the forearm *circumference*, in centimetres, of 252 people.

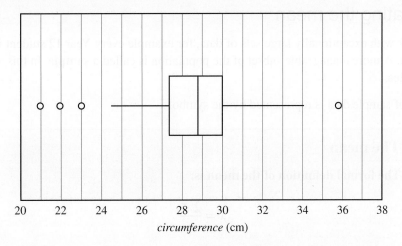

circumference (cm)

The five-number summary for the forearm *circumference* of these 252 people is closest to

   A. 21, 27.4, 28.7, 30, 34   **B.** 21, 27.4, 28.7, 30, 35.9   **C.** 24.5, 27.4, 28.7, 30, 34

   D. 24.5, 27.4, 28.7, 30, 35.9   **E.** 24.5, 27.4, 28.7, 30, 36

▷ **Question 3 (1 mark)**

Source: VCE 2015, Further Mathematics Exam 1, Section A, Q6; © VCAA

MC In New Zealand, rivers flow into either the Pacific Ocean (the Pacific rivers) or the Tasman Sea (the Tasman rivers).

The boxplots below can be used to compare the distribution of the lengths of the Pacific rivers and the Tasman rivers.

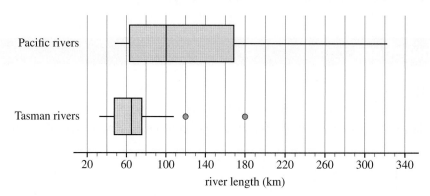

The five-number summary for the lengths of the Tasman rivers is closest to
  A. 32, 48, 64, 76, 108
  B. 32, 48, 64, 76, 180
  C. 32, 48, 64, 76, 322
  D. 48, 64, 97, 169, 180
  E. 48, 64, 97, 169, 322

More exam questions are available online.

1.8 The mean of a sample

LEARNING INTENTION

At the end of this subtopic you should be able to:
- calculate the mean of data sets
- calculate the mean of non-symmetric data
- identify rules for significant figures.

1.8.1 Calculating the mean

We sometimes work with exceptionally large sets of data, for example every Year 12 student in Australia. This is called a **population**. A more manageable subset of the population is called a **sample**. In this section we will be working with samples.

The **mean** of a set of sample data is represented by the symbol $\bar{x}$.

The mean

The formal definition of the mean is:

$$\bar{x} = \frac{\sum x}{n}$$

where:
$\sum x$ **represents the sum of all of the observations in the data set**
n **represents the number of observations in the data set.**

Note that the symbol $\sum$ is the Greek letter sigma, which represents 'the sum of'.

For the set of data {4, 7, 9, 12, 18}:

$$\bar{x} = \frac{4 + 7 + 9 + 12 + 18}{5}$$
$$= 10$$

The mean is also referred to as a *summary statistic* and is a measure of the centre of a distribution. The mean is the point about which the distribution 'balances'.

Consider the masses of seven fruits, given in grams, in the photograph.

The mean is 145 g. The observations 130 g and 160 g 'balance' each other since they are each 15 g from the mean. Similarly, the observations 120 g and 170 g 'balance' each other since they are each 25 g from the mean, as do the observations 100 g and 190 g. Note that the median is also 145 g. That is, for this set of data the mean and the median give the same value for the centre. This is because the distribution is symmetric.

The distribution of non-symmetric data

Case 1

Consider the masses of a set of 7 potatoes, given in grams below.

100, 105, 110, 115, 120, 160, 200

The median of this distribution is 115 g and the mean is 130 g. There are five observations that are less than the mean and only two that are more. In other words, the mean does not give us a good indication of the *centre* of the distribution. However, there is still a 'balance' between observations below the mean and those above, in terms of the spread of all the observations from the mean.

Therefore, the mean is still a useful measure of the central tendency of the distribution, but in cases where the distribution is skewed, the median gives a better indication of the centre.

For a positively skewed distribution, as in the previous case, the mean will be greater than the median. For a negatively skewed distribution, the mean will be less than the median.

Case 2

Consider the data below, showing the weekly income (to the nearest $10) of 10 families living in a suburban street.

$600, $1340, $1360, $1380, $1400, $1420, $1420, $1440, $1460, $1500

In this case, $\bar{x} = \dfrac{13\,320}{10} = \$1332$, and the median is $1410.

One of the values in this set, $600, is clearly an outlier. As a result, the value of the mean is below the weekly income of the other 9 households.

In such a case the mean is not very useful in establishing the centre; however, the 'balance' still remains for this negatively skewed distribution.

As the mean is calculated using the values of the observations, it becomes a less reliable measure of the centre of the distribution when the distribution is skewed or contains an outlier. Because the median is based on the order of the observations rather than their value, it is a better measure of the centre of such distributions.

WORKED EXAMPLE 19 Calculating the mean

Calculate the mean of the set of data shown.

$$10, \quad 12, \quad 15, \quad 16, \quad 18, \quad 19, \quad 22, \quad 25, \quad 27, \quad 29$$

THINK	WRITE
1. Write the formula for calculating the mean.	$\bar{x} = \dfrac{\sum x}{n}$
2. Substitute the values into the formula and evaluate.	$= \dfrac{10 + 12 + 15 + 16 + 18 + 19 + 22 + 25 + 27 + 29}{10}$ $\bar{x} = 19.3$
3. Write the answer in a sentence.	The mean of the given set of data is 19.3.

TI | THINK	DISPLAY/WRITE	CASIO | THINK	DISPLAY/WRITE
1. On a Lists & Spreadsheet page, enter the data into Column A and rename the column 'data'.	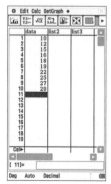	1. On a Statistics screen, enter the data into List 1 and rename the list 'data'.	

2. Press MENU and select: 4: Statistics 1: Stat Calculations 1: One-Variable Statistics Complete the field: Num of Lists: 1 Press ENTER. Complete the fields under One-Variable Statistics as: X1List: data Frequency List: 1 Leave other fields blank. Press ENTER or click OK. The One-Variable Statistics are displayed in Column D.	2. Tap: Calc One-variable Complete the fields under Set Calculation as: XList: main\data Freq: 1 Tap OK.
3. The answer is displayed on the screen. The mean, $\bar{x} = 19.3$	3. The answer is displayed on the screen. The mean, $\bar{x} = 19.3$

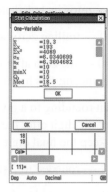

When calculating the mean of a data set, sometimes the answer will contain a long stream of digits after the decimal point.

For example, if we calculate the mean of the data set:

$$44, 38, 55, 61, 48, 32, 49$$

then the mean would be:

$$\bar{x} = \frac{\sum x}{n}$$
$$= \frac{327}{7}$$
$$= 46.714\,285\,71\ \ldots$$

In this case, it makes sense to round the answer to either a given number of decimal places or a given number of **significant figures**.

Rounding to a given number of significant figures

When rounding to a given number of significant figures, we round to the digits in a number that are regarded as 'significant'.

> ### Rules for significant figures
>
> **To determine which digits are significant, we can observe the following rules:**
> - **All digits greater than zero are significant.**
> - **Leading zeros can be ignored. (They are placeholders and are not significant.)**
> - **Zeros included between other digits are significant.**
> - **Zeros included after decimal digits are significant.**
> - **Trailing zeros for integers are not significant (unless specified otherwise).**

The following examples show how these rules work:

0.003 561 — leading digits are ignored, so this has 4 significant figures.

70.036 — zeros between other digits are significant, so this has 5 significant figures.

5.320 — zeros included after decimal digits are significant, so this has 4 significant figures.

450 000 — trailing zeros are not significant, so this has 2 significant figures.

78 000.0 — the zero after the decimal point is considered significant, so the zeros between other numbers are also significant; this has 6 significant figures.

As when rounding to a given number of decimal places, when rounding to a given number of significant figures consider the digit after the specified number of figures. If it is 5 or above, round the final digit up; if it is 4 or below, keep the final digit as is.

5067.37 — rounded to 2 significant figures is 5100

3199.01 — rounded to 4 significant figures is 3199

0.004 931 — rounded to 3 significant figures is 0.004 93

1 020 004 — rounded to 2 significant figures is 1 000 000

1.8 Exercise

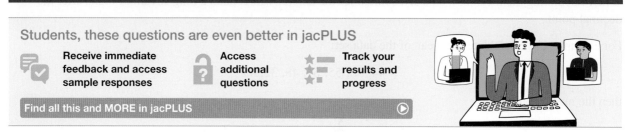
1.  Calculate the mean of the data set shown.

$$9, 12, 14, 16, 18, 19, 20, 25, 29, 33, 35, 36, 39$$

2. Calculate the mean of the data set shown.

$$5.5, 6.3, 7.7, 8.3, 9.7, 6.7, 12.9, 10.5, 9.9, 5.1$$

3. Calculate the mean of each of the following sets of data.
   a. 5, 6, 8, 8, 9 (correct to 2 significant figures)
   b. 3, 4, 4, 5, 5, 6, 7, 7, 7, 8, 8, 9, 9, 10, 10, 12 (correct to 4 significant figures)
   c. 4.3, 4.5, 4.7, 4.9, 5.1, 5.3, 5.5, 5.6 (correct to 5 significant figures)
   d. 11, 13, 15, 15, 16, 18, 20, 21, 22 (correct to 1 decimal place)

4. Calculate the mean of each of the following data sets and explain whether or not it gives a good indication of the centre of the data.
   a. 0.7, 0.8, 0.85, 0.9, 0.92, 2.3
   b. 14, 16, 16, 17, 17, 17, 19, 20
   c. 23, 24, 28, 29, 33, 34, 37, 39
   d. 2, 15, 17, 18, 18, 19, 20

5. The number of people attending sculpture classes at the local TAFE college for each week during the first semester is given.

   15, 12, 15, 11, 14, 8, 14, 15, 11, 10, 7, 11, 12, 14, 15, 14, 15, 9, 10, 11

   Calculate the mean number of people attending each week.
   Express your answer correct to 2 significant figures.

6. **MC** The ages of a group of junior pilots joining an international airline are indicated in the stem plot.

Stem	Leaf
2	1
2	2
2	4 5
2	6 6 7
2	8 8 8 9
3	0 1 1
3	2 3
3	4 4
3	6
3	8

Key: 2 | 1 = 21 years

The mean age of this group of pilots is:

A. 20   B. 28   C. 29   D. 29.15   E. 29.5

7. **MC** The number of people present each week at a 15-week horticultural course is given by the stem plot shown. The mean number of people attending each week was closest to:

A. 17.7
B. 18
C. 19.5
D. 20
E. 21.2

Stem	Leaf
0	4
0*	7
1	2 4
1*	5 5 6 7 8
2	1 2 4
2*	7 7 7

Key: 2 | 4 = 24 people

8. For each of the following, write down whether the mean or the median would provide a better indication of the centre of the distribution.

a. A positively skewed distribution
b. A symmetric distribution
c. A distribution with an outlier
d. A negatively skewed distribution

1.8 Exam questions

Question 1 (1 mark)
Source: VCE 2012, Further Mathematics, Exam 1, Section A, Q3; © VCAA

MC The total weight of nine oranges is 1.53 kg.

Using this information, the mean weight of an orange would be closest to

A. 115 g
B. 138 g
C. 153 g
D. 162 g
E. 170 g

Source: VCE 2012, Further Mathematics, Exam 2, Core, Q3; © VCAA

A weather station records the wind speed and direction each day at 9.00 am.

The wind speed is recorded correct to the nearest whole number.

The parallel boxplots shown were constructed from data that was collected on the 214 days from June to December in 2011.

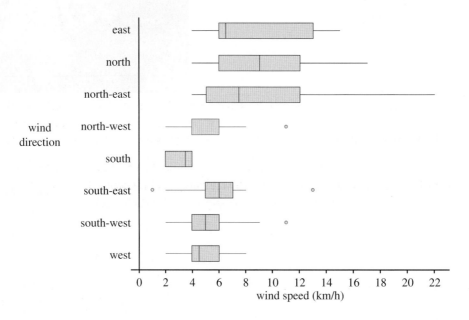

a. Complete the following statements. **(1 mark)**

The wind direction with the lowest recorded wind speed was _____.

The wind direction with the largest range of recorded wind speeds was _____.

b. The wind blew from the south on eight days. **(1 mark)**

Reading from the parallel boxplots above we know that, for these eight wind speeds, the

first quartile $Q_1 = 2 \, \text{km/h}$

median $M = 3.5 \, \text{km/h}$

third quartile $Q_3 = 4 \, \text{km/h}$

Given that the eight wind speeds were recorded to the nearest whole number, write down the eight wind speeds.

⊳ **Question 3 (1 mark)**

MC A golfer has scored an average (mean) of 72.6 shots with his last 15 rounds of golf. He would like to reduce his mean number of shots per round to 71. The total number of shots the golfer can have over the next two rounds to make sure this happens is

A. 71

B. 72.6

C. 118

D. 142

E. 144

More exam questions are available online.

1.9 Standard deviation of a sample

1.9.1 Standard deviation

The **standard deviation** is a measure of how data are spread around the mean. For the set of data {8, 10, 11, 12, 12, 13}, the mean, $\bar{x} = 11$.

The amount that each observation 'deviates' (i.e. differs) from the mean is calculated and shown in the table.

Particular observation (x)	Deviation from the mean ($x - \bar{x}$)
8	$8 - 11 = -3$
10	$10 - 11 = -1$
11	$11 - 11 = 0$
12	$12 - 11 = 1$
12	$12 - 11 = 1$
13	$13 - 11 = 2$

The deviations from the mean are either positive or negative depending on whether the particular observation is lower or higher than the mean. If we added all the deviations from the mean, we would obtain zero.

If we square the deviations from the mean, we overcome the problem of positive and negative deviations cancelling each other out. With this in mind, a quantity known as sample **variance** (s^2) is defined:

$$s^2 = \frac{\sum (x - \bar{x})^2}{n - 1}$$

Variance gives the average of the squared deviations and is also a measure of spread. The standard deviation, which is the square root of variance (s), is more useful as it takes the same unit as the observations (e.g. cm or number of people). Variance would square the units, for example, cm^2 or number of people squared, which is not very practical.

The standard deviation of a sample

In summary,

$$s = \sqrt{\frac{\sum (x - \bar{x})^2}{n - 1}}$$

where:
s **represents sample standard deviation**

$\sum$ **represents 'the sum of'**

x **represents an observation**

$\bar{x}$ **represents the mean**

n **represents the number of observations.**

Note: This is the formula for the standard deviation of a sample. The standard deviation for a population is given by σ (sigma) and is calculated using a slightly different formula, which is outside the scope of this course.

While some of the theory or formulas associated with standard deviation may look complex, the calculation of this measure of spread is straightforward using CAS. Manual computation of standard deviation is therefore rarely necessary; however, application of the formula is required knowledge. Other advantages of calculating the standard deviation will be dealt with later in the topic.

WORKED EXAMPLE 20 Calculating the standard deviation

The price (in cents) per litre of petrol at a service station was recorded each Friday over a 15-week period. The data are given below.

152.4, 160.2, 159.6, 168.6, 161.4, 156.6, 164.8, 162.6,
161.0, 156.4, 159.0, 160.2, 162.6, 168.4, 166.8

Calculate the standard deviation for this set of data, correct to 2 decimal places.

THINK	WRITE
1. Enter the data into CAS to determine the sample statistics.	$s = 4.51592$ $= 4.52 \text{ cents/L}$
2. Round the value correct to 2 decimal places.	

TI | THINK	DISPLAY/WRITE	CASIO | THINK	DISPLAY/WRITE
1. On a Lists & Spreadsheet page, enter the data into Column A and rename the column '*price*'.		1. On a Statistics screen, enter the data into List 1 and rename the list '*price*'.	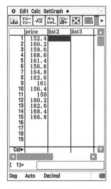
2. Press MENU and select: 4: Statistics 1: Stat Calculations 1: One-Variable Statistics Complete the field: Num of Lists: 1 Press ENTER. Complete the fields under One-Variable Statistics as: X1List: price Frequency List: 1 Leave other fields blank. Press ENTER or click OK. The One-Variable Statistics are displayed in Column D.		2. Tap: Calc One-variable Complete the fields under Set Calculation as: XList: main\price Freq: 1 Tap OK.	

| 3. The answer is displayed on the screen. | The standard deviation, correct to 2 decimal places, is $s = 4.52$ cents/L. | 3. The answer is displayed on the screen. | The standard deviation, correct to 2 decimal places, is $s = 4.52$ cents/L. |

tlvd-3538

WORKED EXAMPLE 21 Calculating the standard deviation

The number of students attending school assemblies during the term is given in the stem plot shown.

Stem	Leaf
0	4
0*	8 8
1	1 3 4
1*	5 8
2	3
2*	5

Key: 1|4 = 14 students

Calculate the standard deviation for this set of data, correct to 4 significant figures, by using the formula $s = \sqrt{\dfrac{\sum(x - \bar{x})^2}{n - 1}}$.

THINK

1. Calculate the value of the mean $(\bar{x})$.

2. Set up a table to calculate the values of $(x - \bar{x})^2$.

WRITE

$\bar{x} = \dfrac{\sum x}{n}$

$= \dfrac{4 + 8 + 8 + 11 + 13 + 14 + 15 + 18 + 23 + 25}{10}$

$= 13.9$

x	$x - \bar{x}$	$(x - \bar{x})^2$
4	−9.9	98.01
8	−5.9	34.81
8	−5.9	34.81
11	−2.9	8.41
13	−0.9	0.81
14	0.1	0.01
15	1.1	1.21
18	4.1	16.81
23	9.1	82.81
25	11.1	123.21
$\sum(x - \bar{x})^2 = 400.9$		

3. Enter the values into the formula to calculate the standard deviation (s).

$$s = \sqrt{\frac{\sum(x - \bar{x})^2}{n-1}}$$

$$= \sqrt{\frac{400.9}{9}}$$

$$= 6.6741 \ldots$$

4. Round the value correct to 4 significant figures. $s = 6.674$ (correct to 4 significant figures)

The standard deviation is a measure of the spread of data from the mean. Consider the two sets of data shown. Each set of data has a mean of 10.

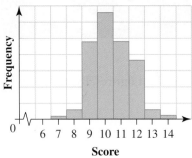

The data has a standard deviation of 1.

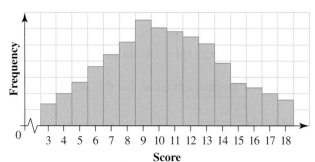

The data has a standard deviation of 3.

As we can see, the larger the standard deviation, the more spread the data are from the mean.

 Resources

 Interactivity The mean and the standard deviation (int-6246)

1.9 Exercise

1. **WE20** The Australian dollar is often compared to the US dollar. The value of the Australian dollar compared to the US dollar each week over a six-month period is shown.

   97.2, 96.8, 98.0, 98.3, 97.5, 95.9, 96.1, 95.6, 95.0, 94.8, 94.6, 94.9, 93.9, 93.3, 92.9, 90.6, 90.4, 89.6, 88.1, 88.3, 89.5, 88.8, 88.5, 88.1

   Calculate the standard deviation for this set of data, correct to 2 decimal places.

2. The test results for a maths class are shown.

$$67, \quad 98, \quad 75, \quad 81, \quad 70, \quad 64, \quad 55, \quad 52, \quad 78, \quad 90, \quad 76,$$
$$92, \quad 78, \quad 80, \quad 83, \quad 59, \quad 67, \quad 45, \quad 78, \quad 48, \quad 82, \quad 62$$

Calculate the standard deviation for the set of data, correct to 2 decimal places.

3. **WE21** Calculate the standard deviation of the data set shown, correct to 4 significant places, by using the formula $s = \sqrt{\dfrac{\sum (x - \bar{x})^2}{n - 1}}$.

Stem	Leaf
1	1 1
1*	6 9
2	4
2*	5 5 8
3	6 9

Key: 2|4 = 24

4. The number of students attending the service learning meetings in preparation for the year's fundraising activities is shown by the stem plot.

Stem	Leaf
1	2 3 3
1*	6 7 8
2	1 4
2*	7
3	
3*	5

Key: 1|3 = 13

Calculate the standard deviation for this set of data, correct to 3 decimal places, by using the formula
$$s = \sqrt{\dfrac{\sum (x - \bar{x})^2}{n - 1}}.$$

5. For each of the following sets of data, calculate the standard deviation correct to 2 decimal places.

a. 3, 4, 4.7, 5.1, 6, 6.2
b. 7, 9, 10, 10, 11, 13, 13, 14
c. 12.9, 17.2, 17.9, 20.2, 26.4, 28.9
d. 41, 43, 44, 45, 45, 46, 47, 49

6. **MC** A new legal aid service has been operational for only 5 weeks. The number of people who have used the service each day during this period is shown. The standard deviation (correct to 2 decimal places) of the data is:

A. 6.00
B. 6.34
C. 6.47
D. 15.44
E. 16.00

Stem	Leaf
0	2 4
0*	7 7 9
1	0 1 4 4 4 4
1*	5 6 6 7 8 8 9
2	1 2 2 3 3 3
2*	7

Key: 1|0 = 10 people

7. First-quarter profit increases for eight leading companies are given below as percentages.

$$2.3, \ 0.8, \ 1.6, \ 2.1, \ 1.7, \ 1.3, \ 1.4, \ 1.9$$

a. Complete the following table.

x	$x - \bar{x}$	$(x - \bar{x})^2$
2.3	0.6625	☐
0.8	☐	0.7014
1.6	−0.0375	☐
2.1	☐	0.2139
1.7	0.0625	☐
1.3	−0.3375	0.1139
1.4	−0.2375	0.0564
1.9	0.2625	0.0689

b. Hence calculate the standard deviation. Express your answer correct to 2 decimal places.

$$\frac{\sum (x - \bar{x})^2}{n - 1} = \frac{1.5987}{\square}$$

$$s = \square$$

8. The number of outgoing phone calls from an office each day over a 4-week period is shown in the stem plot.

Stem	Leaf
0	8 9
1	3 4 7 9
2	0 1 3 7 7
3	3 4
4	1 5 6 7 8
5	3 8

Key: 2|1 = 21 calls

Calculate the standard deviation for this set of data and express your answer correct to 4 significant figures.

9. The speed of 20 cars (in km/h) is monitored along a stretch of road that is a designated 80 km/h zone.

$$80, \ 82, \ 77, \ 75, \ 80, \ 80, \ 81, \ 78, \ 79, \ 78,$$
$$80, \ 80, \ 85, \ 70, \ 79, \ 81, \ 81, \ 80, \ 80, \ 80$$

Calculate the standard deviation of the data, correct to 2 decimal places.

10. Thirty pens are randomly selected off the conveyor belt at a factory and tested to see how long they will last, in hours. Calculate the standard deviation of the data shown, correct to 3 decimal places.

$$20, \ 32, \ 38, \ 22, \ 25, \ 34, \ 47, \ 31, \ 26, \ 29, \ 30, \ 36, \ 28, \ 40, \ 31,$$
$$26, \ 37, \ 38, \ 32, \ 36, \ 35, \ 25, \ 29, \ 30, \ 40, \ 35, \ 38, \ 39, \ 37, \ 30$$

11. Calculate the standard deviation of the data shown, correct to 2 decimal places, representing the temperature of the soil around 25 germinating seedlings.

$$28.9, \quad 27.4, \quad 23.6, \quad 25.6, \quad 21.1, \quad 22.9, \quad 29.6, \quad 25.7, \quad 27.4,$$
$$23.6, \quad 22.4, \quad 24.6, \quad 21.8, \quad 26.4, \quad 24.9, \quad 25.0, \quad 23.5, \quad 26.1,$$
$$23.7, \quad 25.3, \quad 29.3, \quad 23.5, \quad 22.0, \quad 27.9, \quad 23.6$$

12. Companies often use aptitude tests to help decide who to employ. An employer gave 30 potential employees an aptitude test with a total of 90 marks. The scores achieved are shown.

67, 67, 68, 68, 68, 69, 69, 72, 72, 73, 73, 74, 74, 75, 75,
77, 78, 78, 78, 79, 79, 79, 81, 81, 81, 82, 83, 83, 83, 86

Calculate the mean and standard deviation of the data, correct to 1 decimal place.

13. The number of players attending basketball try-out sessions is shown by the stem plot.

Stem	Leaf
2	1
2	
2	
2	6
2	8 9
3	0 0 1 1 1
3	2 3 3 3 3 3
3	4 4 5 5
3	6 7
3	8

Key: 2|1 = 21

Calculate the standard deviation for this set of data, correct to 4 significant figures.

14. The scores obtained out of 20 at a dancing competition are shown.

18.5, 16.5, 18.0, 12.5, 13.0, 18.0, 15.5, 17.5, 18.5, 19.0,
17.0, 12.5, 16.5, 13.5, 19.0, 20.0, 17.5, 19.5, 16.0, 15.5

Calculate the standard deviation of the scores, correct to 3 decimal places.

1.9 Exam questions

Question 1 (1 mark)
Source: VCE 2017, Further Mathematics Exam 1, Section A, Core, Q3; © VCAA

MC The boxplot shows the distribution of the forearm *circumference*, in centimetres, of 252 people.

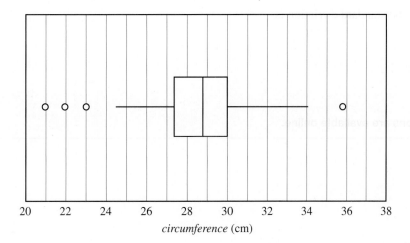

circumference (cm)

The table below shows the forearm *circumference*, in centimetres, of a sample of 10 people selected from this group of 252 people.

Circumference	26.0	27.8	28.4	25.9	28.3	31.5	28.2	25.9	27.9	27.8

The mean, $\bar{x}$, and the standard deviation, s_x, of the forearm *circumference* for this sample of people are closest to

A. $\bar{x} = 1.58$ $s_x = 27.8$
B. $\bar{x} = 1.66$ $s_x = 27.8$
C. $\bar{x} = 27.8$ $s_x = 1.58$
D. $\bar{x} = 27.8$ $s_x = 1.66$
E. $\bar{x} = 27.8$ $s_x = 2.30$

Question 2 (1 mark)

Source: VCE 2015, Further Mathematics Exam 1, Section A, Q3; © VCAA

MC The dot plot displays the difference between female and male life expectancy, in years, for a sample of 20 countries.

difference (years)

The mean ($\bar{x}$) and standard deviation (s) for this data are

A. mean $= 2.32$ standard deviation $= 5.25$
B. mean $= 2.38$ standard deviation $= 5.25$
C. mean $= 5.0$ standard deviation $= 2.0$
D. mean $= 5.25$ standard deviation $= 2.32$
E. mean $= 5.25$ standard deviation $= 2.38$

Question 3 (1 mark)

MC The dot plot below shows the distribution of the number of communicating devices in 15 households.

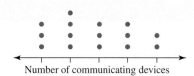

Number of communicating devices

In this distribution, the standard deviation is

A. 1.33
B. 1.37
C. 1.41
D. 1.58
E. 2.03

More exam questions are available online.

1.10 The 68–95–99.7% rule and z-scores

LEARNING INTENTION

At the end of this subtopic you should be able to:
- calculate the normal model and the 68–95–99.7% rule
- determine standardised values (z-scores)
- use z-scores to compare scores from different data sets.

1.10.1 The 68–95–99.7% rule

The heights of students at a graduation ceremony are shown in the histogram.

This set of data is approximately symmetric and has what is termed a *bell* shape. Many sets of data fall into this category and are often referred to as **normal distributions**. Examples are birth weights and people's heights. Data that are normally distributed have their symmetrical, bell-shaped distribution centred on the mean value, $\bar{x}$.

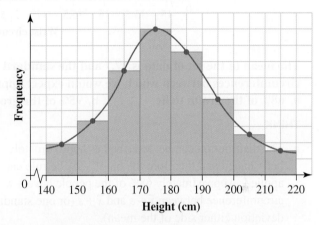

A feature of this type of distribution is that we can predict what percentage of the data lie 1, 2 or 3 standard deviations (s) either side of the mean using what is termed the 68–95–99.7% rule.

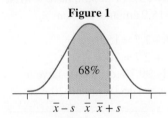

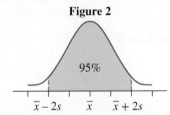

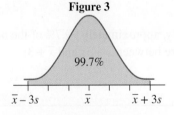

Figure 1	Figure 2	Figure 3
68% of the data lie between the value that is 1 standard deviation below the mean $(\bar{x} - s)$, and the value that is 1 standard deviation above the mean $(\bar{x} + s)$.	95% of the data lie between the value that is 2 standard deviations below the mean $(\bar{x} - 2s)$, and the value that is 2 standard deviations above the mean $(\bar{x} + 2s)$.	99.7% of the data lie between the value that is 3 standard deviations below the mean $(\bar{x} - 3s)$, and the value that is 3 standard deviations above the mean $(\bar{x} + 3s)$.

> ## The 68–95–99.7% rule
>
> **The 68–95–99.7% rule for a bell-shaped curve states that approximately:**
> 1. **68% of data lie within 1 standard deviation either side of the mean**
> 2. **95% of data lie within 2 standard deviations either side of the mean**
> 3. **99.7% of data lie within 3 standard deviations either side of the mean.**

Note: For the purpose of this course, we will use s to denote standard deviations, but for large populations the symbol for standard deviation is s.

The wrist circumferences of a group of people were recorded and the results are shown in the histogram.

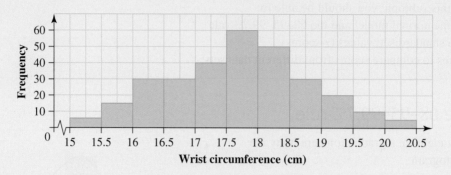

The mean of the set of data is 17.7 and the standard deviation is 0.9. Determine the wrist circumferences between which we would expect approximately:

a. 68% of the group to lie   b. 95% of the group to lie   c. 99.7% of the group to lie.

THINK

a. The distribution can be described as approximately bell-shaped and therefore the 68–95–99.7% rule can be applied. Approximately 68% of the people have a wrist circumference between $\bar{x} - s$ and $\bar{x} + s$ (or one standard deviation either side of the mean).

b. Similarly, approximately 95% of the people have a wrist size between $\bar{x} - 2s$ and $\bar{x} + 2s$.

c. Similarly, approximately 99.7% of the people have a wrist size between $\bar{x} - 3s$ and $\bar{x} + 3s$.

WRITE

a. $\bar{x} - s = 17.7 - 0.9 = 16.8$
$\bar{x} + s = 17.7 + 0.9 = 18.6$
So approximately 68% of the people have a wrist size between 16.8 cm and 18.6 cm.

b. $\bar{x} - 2s = 17.7 - 1.8 = 15.9$
$\bar{x} + 2s = 17.7 + 1.8 = 19.5$
Approximately 95% of people have a wrist size between 15.9 cm and 19.5 cm.

c. $\bar{x} - 3s = 17.7 - 2.7 = 15.0$
$\bar{x} + 3s = 17.7 + 2.7 = 20.4$
Approximately 99.7% of people have a wrist size between 15.0 cm and 20.4 cm.

Using the 68–95–99.7% rule, we can work out the various percentages of the distribution that lie between the mean and 1 standard deviation from the mean and between the mean and 2 standard deviations from the mean and so on. The diagram shown summarises this.

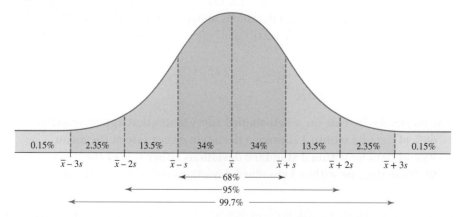

Note: 50% of the data lie below the mean and 50% above due to the symmetry of the distribution about the mean.

tlvd-3539

WORKED EXAMPLE 23 Calculating percentage amounts

The distribution of the masses of packets of Fibre-fill breakfast cereal is known to be bell-shaped with a mean of 250 g and a standard deviation of 5 g. Calculate the percentage of Fibre-fill packets with a mass that is:

a. **less than 260 g**
b. **less than 245 g**
c. **more than 240 g**
d. **between 240 g and 255 g.**

THINK

1. Draw the bell-shaped curve. Label the axis.
$\bar{x} = 250$,
$\bar{x} + s = 255$,
$\bar{x} + 2s = 260$, and so on

WRITE/DRAW

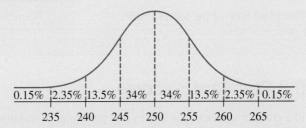

a. 260 g is 2 standard deviations above the mean. Using the summary diagram, we can find the percentage of data that is less than 260 g.

a. Mass of 260 g is 2 standard deviations above the mean. Percentage of distribution less than 260 g is:

$13.5\% + 34\% + 34\% + 13.5\% + 2.35\% + 0.15\% = 97.5\%$
or $13.5\% + 34\% + 50\% = 97.5\%$

b. 245 g is 1 standard deviation below the mean.

b. Mass of 245 g is 1 standard deviation below the mean. Percentage of distribution less than 245 g is:

$13.5\% + 2.35\% + 0.15\% = 16\%$
or $50\% - 34\% = 16\%$

c. 240 g is 2 standard deviations below the mean.

c. Mass of 240 g is 2 standard deviations below the mean. Percentage of distribution more than 240 g is:

$13.5\% + 34\% + 34\% + 13.5\% + 2.35\% + 0.15\% = 97.5\%$
or $13.5\% + 34\% + 50\% = 97.5\%$

d. 240 g is 2 standard deviations below the mean, while 255 g is 1 standard deviation above the mean.

d. Mass of 240 g is 2 standard deviations below the mean. Mass of 255 g is 1 standard deviation above the mean. Percentage of distribution between 240 g and 255 g is:

$13.5\% + 34\% + 34\% = 81.5\%$

The number of matches in a box is not always the same. When a sample of boxes was studied, the number of matches in a box approximated a normal (bell-shaped) distribution with a mean number of matches of 50 and a standard deviation of 2.

In a sample of 200 boxes, determine how many would be expected to have more than 48 matches.

THINK	WRITE
1. Calculate the percentage of boxes with more than 48 matches. Since $48 = 50 - 2$, the score of 48 is 1 standard deviation below the mean.	48 matches is 1 standard deviation below the mean. Percentage of boxes with more than 48 matches is: $34\% + 50\% = 84\%$
2. Calculate 84% of the total sample.	Number of boxes $= 84\%$ of 200 $= 168$ boxes

1.10.2 Standard z-scores

To find a comparison between scores in a particular distribution or in different distributions, we use the z-**score**. The z-score (also called the *standardised score*) indicates the position of a certain score in relation to the mean.

The z-score measures the distance from the mean in terms of the standard deviation. The features of the z-score are summarised in the following diagram.

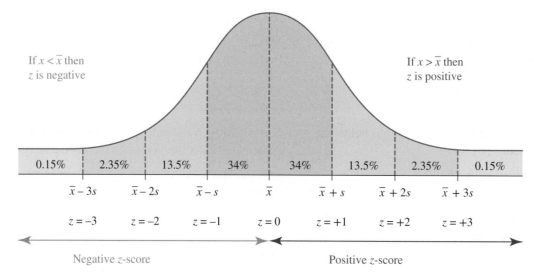

Formula to calculate standard z-scores

$$z = \frac{x - \bar{x}}{s}$$

where:
$x =$ **the score**
$\bar{x} =$ **the mean**
$s =$ **the standard deviation.**

In an IQ test, the mean IQ is 100 and the standard deviation is 15. Dale's test results give an IQ of 130. Calculate this as a z-score.

THINK	WRITE
1. Write the formula.	$z = \dfrac{x - \bar{x}}{s}$
2. Substitute for x, $\bar{x}$ and s.	$= \dfrac{130 - 100}{15}$
3. Calculate the z-score.	$= 2$

Note: Dale's z-score is 2, meaning that his IQ is exactly two standard deviations above the mean.

Not all z-scores will be whole numbers; in fact most will not be. A whole number indicates only that the score is an exact number of standard deviations above or below the mean.

Using Worked example 25, an IQ of 88 would be represented by a z-score of -0.8, as shown.

$$z = \frac{x - \bar{x}}{s}$$
$$= \frac{88 - 100}{15}$$
$$= -0.8$$

The negative value indicates that the IQ of 88 is below the mean.

1.10.3 Comparing data using z-scores

An important use of z-scores is to compare scores from different data sets. Suppose that you scored 74 on your Maths exam and in English your result was 63. In which subject did you achieve the better result?

At first glance, it may appear that the Maths result is better, but this does not take into account the difficulty of the test. A mark of 63 on a difficult English test may be a better result than 74 on an easy Maths test.

The only way that we can fairly compare the results is by comparing each result with its mean and standard deviation. This is done by converting each result to a z-score.

Mathematics	English
$x = 74$, $\bar{x} = 60$, $s = 12$ That implies, $z = \dfrac{x - \bar{x}}{12}$ $= \dfrac{74 - 60}{12}$ $= 1.17$	$x = 63$, $\bar{x} = 50$, $s = 12$ That implies, $z = \dfrac{x - \bar{x}}{12}$ $= \dfrac{63 - 50}{12}$ $= 1.625$
A score of $z = 1.17$ represents a result that is 1.17 standard deviations above the mean.	A score of $z = 1.625$ represents a result that is 1.625 standard deviations above the mean.

The English result is better because the higher z-score shows that the 63 is higher in comparison to the mean of each subject.

Janine scored 82 in her Physics exam and 78 in her Chemistry exam. In Physics, $\bar{x} = 62$ and $s = 10$, while in Chemistry, $\bar{x} = 66$ and $s = 5$.

a. **Write both results as a standardised score.**

b. **Determine which is the better result. Explain your answer.**

THINK

a. 1. Write the formula for each subject.

2. Substitute for x, $\bar{x}$ and s.

3. Calculate each z-score.

b. Explain that the subject with the highest z-score is the better result.

WRITE

a. Physics: $z = \dfrac{x - \bar{x}}{s}$

$= \dfrac{82 - 62}{10}$

$= 2$

Chemistry: $z = \dfrac{x - \bar{x}}{s}$

$= \dfrac{78 - 66}{5}$

$= 2.4$

b. The Chemistry result is better because of the higher z-score.

In each example the circumstances must be analysed carefully to see whether a higher or lower z-score is better. For example, if we were comparing times for runners over different distances, a lower z-score would be better.

 Resources

 Interactivity 68–95–99.7% rule in a normal distribution (int-6247)

1.10 Exercise

Students, these questions are even better in jacPLUS

 Receive immediate feedback and access sample responses

 Access additional questions

Track your results and progress

Find all this and MORE in jacPLUS

1. **WE22** The concentration ability of a group of adults was tested during a short task that they were asked to complete. The length of the concentration span of those involved during the task is shown.

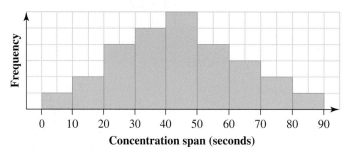

Concentration span (seconds)

The mean, $\bar{x}$, is 49 seconds and the standard deviation, s, is 14 seconds.
Calculate the values between which we would expect approximately:

a. 68% of the group's concentration spans to fall

b. 95% of the group's concentration spans to fall

c. 99.7% of the group's concentration spans to fall.

2. The monthly rainfall in Mathmania Island follows a bell-shaped curve with a mean of 45 mm and a standard deviation of 1.7 mm. Calculate the rainfall range in which we would expect approximately:

   a. 68% of the monthly rainfall totals to lie
   b. 95% of the monthly rainfall totals to lie
   c. 99.7% of the monthly rainfall totals to lie.

3. Copy and complete the entries on the horizontal scale of the following distributions, given that $\bar{x} = 10$ and $s = 2$.

   a.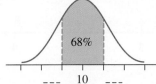
   b.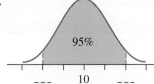
   c.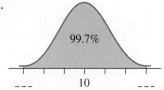

4. Copy and complete the entries on the horizontal scale of the following distributions, given that $\bar{x} = 5$ and $s = 1.3$.

   a.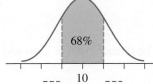
   b.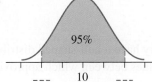
   c.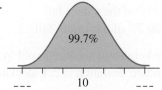

5. In each of the following, decide whether or not the distribution is approximately bell-shaped.

   a.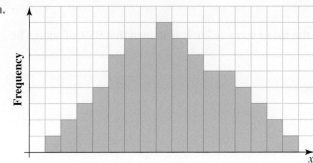
   b.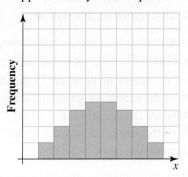

   c.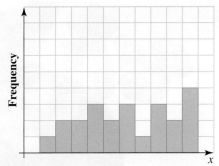
   d.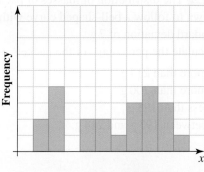

   e.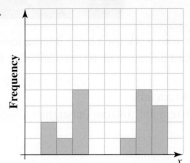
   f.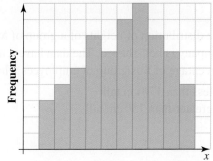

6. A research scientist measured the rate of hair growth in a group of hamsters. The findings are shown in the histogram.

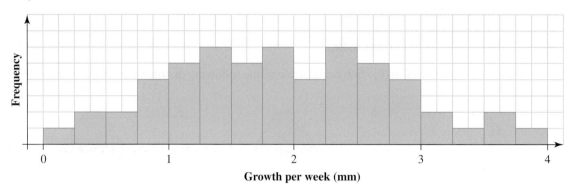

The mean growth per week was 1.9 mm and the standard deviation was 0.6 mm. Calculate the hair growth rates between which approximately:

a. 68% of the values fall
b. 95% of the values fall
c. 99.7% of the values fall.

7. The heights of seedlings sold in a nursery have a bell-shaped distribution. The mean height is 7 cm and the standard deviation is 2. Calculate the values between which approximately:

a. 68% of seedling heights will lie
b. 95% of seedling heights will lie
c. 99.7% of seedling heights will lie.

8. **MC** A distribution of scores is bell-shaped and the mean score is 26. It is known that 95% of scores lie between 21 and 31. It is true to say that:

A. 68% of the scores lie between 23 and 28.
B. 97.5% of the scores lie between 23.5 and 28.5.
C. the standard deviation is 2.5.
D. 99.7% of the scores lie between 16 and 36.
E. the standard deviation is 5.

9. **WE23** The distribution of masses of packets of potato chips is known to follow a bell-shaped curve with a mean of 200 g and a standard deviation of 7 g.
Calculate the percentage of the packets with a mass that is:

a. more than 214 g
b. more than 200 g
c. less than 193 g
d. between 193 g and 214 g.

10. The distribution of heights of a group of Melbourne-based employees who work for a large international company is bell-shaped. The data have a mean of 160 cm and a standard deviation of 10 cm. Calculate the percentage of this group of employees who are:

a. less than 170 cm tall
b. less than 140 cm tall
c. greater than 150 cm tall
d. between 130 cm and 180 cm in height.

11. The number of days taken off in a year by employees of a large company has a distribution that is approximately bell-shaped. The mean and standard deviation of this data are:

    Mean = 9 days   Standard deviation = 2 days

    Calculate the percentage of employees of this company who, in a year, take off:

    a. more than 15 days   b. fewer than 5 days   c. more than 7 days
    d. between 3 and 11 days   e. between 7 and 13 days.

12. **WE24** The number of marbles in a bag is not always the same. From a sample of boxes, it was found that the number of marbles in a bag approximated a normal distribution with a mean number of 20 and a standard deviation of 1.

    In a sample of 500 bags, determine how many would be expected to have more than 19 marbles.

13. The volume of fruit juice in a certain type of container is not always the same. When a sample of these containers was studied, it was found that the volume of juice they contained approximated a normal distribution with a mean of 250 mL and a standard deviation of 5 mL. In a sample of 400 containers, determine how many would be expected to have a volume of:

    a. more than 245 mL

    b. less than 240 mL

    c. between 240 and 260 mL.

14. A particular bolt is manufactured such that the length is not always the same. The distribution of the lengths of the bolts is approximately bell-shaped with a mean length of 2.5 cm and a standard deviation of 1 mm. *Note:* 1 mm is equal to 0.1 cm.

    a. In a sample of 2000 bolts, determine how many would be expected to have a length:

       i. between 2.4 cm and 2.6 cm   ii. less than 2.7 cm   iii. between 2.6 cm and 2.8 cm.

    b. The manufacturer rejects bolts that have a length of less than 2.3 cm or a length of greater than 2.7 cm. In a sample of 2000 bolts, determine how many the manufacturer would expect to reject.

15. **WE25** In a Physics test on electric power, the mean result for the class was 76% with a standard deviation of 9%. Drew's result was 97%. Calculate his mark as a z-score.

16. In a Maths exam, the mean score was 60 and the standard deviation was 12. A student gets 96 marks in their Maths exam. Calculate their mark as a z-score.

17. **WE26** A student received 83 marks in Specialist Mathematics and 88 marks in English. In Specialist Maths the mean was 67 with a standard deviation of 9, while in English the mean was 58 with a standard deviation of 14.

    a. Convert the marks in each subject to a z-score.
    b. Determine which subject is the better result for them. Explain.

18. Ken's English mark was 75 and his Maths mark was 72. In English, the mean was 65 with a standard deviation of 8, while in Maths the mean mark was 56 with a standard deviation of 12.

    a. Convert the mark in each subject to a z-score.
    b. Determine which subject Ken performed better in? Explain your answer.

19. In a major exam, every subject has a mean score of 60 and a standard deviation of 12.5. Clarissa obtains the following marks on her exams. Express each as a z-score.

    a. English 54   b. Maths 78   c. Biology 61   d. Geography 32   e. Art 95

20. The table shows the average number of eggs laid per week by a random sample of chickens with three different types of living conditions.

Number of eggs per week		
Cage chickens	**Barn chickens**	**Free-range chickens**
5.0	4.8	4.2
4.9	4.6	3.8
5.5	4.3	4.1
5.4	4.7	4.0
5.1	4.2	4.1
5.8	3.9	4.4
5.6	4.9	4.3
5.2	4.1	4.2
4.7	4.0	4.3
4.9	4.4	3.9
5.0	4.5	3.9
5.1	4.6	4.0
5.4	4.1	4.1
5.5	4.2	4.1

a. Copy and complete the following table by calculating the mean and standard deviation of barn chickens and free-range chickens correct to 1 decimal place.

Living conditions	Cage	Barn	Free-range
Mean	5.2		
Standard deviation	0.3		

b. A particular free-range chicken lays an average of 4.3 eggs per week. Calculate the z-score relative to this sample correct to 3 significant figures.

The number of eggs laid by free-range chickens is normally distributed. A free-range chicken has a z-score of 1.

c. Approximately what percentage of chickens lay fewer eggs than this chicken?
d. Referring to the table showing the number of eggs per week, prepare five-number summaries for each set of data.

   i. State the median of each set of data.
   ii. Comment on the egg-producing capabilities of chickens in different living conditions.

1.10 Exam questions

Question 1 (1 mark)
Source: VCE 2021, Further Mathematics Exam 1, Section A Core, Q7; © VCAA

MC 800 participants auditioned for a stage musical. Each participant was required to complete a series of ability tests for which they received an overall score.

The overall scores were approximately normally distributed with a mean score of 69.5 points and a standard deviation of 6.5 points.

Only the participants who scored at least 76.0 points in the audition were considered successful.

Using the 68–95–99.7% rule, how many of the participants were considered unsuccessful?
    A. 127   **B.** 128   **C.** 272   **D.** 672   **E.** 673

Source: VCE 2021, Further Mathematics Exam 1, Section A Core, Q8; © VCAA

MC 800 participants auditioned for a stage musical. Each participant was required to complete a series of ability tests for which they received an overall score.

The overall scores were approximately normally distributed with a mean score of 69.5 points and a standard deviation of 6.5 points.

To be offered a leading role in the stage musical, a participant must achieve a standardised score of at least 1.80

Three participants' names and their overall scores are given in the table below.

Participant	Overall score
Amy	81.5
Brian	80.5
Cherie	82.0

Which one of the following statements is true?
- **A.** Only Amy was offered a leading role.
- **B.** Only Cherie was offered a leading role.
- **C.** Only Brian was not offered a leading role.
- **D.** Both Brian and Cherie were offered leading roles.
- **E.** All three participants were offered leading roles.

(▶) **Question 3 (1 mark)**

Source: VCE 2021, Further Mathematics Exam 1, Section A Core, Q9; © VCAA

MC The heights of females living in a small country town are normally distributed:
- 16% of the females are more than 160 cm tall.
- 2.5% of the females are less than 115 cm tall.

The mean and the standard deviation of this female population, in centimetres, are closest to
- **A.** mean = 135   standard deviation = 15
- **B.** mean = 135   standard deviation = 25
- **C.** mean = 145   standard deviation = 15
- **D.** mean = 145   standard deviation = 20
- **E.** mean = 150   standard deviation = 10

More exam questions are available online.

1.11 Review

1.11.1 Summary

doc-38032

1.11 Exercise

Multiple choice

1. **MC** The best distances that a group of twenty 16-year-old competitors achieved in the long jump event at an athletics meeting are recorded.
   This is an example of:
   A. discrete, numerical data.
   B. continuous, numerical data.
   C. categorical data.
   D. discrete, categorical data.
   E. continuous, categorical data.

2. **MC** The observations shown on the stem-and-leaf plot are:

   A. 20, 21, 26, 27, 28, 29, 30, 31, 35
   B. 20, 20, 21, 26, 27, 28, 29, 30, 31, 31, 35
   C. 20, 21, 26, 27, 28, 29, 30, 31, 35
   D. 20, 20, 21, 26, 27, 28, 29, 29, 30, 31, 31, 35
   E. 20, 21, 26, 27, 28, 29, 29, 30, 31, 35

Stem	Leaf
2	0 0 1
2*	6 7 8 9 9
3	0 1 1
3*	5

   Key: $2|1 = 21$

3. **MC** The number of people attending 25 sessions at an outside court, which has a seating capacity of 150, during the Australian Open Tennis Tournament, are displayed in the stem plot. Determine which of the following statements is untrue.

Stem	Leaf
8	5 9
9	2 3 4 9
10	5 5 8
11	0 1 6 6 7
12	4 7 7 8 8
13	5 7 9 9
14	0 2

   Key: $9|2 = 92$

   A. The smallest number of people attending was 85.
   B. Only during six sessions did attendance fall below 100.
   C. The largest number of people attending was 140.
   D. On six occasions the number of people attending was more than 130.
   E. On one occasion the number of people attending was only eight less than the seating capacity.

4. **MC** Determine which of the following frequency tables accurately summarises the scores shown.

$$7, \quad 9, \quad 6, \quad 4, \quad 3, \quad 8, \quad 7, \quad 9, \quad 2,$$
$$1, \quad 3, \quad 4, \quad 7, \quad 6, \quad 2, \quad 8, \quad 9, \quad 4,$$
$$3, \quad 8, \quad 1, \quad 2, \quad 7, \quad 6, \quad 5, \quad 4, \quad 9$$

A.

Score	Frequency
1	2
2	3
3	2
4	5
5	2
6	4
7	6
8	3
9	2

B.

Score	Frequency
1	3
2	2
3	3
4	1
5	5
6	2
7	3
8	4
9	3

C.

Score	Frequency
1	3
2	2
3	1
4	3
5	1
6	5
7	3
8	2
9	4

D.

Score	Frequency
1	2
2	3
3	3
4	4
5	1
6	3
7	4
8	3
9	4

E.

Score	Frequency
1	3
2	4
3	3
4	2
5	2
6	1
7	3
8	2
9	1

5. **MC** The distribution of data shown in the stem plot could best be described as:

A. negatively skewed.
B. negatively skewed with one outlier.
C. positively skewed.
D. positively skewed with one outlier.
E. symmetric.

Stem	Leaf
2	3 4
2*	5 6 8
3	0 1 2 3 4 4
3*	5 5 7 9 9
4	0 1 3 3
4*	6 8 8
5	0 1
5*	6
6	
6*	9

Key: 3 | 1 = 31

6. **MC** The distribution of the data shown in the histogram could best be described as:

A. negatively skewed.
B. negatively skewed with one outlier.
C. positively skewed.
D. positively skewed with one outlier.
E. symmetric.

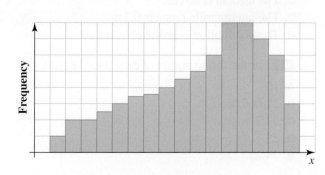

7. **MC** A set of data contains seven observations and has a median of 5 and a range of 3. The set of data could be:

A. 4 4 5 6 7
B. 1 1 2 3 4 5 6
C. 4 5 5 5 6 7 7
D. 1 3 5 5 5 6 7
E. 3 5 7

8. **MC** The median of the set of data shown in the stem plot is:

A. 5
B. 7
C. 9
D. 9.5
E. 37

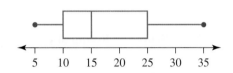

Stem	Leaf
1	2 3
2	0 4 5 7
3	1 2 5 9
4	1 3 6 7
5	2 9 9
6	3

Key: 2 | 4 = 24

9. **MC** For the distribution shown in this boxplot, it is true to say that:

A. the range is 35.
B. the interquartile range is 10.
C. the median is 20.
D. the interquartile range is 25.
E. the median is equal to the interquartile range.

10. **MC** A distribution has a range of 80, an interquartile range of 30 and a median of 50. Determine which one of the following boxplots could represent this distribution.

A.
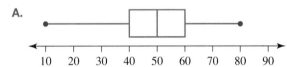

B.
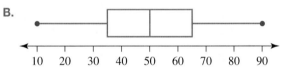

C.
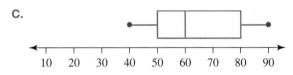

D.
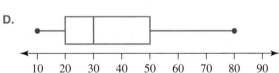

E.
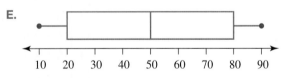

11. **MC** The boxplot represents the lengths of barracuda caught by fishing boats during one day. Determine which one of the following statements is not true about the data.

A. The data contain an outlier.
B. The shortest length is 0.4 m.
C. The median is 60 cm.
D. The interquartile range is 0.2 m.
E. The distribution is positively skewed.

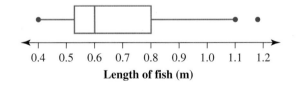

Length of fish (m)

12. **MC** For the following set of data: 14, 18, 20, 21, 23, 23, 24, 25, 29, 30, the mean is:

A. 10
B. 22.666 666 ...
C. 22.7
D. 23
E. 24.222 222 ...

13. **MC** The ages of a group of students entering university for the first time is shown on the stem plot. Calculate the mean age.

A. 18
B. 18.9
C. 19
D. 20.9
E. 21

Stem	Leaf
1	
1*	5 7 7 8 8 8 8 8 8 9
2	0 0 0 1 1
2*	6 8
3	1
3*	5

Key: 1* | 5 = 15 years

14. **MC** Determine in which case below you would expect the mean to be greater than the median.

A.

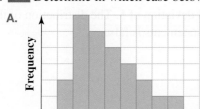

B.

Stem	Leaf
1	
1*	5
2	1 4
2*	5 7 8 8 9
3	1 2 2 3 3 3 4 4
3*	5 5 5 6 Key: 2|4 = 24
	Key: 2|4 = 24
4	3 4
4*	

C. The data: 11, 13, 16, 17, 18, 18, 19, 20

D.

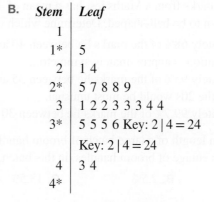

E.

15. **MC** Determine in which case in question 14 the median is not necessarily the better measure of the centre of the data.

16. **MC** The Millers obtained a number of quotes on the price of having their home painted. The quotes, to the nearest hundred dollars, were:

$4200, $4800, $5100, $5000, $4700, $4700, $4600, $4900

The standard deviation for this set of data, to the nearest whole dollar, is:

A. $277   B. $278   C. $324   D. $325   E. $4750

17. **MC** The number of Year 12 students who spent their spare periods studying in the resource centre during each week of terms 3 and 4 is shown on the stem plot.
The standard deviation for this set of data, to the nearest whole number, is:

A. 10
B. 12
C. 14
D. 17
E. 35

Stem	Leaf
0	8
1	
2	5 6 6 7
3	0 2 3 6 9
4	7 9
5	6
6	1

Key: 2 | 5 = 25 students

18. **MC** The lifetime (in hours) of a particular type of battery is known to have a distribution that is bell-shaped. A large number of batteries of this type are sampled and are found to have a mean lifetime of 1200 hours and a standard deviation of 10 hours.

We would expect that approximately 95% of the batteries in the sample would have a lifetime (in hours) between:

A. 10 and 1200
B. 1170 and 1230
C. 1200 and 1210
D. 1180 and 1220
E. 1190 and 1210

19. **MC** A set of marks from a Maths test has a mean of 45 and a standard deviation of 5. The distribution of the marks is known to be bell-shaped. Determine which of the following statements is false.

A. Approximately 68% of the marks lie between 40 and 50.
B. The distribution is approximately symmetric.
C. Approximately 95% of the marks lie between 35 and 50.
D. A mark in the 20s would be most unusual.
E. Approximately 99.7% of the marks lie between 30 and 60.

20. **MC** The mean length of a large batch of broom handles is 120 cm. The data have a standard deviation of 3 cm. The percentage of broom handles, in this batch, that are shorter than 114 cm is:

A. 0.15%   B. 2.5%   C. 13.5%   D. 16%   E. 34%

Short answer

21. Write an example of a variable that produces:

a. categorical data
b. numerical data that are:
    i. discrete   ii. continuous.

22. The money (rounded to the nearest whole dollar) raised by fifteen Year 12 students is shown below.

$$78, \ 84, \ 61, \ 73, \ 71, \ 83, \ 87, \ 65, \ 60, \ 67, \ 71, \ 82, \ 84, \ 79, \ 78$$

Construct a stem plot for the amount raised using:

a. the stems 6, 7 and 8
b. the stems 6, 7 and 8 split into halves
c. the stems 6, 7 and 8 split into fifths.

Discuss your results.

23. Determine the range, median, mode and interquartile range of this set of data.

Stem	Leaf
0	2
0*	5 6 6 8 9
1	0 2 2 4 4 4
1*	5 5 7 8 8 9
2	1 3
2*	6

Key: 1 | 4 = 14

24. The following frequency table shows the speeds of cars recorded by police. The cars were travelling through a 60 km/h zone.

    a. Construct a histogram to display the data.

Class interval	Frequency
50–51.9	3
52–53.9	5
54–55.9	6
56–57.9	7
58–59.9	9
60–61.9	10
62–63.9	9
64–65.9	10
66–67.9	8
68–69.9	5
70–71.9	3
72–73.9	4
74–75.9	2

    Check your work using a CAS calculator.
    b. Draw two conclusions about these data.

25. The money raised (correct to the nearest whole dollar) by each student in a Year 3 class on the school walkathon is shown in the stem plot.

Stem	Leaf
0	8 9
1	2 3 4 7
2	1 2 2 3 5 7 9
3	0 1 4 5 8
4	3 5 6 7
5	1 3 5
6	4 6
7	6

    Key: 0 | 8 = $8

    a. Describe the shape of the distribution of the data.
    b. Describe how this distribution would need to change for it to become symmetric.

26. a. For this set of data, construct a boxplot to display the distribution.

    2, 5, 4, 6, 3, 7, 9, 8, 5, 3,
    1, 4, 6, 8, 7, 5, 2, 9, 5, 6

    b. Describe the shape of the distribution.

27. A local council monitored the use of the public swimming pool over a 12-day period. The attendance is recorded in the table.

Attendance	275	180	201	345	480	150	500	155	302	650	362	410

    a. Calculate the range.
    b. Calculate the percentage of days where the attendance greater than 300. Give your answer correct to 1 decimal place.

28. A chemical component is added to a filtering system on a weekly basis. The amount of chemical component required each week varies. The amounts required (in mL) over the past 20 weeks are shown in the stem plot.

Stem	Leaf
2	1
2	2 2
2	4 4 4 5
2	6 6
2	8 8 9 9
3	0
3	2 2
3	4 5
3	6
3	8

Key: $3\,|\,8 = 0.38$ mL

Calculate, correct to 2 decimal places, the standard deviation of the amount of chemical used.

29. The life spans of dogs of a particular breed follow a bell-shaped distribution. A group of this particular breed at a dog club was found to have a mean life span of 12 years with a standard deviation of 1.2 years.

a. For this group, calculate the expected values between which the life spans of approximately:

   i. 68% of the dogs would lie

   ii. 95% of the dogs would lie

   iii. 99.7% of the dogs would lie

b. Comment on what this information suggests about this breed.

30. Ricardo scored 85 on an entrance test for a job. The test has a mean score of 78 and a standard deviation of 8. Kory sits a similar test and scores 27. In this test, the mean is 18 and the standard deviation is 6. Based on these tests, determine who is the better candidate for the job. Explain your answer.

31. An experimental physicist recorded two measurements using log (base 10) scale. The measurements were 8.2 and 6.6. Determine how many times greater the first measurement was compared to the second. Give your answer correct to the nearest whole number.

32. The quarterly incomes (in dollars) of a sample of 12 local businesses was recorded as follows:

$45 000, $20 000, $360 000, $750 000, $3 200, $1 048 500,
$34 590, $37 250, $65 710, $290 060, $59 070, $8 450

Complete the frequency table below for log (base 10) of incomes using class intervals of width 1.

$\log_{10}$ (income)	Frequency

33. Mr Fahey gives the same test to the two Year 10 classes that he teaches, 10C and 10E. This test is out of 20. The results in 10C are:

4,   7,   7,   9,   9,   10,   10,   11,   12,   13,   14,   14,   15,

15,   15,   16,   17,   17,   17,   17,   18,   18,   18,   19,   19

The results in 10E are:

8,   9,   10,   11,   11,   12,   12,   12,   13,   13,   13,   13,   13,

14,   14,   14,   14,   14,   14,   15,   15,   15,   16,   16,   19

a. For each of these sets of data:

  i. display the data using a histogram or a stem plot. Give reasons for your choices. Also describe the shapes of the distributions

  ii. calculate the median, the interquartile range, the range and the mode

  iii. represent each set of data using a boxplot

  iv. calculate the mean and the standard deviation

  v. state whether the mean or the median is a better measure of the centre of each distribution

  vi. comment as to whether the 68–95–99.7% rule can be applied to either of the distributions.

b. Using the summary statistics you have calculated, comment on and compare the performances of 10C and 10E on the test.

34. A group of office workers and a group of sports instructors were asked to complete 5 minutes of exercise as part of a study of heart rates.
Following the exercise, participants rested for 2 minutes before their pulse rates were measured. The results are set out in the stem plots.

Pulse rates for office workers (beats/min)

Stem	Leaf
7	6
8	
9	5
10	6 7
11	0 2
12	0 1 2 4 6 7 9
13	0 0 4

Pulse rates for sports instructors (beats/min)

Stem	Leaf
6	2 4 8 8 9
7	2 2 3 5 7 9
8	2 8
9	6
10	8

Key: 12 | 4 = 124 beats/min

a. Describe the shape of each distribution.

b. Calculate the median, the interquartile range, the mode and the range for both.

c. Represent each set of data using a boxplot.

d. Calculate the mean and the standard deviation for both sets of data.

e. Comment as to whether the 68–95–99.7% rule can be applied to either of the distributions.

f. Use the summary statistics you have calculated to comment on the pulse rates of each group, noting any differences between the two.

35. A hatch of Atlantic salmon has been reared in a coastal environment over a period of 12 months. The lengths (to the nearest cm) of a sample of 20, out of the total population of 10 000 fish, are shown below.

13, 16, 17, 14, 16, 19, 15, 17, 16, 15,
16, 18, 16, 13, 17, 14, 18, 15, 19, 16

a. Describe the type of data that the variable produces.
b. Construct an appropriate stem plot from the data and use it to describe the shape of the distribution.
c. Using your stem plot, calculate the five-number summary statistics and then draw a boxplot.
d. Describe the shape of the distribution from the boxplot.
e. Determine whether the stem plot or boxplot gives a better indication of the distribution's shape.
f. For a symmetric distribution the mean is the same as the median. Determine whether that is the case here.

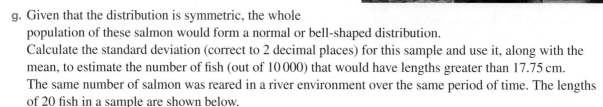

g. Given that the distribution is symmetric, the whole population of these salmon would form a normal or bell-shaped distribution.
Calculate the standard deviation (correct to 2 decimal places) for this sample and use it, along with the mean, to estimate the number of fish (out of 10 000) that would have lengths greater than 17.75 cm.
The same number of salmon was reared in a river environment over the same period of time. The lengths of 20 fish in a sample are shown below.

18, 20, 17, 19, 16, 19, 19, 17, 16, 18,
19, 18, 12, 18, 17, 14, 18, 15, 19, 17

h. Use an appropriate method to help you describe the shape of this distribution.
i. Determine how many of this population of 10 000 salmon would have a length greater than 19.25 cm (calculate the standard deviation of the sample correct to 2 decimal places).
j. Comment on the growth of each hatch of salmon over the 12 months.

1.11 Exam questions

Question 1 (1 mark)
Source: VCE 2021, Further Mathematics Exam 1, Section A Core, Q5; © VCAA
MC The stem plot below shows the *height*, in centimetres, of 20 players in a junior football team.

key: 14|2 = 142 cm		$n = 20$

14	2 2 4 7 8 8 9
15	0 0 1 2 5 5 6 8
16	0 1 1 2
17	9

A player with a height of 179 cm is considered an outlier because 179 cm is greater than
  A. 162 cm   **B.** 169 cm   **C.** 172.5 cm   **D.** 173 cm   **E.** 175.5 cm

Question 2 (7 marks)
Source: VCE 2021, Further Mathematics Exam 2, Section A Core, Q1; © VCAA

In the sport of heptathlon, athletes compete in seven events.

These events are the 100 m hurdles, high jump, shot-put, javelin, 200 m run, 800 m run and long jump.

Fifteen female athletes competed to qualify for the heptathlon at the Olympic Games.

Their results for three of the heptathlon events – high jump, shot-put and javelin – are shown in Table 1.

Table 1

Athlete number	High jump (metres)	Shot-put (metres)	Javelin (metres)
1	1.76	15.34	41.22
2	1.79	16.96	42.41
3	1.83	13.87	46.53
4	1.82	14.23	40.62
5	1.87	13.78	45.64
6	1.73	14.50	42.33
7	1.68	15.08	40.88
8	1.82	13.13	39.22
9	1.83	14.22	42.51
10	1.87	13.62	42.75
11	1.87	12.01	38.12
12	1.80	12.88	42.65
13	1.83	12.68	45.68
14	1.87	12.45	41.32
15	1.78	11.31	42.88

a. Write down the number of numerical variables in Table 1. **(1 mark)**

b. Complete Table 2 below by calculating the mean height jumped for the high jump, in metres, by the 15 athletes. Write your answer in the space provided in the table. **(1 mark)**

Table 2

Statistic	High jump (metres)	Shot-put (metres)
mean		13.74
standard deviation	0.06	1.43

c. In shot-put, athletes throw a heavy spherical ball (a shot) as far as they can.
Athlete number six, Jamilia, threw the shot 14.50 m.
Calculate Jamilia's standardised score (z).
Round your answer to one decimal place. **(1 mark)**

d. In the qualifying competition, the heights jumped in the high jump are expected to be approximately normally distributed.
Chara's jump in this competition would give her a standardised score of $z = -1.0$
Use the 68−95−99.7% rule to calculate the percentage of athletes who would be expected to jump higher than Chara in the qualifying competition. **(1 mark)**

e. The boxplot below was constructed to show the distribution of high jump heights for all 15 athletes in the qualifying competition.

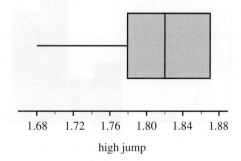

high jump

Explain why the boxplot has no whisker at its upper end. **(1 mark)**

f. For the javelin qualifying competition (refer to Table 1 on previous page), another boxplot is used to display the distribution of athletes' results.

An athlete whose result is displayed as an outlier at the upper end of the plot is considered to be a potential medal winner in the event.

What is the minimum distance that an athlete needs to throw the javelin to be considered a potential medal winner? **(2 marks)**

Use the following information to answer Questions 3 and 4.

▷ Question 3 (1 mark)
Source: VCE 2019, Further Mathematics Exam 1, Section A, Q2; © VCAA

MC The histogram below shows the distribution of the *population size* of 48 countries in 2018.

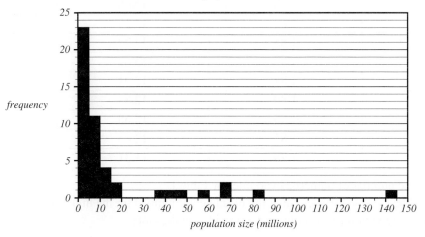

The shape of this histogram is best described as
  A. positively skewed with no outliers.
  B. positively skewed with outliers.
  C. approximately symmetric.
  D. negatively skewed with no outliers.
  E. negatively skewed with outliers.

▷ Question 4 (1 mark)
Source: VCE 2019, Further Mathematics Exam 1, Section A, Q3; © VCAA

MC The histogram below shows the *population size* for these 48 countries plotted on a $\log_{10}$ scale.

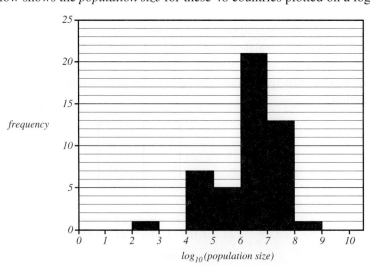

Based on this histogram, the number of countries with a *population size* that is less than 100 000 people is
  A. 1   **B.** 5   **C.** 7   **D.** 8   **E.** 48

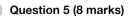 **Question 5 (8 marks)**

Source: VCE 2018, Further Mathematics Exam 2, Section A, Q1; © VCAA

Table 1

City	Congestion level	Size	Increase in travel time (minutes per day)
Belfast	high	small	52
Edinburgh	high	small	43
London	high	large	40
Manchester	high	large	44
Brighton and Hove	high	small	35
Bournemouth	high	small	36
Sheffield	medium	small	36
Hull	medium	small	40
Bristol	medium	small	39
Newcastle-Sunderland	medium	large	34
Leicester	medium	small	36
Liverpool	medium	large	29
Swansea	low	small	30
Glasgow	low	large	34
Cardiff	low	small	31
Nottingham	low	small	31
Birmingham-Wolverhampton	low	large	29
Leeds-Bradford	low	large	31
Portsmouth	low	small	27
Southampton	low	small	31
Reading	low	small	30
Stoke-on-Trent	low	small	29

The data in Table 1 relates to the impact of traffic congestion in 2016 on travel times in 23 cities in the United Kingdom (UK).

The four variables in this data set are:
- *city* – name of city
- *congestion level* – traffic congestion level (high, medium, low)
- *size* – size of city (large, small)
- *increase in travel time* – increase in travel time due to traffic congestion (minutes per day).

a. How many variables in this data set are categorical variables? **(1 mark)**

b. How many variables in this data set are ordinal variables? **(1 mark)**

c. Name the large UK cities with a medium level of traffic congestion. **(1 mark)**

d. Use the data in Table 1 to completed the following two-way frequency table, Table 2. **(2 marks)**

Table 2

Congestion level	City size	
	Small	Large
high	4	
medium		
low		
Total	16	

e. What percentage of the small cities have a high level of traffic congestion? **(1 mark)**

f. Traffic congestion can lead to an increase in travel times in cities. The dot plot and boxplot below both show the increase in travel time due to traffic congestion, in minutes per day, for the 23 UK cities. **(1 mark)**

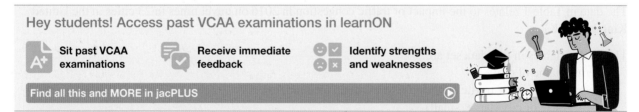

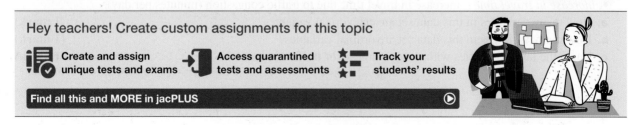

Describe the shape of the distribution of the increase in travel time for the 23 cities.

g. The date value 52 is below the upper fence and is not an outlier. **(1 mark)**
Determine the value of the upper fence.

More exam questions are available online.

Answers

Topic 1 Investigating data distributions

1.2 Types of data

1.2 Exercise

1. D
2. D
3. Numerical: **a, b, c**
   Categorical: **d, e, f**
4. Categorical: **c, d, e, f, g**
5. C
6. B
7. Discrete: **c**
   Continuous: **a, b**
8. C
9. C
10. Categorical and nominal
11. B
12. A
13. Categorical and ordinal
14. Discrete
15. Ordinal

1.2 Exam Questions

Note: Mark allocations are available with the fully worked solutions online.
1. A
2. A
3. E

1.3 Stem plots

1.3 Exercise

1.

Stem	Leaf
1	6 8
2	1 5 8 8 9
3	0 1 3 5 8 9
4	2 8 9

Key: 1 | 6 = 16

2.

Stem	Leaf
0	5
1	1 8 9
2	3 7 9
3	1 2 5 6 7 9
4	1 2 3 5
5	2

Key: 0 | 5 = $5
The busker's earnings are inconsistent.

3.

Stem	Leaf
86	8
87	7
88	0
89	8
90	2 4 8 9
91	
92	0 1 2 6
93	
94	3 9
95	
96	
97	0
98	3 5 9

Key: 86 | 8 = 86.8%

4.

Stem	Leaf
18	5 7 9
19	1 5 6 6 7 9
20	1 3 3 5 9
21	7
22	1

Key: 18 | 5 = 1.85 cm

5.

Stem	Leaf
3	7 9
4	2 9 9
5	1 1 2 3 7 8 9
6	1 3 3 8

Key: 3 | 7 = 37 years
It seems to be an activity for older people.

6. C

7.

Stem	Leaf
1*	9
2	
2*	5 8 8 9 9 9
3	0 0 2 2 2 3 3 4
3*	5 5 7 8 9

Key: 2 | 5 = 25 years
More than half of the parents are 30 or older with a considerable spread of ages, so this statement is not very accurate.

8.

Stem	Leaf
1	8 9
2	7 7 8 9
3	3 6 6 9
4	0 1 6
5	0 2 4 5
6	
7	7

The three highest performing teams are the GWS Giants, Gold Coast Suns and Port Adelaide Power.

9.

Stem	Leaf
33	0
34	
35	0 0 1
36	5
37	3
38	0 0
39	0 0 5
40	0 6
41	0 5
42	1 3
43	0 0
44	
45	0

Key: 33|0 = $330

The stem plot shows a fairly even spread of rental prices with no obvious outliers.

10. a.

Stem	Leaf
2	0 2 2
2*	5 6 8 8
3	3 3 3
3*	7 7 8 9 9 9

Key: 2|0 = 20 points

b.

Stem	Leaf
2	0
2	2 2
2	5
2	6
2	8 8
3	
3	3 3 3
3	
3	7 7
3	8 9 9 9

Key: 2|0 = 20 cm

11. a.

Stem	Leaf
4	3 7 7 8 8 9 9 9
5	0 0 0 0 1 2 2 3

Key: 4|3 = 43 cm

b.

Stem	Leaf
4	3
4*	7 7 8 8 9 9 9
5	0 0 0 0 1 2 2 3
5*	

Key: 4|3 = 43 cm

c.

Stem	Leaf
4	
4	3
4	
4	7 7
4	8 8 9 9 9
5	0 0 0 0 1
5	2 2 3
5	
5	
5	

Key: 4|3 = 43 cm

12. a. 1, 2, 5, 8, 12, 13, 13, 16, 16, 17, 21, 23, 24, 25, 25, 26, 27, 30, 32

b. 10, 11, 23, 23, 30, 35, 39, 41, 42, 47, 55, 62

c. 101, 102, 115, 118, 122, 123, 123, 136, 136, 137, 141, 143, 144, 155, 155, 156, 157

d. 50, 51, 53, 53, 54, 55, 55, 56, 56, 57, 59

e. 1, 4, 5, 8, 10, 12, 16, 19, 19, 21, 21, 25, 29

13. a.

Stem	Leaf
1	5 6 7 7 7 8 9 9 9
2	0 0 0 1 1 1 2 3 3 3

Key: 1|5 = 15 mm

b.

Stem	Leaf
1	
1*	5 6 7 7 7 8 9 9 9 9
2	0 0 0 1 1 1 2 3 3 3
2*	

Key: 1|5 = 15 mm

c.

Stem	Leaf
1	
1	
1	5
1	6 7 7 7
1	8 9 9 9 9
2	0 0 0 1 1 1
2	2 3 3 3
2	
2	
2	

Key: 1|5 = 15 mm

Values are bunched together; they vary little.

14.

Stem	Leaf
7	2 8
8	3 3 5 7 8 8
9	0 1 2 2 3 4 8 9
10	0 2 4
11	2

Key: 7 | 2 = 72 shots

15.

Stem	Leaf
6	
6*	7 8 9
7	1 1 2 2 3 3 3 3 4 4
7*	5 5 5 6 6 7
8	
8*	6

Key: 7 | 1 = 71 net score

The handicapper has done a good job as most of the net scores are around the same scores; that is, in the 70s.

16. a.

Stem	Leaf
6	0 3 9
7	0 1 3 5 6 7 8
8	0 1 3 4 7 8 9 9
9	1 3 7 8

Key: 6 | 0 = 60%

b.

Stem	Leaf
6	0 3
6*	9
7	0 1 3
7*	5 6 7 8
8	0 1 3 4
8*	7 8 9 9
9	1 3
9*	7 8

Key: 6 | 0 = 60%

17. a.

Computer 1	Stem	Computer 2
5	34	0 2 6 8
8 2	35	2 3 5 5 7 8
6 3	36	1 2
6	37	
1 0	38	
2 1	39	
5	40	
0	41	

Key: 34 | 0 = 340 minutes

b. Computer 1 lasts longer but is not as consistent. Computer 2 is more consistent but doesn't last as long.

18. a.

Year 8	Stem	Year 10
9 8	14	
7 5 5 5 3 1 0	15	2 4 6 8 9
8 6 5 4 3 2 1 0	16	0 4 5 7 7 9
5 2 1	17	2 3 4 6 7 8 8
	18	2 5

Key: 14 | 8 = 148 cm

b. As you would expect the Year 10 students are generally taller than the Year 8 students; however, there is a large overlap in the heights.

1.3 Exam questions

Note: Mark allocations are available with the fully worked solutions online.

1. A
2. C
3. C

1.4 Dot plots, frequency tables, histograms, bar charts and logarithmic scales

1.4 Exercise

1.

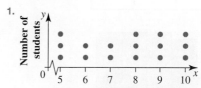

2. See graph at the bottom of the page*

3.

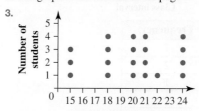

4.

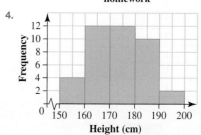

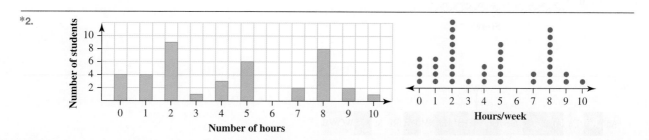

*2.

5. a.

Class	Frequency
1–	1
2–	2
3–	2
4–	6
5–	5
6–	1

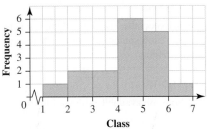

b.

Class interval	Frequency
10–	3
15–	9
20–	10
25–	10
30–	10
35–	1

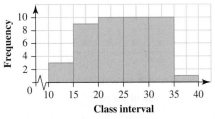

c.

Score	Frequency
0.3	1
0.4	2
0.5	1
0.6	1
0.7	1
0.8	2
0.9	2
1.0	2
1.1	1
1.2	1
1.3	1

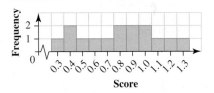

6. a.

Number of dogs	Tally	Frequency
2	||	2
3	|||	3
4	||	2
5	||	2
6	|	1
7	|||||	5
8	||	2
9	|||| |||	8
10	|||	3
11		0
12	||	2
Total		**30**

b.

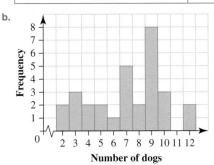

c. Check your histogram against that shown in the answer to **b.**

7.

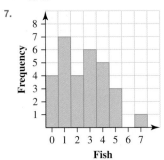

8.

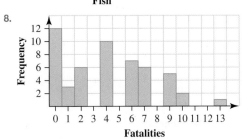

9. See segmented bar chart at the bottom of the page*

---

*9.

GWS

| Ad | Br | Ca | Co | Es | Fr | Ge | GC | Ha | Me | NM | PA | Ri | St | Syd | WCWB |

10. Participation in activities

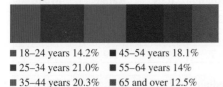

- ■ 18–24 years 14.2% ■ 45–54 years 18.1%
- ■ 25–34 years 21.0% ■ 55–64 years 14%
- ■ 35–44 years 20.3% ■ 65 and over 12.5%

The statement seems untrue as there are similar participation rates for all ages. However, the data don't indicate types of activities.

11.

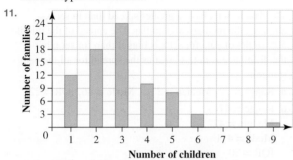

12.

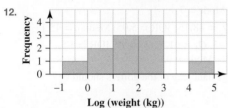

13. D

14. D

15. a. B   **b.** A   **c.** D

16. A

17. B

18. 16

19. 5 times

20. a.

NZ	24.5%
US	13.1%
UK	11.2%
Indonesia	10.1%
China	7.4%
Thailand	10.8%
Fiji	6.3%
Singapore	5.8%
Hong Kong	5.7%
Malaysia	5.1%

b.

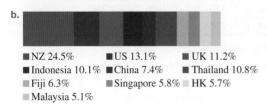

- ■ NZ 24.5%   ■ US 13.1%   ■ UK 11.2%
- ■ Indonesia 10.1% ■ China 7.4%   ■ Thailand 10.8%
- ■ Fiji 6.3%   ■ Singapore 5.8%   ■ HK 5.7%
- ■ Malaysia 5.1%

1.4 Exam questions

Note: Mark allocations are available with the fully worked solutions online.

1. D

2. D

3. E

1.5 Describing stem plots and histograms

1.5 Exercise

1. Positively skewed

2. Negatively skewed

3. E

4. C

5. a. Approximately symmetric
  b. Negatively skewed
  c. Positively skewed
  d. Approximately symmetric
  e. Approximately symmetric
  f. Positively skewed

6. Negatively skewed

7. a. Approximately symmetric, no outliers
  b. Approximately symmetric, no outliers
  c. Approximately symmetric, no outliers
  d. Negatively skewed, no outliers
  e. Negatively skewed, no outliers
  f. Positively skewed, no outliers

8. Positively skewed. This tells us that most of the flight attendants in this group spend a similar number of nights (between 2 and 5) interstate per month. A few stay away more than this and a very few stay away a lot more.

9. a. Approximately symmetric
  b. This tells us that there are few low-weight dogs and few heavy dogs but most dogs have a weight in the range of 10 to 19 kg.

10. a. Approximately symmetric
  b. Most students receive about $8 (give or take $2).

11. a. Positively skewed
  b. i. 15   **ii.** 85%

12. a. Positively skewed
  b. Since most of the data is linked to the lower stems, this suggests that some students do little exercise, but those students who exercise, do quite a bit each week. This could represent the students in teams or in training squads.

13. a. Club A: negatively skewed
    Club B: positively skewed
  b. Since Club A has more members of its bowling team at the higher stems as compared to Club B; you could say Club A has the older team as compared to Club B.
  c. i. Club A: 11 members over 70 years of age
    ii. Club B: 4 members over 70 years of age.

14. a.

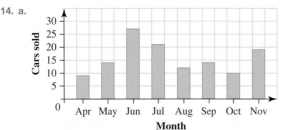

b. Positively skewed

c. June, July and November represent the months with the highest number of sales.

d. This is when the end of financial year sales occur.

1.5 Exam questions

Note: Mark allocations are available with the fully worked solutions online.

1. B
2. B
3. E

1.6 Summary statistics

1.6 Exercise

1. Median $= 33$
2. Median $= 36.5$ goals
3. IQR $= 14$
4. IQR $= 8$
5. IQR $= 6.5$
6. IQR $= 3.3$

7.

	Median	Range	Mode
a	37	56	38, 49
b	5	17	5
c	11	18	8, 11
d	42.5	18	43

8.

	Median	Range
a	6	7
b	17	9
c	6	6
d	10	13
e	18.5	14
f	4	7

9. a. 10

b. 8

c. The IQRs (middle 50%) are similar for the two restaurants, but they don't give any indication about the number of cars in each data set.

10. An example is 2, 3, 6, 8, 9. There are many others.

11. a. The lowest score occurs several times. An example is 5, 5, 5, 5, 6, 9, 10.

b. There are several data points that have the median value. An example is 3, 5, 5, 5, 5, 5, 7.

12. C

13.

	Median	IQR	Range	Mode
a	21	18	45	15, 23, 32
b	27.5	8	20	29
c	3.7	3.0	5.9	3.7

14.

	Median	IQR	Range	Mode
a	42	21	91	46
b	32	7	30	34

The data in set **a** have a greater spread than in set **b**, although the medians are similar. The spread of the middle 50% (IQR) of data for set **a** is bigger than for set **b** but the difference is not as great as the spread for all the data (range).

15. a. Range $= 72$, Median $= 37.5$, Mode $= 46$, IQR $= 22$

b. Range $= 47$
Median $= 422$
Mode $= 411$
IQR $= 20$

16. Median $= 7$, Mode $= 7$

17. $Q_1 = 42.2$, $Q_3 = 48.15$, IQR $= 5.95$, Median $= 45.1$

18. a. Median $= 93$, $Q_1 = 91.5$, $Q_3 = 97$, IQR $= 5.5$, Range $= 30$, Mode $= 93$

b. The average handicap of the golfers should be around 21.

1.6 Exam questions

Note: Mark allocations are available with the fully worked solutions online.

1. C
2. D
3. C

1.7 The five-number summary and boxplots

1.7 Exercise

1. Range $= 39$
Median $= 25$
IQR $= 19$

2. Range $= 3$
Median $= 7.5$
IQR $= 1.4$

3.

	Range	IQR	Median
a	12	6	8
b	7	2	5
c	350	100	250
d	100	30	65

4. They could represent the same data.

5. They could represent the same data.

6. a. iii   b. iv   c. i   d. ii

7.

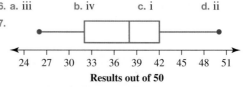

Results out of 50

Negatively skewed; 50% of results are between 32 and 42.

8.

Fairly symmetrical.

9. The boxplots should show the following:

	Min	Q_1	Med	Q_3	Max
a	3	6	8.5	14	18
b	3	5	7	9	12
c	4.3	4.6	5	5.4	5.6
d	11	15.5	18	20	22

10. D

11.

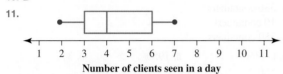

Number of clients seen in a day

12. See graph at the bottom of the page*

13. a.

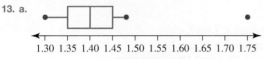

Height jumped

b. The data is symmetrical and 1.75 is an outlier.

14.

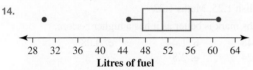

Litres of fuel

30.3 is an outlier.

15. a.

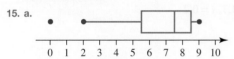

Number of rides

The data are negatively skewed with an outlier on the lower end. The reason for the outlier may be that the person wasn't at the show for long or possibly didn't like the rides.

16. a. Two similar properties: both sets of data have the same minimum value and similar IQR value.

b. Year 12s IQR = 16
Year 11s IQR = 16.5

c. The reason for an outlier in the Year 12s' data may be that the student did not understand how to do the test, or they stopped during the test rather than working continuously.

17. Median = 5

$Q_1 = 4.5$

$Q_3 = 7$

$Min_x = 2$, $Max_x = 15$
IQR = 2.5

15 is an outlier.

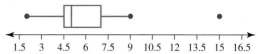

Number of times perform used per week

18. a. Median = 25

$Q_1 = 17.5$

$Q_3 = 32$

$Min_x = 11$, $Max_x = 39$
IQR = 14.5

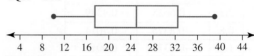

b. No outliers

c. Check your boxplot against that shown in the answer to part **a.**

1.7 Exam questions

Note: Mark allocations are available with the fully worked solutions online.

1. D

2. B

3. B

1.8 The mean of a sample

1.8 Exercise

1. 23.46

2. 8.26

3. a. 7.2   **b.** 7.125   **c.** 4.9875   **d.** 16.8

4. a. 1.0783 No, because of the outlier.

b. 17 Yes

c. 30.875 Yes

d. 15.57 No, because of the outlier.

5. 12

6. D

7. A

8. a. Median   **b.** Mean   **c.** Median   **d.** Median

1.8 Exam questions

Note: Mark allocations are available with the fully worked solutions online.

1. E

2. a. SE, NE   **b.** 2, 2, 2, 3, 4, 4, 4, 4

3. C

***12.**

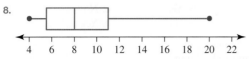

Temperature (°C)

1.9 Standard deviation of a sample

1.9 Exercise

1. 3.54 cents
2. 14.28 %
3. 9.489
4. 7.306
5. a. 1.21   b. 2.36   c. 6.01   d. 2.45
6. C
7. a.

x	$x - \bar{x}$	$(x - \bar{x})^2$
2.3	0.6625	0.4389
0.8	−0.8375	0.7014
1.6	−0.0375	0.0014
2.1	0.4625	0.2139
1.7	0.0625	0.0039
1.3	−0.3375	0.1139
1.4	−0.2375	0.0564
1.9	0.2625	0.0689

   b. 0.48
8. 15.49
9. 2.96 km/h
10. 6.067 pens
11. 2.39°C
12. $\bar{x} = 75.7$, $s = 5.6$
13. 3.786 players
14. 2.331

1.9 Exam questions

Note: Mark allocations are available with the fully worked solutions online.
1. D
2. E
3. B

1.10 The 68–95–99.7% rule and z-scores

1.10 Exercise

1. a. 68% of group's concentration span falls between 35 secs and 63 secs.
   b. 95% of group's concentration span falls between 21 secs and 77 secs.
   c. 99.7% of group's concentration span falls between 7 secs and 91 secs.
2. a. 68% of the monthly rainfall totals lie between 43.3 mm and 46.7 mm
   b. 95% of the monthly rainfall totals lie between 41.6 mm and 48.4 mm
   c. 99.7% of the monthly rainfall totals lie between 39.9 mm and 50.1 mm
3. a. 8 and 12   b. 6 and 14   c. 4 and 16
4. a. 3.7 and 6.3   b. 2.4 and 7.6   c. 1.1 and 8.9

5. a. Yes   b. Yes   c. No
   d. No   e. No   f. Yes
6. a. 1.3 mm and 2.5 mm
   b. 0.7 mm and 3.1 mm
   c. 0.1 mm and 3.7 mm
7. a. 5 and 9   b. 3 and 11   c. 1 and 13
8. C
9. a. 2.50%   b. 50%   c. 16%   d. 81.5%
10. a. 84%   b. 2.5%   c. 84%   d. 97.35%
11. a. 0.15%   b. 2.5%   c. 84%
    d. 83.85%   e. 81.5%
12. 420 bags
13. a. 336 containers
    b. 10 containers
    c. 380 containers
14. a. i. 1360   ii. 1950   iii. 317
    b. 100
15. 2.33
16. 3
17. a. Specialist: $\bar{x} = 67$, $s = 9$
    English: $\bar{x} = 58$, $s = 14$
    $z_s = 1.78$, $z_e = 2.14$
    b. English has the higher result as it has the higher z-score.
18. a. English 1.25, Maths 1.33
    b. Maths mark is better as it has a higher z-score.
19. a. −0.48   b. 1.44   c. 0.08
    d. −2.24   e. 2.8
20. a. Barn: $\bar{x} = 4.4$, $s = 0.3$
    FR: $\bar{x} = 4.1$, $s = 0.2$
    b. 1.18
    c. 84%
    d.

	Cage	Barn	Free range
Min_x	4.7	3.9	3.8
Q_1	5	4.1	4
med	5.15	4.35	4.1
Q_3	5.5	4.6	4.2
Max_x	5.8	4.9	4.4

   i. Cage: 5.15
   Barn: 4.35
   Free: 4.1
   ii. It could be concluded that the more space a chicken has, the fewer eggs it lays because the median is greatest for cage eggs.

1.10 Exam questions

Note: Mark allocations are available with the fully worked solutions online.
1. D
2. C
3. C

1.11 Review

1.11 Exercise

Multiple choice

1. B
2. D
3. C
4. D
5. C
6. A
7. C
8. E
9. E
10. B
11. D
12. C
13. D
14. A
15. B
16. B
17. C
18. D
19. C
20. B

Short answer

21. Many answers possible

22. a.

Stem	Leaf
6	0 1 5 7
7	1 1 3 8 8 9
8	2 3 4 4 7

Key: 6|0 = $6

b.

Stem	Leaf
6	0 1
6*	5 7
7	1 1 3
7*	8 8 9
8	2 3 4 4
8*	7

Key: 6|0 = $6

c.

Stem	Leaf
6	0 1
6	
6	5
6	7
6	
7	1 1
7	3
7	
7	
7	8 8 9
8	
8	2 3
8	4 4
8	7
8	

Key: 6|0 = $6

Stem plot **b** is probably an appropriate display. No real need for stems in fifths.

23. Range = 24, median = 14, mode = 14, interquartile range = 9.5

24. a. See graph at the bottom of the page*

b. The data are approximately symmetrical. More than half the drivers exceeded the limit. The fastest drivers were about 15 km/h over the limit. Many other conclusions are possible.

25. a. Positively skewed

b. There would need to be a shift of some of the amounts in the twenties to the thirties and forties.

26. a.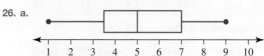

b. Approximately symmetric

27. a. 500   b. 58.3%

28. 0.05 mL

29. a. i. 10.8 and 13.2 years

ii. 9.6 and 14.4 years

iii. 8.4 and 15.6 years

b. There is a large range of life spans for these dogs. The oldest dog is almost twice as old as the youngest.

30. Kory is the better candidate as he has a greater z-score (1.5 compared with 0.875).

31. 40 times

*24. a.
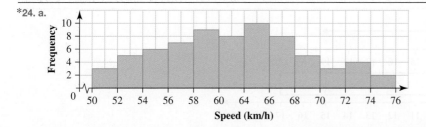

32.

$\log_{10}$ (income)	Frequency
$3 - <4$	2
$4 - <5$	6
$5 - <6$	3
$6 - <7$	1

Extended response

33. a. i. A stem plot is more appropriate since there are only 25 observations in each set.

	10C			**10E**
Stem	*Leaf*		*Stem*	*Leaf*
0			0	
0			0	
0	4		0	
0	7 7		0	
0	9 9		0	8 9
1	0 0 1		1	0 1 1
1	2 3		1	2 2 2 3 3 3 3 3
1	4 4 5 5 5		1	4 4 4 4 4 4 5 5 5
1	6 7 7 7 7		1	6 6
1	8 8 8 9 9		1	9

Key: 1 | 3 = 13

The distribution of 10C is negatively skewed with no outliers.

The distribution of 10E is symmetric with no outliers.

ii.

Class	Median	IQR	Range	Mode
10C	15	7	15	17
10E	13	2.5	11	14

iii. See graph at the bottom of the page*

iv.

Class	Mean	Standard deviation
10C	13.64	4.24
10E	13.2	2.35

v. For 10C the median provides a better indication, whereas for 10E the mean and the median are close anyway.

vi. The $68 - 95 - 99.7\%$ rule can be applied to the distribution of 10E's data since it is approximately bell-shaped.

b. We use the median of the 10C scores to give us an indication of the centre of the distribution. The median of the 10C scores is 15. For 10E, the distribution is symmetric and hence we use the mean to give us an indication of the centre of the distribution. The mean of 10E scores is 13.2. The range of the 10C scores is 15 whereas for 10E the range is 11. Also, the standard deviation for 10C is 4.24 and for 10E it is 2.35. This

means that the scores in 10C are more spread out than those in 10E, which are relatively bunched. So, while in 10C there are more students with higher marks than in 10E, the range of marks in 10C is greater and this would make it a more challenging class to teach.

34. a. Office workers: negatively skewed with outlier. Sports instructors: positively skewed with outlier.

b.

	Office workers	**Sports instructors**
Median	121.5 beats/min	73 beats/min
IQR	19.5 beats/min	14 beats/min
Range	58 beats/min	46 beats/min
Mode	130 beats/min	68, 72 beats/min

c.

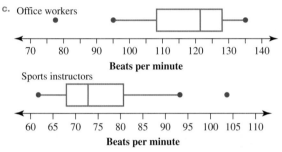

d.

Office workers	**Sports instructors**
$\bar{x} = 116.8$ beats/min	$\bar{x} = 76.9$ beats/min
$s = 15.3$ beats/min	$s = 12.4$ beats/min

e. Not used since the distributions are not bell-shaped.

f. Office workers: Pulse rates are generally very high, clustered around $120 - 130$ beats/min. Also, there is one person whose rate was much lower than the rest. This outlier (76) produces a large range and makes the mean slightly lower than the median. As a result, the median is a more appropriate measure of the centre of the data rather than the mean.

Sports instructors: Pulse rates are generally low, clustered around $60 - 70$ beats/min, although there are a few people with rates much higher, which makes the mean slightly higher than the median and also produces quite a large range. As a result of the skewed distribution the median is the more appropriate measure of the centre of the data rather than the mean, although there is little difference between these values.

*33. a. iii.

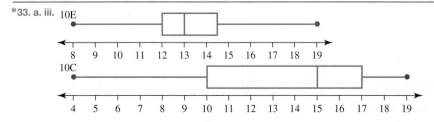

35. a. Discrete, numerical data

b.

Stem	Leaf
1	
1	3 3
1	4 4 5 5 5
1	6 6 6 6 6 6 7 7 7
1	8 8 9 9

Key: $1\,|\,3 = 13$ cm

Looks to be slightly negatively skewed.

c. $Q_2 = 16$ cm
$Q_1 = 15$ cm
$Q_3 = 17$ cm
Lowest score $= 13$ cm
Highest score $= 19$ cm

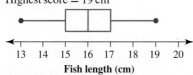

Fish length (cm)

d. Symmetric

e. Given the data itself, the boxplot does.

f. Mean $= 16$ cm $=$ median

g. $s = 1.75$ cm, 1600

h.

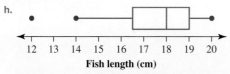

Fish length (cm)

Negatively skewed with an outlier at the lower end

i. 1600

j. The river fish seem to be larger overall. Only 1600 of the coastal fish lie above 17.75 cm, whereas 1600 of the river fish lie above 19.25 cm. All the quartiles for the river fish are higher than those for the coastal ones. It would seem that the river fish have grown more than the coastal fish.

1.11 Exam questions

Note: Mark allocations are available with the fully worked solutions online.

1. E

2. a. Three

b. 1.81 m

c. 0.5

d. 84%

e. Q_3 and the maximum are the same value of 1.87.

f. 45.88 m

3. B

4. D

5. a. There are three categorical variables: City, Congestion level and Size.

b. There are two categories that are ordinal: Congestion level and Size.

c. The large cities with medium traffic congestion levels are Newcastle-Sunderland and Liverpool.

d.

Congestion level	City size	
	Small	Large
high	4	2
medium	4	2
low	8	3
Total	16	7

e. 25%

f. The distribution of the increase in travel time is positively skewed.

g. The value of the upper fence is 52.5.

2 Investigating associations between two variables

LEARNING SEQUENCE

Fully worked solutions for this topic are available online.

2.1 Overview

2.1.1 Introduction

Recent research by the University of Sydney using data from 80 000 adults over the age of 30 determined that strength-bearing exercise and cardio exercise are equally important in the prevention of premature death, but strength-bearing exercise may be more effective when applied to premature death from cancer.

In this study, researchers considered the relationships between many variables that pertained to the 80 000 people in their study. These variables would have included their age, diet and pre-existing health issues. After taking these variables into consideration, mortality rates could be compared based on the type of exercise undertaken.

Detailed studies such as this require researchers with excellent skills in data science. Data scientists, engineers and statisticians are needed in every industry, university and government department to help set up data collection infrastructure and to provide in-depth analysis of the data to assist decision makers. Some of these roles require programming skills, statistical skills, technical skills, written and oral communication skills or any combination of these. Small and large businesses also employ consultants to provide timely analysis of available data to grow and sustain their businesses. There is a vast range of careers within the data science field, and it is an area that will continue to provide employment opportunities into the future.

KEY CONCEPTS

This topic covers the following key concepts from the VCE Mathematics Study Design:
- response and explanatory variables and their role in investigating associations between variables
- contingency (two-way) frequency tables, their associated bar charts (including percentage segmented bar charts) and their use in identifying and describing associations between two categorical variables
- back-to-back stem plots, parallel dot plots and boxplots and their use in identifying and describing associations between a numerical variable and a categorical variable
- scatterplots and their use in identifying and qualitatively describing the association between two numerical variables in terms of direction (positive/negative), form (linear/non-linear) and strength (strong/moderate/weak)
- answering statistical questions that require a knowledge of the associations between pairs of variables
- Pearson correlation coefficient, r, and its calculation and interpretation
- cause and effect; the difference between observation and experimentation when collecting data and the need for experimentation to definitively determine cause and effect.

Source: VCE Mathematics Study Design (2023–2027) extracts © VCAA; reproduced by permission

2.2 Response and explanatory variables

2.2.1 The definition of response and explanatory variables

Many sets of data involve two variables, where one variable affects the other. If the values of one variable 'respond' to the values of another variable, then the former variable is referred to as the **response** (dependent) **variable**. An **explanatory** (independent) **variable** is a factor that influences the response (dependent) variable and vice versa.

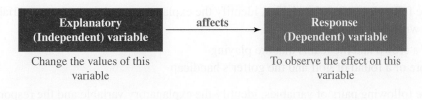

For example:

Change the level of education ——————→ To observe the effect on the salary

Change the daily temperature ——————→ To observe the effect on the sales of air conditioners

Change the amount of sunlight ——————→ To observe the effect on the height of a plant

Change the homework hours ——————→ To observe the effect on the test result

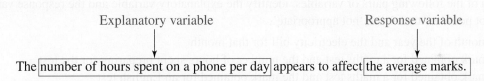

tlvd-3542

WORKED EXAMPLE 1 Identifying response and explanatory variables

For each of the following pairs of variables, identify the explanatory variable and the response variable. If this is not possible, then write 'not appropriate'.
a. The number of visitors at a local swimming pool and the daily temperature
b. The blood group of a person and his or her favourite streaming service

THINK	WRITE
a. It is reasonable to expect that the number of visitors at the swimming pool on any day will respond to the temperature on that day (and not the other way around).	a. Daily temperature is the explanatory variable; number of visitors at a local swimming pool is the response variable.
b. Common sense suggests that the blood type of a person does not respond to the person's streaming service preferences. Similarly, the choice of a streaming service does not respond to a person's blood type.	b. Not appropriate

1. **WE1** For each of the following pairs of variables, identify the explanatory and the response variable. If this is not possible, then write 'not appropriate'.

   a. The number of air conditioners sold and the daily temperature
   b. The age of a person and their favourite colour

2. For each of the following pairs of variables, identify the explanatory and the response variable. If this is not possible, then write 'not appropriate'.

   a. The size of a crowd and the teams that are playing
   b. The net score of a round of golf and the golfer's handicap

3. For each of the following pairs of variables, identify the explanatory variable and the response variable. If this is not possible, then write 'not appropriate'.

   a. The age of an AFL footballer and their annual salary
   b. The growth of a plant and the amount of fertiliser it receives
   c. The number of books read in a week and the eye colour of the readers
   d. The voting intentions of a person and their weekly consumption of red meat
   e. The number of members in a household and the size of the house

4. For each of the following pairs of variables, identify the explanatory variable and the response variable. If this is not possible, then write 'not appropriate'.

   a. The month of the year and the electricity bill for that month
   b. The mark obtained for a maths test and the number of hours spent preparing for the test
   c. The mark obtained for a maths test and the mark obtained for an English test
   d. The cost of grapes (in dollars per kilogram) and the season of the year

5. **MC** In a scientific experiment, the explanatory variable was the amount of sleep (in hours) a new parent got per night during the first month following the birth of their baby. The response variable would most likely have been:

   A. the number of times (per night) the baby woke up for a feed.
   B. the blood pressure of the baby.
   C. the parent's reaction time (in seconds) to a certain stimulus.
   D. the level of alertness of the baby.
   E. the amount of time (in hours) spent by the parent on reading.

6. A paediatrician investigated the relationship between the amount of time children aged two to five spend outdoors and their vitamin D levels. 'The amount of time children spend outdoors is the explanatory variable.' Determine whether this statement is true or false.

7. Alex works as a personal trainer at the local gym. He wishes to analyse the relationship between the number of weekly training sessions and the weekly weight loss of his clients.
'The weekly weight loss of his clients is the explanatory variable.' Determine whether this statement is true or false.

8. Two variables investigated are the number of minutes on a basketball court and the number of points scored:
   a. Determine which is the explanatory variable.
   b. Determine which is the response variable.

9. Callum decorated his house with Christmas lights for everyone to enjoy. He investigated two variables, the number of Christmas lights he has and the size of his electricity bill.
   a. Determine which is the response variable.
   b. Determine which is the explanatory variable.

10. In a study, the two variables investigated are the number of hours a child spends playing an imaginary game with an adult and the vocabulary of the child.
   a. Determine which is the explanatory variable.
   b. Determine which is the response variable.

2.2 Exam questions

Question 1 (1 mark)

MC A survey of 103 students who recently joined the workforce was taken. The focus of the survey was to determine the area of employment the students had joined. The results are in the table.

		Age	
		20–25	26–30
Area of employment	Emergency services	12	7
	Health services	13	19
	Hospitality	24	20
	Clerical	7	34
	Retail	25	22

The two variables of gender and employment area would be defined as
   A. both numerical.
   B. both categorical.
   C. numerical and categorical respectively.
   D. categorical and numerical respectively.
   E. categorical and ordinal respectively.

Question 2 (1 mark)

MC The relationship between pulse rate (beats per minute) and exercise regime (walking, jogging, running) is best displayed using
   A. a histogram.
   B. a scatterplot.
   C. a back-to-back stem plot.
   D. a time series plot.
   E. parallel plots.

Question 3 (1 mark)

MC Paul owns a gym specialising in weight loss. He has collected data on each client in order to determine whether there is a relationship between the number of weekly training sessions and the weekly weight loss. Paul wants to graph the data so that he can more easily interpret the result. Determine which of the following statements describes how Paul should graph the data.

- **A.** When graphing, the number of weekly training sessions should be on the vertical axis as it is the response variable.
- **B.** When graphing, the weekly loss should be on the vertical axis because it is the explanatory variable.
- **C.** When graphing, the weekly weight loss should be on the horizontal axis because it is the explanatory variable.
- **D.** When graphing, the weekly training sessions should be on the horizontal axis because it is the explanatory variable.
- **E.** When graphing, a scatterplot should be used because there is neither an independent nor response variable.

More exam questions are available online.

2.3 Contingency (two-way) frequency tables and segmented bar charts

LEARNING INTENTION

At the end of this subtopic you should be able to:
- construct contingency frequency tables and segmented bar charts
- convert the numbers in a table to percentages.

2.3.1 Contingency (two-way) frequency tables

Contingency (two-way) tables are an excellent tool, when we examine the relationship between two categorical variables.

Once a two-way table is formed, **marginal distributions** and **conditional distributions** can both be found. Marginal distributions are the sums (totals) of the row or column and are found in the margins of the table. A conditional distribution is a sub-population (sample) and this is found in the middle of the table.

If we look at mobile phone preference, as shown in the table, the marginal distributions are the totals — the numbers in the two green rectangular boxes.

		Phone			Total
		Apple	Samsung	Nokia	
Age group	20–50	13	9	3	25
	Above 50	17	7	1	25
	Total	30	16	4	50

Marginal distributions

A conditional distribution is a sub-population, so if we look at people who prefer Samsung, the conditional distribution is shown by the numbers in the pink rectangular box.

		Phone			Total
		Apple	Samsung	Nokia	
Age group	20–50	13	9	3	25
	Above 50	17	7	1	25
Total		30	16	4	50

↓
Conditional distribution

tlvd-3543

WORKED EXAMPLE 2 Constructing a two-way table

At a local shopping centre, 34 voters aged 18 to 45 and 23 voters aged over 45 were asked which of the two major political parties they preferred. Eighteen voters aged 18 to 45 and 12 voters aged over 45 preferred Labor.
Display these data in a two-way (contingency) table, and calculate the party preference for voters aged over 45 and voters aged 18 to 45.

THINK

1. Draw a table. Record the respondents' gender in the columns and party preference in the rows of the table.

WRITE/DRAW

		Voter's age		Total
		18–45	Over 45	
Party preference	Labor			
	Liberal			
Total				

2. We know that 34 voters aged 18 to 45 and 23 voters aged over 45 were asked. Put this information into the table and fill in the total.

   We also know that 18 voters aged 18 to 45 and 12 voters aged over 45 preferred Labor. Put this information in the table and calculate the total of people who preferred Labor.

		Voter's age		Total
		18–45	Over 45	
Party preference	Labor	18	12	30
	Liberal			
Total		34	23	57

▶

3. Fill in the remaining cells. For example, to calculate the number of voters aged 18 to 45 who preferred the Liberals, subtract the number of voters aged 18 to 45 preferring Labor from the total number of voters aged 18 to 45: $34 - 18 = 16$.

		Voter's age		Total
		18–45	Over 45	
Party preference	Labor	18	12	30
	Liberal	16	11	27
	Total	34	23	57

In Worked example 2, we have a very clear breakdown of data. We know how many voters aged 18 to 45 preferred Labor, how many voters aged 18 to 45 preferred the Liberals, how many voters aged over 45 preferred Labor and how many voters aged over 45 preferred the Liberals.

If we wish to compare the number of voters aged 18 to 45 who prefer Labor with the number of voters aged over 45 who prefer Labor, we must be careful. While 12 voters aged over 45 preferred Labor compared to 18 voters aged 18 to 45, there were fewer voters aged over 45 than voters aged 18 to 45 in the survey. That is, only 23 voters aged over 45 were asked for their opinion, compared to 34 voters aged 18 to 45.

To overcome this problem, we can express the figures as percentages.

WORKED EXAMPLE 3 Converting the numbers in a table to percentages

Fifty-seven people in a local shopping centre were asked whether they preferred the Australian Labor Party or the Liberal Party. The results are as shown.

		Voter's age		Total
		18–45	Over 45	
Party preference	Labor	18	12	30
	Liberal	16	11	27
	Total	34	23	57

Convert the numbers in this table to percentages.

THINK

1. Draw the table, omitting the 'total' column.

2. Fill in the table by expressing the number in each cell as a percentage of its column's total. For example, to obtain the percentage of voters aged over 45 who prefer Labor, divide the number of voters aged over 45 who prefer Labor by the total number of voters aged over 45 and multiply by 100%. $\dfrac{12}{23} \times 100\% = 52.2\%$ (correct to 1 decimal place).

WRITE/DRAW

		Voter's age	
		18–45	Over 45
Party preference	Labor	52.9	52.2
	Liberal	47.1	47.8
	Total	100.0	100.0

We could have also calculated percentages from the table rows, rather than columns. To do that we would have divided the number of voters aged 18 to 45 who preferred Labor (18) by the total number of people who preferred Labor (30) and so on. The table shows this:

| | | Voter's age | | Total |
		18–45	Over 45	
Party preference	Labor	60.0	40.0	100.0
	Liberal	59.3	40.7	100.0

By doing this, we have obtained the percentage of people who were female and preferred Labor (60%) and the percentage of people who were male and preferred Labor (40%), and so on. This highlights different facts from those shown in the previous table. In other words, different results can be obtained by calculating percentages from a table in different ways.

In the above example, the respondent's age is referred to as the explanatory variable and the party preference as the response variable.

The explanatory variable and percentages

As a general rule, when the explanatory variable is placed in the columns of the table, the percentages should be calculated in columns.

Comparing percentages in each row of a two-way table allows us to establish whether a relationship exists between the two categorical variables that are being examined. As we can see from the table in Worked example 3, the percentage of voters aged 18 to 45 who preferred Labor is about the same as that of voters aged over 45. Likewise, the percentage of voters aged 18 to 45 and voters aged over 45 preferring the Liberal Party are almost equal. This indicates that for the group of people participating in the survey, party preference is not related to age.

2.3.2 Segmented bar charts

When comparing two categorical variables, it can be useful to represent the results from a two-way table (in percentage form) graphically. We can do this using **segmented bar charts**.

A segmented bar chart consists of two or more columns, each of which matches one column in the two-way table. Each column is subdivided into segments, corresponding to each cell in that column.

For example, the data from Worked example 3 can be displayed using the segmented bar chart shown.

The segmented bar chart is a powerful visual aid for comparing and examining the relationship between two categorical variables.

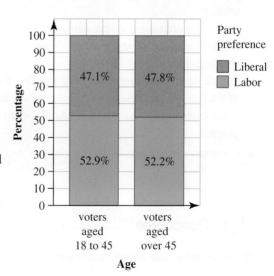

Sixty-seven primary school students and 47 secondary school students were asked about their attitude to the number of school holidays that should be given. They were asked whether there should be fewer, the same number, or more school holidays.

Five primary students and 2 secondary students wanted fewer holidays, 29 primary and 9 secondary students thought they had enough holidays (that is, they chose the same number) and the rest thought they should be given more holidays.

Present the data in percentage form in a two-way table and a segmented bar chart. Compare the opinions of the primary and secondary students.

THINK

1. Put the data in a table. First, fill in the given information, then determine the missing information by subtracting the appropriate numbers from the totals.

2. Calculate the percentages. Since the explanatory variable (the level of the student: primary or secondary) has been placed in the columns of the table, we calculate the percentages in columns. For example, to obtain the percentage of primary students who wanted fewer holidays, divide the number of such students by the total number of primary students and multiply by 100%.

That is, $\dfrac{5}{67} \times 100\% = 7.5\%$.

3. Rule out a set of axes. (The vertical axis shows percentages from 0 to 100, while the horizontal axis represents the categories from the columns of the table.)
Draw two columns to represent each category — primary and secondary. Columns must be the same width and height (up to 100%).
Divide each column into segments so that the height of each segment is equal to the percentage in the corresponding cell of the table. Add a legend to the graph.

4. Comment on the results.

WRITE/DRAW

		School		Total
		Primary	Secondary	
Attitude	Fewer	5	2	7
	Same	29	9	38
	More	33	36	69
Total		67	47	114

		School	
		Primary	Secondary
Attitude	Fewer	7.5%	4.3%
	Same	43.3%	19.1%
	More	49.2%	76.6%
Total		100%	100%

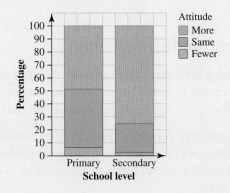

Secondary students were much keener on having more holidays than were primary students.

 Resources

 Interactivity Two-way tables and segmented bar graphs (int-6249)

1. **WE2** A group of 60 people, 38 people aged 15 to 30 and 22 people aged over 30, were asked whether they preferred an Apple or Samsung phone. Twenty-three people aged 15 to 30 and 15 people aged over 30 said they preferred an Apple phone.
Display this data in a two-way (contingency) table and calculate the marginal distribution for phone preference for these age groups.

2. A group of 387 teenagers and 263 adults were asked their preference from Coke and Pepsi. Two hundred and twenty-one teenagers preferred Coke, whereas 108 adults preferred Pepsi.
Display this data in a two-way (contingency) table and calculate the conditional distribution of drink preference among teenagers.

3. In a survey, 139 inner-city residents and 102 outer suburban residents were asked whether they approved or disapproved of a proposed freeway. Thirty-seven inner-city residents and 79 outer suburban residents approved of the freeway.
Display these data in a two-way table (not as percentages), and calculate the approval or disapproval of the proposed freeway for inner-city and outer suburban residents.

4. Students at a secondary school were asked whether the length of lessons should be 45 minutes or 1 hour. Ninety-three senior students (Years 10–12) were asked and 60 preferred 1-hour lessons, whereas of the 86 junior students (Years 7–9), 36 preferred 1-hour lessons.
Display these data in a two-way table (not as percentages), and calculate the conditional distribution on length of lessons and senior students.

5. **WE3** A group of 60 people were asked their preferences on phones. The results are shown.

		Age group		Total
		Teenagers	Adults	
Phone types	Apple	23	15	38
	Samsung	15	7	22
	Total	38	22	60

Convert the numbers in this table to percentages.

6. A group of 650 people were asked their preferences on soft drink. The results are shown.

		Age group		Total
		Teenagers	Adults	
Drink types	Pepsi	221	155	376
	Coke	166	108	274
	Total	387	263	650

Convert the numbers in this table to percentages.

7. For each of the following two-way frequency tables, complete the missing entries.

a.

		Grades		Total
		Year 12	Year 11	
Transport preference	Tram	25	i	47
	Train	ii	iii	iv
	Total	51	v	92

b.

		Voters' ages		
		18 to 45	over 45	
Party preference	Labor	i	42%	
	Liberal	53%	ii	
	Total	iii	iv	

8. Sixty single people were asked whether they preferred to rent by themselves or to share accommodation with friends. The results are shown.

		Age group		Total
		18 to 30	31 to 40	
Preference	Rent by themselves	12	23	35
	Share with friends	9	16	25
	Total	21	39	60

Convert the numbers in this table to percentages.

The information in the following two-way frequency table relates to questions 9 and 10.

The data show the reactions of administrative staff and technical staff to an upgrade of the computer systems at a large corporation.

		Staff type		Total
		Administrative	Technical	
Attitude	For	53	98	151
	Against	37	31	68
	Total	90	129	219

9. **MC** From the previous table, we can conclude that:
   A. 53% of administrative staff were for the upgrade.
   B. 37% of administrative staff were for the upgrade.
   C. 37% of administrative staff were against the upgrade.
   D. 59% of administrative staff were for the upgrade.
   E. 54% of administrative staff were against the upgrade.

10. **MC** From the previous table, we can conclude that:
   A. 98% of technical staff were for the upgrade.
   B. 65% of technical staff were for the upgrade.
   C. 76% of technical staff were for the upgrade.
   D. 31% of technical staff were against the upgrade.
   E. 14% of technical staff were against the upgrade.

11. **WE4** Sixty-one adults and 58 children were asked which they preferred off a restaurant menu: entrée, main or dessert. Eight children and 18 adults preferred the entrée, while 31 children and 16 adults said they preferred the main course, with the remainder having dessert as their preference.

Present these data in percentage form in a two-way table and a segmented bar chart.

Compare the preferences of children and adults.

12. Ninety-three people less than 40 years of age and 102 people aged 40 and over were asked where their priority financially was, given the three options 'mortgage', 'superannuation' or 'investing'. Eighteen people in the 40 and over category and 42 people in the less than 40 category identified mortgage as their priority, whereas 21 people under 40 years of age and 33 people aged 40 and over said investment was most important. The rest suggested superannuation was their priority.

Present these data in percentage form in a two-way table and segmented bar chart.

Compare the financial priorities of the under 40s to the people aged 40 and over.

13. Delegates at the respective Liberal Party and Australian Labor Party conferences were surveyed on whether or not they believed that marijuana should be legalised. Sixty-two Liberal delegates were surveyed and 40 of them were against legalisation. Seventy-one Labor delegates were surveyed and 43 were against legalisation.

   a. Present the data in percentage form in a two-way frequency table.
   b. Comment on any differences between the reactions of the Liberal and Labor delegates.
   c. Use the results in a to draw a segmented bar chart.

The information in the following table relates to questions 14–16.

The amount of waste recycled by 100 townships across Australia was rated as low, medium or high and the size of the town as small, mid-sized or large. The results of the ratings are shown in the table.

		Type of town		
		Small	**Mid-sized**	**Large**
Amount of waste recycled	Low	6	7	4
	Medium	8	31	5
	High	5	16	18

14. **MC** The percentage of mid-sized towns rated as having a high level of waste recycling is closest to:

   A. 41%   **B.** 25%   **C.** 30%   **D.** 17%   **E.** 50%

15. Calculate the conditional distribution for amount of waste and large towns.

16. Calculate the percentage of small towns rated as having a high level of waste recycling.

Question 1 (1 mark)

Source: VCE 2017, Further Mathematics Exam1, Section A, Core, Q6; © VCAA

`MC` A study was conducted to investigate the association between the number of moths caught in a moth trap (less than 250, 250–500, more than 500) and the trap type (sugar, scent, light) The results are summarised in the percentage segmented bar chart below.

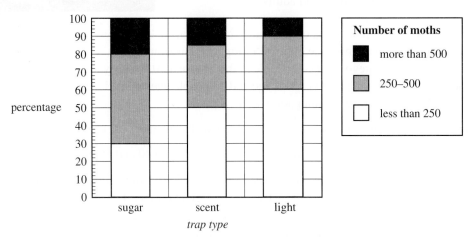

The data displayed in the percentaged segmented bar chart supports the contention that there is an association between the number of moths caught in a moth trap and the trap type because

A. most of the light traps contained less than 250 moths.

B. 15% of the scent traps contained 500 or more moths.

C. the percentage of sugar traps containing more than 500 moths was greater than the percentage of scent traps containing less than 500 moths.

D. 20% of sugar traps cantained more than 500 moths, while 50% of light traps contained less than 250 moths.

E. 20% of sugar traps contained more than 500 moths while 10% of light traps contained more than 500 moths.

Question 2 (1 mark)

Source: VCE 2016, Further Mathematics Exam1, Section A, Core, Q1; © VCAA

`MC` The blood pressure (low, normal, high) and the age (under 50 years, 50 years or over) of 110 adults were recorded. The results are displayed in the two-way frequency table.

	Age	
Blood pressure	**Under 50 years**	**50 years or over**
Low	15	5
Normal	32	24
High	11	23
Total	58	52

The percentage of adults under 50 years of age who have high blood pressure is closest to

A. 11%   **B.** 19%   **C.** 26%   **D.** 44%   **E.** 58%

Question 3 (1 mark)

Source: VCE 2012, Further Mathematics Exam1, SectionA, Core, Q7; © VCAA

MC A study was conducted to investigate the association between the number of moths caught in a moth trap (less than 250, 250–500, more than 500) and the *trap type* (sugar, scent, light). The results are summarised in the percentaged segmented bar chart.

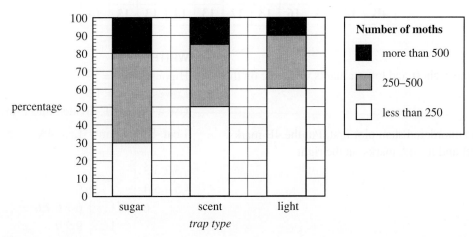

There were 300 sugar traps.

The number of sugar traps that caught less than 250 moths is closest to

  A. 30   **B.** 90   **C.** 250   **D.** 300   **E.** 500

More exam questions are available online.

2.4 Back-to-back stem plots

LEARNING INTENTION

At the end of this subtopic you should be able to:
- display data on back-to-back stem plots and use them to compare data distribution.

2.4.1 Constructing back-to-back stem plots

We can create a stem plot that displays the relationship between a numerical variable and a categorical variable. We will limit ourselves in this section to categorical variables with just two categories, for example, gender. The two categories are used to provide the two back-to-back leaves of the stem plot.

Back-to-back stem plot

A **back-to-back stem plot** is used to display two sets of data, involving a numerical variable and a categorical variable with two categories.

tlvd-3544

WORKED EXAMPLE 5 Displaying data on a back-to-back stem plot

The children in classes 4A and 4B at Kingston Primary School submitted projects on the Olympic Games. The marks they obtained out of 20 are shown. Display the data on a back-to-back stem plot.

4A	16	17	19	15	12	16	17	19	19	16
4B	14	15	16	13	12	13	14	13	15	14

THINK

1. Identify the highest and lowest marks to decide on the stems.

2. Create an unordered stem plot first. Put the 4B marks on the left and the 4A marks on the right.

3. Now order the stem plot. The marks on the left should increase in value from right to left, while the marks on the right should increase in value from left to right.

WRITE

Highest score = 19
Lowest score = 12
Use a stem of 1, divided into fifths.

Leaf 4B	Stem	Leaf 4A
	1	
3 2 3 3	1	2
4 5 4 5 4	1	5
6	1	6 7 6 7 6
	1	9 9 9

Key: 1 | 2 = 12

Leaf 4B	Stem	Leaf 4A
3 3 3 2	1	2
5 5 4 4 4	1	5
6	1	6 6 6 7 7
	1	9 9 9

Key: 1 | 2 = 12

The back-to-back stem plot allows us visually compare the two distributions. In Worked example 5, the centre of the distribution for class 4A is higher than the centre of the distribution for class 4B. The spread of each of the distributions seems to be about the same. For class 4B, the scores are grouped around the 12−15 mark; for class 4A, they are grouped around the 16−19 mark. On the whole, we can conclude that class 4A obtained better scores than class 4B did.

To get a more precise picture of the centre and spread of each of the distributions, we can use the summary statistics discussed in topic 1. Specifically, we are interested in:
1. the mean and median (to measure the centre of the distributions)
2. the interquartile range and standard deviation (to measure the spread of the distributions).

WORKED EXAMPLE 6 Comparing distributions

The number of how-to-vote cards handed out by various Australian Labor Party and Liberal Party volunteers during the course of a polling day is shown.

Labor	180	233	246	252	263	270	229	238	226	211
	193	202	210	222	257	247	234	226	214	204
Liberal	204	215	226	253	263	272	285	245	267	275
	287	273	266	233	244	250	261	272	280	279

Display the data using a back-to-back stem plot and use this, together with summary statistics, to compare the distributions of the number of cards handed out by the Labor and Liberal volunteers.

THINK

WRITE

1. Construct the stem plot.

Leaf Labor	Stem	Leaf Liberal
0	18	
3	19	
4 2	20	4
4 1 0	21	5
9 6 6 2	22	6
8 4 3	23	3
7 6	24	4 5
7 2	25	0 3
3	26	1 3 6 7
0	27	2 2 3 5 9
	28	0 5 7

Key: $18 \mid 0 = 180$

2. Use CAS to obtain summary statistics for each party. Record the mean, median, IQR and standard deviation in the table. (IQR $= Q_3 - Q_1$, which is determined by assigning the middle score of the lower half of the data to Q_1 and the middle score of the upper half of the data to Q_3).

	Labor	Liberal
Mean	227.9	257.5
Median	227.5	264.5
IQR	36	29.5
Standard deviation	23.9	23.4

3. Comment on the relationship.

From the stem plot we see that the Labor distribution is symmetric and therefore the mean and the median are very close, whereas the Liberal distribution is negatively skewed.

Since the distribution is skewed, the median is a better indicator of the centre of the distribution than the mean. Comparing the medians, therefore, we have the median number of cards handed out for Labor at 228 and for Liberal at 265, which is a big difference. The standard deviations were similar, as were the interquartile ranges. There was not a lot of difference in the spread of the data.

In essence, the Liberal party volunteers handed out more how-to-vote cards than the Labor party volunteers did.

 Resources

Interactivity: Back-to-back stem plots (int-6252)

2.4 Exercise

1. **WE5** Two classes submitted an assignment on the history of the ANZACs. The results out of 40 are shown.

4A's results	30	35	31	32	38	33	35	30
4B's results	34	33	37	39	31	32	39	36

Display the data on a back-to-back stem plot.

2. The marks obtained out of 50 by students in Physics and Chemistry are shown.

Physics	32	45	48	32	24	30	41	29	44	45	36	34	28	49
Chemistry	46	31	38	28	45	49	34	45	47	33	30	21	32	28

Display the data on a back-to-back stem plot.

3. The marks out of 50 obtained for the end-of-term test by the students in Chinese and French classes are shown.

Chinese	20	38	45	21	30	39	41	22	27	33	30	21	25	32	37	42	26	31	25	37
French	23	25	36	46	44	39	38	24	25	42	38	34	28	31	44	30	35	48	43	34

Display the data on a back-to-back stem plot.

4. The masses of 10 one-month-old lion cubs and 10 one-month-old tiger cubs (in kilograms, correct to the nearest 100 grams) are recorded in the table.

Lion cubs	3.4	5.0	4.2	3.7	4.9	3.4	3.8	4.8	3.6	4.3
Tiger cubs	3.0	2.7	3.7	3.3	4.0	3.1	2.6	3.2	3.6	3.1

Display the data on a back-to-back stem plot.

5. **WE6** The number of promotional pamphlets handed out for company A and company B by a number of their reps is shown.

Company A	144	156	132	138	148	160	141	134	132	142	132	134	168	149
Company B	146	131	138	155	145	153	134	153	138	133	130	162	148	160

Display the data using a back-to-back stem plot and use this, together with summary statistics, to compare the number of pamphlets handed out by each company.

6. Student achievements (out of 100) in History and English were recorded and the results are shown.

History	75	78	42	92	59	67	78	82	84	64	77	98
English	78	80	57	96	58	71	74	87	79	62	75	100

a. Draw a back-to-back stem plot.
b. Use summary statistics and the stem plot to comment on the results in the two subjects.

7. The number of delivery trucks making deliveries to a supermarket each day over a 2-week period was recorded for two neighbouring supermarkets — supermarket A and supermarket B. The data are shown in the table.

A	11	15	20	25	12	16	21	27	16	17	17	22	23	24
B	10	15	20	25	30	35	16	31	32	21	23	26	28	29

a. Display the data on a back-to-back stem plot.
b. Use the stem plot, together with some summary statistics, to compare the distributions of the number of trucks delivering to supermarkets A and B.

8. The marks out of 20 obtained by two Year 10 classes for a science test are given.

10A	12	13	14	14	15	15	16	17
10B	10	12	13	14	14	15	17	19

a. Display the data on a back-to-back stem plot.
b. Use the stem plot, together with some summary statistics, to compare the distributions of the marks of the two classes.

9. The end-of-year English marks for 10 students in an English class were compared over two years. The marks for 2021 and for the same students in 2022 are shown.

2021	30	31	35	37	39	41	41	42	43	46
2022	22	26	27	28	30	31	31	33	34	36

a. Display the data on a back-to-back stem plot.
b. Use the stem plot, together with some summary statistics, to compare the distributions of the marks obtained by the students in 2021 and 2022.

10. The age and species of gum trees in a forest were recorded.

Species 1	23	24	25	26	27	28	30	31
Species 2	22	25	30	31	36	37	42	46

a. Display the data on a back-to-back stem plot.
b. Use the stem plot, together with some summary statistics, to compare the distributions of the ages of the Species 1 gum trees with the ages of the Species 2 gum trees.

11. The scores on a board game for a group of kindergarten children and for a group of children in a preparatory school are shown.

Kindergarten	3	13	14	25	28	32	36	41	47	50
Prep. school	5	12	17	25	27	32	35	44	46	52

a. Display the data on a back-to-back stem plot.
b. Use the stem plot, together with some summary statistics, to compare the distributions of the scores of the kindergarten children with the scores of the preparatory school children.

12. **MC** A pair of variables that could be displayed on a back-to-back stem plot is:
A. the height of a student and the number of people in the student's household.
B. the time put into completing an assignment and a pass or fail score on the assignment.
C. the weight of a businessman and his age.
D. the religion of an adult and the person's head circumference.
E. the income of an employee and the time the employee has worked for the company.

13. **MC** A back-to-back stem plot is a useful way of displaying the relationship between:
    A. the proximity to markets in kilometres and the cost of fresh foods on average per kilogram.
    B. height and head circumference.
    C. age and attitude to gambling (for or against).
    D. weight and age.
    E. the money spent during a day of shopping and the number of shops visited on that day.

14. The scores out of 100 that a group of Year 12 and Year 11 students received when going for their licence are shown.

Year 12	86	92	100	90	94	82	72	90	88	94	76	80
Year 11	94	96	72	80	84	92	83	88	90	70	81	83

Construct a back-to-back stem plot of the data.

15. **MC** A back-to-back stem plot is used to display two sets of data. Name the two variables involved.
    A. Increasing variables
    B. Discrete and numerical variables
    C. Continuous and categorical variables
    D. Numerical and categorical variables
    E. Numerical and continuous variables

16. The study scores (out of 50) of students who studied both Mathematical Methods and General Mathematics are shown.

Methods	28	34	41	36	33	39	44	40	39	42	36	31	29	44
General	30	37	38	41	35	43	44	46	43	48	37	31	28	48

a. Display the data in a back-to-back stem plot.
b. Use the stem plot, together with some summary statistics, to compare the distributions of the scores for Mathematical Methods with the scores for General Mathematics.

2.4 Exam questions

Question 1 (1 mark)
Source: VCE 2014, Further Mathematics Exam1, Section A, Q8; © VCAA

MC A single back-to-back stem plot would be an appropriate graphical tool to investigate the association between a car's speed, in kilometres per hour, and the
    A. driver's age, in years.
    B. car's colour (white, red, grey, other).
    C. car's fuel consumption, in kilometres per litre.
    D. average distance travelled, in kilometres.
    E. driver's sex (female, male).

MC To ensure the IQR value for both the males and females is equal, the missing value from the female data must be

Male	Stem	Female
3 3 2	2	
9 8 6 5	3	
4 3	4	
7	5	1 2 5
	6	5 7 7
	7	4 6
	8	2

Key: 5|2 = 52

A. 43   **B.** 45   **C.** 74   **D.** 75   **E.** 84

Question 3 (19 marks)

The following stem-and-leaf plot details the age of 26 offenders who were caught drink driving during a Friday night blitz.

Male offenders (Leaf)	Stem	Female offenders (Leaf)
7 7 7 7 8 8 8 8 8 9	1	8 8 8
1 1 1 1 1 1	2	1 1 2 4
7	3	
	4	2 5

Key: 1|8 = 18 years old

Note: Give answers to 1 decimal place where appropriate.

  a. Calculate the mean, median, mode and standard deviation for the male offenders.   **(4 marks)**
  b. Calculate the five-figure summary, and hence the range and IQR, for the male offenders.   **(3 marks)**
  c. Calculate the mean, median, mode and standard deviation for the female offenders.   **(4 marks)**
  d. Calculate the five-figure summary, and hence the range and IQR, for the female offenders.   **(3 marks)**
  e. Using statistics, write a paragraph about the results of the Friday night drink driving blitz.   **(5 marks)**

More exam questions are available online.

2.5 Parallel boxplots and dot plots

LEARNING INTENTION

At the end of this subtopic you should be able to:
- construct parallel boxplots.

2.5.1 Constructing parallel boxplots

When we want to display a relationship between a numerical variable and a categorical variable with two or more categories, **parallel boxplots** or **parallel dot plots** can be used.

Parallel boxplots and dot plots

Parallel boxplots are obtained by constructing individual boxplots for each distribution and positioning them on a common scale.

Parallel dot plots are obtained by constructing individual dot plots for each distribution and positioning them on a common scale.

Construction of individual boxplots was discussed in detail in topic 1. In this section, we concentrate on comparing distributions represented by a number of boxplots (that is, on the interpretation of parallel boxplots).

tlvd-3545

WORKED EXAMPLE 7 Constructing and interpreting parallel boxplots.

The four Year 7 classes at Western Secondary College complete the same end-of-year maths test. The marks, expressed as percentages, are given.

7A	40	43	45	47	50	52	53	54	57	60	69	63	63	68	70	75	80	85	89	90
7B	60	62	63	64	70	73	74	76	77	77	78	82	85	87	89	90	92	95	97	97
7C	50	51	53	55	57	60	63	65	67	69	70	72	73	74	76	80	82	82	85	89
7D	40	42	43	45	50	53	55	59	60	61	69	73	74	75	80	81	82	83	84	90

Display the data using parallel boxplots. Use these to describe any similarities or differences in the distributions of marks between the four classes.

THINK

1. Use CAS to determine the five-number summary for each data set.

WRITE/DRAW

	7A	7B	7C	7D
Minimum	40	60	50	40
Q_1	51	71.5	58.5	51.5
Median $= Q_2$	61.5	77.5	69.5	65
Q_3	72.5	89.5	78	80.5
Maximum	90	97	89	90

2. Draw the boxplots, labelling each class. All four boxplots share a common scale.

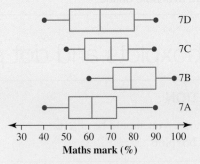

3. Describe the similarities and differences between the four distributions.

Class 7B had the highest median mark and the range of the distribution was only 37. The lowest mark in 7B was 60.

We notice that the median of 7A's marks is 61.5. So, 50% of students in 7A received less than 61.5. This means that about half of 7A had scores that were less than the lowest score in 7B.

The range of marks in 7A was the same as that of 7D with the highest scores in each equal (90) and the lowest scores equal (40). However, the median mark in 7D (65) was slightly higher than the median mark in 7A (61.5), so, despite a similar range, more students in 7D received a higher mark than in 7A. While 7D had a top score that was higher than that of 7C, the median score in 7C (69.5) was higher than that of 7D and almost 25% of scores in 7D were less than the lowest score in 7C. In summary, 7B did best, followed by 7C, then 7D and finally 7A.

Resources

Interactivity Parallel boxplots (int-6248)

2.5 Exercise

Students, these questions are even better in jacPLUS

Receive immediate feedback and access sample responses

Access additional questions

Track your results and progress

Find all this and MORE in jacPLUS

1. **WE7** The times run for a 100 m race in Year 6 are shown for two classes. The times are expressed in seconds.

6A	15.5	16.1	14.5	16.9	18.1	14.3	13.8	15.9	16.4	17.3	18.8	17.9	16.1
6B	16.7	18.4	19.4	20.1	16.3	14.8	17.3	20.3	19.6	18.4	16.5	17.2	16.0

Display the data using parallel boxplots and use these to describe any similarities or differences between the 6A and 6B performances.

2. A teacher taught two Year 10 maths classes and wanted to see how they compared on the end-of-year examination. The marks are expressed as percentages.

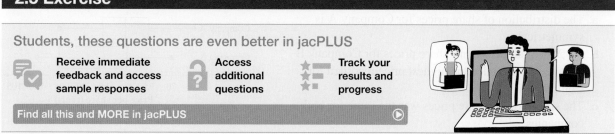

10A	67	73	45	59	67	89	42	56	68	75	87	94	80	98
10D	76	82	62	58	40	55	69	71	89	95	100	84	70	66

Display the data using parallel boxplots and parallel dot plots. Use these to describe any similarities or differences between the two classes.

3. The heights (in cm) of students in 9A, 10A and 11A were recorded and are shown in the table.

9A	120	126	131	138	140	143	146	147	150
10A	140	143	146	147	149	151	153	156	162
11A	151	153	154	158	160	163	164	166	167

9A	156	157	158	158	160	162	164	165	170
10A	164	165	167	168	170	173	175	176	180
11A	169	169	172	175	180	187	189	193	199

a. Construct parallel boxplots to show the data.

b. Use the boxplots to compare the distributions of height for the three classes.

4. The amounts of money contributed annually to superannuation schemes by people in three different age groups are shown.

20–29	2000	3100	5000	5500	6200	6500	6700	7000	9200	10 000
30–39	4000	5200	6000	6300	6800	7000	8000	9000	10 300	12 000
40–49	10 000	11 200	12 000	13 300	13 500	13 700	13 900	14 000	14 300	15 000

a. Construct parallel boxplots to show the data.

b. Use the boxplots to comment on the distributions.

5. The daily share price of two companies was recorded over a period of one month. The results are presented as parallel boxplots.
State whether each of the following statements is true or false.

a. The distribution of share prices for Company A is symmetrical.

b. On 25% of all occasions, share prices for Company B equalled or exceeded the highest price recorded for Company A.

c. The spread of the share prices was the same for both companies.

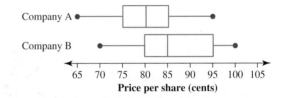

d. 75% of share prices for Company B were at least as high as the median share price for Company A.

6. Last year, the spring season at the Sydney Opera House included two major productions: *The Pearlfishers* and *Orlando*. The number of A-reserve tickets sold for each performance of the two operas is shown as parallel boxplots.

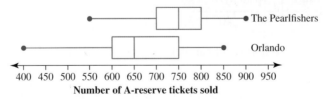

a. Determine which of the two productions proved to be more popular with the public, assuming A-reserve ticket sales reflect total ticket sales. Explain your answer.

b. Determine which production had a larger variability in the number of patrons purchasing A-reserve tickets. Support your answer with the necessary calculations.

7. **MC** The results in two exams, one in Year 8 and the other in Year 10, are given by the parallel boxplots.

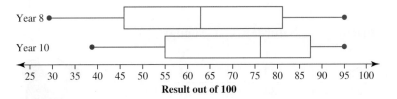

Result out of 100

The percentage of Year 10 students who obtained a mark greater than 87 was:

A. 2%   **B.** 5%   **C.** 20%   **D.** 25%   **E.** 75%

8. **MC** From the parallel boxplots in question 7, it can be concluded that:
   A. the Year 8 results were similar to the Year 10 results.
   B. the Year 8 results were lower than the Year 10 results and less variable.
   C. the Year 8 results were lower than the Year 10 results and more variable.
   D. the Year 8 results were higher than the Year 10 results and less variable.
   E. the Year 8 results were higher than the Year 10 results and more variable.

9. The scores of 10 competitors on two consecutive days of a diving competition were recorded.

Day 1	5.4	4.1	5.4	5.6	4.9	5.6	5.4	6.0	5.8	6.0
Day 2	4.9	5.1	5.3	5.8	5.7	5.2	5.8	5.4	5.5	6.0

Construct parallel dot plots to show the data and comment on the divers' results over the two days.

The following figure relates to questions 10 to 12.
The ages of customers in different areas of a department store are shown.

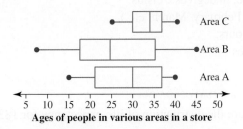

Ages of people in various areas in a store

10. Determine which area has the largest range from Q_2 (the median) to Q_3.

11. Determine which area has the largest range.

12. Determine which area has the highest median age.

13. The numbers of jars of vitamins A, B, C and multi-vitamins sold per week by a local chemist are shown in the table.

Vitamin A	5	6	7	7	8	8	9	11	13	14
Vitamin B	10	10	11	12	14	15	15	15	17	19
Vitamin C	8	8	9	9	9	10	11	12	12	13
Multi-vitamins	12	13	13	15	16	16	17	19	19	20

Construct parallel boxplots to display the data and use them to compare the distributions of sales for the four types of vitamins.

14. Eleven golfers in a golf tournament play 18 holes each day. The scores for each of the golfers on the four days are given.

Thursday	Friday	Saturday	Sunday
70	77	81	70
71	78	83	81
75	81	84	81
79	82	84	88
80	83	86	88
81	83	86	89
83	85	87	90
83	85	87	90
84	85	87	91
85	88	88	93
90	89	89	94

Display this data using parallel boxplots.

2.5 Exam questions

Question 1 (1 mark)

Source: VCE 2016, Further Mathematics Exam1, Section A, Q8; © VCAA

MC Parallel boxplots would be an appropriate graphical tool to investigate the association between the monthly median rainfall, in millimetres, and the
- **A.** monthly median wind speed, in kilometres per hour.
- **B.** monthly median temperature, in degrees Celsius.
- **C.** month of the year (January, February, March, etc.).
- **D.** monthly sunshine time, in hours.
- **E.** annual rainfall, in millimetres.

Question 2 (3 marks)

Source: VCE 2015, Further Mathematics Exam 2, Core, Q2; © VCAA

The parallel boxplots below compare the distribution of life expectancy for 183 countries for the years 1953, 1973 and 1993.

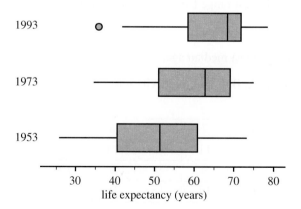

- **a.** Describe the shape of the distribution of life expectancy for 1973. **(1 mark)**
- **b.** Explain why life expectancy for these countries is associated with the year. Refer to specific statistical values in your answer. **(2 marks)**

▶ **Question 3 (1 mark)**

MC The time taken, measured in seconds, for a sample of people to type a 250 character SMS was recorded. The results were grouped into the age brackets of 0–15, 16–35 and 36–70 and displayed in the boxplots shown.

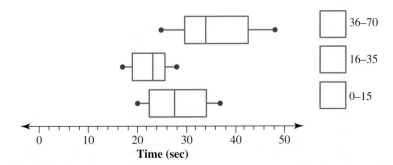

Compare the information given in the boxplots to decide which of the following is inaccurate.

- **A.** At least 75% of the times recorded by the 0–15 age bracket were less than the median of the 36–70 age bracket.
- **B.** At least 75% of the times recorded by the 36–70 age bracket were more than the median of the 0–15 age bracket.
- **C.** At least 25% of the times recorded by the 36–70 age bracket were more than the maximum recorded time of the 0–15 age bracket.
- **D.** 100% of the times recorded by the 16–35 age bracket were less than the median of the 0–15 age bracket.
- **E.** At least 25% of the times recorded by the 16–35 age bracket were less than the minimum recorded time of the 0–15 age bracket.

More exam questions are available online.

2.6 Scatterplots

LEARNING INTENTION

At the end of this subtopic you should be able to:
- construct and describe scatterplots.

2.6.1 Constructing and describing scatterplots

A **scatterplot** gives a visual display of the relationship between two variables. Recall that when we display bivariate data as a scatterplot, the explanatory variable is placed on the horizontal axis and the response variable is placed on the vertical axis.

Consider the data obtained from last year's 12B class at Northbank Secondary College.

Each student in this class of 29 was asked to give an estimate of the average number of hours they studied per week during Year 12. They were also asked for the ATAR score they obtained.

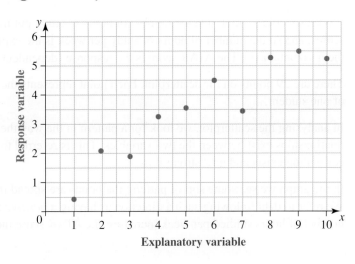

Average hours of study	ATAR score
18	59
16	67
22	74
27	90
15	62
28	89
18	71
19	60
22	84
30	98
14	54
17	72
14	63
19	72
20	58

Average hours of study	ATAR score
10	47
28	85
25	75
18	63
19	61
17	59
16	76
14	59
29	89
30	93
30	96
23	82
26	35
22	78

The figure below shows the data plotted on a scatterplot.

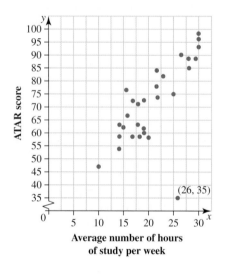
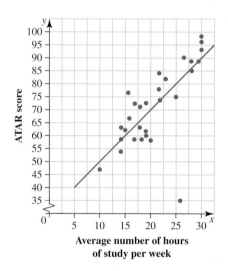

It is reasonable to think that the number of hours of study put in each week by students would affect their ATAR scores, and so the number of hours of study per week is the explanatory (independent) variable and appears on the horizontal axis. The ATAR score is the response (dependent) variable and appears on the vertical axis.

There are 29 points on the scatterplot. Each point represents the number of hours of study and the ATAR score of one student.

In analysing the scatterplot, we look for a pattern in the way the points lie. Certain patterns tell us that certain relationships exist between the two variables. This is referred to as **correlation**. We look at what type of correlation exists and how strong it is.

In the graph we see some sort of pattern: the points are spread in a rough corridor from bottom left to top right. We refer to data following such a direction as having a *positive relationship*. This tells us that as the average number of hours studied per week increases, the ATAR score increases.

The point (26, 35) is an outlier. It stands out because it is well away from the other points and clearly is not part of the 'corridor' referred to previously. This outlier may have occurred because a student exaggerated the number of hours he or she worked in a week or perhaps there was a recording error. This needs to be checked.

We could describe the rest of the data as having a **linear relationship** as the straight line in the second graph indicates.

When describing the relationship between two variables displayed on a scatterplot, we need to comment on:
a. the direction — whether it is positive or negative
b. the form — whether it is linear or non-linear
c. the strength — whether it is strong, moderate or weak
d. possible outliers.

Describing relationships on scatterplots

Here is a gallery of scatterplots showing the various patterns.

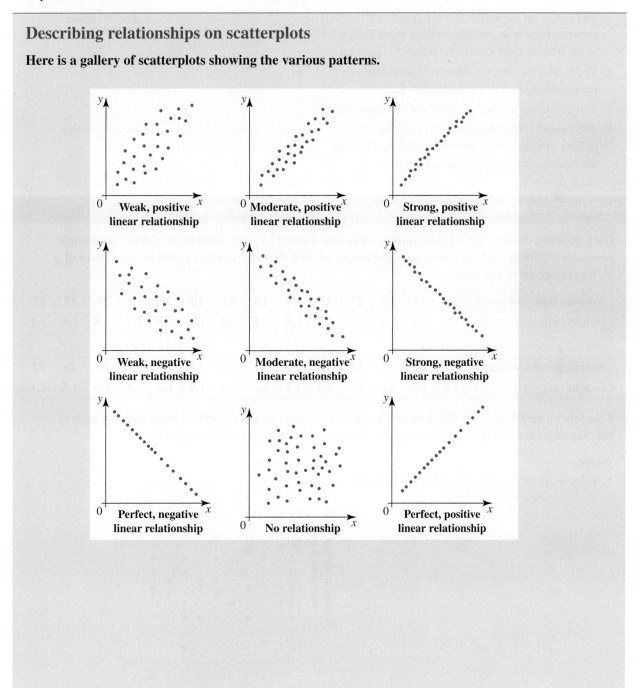

WORKED EXAMPLE 8 Describing a scatterplot

The scatterplot shows the number of hours people spend at work each
week and the number of hours people get to spend on recreational
activities during the week.

Decide whether or not a relationship exists between the variables
and, if it does, comment on whether it is positive or negative; weak,
moderate or strong; and whether or not it has a linear form.

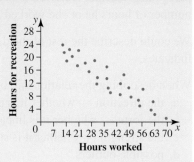

THINK	WRITE
1. The points on the scatterplot are spread in a certain pattern, namely in a rough corridor from the top left to the bottom right corner. Interpret this pattern.	As work hours increase, the recreation hours decrease.
2. The corridor is straight (that is, it would be reasonable to fit a straight line into it).	
3. The points are neither too tight nor too dispersed.	
4. The pattern resembles the central diagram in the gallery of scatterplots shown previously. Describe the relationship between the variables.	There is a moderate, negative linear relationship between the two variables.

WORKED EXAMPLE 9 Constructing and describing a scatterplot

Data showing the average weekly number of hours studied by each student in 12B at Northbank
Secondary College and the corresponding height of each student (correct to the nearest tenth of a
metre) are given in the table.

Average hours of study	18	16	22	27	15	28	18	20	10	28	25	18	19	17
Height (m)	1.5	1.9	1.7	2.0	1.9	1.8	2.1	1.9	1.9	1.5	1.7	1.8	1.8	2.1

Average hours of study	19	22	30	14	17	14	19	16	14	29	30	30	23	22
Height (m)	2.0	1.9	1.6	1.5	1.7	1.8	1.7	1.6	1.9	1.7	1.8	1.5	1.5	2.1

Construct a scatterplot for the data and use it to comment on the direction, form and strength of any
relationship between the number of hours studied and the height of the students.

THINK

1. CAS technology can be used to assist you in
   drawing a scatterplot.

WRITE/DRAW

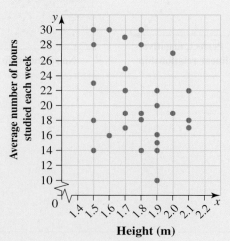

2.	Comment on the direction of any relationship.	There is no relationship; the points appear to be randomly placed.
3.	Comment on the form of the relationship.	There is no form, no linear trend, no quadratic trend, just a random placement of points.
4.	Comment on the strength of any relationship.	Since there is no relationship, strength is not relevant.
5.	Draw a conclusion.	Clearly, from the graph, the number of hours spent studying has no relation to how tall you are.

TI	THINK	DISPLAY/WRITE	CASIO	THINK	DISPLAY/WRITE
1.	On a Lists & Spreadsheet page, enter the data into the lists named 'hours' and 'height'. Height is the explanatory variable and hours is the response variable.	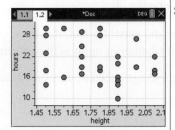	1.	On a Statistics screen, enter the data into the lists named 'hours' and 'height'. Height is the explanatory variable and hours is the response variable.	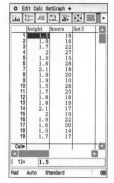
2.	Press CTRL + I and select: 5: Add Data & Statistics Use the tab key to move to each axis and press the click key or ENTER to place 'height' on the horizontal axis and 'hours' on the vertical axis.		2.	Tap: SetGraph Setting Draw: On Type: Scatter X List: main\height Y List: main\hours Freq: 1 Mark: square Tap Set to lock your selection. Tap Resize to enlarge the scatterplot.	

 Resources

 Interactivity Scatterplots (int-6250)

Create scatterplots (int-6497)

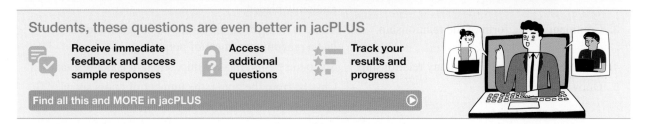
1. **WE8** The scatterplot shown represents the number of hours of basketball practice each week and a player's shooting percentage.

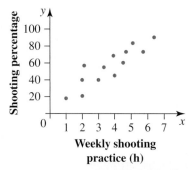

Decide whether or not a relationship exists between the variables and, if it does, comment on whether it is positive or negative; weak, moderate or strong; and whether or not it is linear form.

2. The scatterplot shows the hours after 5 pm and the average speed of cars on a freeway.

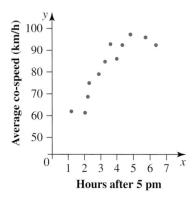

Explain the direction, form and strength of the relationship of the two variables.

3. For each of the following pairs of variables, determine whether or not you would reasonably expect a relationship to exist between the pair and, if so, comment on whether it would be a positive or negative association.

   a. Time spent in a supermarket and total money spent
   b. Income and value of car driven
   c. Number of children living in a house and time spent cleaning the house
   d. Age and number of hours of competitive sport played per week
   e. Amount spent on petrol each week and distance travelled by car each week
   f. Number of hours spent in front of a computer each week and time spent playing the piano each week
   g. Amount spent on weekly groceries and time spent gardening each week

4. For each of the scatterplots, describe whether or not a relationship exists between the variables and, if it does, comment on whether it is positive or negative; weak, moderate or strong; and whether or not it has a linear form.

a.
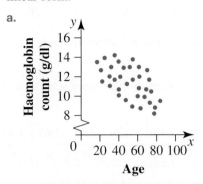
Age

b.
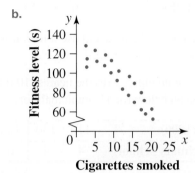
Cigarettes smoked

c.
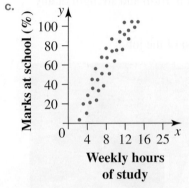
Weekly hours
of study

d.
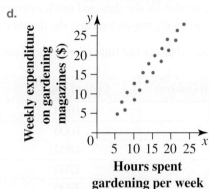
Hours spent
gardening per week

e.
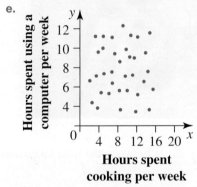
Hours spent
cooking per week

f.
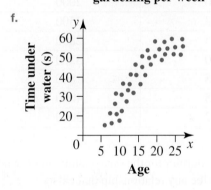
Age

5. **MC** From the scatterplot shown, it would be reasonable to observe that:

A. as the value of x increases, the value of y increases.
B. as the value of x increases, the value of y decreases.
C. as the value of x increases, the value of y remains the same.
D. as the value of x remains the same, the value of y increases.
E. there is no relationship between x and y.

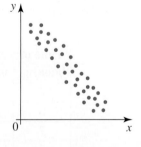

6. **WE9** Data on the height of a person and the length of their hair is shown.

Height (cm)	158	164	184	173	194	160	198	186	166
Hair length (cm)	18	12	5	10	7	3	10	6	14

Construct a scatterplot for the data and use it to comment on the direction, form and strength of any relationship between the height of a person and the length of their hair.

7. The following table shows data on hours spent watching television per week and a person's age.

Age (yr)	12	25	61	42	18	21	33	15	29
TV per week (h)	23	30	26	18	12	30	20	19	26

Construct a scatterplot for the data and use it to comment on the direction, form and strength of any relationship between the two variables.

8. The population of a municipality (to the nearest 10 000) together with the number of primary schools in that municipality are given below for 11 municipalities.

Population (×1000)	110	130	130	140	150	160	170	170	180	180	190
Number of primary schools	4	4	6	5	6	8	6	7	8	9	8

Construct a scatterplot for the data and use it to comment on the direction, form and strength of any relationship between the population and the number of primary schools.

9. The table contains data for the time taken to do a paving job and the cost of the job.

Time taken (hours)	Cost of job ($)
5	1000
7	1000
5	1500
8	1200
10	2000
13	2500
15	2800
20	3200
18	2800
25	4000
23	3000

Construct a scatterplot for the data. Comment on whether a relationship exists between the time taken and the cost. Describe any relationship that exists.

10. The table shows the time of booking (how many days in advance) of the tickets for a musical performance and the corresponding row number in A-reserve seating.

Time of booking	Row number
5	15
6	15
7	15
7	14
8	14
11	13
13	13

Time of booking	Row number
14	12
14	10
17	11
20	10
21	8
22	5
24	4

Time of booking	Row number
25	3
28	2
29	2
29	1
30	1
31	1

Construct a scatterplot for the data. Comment on whether a relationship exists between the time of booking and the number of the row and, if there is a relationship, describe it.

11. **MC** The correlation of this scatterplot is:

A. weak, positive, linear.
B. no correlation.
C. strong, positive linear.
D. weak, negative, linear.
E. strong, negative, linear.

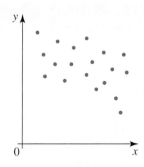

12. Draw a scatterplot to display the following data:

a. by hand
b. using CAS.

Number of dry-cleaning items	1	2	3	4	5	6	7
Cost ($)	12	16	19	20	22	24	25

13. Draw a scatterplot to display the following data:

a. by hand
b. using CAS.

Maximum daily temperature (°C)	26	28	19	17	32	36	33	23	24	18
Number of drinks sold	135	156	98	87	184	133	175	122	130	101

14. Describe the correlation between:

a. the number of dry-cleaning items and the cost in question 12
b. the maximum daily temperature and the number of drinks sold in question 13.

15. Draw a scatterplot and describe the correlation for the following data.

	NSW	VIC	QLD	SA	WA	TAS	NT	ACT
Population	7 500 600	5 821 000	4 708 000	1 682 000	2 565 000	514 000	243 000	385 000
Area of land (km²)	800 628	227 010	1 723 936	978 810	2 526 786	64 519	1 335 742	2358

16. The table shows data giving the time taken to engineer a finished product from the raw recording (of a song) and the length of the finished product.

Time spent engineering studio (hours)	1	2	3	4	5	6	7	8	9	10	11	12	13	14	15
Finished length of recording (minutes)	3	4	10	12	20	16	18	25	30	28	35	36	39	42	45

a. Construct a scatterplot for the data.
b. Comment on whether a relationship exists between the time spent engineering and the length of the finished recording.

2.6 Exam questions

Question 1 (1 mark)

MC The best description for the relationship shown in the scatterplot would be

    A. moderate, positive and linear.
    B. strong, negative and linear.
    C. moderate, negative and non-linear.
    D. strong, positive and linear.
    E. weak, positive and linear.

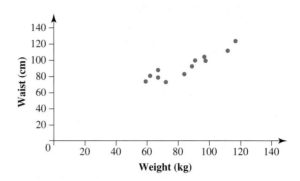

Question 2 (1 mark)

MC Determine which of the following conclusions can be drawn from the scatterplot shown.

    A. As the number of cigarettes smoked increases, fitness level increases.
    B. As the number of cigarettes smoked increases, fitness level decreases.
    C. As the number of cigarettes smoked decreases, fitness level remain unchanged.
    D. As fitness level decreases, there is little change in the number of cigarettes smokes.
    E. There is no relationship between the number of cigarettes smoked and fitness level.

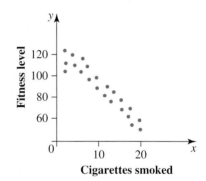

Question 3 (1 mark)

MC Determine which of the following conclusions can be drawn from the scatterplot shown.

    A. There is strong, linear, positive relationship between the number of hours worked and the number of hours for recreation.
    B. There is a moderate, linear, positive relationship between the number of hours worked and the number of hours for recreation.
    C. There is a strong, linear, negative relationship between the number of hours worked and the number of hours for recreation.
    D. There is a moderate, linear, negative relationship between the number of hours worked and the number of hours for recreation.
    E. There is no relationship between the number of hours worked and the number of hours for recreation.

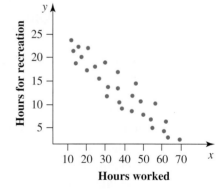

More exam questions are available online.

2.7 Estimating and interpreting Pearson's product–moment correlation coefficient

---

LEARNING INTENTION

At the end of this subtopic you should be able to:
- describe correlation and strength on scatterplots
- estimate and interpret Pearson's product–moment correlation coefficient, r.

2.7.1 Pearson's product–moment correlation coefficient

In the previous section, we estimated the strength of association by looking at a scatterplot and forming a judgement about whether the correlation between the variables was positive or negative and whether the correlation was weak, moderate or strong.

A more precise tool for measuring correlation between two variables is **Pearson's product–moment correlation coefficient**. This coefficient is used to measure the strength of *linear relationships* between variables. The symbol for Pearson's product–moment correlation coefficient is r. The value of r ranges from -1 to 1, that is, $-1 \leq r \leq 1$. The two extreme values of r (1 and -1) are shown in the first two diagrams in the following pink box.

Describing correlation on scatterplots

Following is a gallery of scatterplots with the corresponding value of r for each.

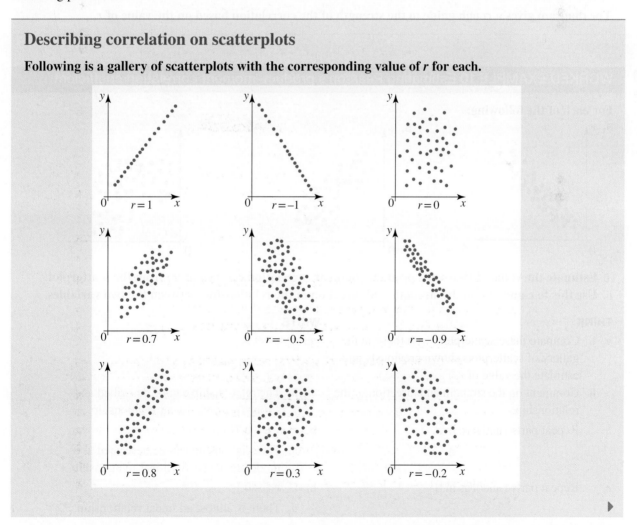

From these diagrams, we can see that a value of $r = 1$ or -1 means that there is perfect linear association between the variables.

The value of the Pearson's product–moment correlation coefficient indicates the strength of the linear relationship between two variables.

Describing strength of the correlation

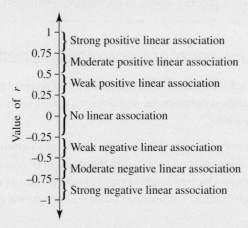

The diagram gives a rough guide to the strength of the correlation based on the value of r.

WORKED EXAMPLE 10 Estimating Pearson's product–moment correlation coefficient

tlvd-3546

For each of the following:

a.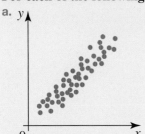

b.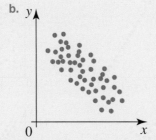

c.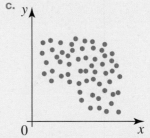

i. **Estimate the value of Pearson's product–moment correlation coefficient (r) from the scatterplot.**
ii. **Use this to comment on the strength and direction of the relationship between the two variables.**

THINK	WRITE
a. i. Compare these scatterplots with those in the gallery of scatterplots shown previously and estimate the value of r.	a. i. $r \approx 0.9$
ii. Comment on the strength and direction of the relationship.	ii. The relationship can be described as a strong, positive, linear relationship.
b. Repeat parts i and ii as in a.	b. i. $r \approx -0.7$
	ii. The relationship can be described as a moderate, negative, linear relationship.
c. Repeat parts i and ii as in a.	c. i. $r \approx -0.1$
	ii. There is almost no linear relationship.

In completing Worked example 10, we notice that estimating the value of *r* from a scatterplot is rather like making an informed guess. In the next section, we will see how to obtain the actual value of *r*.

 Resources

 Interactivity Pearson's product–moment correlation coefficient and the coefficient of determination (int-6251)

2.7 Exercise

Students, these questions are even better in jacPLUS

Receive immediate feedback and access sample responses

Access additional questions

Track your results and progress

Find all this and MORE in jacPLUS

1. **WE10** For each of the following:

   a.

   b.

   i. estimate the Pearson's product–moment correlation coefficient (*r*) from the scatterplot
   ii. use this to comment on the strength and direction of the relationship between the two variables.

2. Determine what type of linear relationship each of the following values of *r* suggests.

   a. 0.85   b. −0.3

3. Determine what type of linear relationship each of the following values of *r* suggests.

   a. 0.21   b. 0.65   c. −1   d. −0.78

4. Determine what type of linear relationship each of the following values of *r* suggests.

   a. 1   b. 0.9   c. −0.34   d. −0.1

5. For each of the following:

   a. 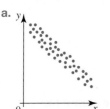   b. 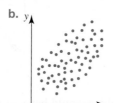   c. 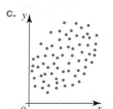   d.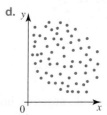

   i. estimate the value of Pearson's product–moment correlation coefficient (*r*) from the scatterplot
   ii. use this to comment on the strength and direction of the relationship between the two variables.

6. For each of the following:

a.

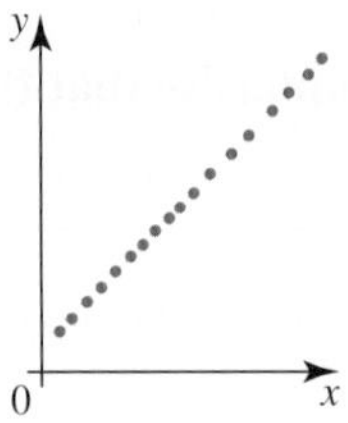

b.

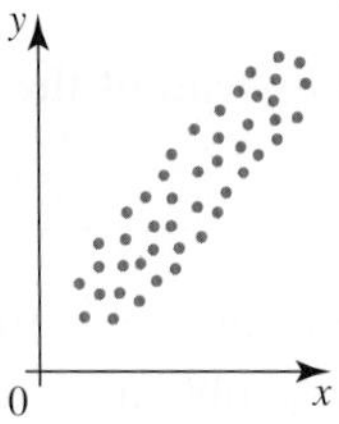

c.

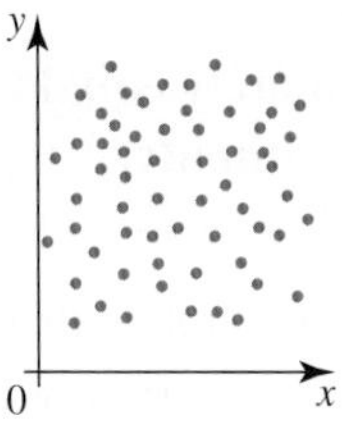

d.

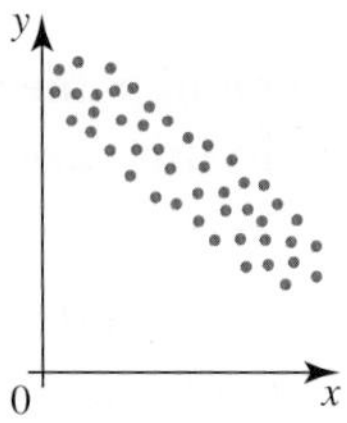

i. estimate the value of Pearson's product–moment correlation coefficient (r) from the scatterplot
ii. use this to comment on the strength and direction of the relationship between the two variables.

7. **MC** A set of data relating the variables x and y is found to have an r value of 0.62. The scatterplot that could represent the data is:

A.

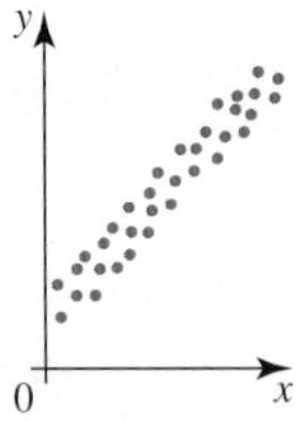

B.

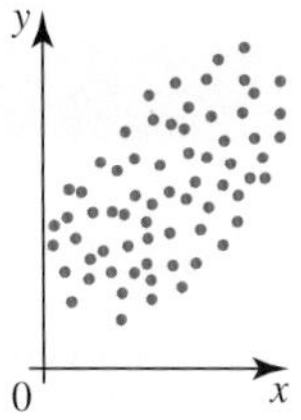

C.

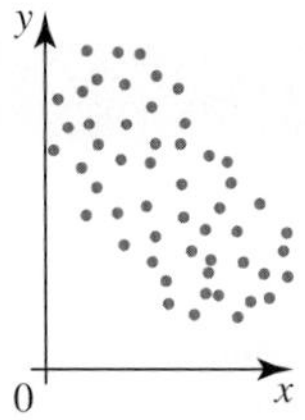

D.

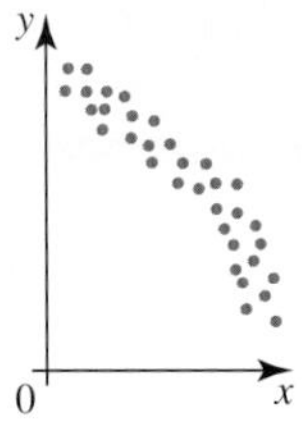

E.

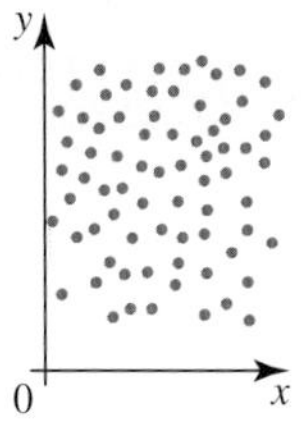

8. **MC** A set of data relating the variables x and y is found to have an r value of −0.45. A true statement about the relationship between x and y is:

A. There is a strong linear relationship between x and y and when the x-values increase, the y-values tend to increase also.
B. There is a moderate linear relationship between x and y and when the x-values increase, the y-values tend to increase also.
C. There is a moderate linear relationship between x and y and when the x-values increase, the y-values tend to decrease.
D. There is a weak linear relationship between x and y and when the x-values increase, the y-values tend to increase also.
E. There is a weak linear relationship between x and y and when the x-values increase, the y-values tend to decrease.

9. From the scatterplots shown estimate the value of r and comment on the strength and direction of the relationship between the two variables.

a.

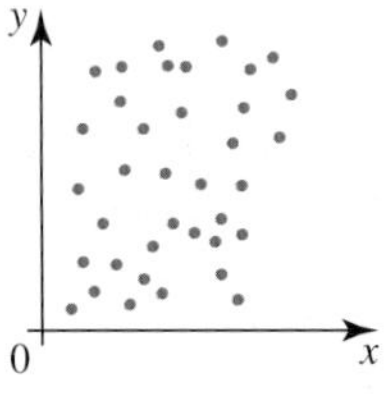

b.

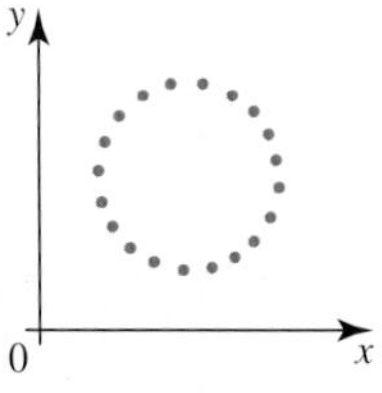

c.

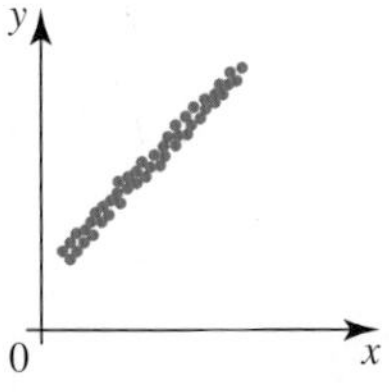

10. **MC** A weak, negative, linear association between two variables would have an r value closest to:

   A. −0.55   **B.** 0.55   **C.** −0.65   **D.** −0.45   **E.** 0.45

11. **MC** Determine which of the following is *not* a Pearson product–moment correlation coefficient.

    A. 1.0   **B.** 0.99   **C.** −1.1   **D.** −0.01   **E.** 0

12. Draw a scatterplot that has a Pearson product–moment correlation coefficient of approximately −0.7.

13. **MC** If two variables have an r value of 1, then they are said to have:

    A. a strong positive linear relationship.   **B.** a strong negative linear relationship.

    C. a perfect positive relationship.   **D.** a perfect negative linear relationship.

    E. a perfect positive linear relationship.

14. **MC** Select which is the correct ascending order of positive values of r.

    A. Strong, moderate, weak, no linear association   **B.** Weak, strong, moderate, no linear association

    C. No linear association, weak, moderate, strong   **D.** No linear association, moderate, weak, strong

    E. Strong, weak, moderate, no linear association

2.7 Exam questions

Question 1 (1 mark)

Source: VCE 2016, Further Mathematics Exam 1, Section A, Q12; © VCAA

MC There is a strong positive association between a country's Human Development Index and its carbon dioxide emissions. From this information, it can be concluded that

    A. increasing a country's carbon dioxide emissions will increase the Human Development Index of the country.

    B. decreasing a country's carbon dioxide emissions will increase the Human Development Index of the country.

    C. this association must be a chance occurrence and can be safely ignored.

    D. countries that have higher human development indices tend to have higher levels of carbon dioxide emissions

    E. countries that have higher human development indices tend to have lower levels of carbon dioxide emissions.

Question 2 (1 mark)

MC A scatterplot cannot be used to display the relationship between which of the following pairs of variables?

    A. The marks obtained by a group of students and their gender.

    B. The masses (kg) of newborn male and female giraffes born in captivity

    C. The number of couriers delivering packages to two nearby factories.

    D. The weights and ages of females attending a gym.

    E. The scores obtained by basketball teams from two neighbouring schools over the course of a season.

Question 3 (1 mark)

MC An estimate of the correlation coefficient for the data shown is

    A. 0.93

    B. −0.93

    C. 0.67

    D. −0.67

    E. 0.28

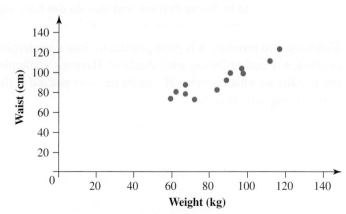

More exam questions are available online.

2.8 Calculating r and the coefficient of determination, r^2

2.8.1 Pearson's product–moment correlation coefficient (r)

The formula for calculating Pearson's correlation coefficient r is as follows:

The rule for 'r'

$$r = \frac{1}{n-1} \sum_{i=1}^{n} \left(\frac{x_i - \bar{x}}{s_x} \right) \left(\frac{y_i - \bar{y}}{s_y} \right)$$

where n is the number of pairs of data in the set

s_x is the standard deviation of the x-values

s_y is the standard deviation of the y-values

$\bar{x}$ is the mean of the x-values

$\bar{y}$ is the mean of the y-values.

The calculation of r is often done using CAS.

There are two important limitations on the use of r.

First, since r measures the strength of a linear relationship, it would be inappropriate to calculate r for data that are not linear — for example, data that a scatterplot shows to be in a quadratic form.

Second, outliers can bias the value of r. Consequently, if a set of linear data contains an outlier, then r is not a reliable measure of the strength of that linear relationship.

When to use 'r'

The calculation of r is applicable to sets of bivariate data that are known to be linear in form and that do not have outliers.

With those two provisos, it is good practice to draw a scatterplot for a set of data to check for a linear form and an absence of outliers before r is calculated. Having a scatterplot in front of you is also useful because it enables you to estimate what the value of r might be — as you did in the previous exercise, and thus you can check that your workings are correct.

The heights (in centimetres) of 21 football players were recorded against the number of marks they took in a game. The data are shown in the following table.

a. Construct a scatterplot for the data.

b. Comment on the correlation between the heights of players and the number of marks that they take, and estimate the value of r.

c. Calculate r and use it to comment on the relationship between the heights of players and the number of marks they take in a game.

Height (cm)	Number of marks taken
184	6
194	11
185	3
175	2
186	7
183	5
174	4
200	10
188	9
184	7
188	6

Height (cm)	Number of marks taken
182	7
185	5
183	9
191	9
177	3
184	8
178	4
190	10
193	12
204	14

THINK

a. Height is the explanatory variable, so plot it on the *x*-axis; the number of marks is the response variable, so show it on the *y*-axis.

b. Comment on the correlation between the variables, and estimate the value of r.

c. 1. Because there is a linear form and no outliers, the calculation of r is appropriate. Use CAS to find the value of r. Round correct to 2 decimal places.

  2. The value of $r = 0.86$ indicates a strong positive linear relationship.

WRITE/DRAW

a.

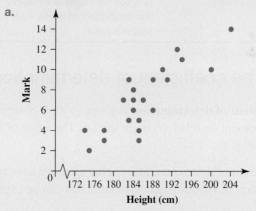

b. The data show what appears to be a linear form of moderate strength. We might expect $r \approx 0.8$.

c. $r = 0.859\,311...$
  ≈ 0.86

  $r = 0.86$. This indicates a strong positive linear association between the height of a player and the number of marks they take in a game.

TI	THINK	WRITE	CASIO	THINK	WRITE
1.	On a Lists & Spreadsheet page, enter the data into the lists named 'height' and 'marks'.	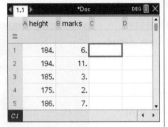	1.	On a Statistics screen, enter the data into the lists named 'height' and 'marks'.	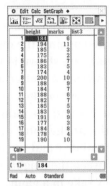
2.	Press CTRL + I and select: 1: Add Calculator Complete the entry line as:corrMat(height, marks) Then press ENTER.		2.	Tap: Calc Linear reg On the Set Calculation screen, select the following: X List: main\height Y List: main\marks Freq: 1 Copy Formula: y1 OK	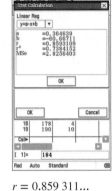
3.	Read r, the correlation coefficient, from the matrix on the screen and round correct to 2 decimal places.	$r = 0.859\,311...$ ≈ 0.86		Read r, the correlation coefficient, from the screen and round correct to 2 decimal places.	$r = 0.859\,311...$ ≈ 0.86
	The value of $r = 0.86$ indicates a strong positive linear relationship.	$r = 0.86$. This indicates a strong positive linear association between the height of a player and the number of marks they take in a game.		The value of $r = 0.86$ indicates a strong positive linear relationship.	$r = 0.86$. This indicates a strong positive linear association between the height of a player and the number of marks they take in a game.

2.8.2 The coefficient of determination $\left(r^2\right)$

The **coefficient of determination** is given by r^2. It is very easy to calculate; we merely square Pearson's product–moment correlation coefficient (r). The value of the coefficient of determination ranges from 0 to 1, that is, $0 \leq r^2 \leq 1$.

The coefficient of determination is useful when we have two variables that have a linear relationship. It tells us the proportion of variation in one variable that can be explained by the variation in the other variable.

The coefficient of determination

The coefficient of determination provides a measure of how well the linear rule linking the two variables (x and y) predicts the value of y when we are given the value of x.

tlvd-3547

WORKED EXAMPLE 12 Calculating the coefficient of determination

A set of data giving the number of police traffic patrols on duty and the number of drink-driving charges for the region was recorded and a correlation coefficient of $r = -0.8$ was found. Calculate the coefficient of determination and interpret its value.

THINK	WRITE
1. Calculate the coefficient of determination by squaring the given value of r.	Coefficient of determination $= r^2$ $= (-0.8)^2$ $= 0.64$
2. Interpret your result.	We can conclude from this that 64% of the variation in the number of drink-driving charges can be explained by the variation in the number of police traffic patrols on duty. This means that the number of police traffic patrols on duty is a major factor in predicting the number of drink-driving charges.

Note: In Worked example 12, 64% of the variation in the number of drink-driving charges was due to the variation in the number of police cars on duty and 36% was due to other factors, for example, days of the week or hour of the day.

2.8 Exercise

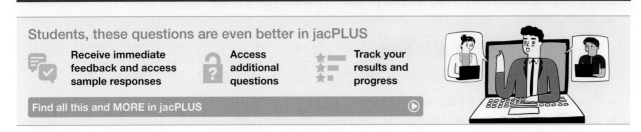

Students, these questions are even better in jacPLUS

Receive immediate feedback and access sample responses

Access additional questions

Track your results and progress

Find all this and MORE in jacPLUS

1. **WE11** The heights (cm) of basketball players were recorded against the number of points they scored in a game. The data are shown in the following table.

Height (cm)	Points scored	Height (cm)	Points scored
194	6	201	13
203	4	196	10
208	18	205	20
198	22	215	14
195	2	203	3

a. Construct a scatterplot of the data.
b. Comment on the correlation between the heights of basketballers and the number of points scored, and estimate the value of r.
c. Calculate the r value and use it to comment on the relationship between heights of players and the number of points they scored in a game.

2. The following table shows the gestation time and the birth mass of 10 babies.

Gestation time (weeks)	31	32	33	34	35	36	37	38	39	40
Birth mass (kg)	1.08	1.47	1.82	2.06	2.23	2.54	2.75	3.11	3.08	3.37

a. Construct a scatterplot of the data.
b. Comment on the correlation between gestation time and birth mass, and estimate the value of r.
c. Calculate the r value and use it to comment on the relationship between gestation time and birth mass.

3. The yearly salary ($\times$ $1000) and the number of votes polled in the Brownlow medal count are given below for 10 footballers.

Yearly salary ($\times$ 1000)	360	400	320	500	380	420	340	300	280	360
Number of votes	24	15	33	10	16	23	14	21	31	28

a. Construct a scatterplot for the data.
b. Comment on the correlation of salary and the number of Brownlow votes, and make an estimate of r.
c. Calculate r and use it to comment on the relationship between yearly salary and number of votes.

4. **WE12** Data on the number of booze buses in use and the number of drivers registering a blood alcohol reading over 0.05 was recorded, and a correlation coefficient of $r = 0.77$ was found.
Calculate the coefficient of determination and interpret its value.

5. A set of data, obtained from 40 smokers, gives the number of cigarettes smoked per day and the number of visits per year to a doctor. The Pearson's correlation coefficient for the data was found to be 0.87. Calculate the coefficient of determination and interpret its value.

6. Data giving the annual advertising budgets ($\times$ $1000) and the yearly profit increases (%) of eight companies are shown below.

Annual advertising budget ($\times$ 1000)	11	14	15	17	20	25	25	27
Yearly profit increase (%)	2.2	2.2	3.2	4.6	5.7	6.9	7.9	9.3

a. Construct a scatterplot for the data.
b. Comment on the correlation between the advertising budget and profit increase, and make an estimate of r.
c. Calculate r.
d. Calculate the coefficient of determination.
e. Write the proportion of the variation in the yearly profit increase that can be explained by the variation in the advertising budget.

7. Data showing the number of people in nine households against weekly grocery costs are given below.

Number of people in household	2	5	6	3	4	5	2	6	3
Weekly grocery costs ($)	60	180	210	120	150	160	65	200	90

a. Construct a scatterplot for the data.
b. Comment on the correlation between the number of people in a household and the weekly grocery costs, and give an estimate of r.
c. Calculate r.
d. Calculate the coefficient of determination.
e. Write the proportion of the variation in the weekly grocery costs that can be explained by the variation in the number of people in a household.

8. An investigation is undertaken with people following the Certain Slim diet to explore the link between weeks of dieting and total weight loss. The data are shown in the table.

Total weight loss (kg)	1.5	4.5	9	3	6	8	3.5	3	6.5	8.5	4	6.5	10	2.5	6
Number of weeks on the diet	1	5	8	3	6	9	4	2	7	10	4	6	9	2	5

a. Display the data on a scatterplot.
b. Describe the association between the two variables in terms of direction, form and strength.
c. Explain whether it is appropriate to use Pearson's correlation coefficient to describe the link between the number of weeks on the Certain Slim diet and total weight loss.
d. Estimate the value of Pearson's correlation coefficient from the scatterplot.
e. Calculate the value of this coefficient.
f. Determine whether the total weight loss is affected by the number of weeks on the diet.
g. Calculate the value of the coefficient of determination.
h. Describe what the coefficient of determination says about the relationship between total weight loss and the number of weeks on the Certain Slim diet.

2.8 Exam questions

Question 1 (1 mark)

Source: *VCE 2013, Further Mathematics, Exam 1, Section A Core, Q8; © VCAA*

MC The table below shows the hourly rate of pay earned by 10 employees in a company in 1990 and in 2010.

Employee	Hourly rate of pay ($)	
	1990	**2010**
Ben	9.53	17.02
Lani	9.15	16.71
Freya	8.88	15.10
Jill	8.60	15.93
David	7.67	14.40
Hong	7.96	13.32
Stuart	6.42	15.40
Mei lien	11.86	19.79
Tim	14.64	23.38
Simon	15.31	25.11

The value of the correlation coefficient, r, for this set of data is closed to
- **A.** 0.74
- **B.** 0.86
- **C.** 0.92
- **D.** 0.93
- **E.** 0.96

Question 2 (1 mark)

Source: VCE 2013, Further Mathematics, Exam 1, Section A Core, Q7; © VCAA

MC For a city, the correlation between:
- population density and distance from the centre of the city is $r = -0.563$
- house size and distance from the centre of the city is $r = 0.357$.
  Given this information, determine which of the following statements is true.
  A. Around 31.7% of the variation observed in house size in the city can be explained by the variation in distance from the centre of the city.
  B. Population density tends to increase as the distance from the centre of the city increases.
  C. House sizes tend to be larger as the distance from the centre of the city decreases.
  D. The slope of a least-squares regression line relating population density to distance from the centre of the city is positive.
  E. Population density is more strongly associated with distance from the centre of the city than is house size.

Question 3 (1 mark)

Source: VCE 2013, Further Mathematics, Exam 1, Section A, Q11; © VCAA

MC For a group of 15-year-old students who regularly played computer games, the correlation between the time spent playing computer games and fitness level was found to be $r = -0.56$.

On the basis of this information, it can be concluded that
  A. 56% of these students were not very fit.
  B. these students would become fitter if they spent less time playing computer games.
  C. these students would become fitter if they spent more time playing computer games.
  D. the students in the group who spent a short amount of time playing computer games tended to be fitter.
  E. the students in the group who spent a large amount of time playing computer games tended to be fitter.

More exam questions are available online.

2.9 Cause and effect

> **LEARNING INTENTION**
>
> At the end of this subtopic you should be able to:
> - distinguish between correlation and causation
> - answer statistical questions that require a knowledge of the associations between pairs of variables.

2.9.1 Observation and experimentation

Correlation and causality

In an area known as Alsace on the eastern border of France, it has long been noted that there is a strong correlation between the number of storks and the number of babies in these villages.

This is an extreme example of why **correlation** does not necessary imply **causation**. The diagram shows why large numbers of storks do not cause the growth in births.

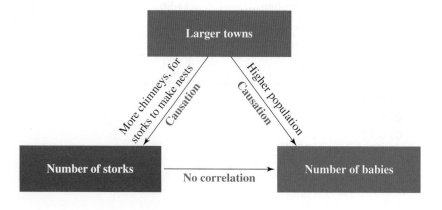

In Worked example 11, there was a high positive correlation between the height of a footballer and the number of marks they take. Does this mean that if a football team needs players with greater marking ability then they just should recruit more tall players?

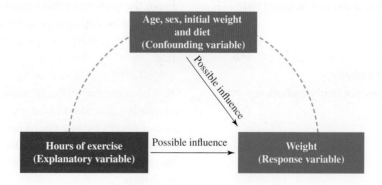

An experienced recruiter knows that height is only one advantage in a marking contest; fitness levels, skill level, hand-eye coordination and ability to read the game are all necessary attributes of a player who can take marks.

Therefore, a high degree of correlation between two variables is not a basis on which to state that one variable *causes* a change in the other variable; further investigation is necessary. In professional research, many controlled experiments must be carried out to determine and identify a causal relationship.

Experimentation

One way to establish causation is to conduct experiments where a control group is used. Scientists researching the effectiveness of a new drug designed to destroy a particular type of cancer cell split the population (those participating in the study) into two groups. One group will be given the drug; the other group is given a placebo and all other variables are the same across both groups. A study of the correlation between the administered drug and a decrease in the presence of the cancer cells can then establish causation. In this example, the administered drug and placebo are the explanatory variables, and the number of cancer cells present is the response variable.

Common response

In some cases, the correlation between two variables can be explained by a common response that provides the association. A common response affects both the explanatory and response variables. For example, a study may show a strong correlation between house sizes and the life expectancy of home owners. While a bigger house will not directly lead to a longer life expectancy, a common response, the income of the house owner provides a direct link to both variables and is more likely to be the underlying cause for the observed correlation.

Confounding variables

A **confounding variable** is an outside influence that changes the relationship between a response variable and an explanatory variable. It could also show a false relationship between them when no true association exists.

For example, a study of a group of people may show a strong association between weight (response variable) and hours of exercise (explanatory variable).

But, there are other factors that can influence weight, such as age, sex, initial weight and diet. These are called confounding variables. A study that does not take these variables into account produces results that cannot show causality.

Sometimes an association between variables can just be a **coincidence**. There are no clear confounding or common response variables. For example, during a discussion over a community breakfast at a time-share resort in Queensland, it was established that of six people sitting together, five had a relative who lived on Phillip Island, in Victoria. Does this suggest a strong correlation between people who own a time-share property and those who have a relative on Phillip Island? No, it is just a coincidence.

When observing an association between two variables, it can never be stated that one variable causes a response in the other variable.

tlvd-3548

WORKED EXAMPLE 13 Establishing correlation

Data showing the number of tourists visiting a small country in a month and the corresponding average monthly exchange rate for the country's currency against the US dollar are given.

Number of tourists (× 1000)	2	3	4	5	7	8	8	10
Exchange rate	1.2	1.1	0.9	0.9	0.8	0.8	0.7	0.6

a. **Construct a scatterplot for the data.**
b. **Comment on the correlation between the number of tourists and the exchange rate.**
c. **Calculate r.**
d. **Based on the value of r obtained in part c, determine whether it is appropriate to conclude that the decrease in tourist numbers is caused by the increasing exchange rate.**

THINK

a. 1. Determine the explanatory and response variables.

2. Sketch a scatterplot.

b. A negative association can be observed.

c. Use CAS to calculate r.
d. Correlation does not imply causation.

WRITE

a. Exchange rate (explanatory)
Tourist numbers (response)

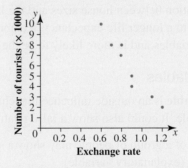

b. From the scatterplot there appears to be a strong negative correlation between the number of tourists visiting the country and the exchange rate.

c. $r = -0.96$
d. Just because $r = -0.96$, it cannot be stated that the increasing tourist numbers are caused by the decreasing exchange rate. There may be other factors that need to be considered. Further investigation is needed before any conclusions can be drawn.

2.9 Exercise

1. **WE13** Data showing the number of people on eight fundraising committees and the annual funds raised are given in the table.

Number of people on a committee	3	6	4	8	5	7	3	6
Annual funds raised ($)	4500	8500	6100	12 500	7200	10 000	4700	8800

   a. Construct a scatterplot for the data.
   b. Comment on the correlation between the number of people on a committee and the annual funds raised.
   c. Calculate r.
   d. Based on the value of r obtained in part c, determine whether it is appropriate to conclude that the increase in funds raised is caused by an increase in the number of people on a committee.

2. Data was collected on the number of books read by each of 50 ten-year olds and their spelling skills. The correlation coefficient for this data was 0.90.
   Explain what can be said about the association between the number of books read and spelling skills for this group of 10-year olds.

The following information applies to questions 3 and 4.

A set of data was gathered from a large group of parents with children under 5 years of age. They were asked the number of hours they worked per week and the amount of money they spent on child care. The results were recorded, and the value of Pearson's correlation coefficient was found to be 0.92.

3. **MC** Determine which of the following statements is *not* true.
   A. There is a positive correlation between the number of working hours and the amount of money spent on child care.
   B. The correlation between the number of working hours and the amount of money spent on child care can be classified as strong.
   C. The relationship between the number of working hours and the amount spent on child care is linear.
   D. The linear relationship between the two variables suggests that as the number of working hours increases, so too does the amount spent on child care.
   E. The increase in the number of hours worked causes the increase in the amount of money spent on child care.

4. **MC** Determine which of the following statements is *not* true.
   A. The coefficient of determination is about 0.85.
   B. The number of working hours is the major factor in predicting the amount of money spent on child care.
   C. About 85% of the variation in the number of hours worked can be explained by the variation in the amount of money spent on child care.
   D. Other factors such as income levels can affect the amount of money spent on child care.
   E. About $\frac{17}{20}$ of the variation in the amount of money spent on child care can be explained by the variation in the number of hours worked.

5. MC Over a 10-week period a strong positive association ($r = 0.91$) was found between the number of baby penguins born on Phillip Island each week and the amount of wildlife killed on the roads on Phillip Island each week.

A strong positive association was also found both between the number of baby penguins born on the island ($r = 0.93$), the amount of wildlife killed on the island ($r = 0.86$), and the number of visitors to the island. Using this information, determine which of the following statements is true.

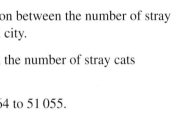

   A. During peak breeding season, drivers get distracted looking at the baby penguins and don't notice the wildlife on the roads.

   B. It is just a coincidence that there is a strong positive association between the number of baby penguins born on the island and the amount of wildlife killed on the roads.

   C. Breeding season attracts larger numbers of visitors and more visitors means more drivers on the roads, which leads to more wildlife being killed.

   D. Everyone knows that tourists are very poor drivers.

   E. Penguins are very aggressive and cause other wildlife to run onto the road.

6. There is a strong positive correlation between rates of diabetes and the number of hip replacements. Does diabetes cause deterioration of the hip joint? Describe the common cause(s) that could link these two variables.

7. A strong positive association has been found between weight gain and the length of holidays. Does this mean that holidays cause weight gain? Describe a common cause that could link these variables.

8. A study of emergency service workers (police, fire and paramedics) showed a strong positive association between levels of stress and hours worked. Does this mean that the length of the shift causes stress? Describe other confounding variables that could also contribute to this association.

2.9 Exam questions

Question 1 (1 mark)

Source: VCE 2017, Further Mathematics Exam 1, Section A, Core, Q12; © VCAA

MC Data collected over a period of 10 years indicated a strong, positive association between the number of stray cats and the number of stray dogs reported each year ($r = 0.87$) in a large regional city.

A positive association was also found between the population of the city and both the number of stray cats ($r = 0.61$) and the number of stray dogs ($r = 0.72$).

During the time the data was collected, the population of the city grew from 34 564 to 51 055.

From this information, we can conclude that

   A. if cat owners paid more attention to keeping dogs off their property, the number of stray cats reported would decrease.

   B. the association between the number of stray cats and stray dogs reported cannot be causal because only a correlation of $+1$ or -1 shows causal relationship.

   C. there is no logical explanation for the association between the number of stray cats and stray dogs reported in the city, so it must be a chance occurrence.

   D. because larger populations tend to have both a larger number of stray cats and stray dogs, the association between these numbers can be explained by a common response to a third variable, which is the increasing population size of the city.

   E. more stray cats were reported because people are no longer as careful about keeping their cats properly contained on their property as they were in the past.

Question 2 (1 mark)

MC Following an investigation into the relationship between the two variables of tissue sales and hot chocolate sales, a correlation value of $r = 0.95$ was found.

Determine which of the following conclusions can be drawn from this result.
- **A.** Strong tissue sales cause an increase in sales of hot chocolate.
- **B.** 90.25% of the variation in hot chocolate sales can be explained by the variation in tissue sales.
- **C.** There is no causal relationship between the two variables.
- **D.** 90.25% of the variation in tissue sales can be explained by the variation in hot chocolate sales.
- **E.** Higher tissue sales are the cause of greater hot chocolate sales.

Question 3 (1 mark)

MC A recent study over a summer period found a strong, positive and linear association between ice-cream sales and house burglaries.

Select which of the following statements is false.
- **A.** House burglaries tended to increase as ice-cream sales increased.
- **B.** Higher ice-cream sales caused more burglaries.
- **C.** A confounding factor is evident.
- **D.** Non-causal reasons are evident in this study.
- **E.** More homes are vacant in hot weather as more people go out to enjoy warmer weather.

More exam questions are available online.

2.10 Review

2.10 Exercise

Multiple choice

1. **MC** A back-to-back stem plot is a useful way of displaying the relationship between:

   A. the number of children attending a day care centre and whether or not the centre has federal funding.
   B. height and wrist circumference.
   C. age and weekly income.
   D. weight and the number of takeaway meals eaten each week.
   E. the age of a car and amount spent each year on servicing it.

The following information is for questions 2 and 3.

The salaries of people working at five different advertising companies are shown on the following parallel boxplots.

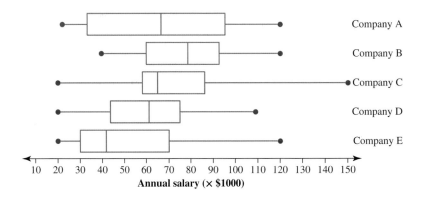

2. **MC** The company with the largest interquartile range is:

   A. Company A
   B. Company B
   C. Company C
   D. Company D
   E. Company E

3. **MC** The company with the lowest median is:

   A. Company A
   B. Company B
   C. Company C
   D. Company D
   E. Company E

Use the following information to answer questions 4 and 5.

Data showing reactions of junior staff and senior staff to a relocation of offices are given below in a contingency table.

	Attitude	Staff type Junior	Staff type Senior	Total
Attitude	For	23	14	37
	Against	31	41	72
	Total	54	55	109

4. **MC** From this table, we can conclude that:

   A. 23% of junior staff were for the relocation.
   B. 42.6% of junior staff were for the relocation.
   C. 31% of junior staff were against the relocation.
   D. 62.1% of junior staff were for the relocation.
   E. 28.4% of junior staff were against the relocation.

5. **MC** From this table, we can conclude that:

   A. 14% of senior staff were for the relocation.
   B. 37.8% of senior staff were for the relocation.
   C. 12.8% of senior staff were for the relocation.
   D. 72% of senior staff were against the relocation.
   E. 74.5% of senior staff were against the relocation.

6. **MC** The relationship between the variables x and y is shown on the scatterplot. The correlation between x and y would be best described as:

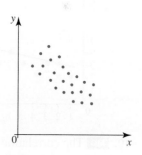

   A. a weak positive association.
   B. a weak negative association.
   C. a strong positive association.
   D. a strong negative association.
   E. non-existent.

7. **MC** A set of data relating the variables x and y is found to have an r-value of -0.83. The scatterplot that would best represent this data set is:

A. 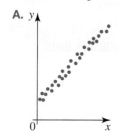   **B.** 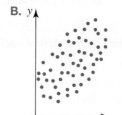   **C.**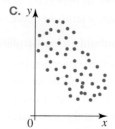

D. 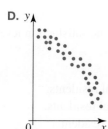   **E.**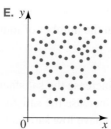

8. **MC** A set of data comparing age with blood pressure is found to have a Pearson's correlation coefficient of 0.86. The coefficient of determination for the data would be closest to:

   A. -0.86 **B.** -0.74 **C.** -0.43 **D.** 0.43   **E.** 0.74

9. **MC** The most appropriate line of best fit for the figure is:

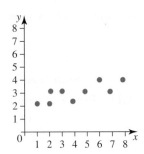

A. 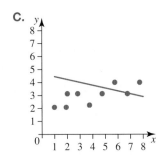 B. C.

D. 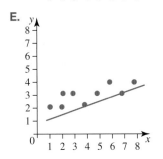 E.

10. **MC** The correlation between two variables x and y is -0.88. Determine which of the following statements is true.

A. As y increases, it causes x to increase.   B. As y increases, it causes x to decrease.

C. There is a poor fit between x and y.   D. As x increases, y tends to increase.

E. As x increases, y tends to decrease.

11. **MC** When calculating a least-squares regression line, a correlation coefficient of -1 indicates that:

A. the y-axis variable depends linearly on the x-axis variable.

B. the y-axis variable increases as the x-axis variable decreases.

C. the y-axis variable decreases as the x-axis variable decreases.

D. all the data lie on the same straight line.

E. the two variables depend upon each other.

12. **MC** There is a strong negative association between the number of tourists in cities and the satisfaction levels of local residents.

From this information it can be concluded that:

A. increasing the number of tourists in a city will increase the satisfaction levels of local residents.

B. decreasing the number of tourists in a city will increase the satisfaction levels of local residents.

C. cities that have large numbers of tourists tend to have low satisfaction levels of local residents.

D. cities that have large numbers of tourists tend to have high satisfaction levels of local residents.

E. this association is a coincidence.

Short answer

13. The number of hours of counselling received by a group of nine full-time firefighters and nine volunteer firefighters after a serious bushfire is given in the table.

Full-time	2	4	3	5	2	4	6	1	3
Volunteer	8	10	11	11	12	13	13	14	15

a. Construct a back-to-back stem plot to display the data.
b. Comment on the distributions of the number of hours of counselling of the full-time firefighters and the volunteers.

14. The IQ of eight players in three different football teams are shown below.

Team A	120	105	140	116	98	105	130	102
Team B	110	104	120	109	106	95	102	100
Team C	121	115	145	130	120	114	116	123

Display the data in parallel boxplots.

15. Delegates at the respective Liberal and Labor Party conferences were surveyed on whether or not they believed that uranium mining should continue. Forty-five Liberal delegates were surveyed and 15 were against continuation. Fifty-three Labor delegates were surveyed and 43 were against continuation.

a. Present the data as percentages in a two-way frequency table and a segmented bar chart.
b. Comment on any difference between the reactions of the Liberal and Labor delegates.

16. a. Construct a scatterplot for the data given in the table below.

Age	15	17	18	16	19	19	17	15	17
Pulse rate	79	74	75	85	82	76	77	72	70

b. Use the scatterplot to comment on any relationship that exists between the variables.

17. For the variables shown on the scatterplot below, give an estimate of the value of r and use it to comment on the nature of the relationship between the two variables.

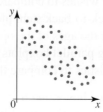

18. The table gives data relating the percentage of lectures attended by students in a semester and the corresponding mark for each student in the exam for that subject.

Lectures attended (%)	70	59	85	93	78	85	84	69	70	82
Exam result (%)	80	62	89	98	84	91	83	72	75	85

a. Identify the explanatory and response variables.
b. Construct a scatterplot for the data.
c. Comment on the correlation between the percentage of lectures attended and the examination results and make an estimate of r.
d. Calculate r.
e. Calculate the coefficient of determination.
f. Write the proportion of the variation in the examination results that can be explained by the variation in the percentage of lectures attended.

Extended response

19. For marketing purposes, the administration of an arts centre needs to compare the ages of people attending two different concerts: a symphony orchestra concert and a jazz concert. Twenty people were randomly selected from each audience and their ages were recorded.

Event	Ages of people attending the event
Symphony orchestra concert	20, 23, 30, 35, 39, 42, 45, 45, 47, 48, 48, 49, 49, 50, 53, 54, 56, 58, 58, 60
Jazz concert	16, 18, 19, 19, 20, 23, 24, 27, 29, 30, 33, 34, 38, 39, 40, 42, 43, 45, 46, 62

   a. Display the data on a back-to-back stem plot.

   b. For each category, calculate the following statistics:

      i. X_{min}

      ii. Q_1

      iii. median

      iv. Q_3

      v. X_{max}

      vi. mean

      vii. interquartile range (IQR)

      viii. standard deviation.

   c. Use the stem plot together with some summary statistics to compare the distributions of the ages of patrons attending the two concerts.

One month later, at the beginning of the opera season, 20 people were again selected (this time from the opera audience) and their ages were recorded.

Event	Ages of people attending the event
Opera	12, 18, 29, 30, 33, 35, 38, 39, 42, 46, 49, 50, 54, 56, 56, 57, 59, 63, 65, 68

The administration of the arts centre now wishes to compare all three age distributions.

   d. Explain why it is not possible to use a back-to-back stem plot for this task.

   e. Calculate the eight summary statistics for the ages of the opera-goers (as in part **b** above).

   f. Display the data for the three events using parallel boxplots.

   g. Use the boxplots and some summary statistics to compare the three distributions.

20. In one study, 380 Year 12 students were asked how often they engaged in any sporting activity outside school. Students were also asked to classify their stress level in relation to their VCE studies. The results obtained are shown in the table.

	Engaged in sporting activity outside school		
Level of stress	Regularly	Sometimes	Never
Low	16	32	36
Medium	12	40	56
High	6	52	130

   a. Determine how many students in this study reported a high level of stress.

   b. Determine how many students were engaged in sport activity outside of school.

   c. Represent the data in a contingency table in percentage form.

   d. Display the data from part **c** using a segmented bar chart.

   e. Comment on any relationship between the stress level and the amount of sporting activity for this group of Year 12 students.

Question 1 (1 mark)

Source: VCE 2021 Further Mathematics, Exam 1, Section A, Q4; © VCAA

MC The boxplots below show the distribution of the length of fish caught in two different ponds, Pond A and Pond B.

Based on the boxplots, it can be said that

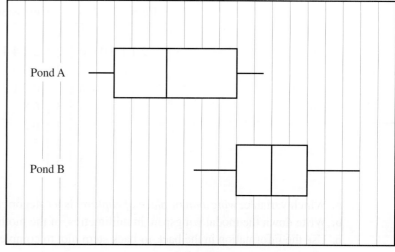

length

A. 50% of the fish caught in Pond A are the same length as the fish caught in Pond B.

B. 50% of the fish caught in Pond B are longer than all of the fish caught in Pond A.

C. 50% of the fish caught in Pond B are shorter than all of the fish caught in Pond A.

D. 75% of the fish caught in Pond A are shorter than all of the fish caught in Pond B.

E. 75% of the fish caught in Pond B are longer than all of the fish caught in Pond A.

Question 2 (1 mark)

Source: VCE 2019, Further Mathematics Exam 1, Section A, Q11; © VCAA

MC A study was conducted to investigate the effect of drinking *coffee* on sleep.

In this study, the amount of *sleep*, in hours, and the amount of *coffee* drunk, in cups, on a given day were recorded for a group of adults.

The following summary statistics were generated.

	Sleep (hours)	*Coffee* (cups)
Mean	7.08	2.42
Standard deviation	1.12	1.56
Correlation coefficient (r)	−0.770	

On average, for each additional cup of *coffee* drunk, the amount of *sleep*

A. decreased by 0.55 hours.

B. decreased by 0.77 hours.

C. decreased by 1.1 hours.

D. increased by 1.1 hours.

E. increased by 2.3 hours.

Question 3 (6 marks)

Source: VCE 2017, Further Mathematics Exam 2, Section A, Core, Q2; © VCAA

The back-to-back stem plot below displays the *wingspan*, in millimetres, of 32 moths and their *place of capture* (forest or grassland).

```
key 1|8 = 18        wingspan (mm)
    forest (n = 13)  |    | grassland (n = 19)
                  6  | 1  | 8
        2 1 1 0 0 0 0 | 2  | 2 2 4 4
                  7  | 2  | 5 5 9
                4 0  | 3  | 0 0 1 2 3 4
                  5  | 3  | 6 8
                     | 4  | 0 3
                     | 4  | 5
                  2  | 5  |
```

a. Which variable, *wingspan* or *place of capture*, is a categorical variable? **(1 mark)**

b. Write down the modal wingspan, in millimetres, of the moths captured in the forest. **(1 mark)**

c. Use the information in the back-to-back stem plot to complete the table below. **(2 marks)**

	Wingspan (mm)				
Place of capture	Minimum	Q_1	Median (M)	Q_3	Maximum
forest		20	21	32	52
grassland	18	24	30		45

d. Show that the moth captured in the forest that had a *wingspan* of 52 mm is an outlier. **(2 marks)**

e. The back-to-back stem plot suggests that *wingspan* is associated with *place of capture*. Explain why, quoting the values of an appropriate statistic. **(2 marks)**

Question 4 (1 mark)

Source: VCE 2016, Further Mathematics Exam 1, Section A, Q12; © VCAA

MC There is a strong positive association between a country's Human Development Index and its carbon dioxide emissions.

From this information, it can be concluded that

A. increasing a country's carbon dioxide emissions will increase the Human Development Index of the country.

B. decreasing a country's carbon dioxide emissions will increase the Human Development Index of the country.

C. this association must be a chance occurrence and can be safely ignored.

D. countries that have higher Human Development Indices tend to have higher levels of carbon dioxide emissions.

E. countries that have higher Human Development Indices tend to have lower levels of carbon dioxide emissions.

The segmented bar chart shows the age distribution of people in three countries, Australia, India and Japan, for the year 2010.

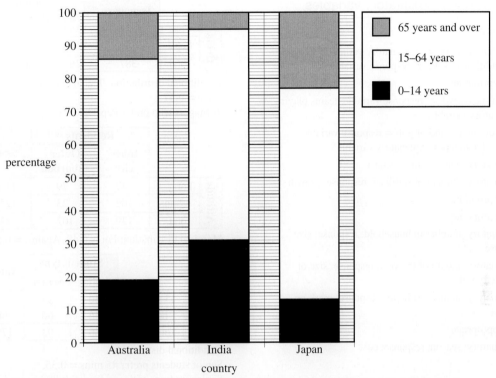

Source: Australian Bureau of Statistics, *3201.0 – Population by Age and Sex, Australian States and Territories*, June 2010

a. Write down the percentage of people in Australia who were aged 0–14 years in 2010.
Write your answer, correct to the nearest percentage. **(1 mark)**

b. In 2010, the population of Japan was 128 000 000.
How many people in Japan were aged 65 years and over in 2010? **(1 mark)**

c. From the graph above, it appears that there is no association between the percentage of people in the $15 - 64$ age group and the country in which they live.
Explain why, quoting appropriate percentages to support your explanation. **(1 mark)**

More exam questions are available online.

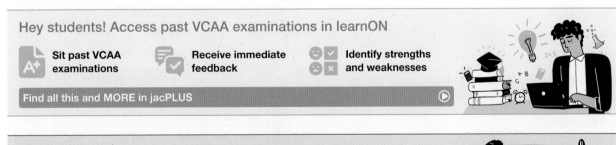

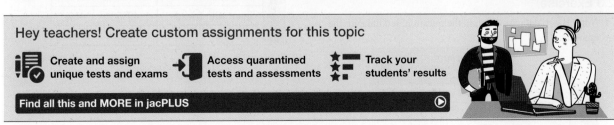

Answers

Topic 2 Investigating associations between two variables

2.2 Response and explanatory variables

2.2 Exercise

1. a. Daily temperature = explanatory variable, air conditioners sold = response variable
   b. Not appropriate
2. a. Size of the crowd = response variable, teams playing = explanatory variable
   b. Net score of a round of golf = response variable, golfer's handicap = explanatory variable
3. a. Explanatory: age, response: salary
   b. Explanatory: amount of fertiliser, response: growth
   c. Not appropriate
   d. Not appropriate
   e. Explanatory: number in household, response: size of house
4. a. Explanatory: month of the year, response: size of electricity bill
   b. Explanatory: number of hours, response: mark on the test
   c. Not appropriate
   d. Explanatory: season, response: cost
5. C
6. True
7. False
8. a. Minutes on the court: explanatory variable
   b. Points scored: response variable
9. a. Response variable: electricity bill
   b. Explanatory variable: number of Christmas lights
10. a. hours spent playing
    b. vocabulary

2.2 Exam questions

Note: Mark allocations are available with the fully worked solutions online.
1. B
2. E
3. D

2.3 Contingency (two-way) frequency tables and segmented bar charts

2.3 Exercise

1. *Note:* Black data are given in the question; red data are the answers.

		Age group		Total
		15 to 30	over 30	
Phone	Apple	23	15	38
	Samsung	15	7	22
	Total	38	22	60

Marginal distribution:
Apple = 0.63 Samsung = 0.37

2. *Note:* Black data are given in the question; red data are the answers.

		Age group		Total
		Teenagers	Adults	
Drink	Coke	221	155	376
	Pepsi	166	108	274
	Total	387	263	650

Conditional distribution: Teenagers who prefer Coke = 0.57
Teenagers who prefer Pepsi = 0.43

3.

		Residents		Total
		Inner city	Outer suburban	
Attitude	For	37	79	116
	Against	102	23	125
	Total	139	102	241

Marginal distribution: For = 0.48, Against = 0.52

4.

		Student type		Total
		Junior	Senior	
Lesson length	45 minutes	50	33	83
	1 hour	36	60	96
	Total	86	93	179

Conditional distribution:
Senior students prefer 45 mins = 0.35,
Senior students prefer an hour = 0.65

5.

		Age group	
		Teenagers	Adults
Phone	Apple	60.5%	68.2%
	Samsung	39.5%	31.8%
	Total	100%	100%

6.

		Age group	
		Teenagers	Adults
Drink	Coke	57.1%	58.9%
	Pepsi	42.9%	41.1%
	Total	100%	100%

7. a. i. 22   ii. 26   iii. 19   iv. 45   v. 41
   b. i. 47%   ii. 58%   iii. 100%   iv. 100%

8.

		Age group	
		18 to 30	31 to 40
Preference	Rent by themselves	57%	59%
	Share with friends	43%	41%
	Total	100%	100%

9. D
10. C

11.

		Age group		Total
		Children	**Adults**	
Choice	Entrée	8	18	26
	Main	31	16	47
	Dessert	19	27	46
	Total	58	61	118

		Age group	
		Children	**Adults**
Choice	Entrée	14%	30%
	Main	53%	26%
	Dessert	33%	44%
	Total	100%	100%

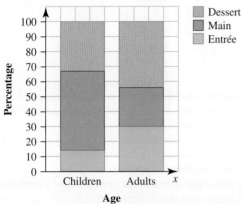

Children enjoy the main meal the most compared to Adults who prefer dessert the most.

12.

		Age group		Total
		< 40	**40+**	
Choice	Mortgage	42	18	60
	Superannuation	30	51	81
	Investment	21	33	54
	Total	93	102	195

		Age group	
		< 40	**40+**
Choice	Mortgage	45%	18%
	Superannuation	32%	50%
	Investment	23%	32%
	Total	100%	100%

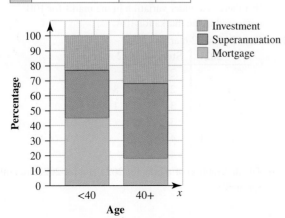

The under 40s have a focus on their mortgage, whereas the 40 and overs prioritise their superannuation.

13. a.

		Delegates	
		Liberal	**Labor**
Attitude	For	35.5%	39.4%
	Against	64.5%	60.6%
	Total	100.0%	100.0%

b. There is not a lot of difference in the reactions.

c.

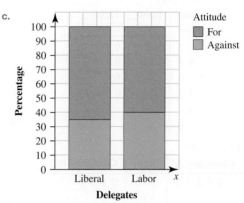

14. C

15. Conditional distribution:
Large town and no waste = 0.15
Large town and medium waste = 0.19
Large town and high waste = 0.67
Note: Rounding causes the total to be greater than 100%.

16. 26.32%

2.3 Exam questions

Note: Mark allocations are available with the fully worked solutions online.

1. E

2. B

3. B

2.4 Back-to-back stem plots

2.4 Exercise

1. Key 3 | 1 = 31

4B	Stem	4A
1	3	0 0 1
3 2	3	2 3
4	3	5 5
7 6	3	
9 9	3	8

2. Key 2 | 4 = 24

Physics	Stem	Chemistry
4	2	1
9 8	2*	8 8
4 2 2 0	3	0 1 2 3 4
6	3*	8
4 1	4	
9 8 5 5	4*	5 5 6 7 9

3. Key: $2 \mid 3 = 23$

Chinese	Stem	French
2 1 1 0	2	3 4
7 6 5 5	2*	5 5 8
3 2 1 0 0	3	0 1 4 4
9 8 7 7	3*	5 6 8 8 9
2 1	4	2 3 4 4
5	4*	6 8

4. Key: $2 * \mid 7 = 2.7 \,(\text{kg})$

Lion cubs	Stem	Tiger cubs
	2*	6 7
4 4	3	0 1 1 2 3
8 7 6	3*	6 7
3 2	4	0
9 8	4*	
0	5	

5. Key $13 \mid 0 = 130$

Company A	Stem	Company B
4 4 2 2 2	13	0 1 3 4
8	13*	8 8
4 2 1	14	
9 8	14*	5 6 8
	15	3 3
6	15*	5
0	16	0 2
8	16*	

	Company A	Company B
Mean	143.57	144.71
Median	141.5	145.5
IQR	$149 - 134 = 15$	$153 - 134 = 19$
Standard deviation	11.42	10.87

They are both positively skewed. The median is a better indicator of the centre of the distribution than the mean. This shows Company B handed out more pamphlets, taking into account that the IQR and the standard deviations are quite similar.

6. a. Key $5 \mid 7 = 57$

History	Stem	English
2	4	
9	5	7 8
7 4	6	2
8 8 7 5	7	1 4 5 8 9
4 2	8	0 7
8 2	9	6
	10	0

b.

	History	English
Mean	74.67	76.42
Median	77.5	76.5
IQR	$83 - 65.5 = 17.5$	$83.5 - 66.5 = 17$
Standard deviation	15.07	13.60

History has a slightly higher median; however, English has a slightly higher mean. Their standard deviations are similar, so overall the results are quite similar.

7. a. Key: $1 \mid 0 = 10$ trucks

Supermarket A	Stem	Supermarket B
2 1	1	0
7 7 6 6 5	1*	5 6
4 3 2 1 0	2	0 1 3
7 5	2*	5 6 8 9
	3	0 1 2
	3*	5

b. For supermarket A, the mean is 19, the median is 18.5, the standard deviation is 4.9 and the interquartile range is 7. The distribution is symmetric.
For supermarket B, the mean is 24.4, the median is 25.5, the standard deviation is 7.2 and the interquartile range is 10. The distribution is symmetric.
The centre and spread of the distribution of supermarket B is higher than that of supermarket A.
There is greater variation in the number of trucks arriving at supermarket B.

8. a. Key: $1 \mid 2 = 12$ marks

10A	Stem	10B
	1	0
3 2	1	2 3
5 5 4 4	1	4 4 5
7 6	1	7
	1	9

b. For the 10A marks, the mean is 14.5, the median is 14.5, the standard deviation is 1.6 and the interquartile range is 2. The distribution is symmetric.
For the 10B marks, the mean is 14.25, the median is 14, the standard deviation is 2.8 and the interquartile range is 3.5. The distribution is symmetric.
The centre of each distribution is about the same. The spread of marks for 10B is greater, however. This means that there is a wider variation in the marks for 10B compared to the marks for 10A.

9. a. Key: $2 \mid 2 = 22$ marks

2021	Stem	2022
	2	2
	2*	6 7 8
1 0	3	0 1 1 3 4
9 7 5	3*	6
3 2 1 1	4	
6	4*	

b. The distribution of marks for 2021 and for 2022 are both symmetric.

For the 2021 marks, the mean is 38.5, the median is 40, the standard deviation is 5.2 and the interquartile range is 7. The distribution is symmetric.

For the 2022 marks, the mean is 29.8, the median is 30.5, the standard deviation is 4.2 and the interquartile range is 6.

The spread of each of the distributions is much the same, but the centre of each distribution is quite different with the centre of the 2022 distribution lower. The work may have become a lot harder.

10. a. Key: 3 | 6 = 36 years old

Species 1	Stem	Species 2
4 3	2	2
8 7 6 5	2*	5
1 0	3	0 1
	3*	6 7
	4	2
	4*	6

b. For the distribution of species 1, the mean is 26.75, the median is 26.5, the standard deviation is 2.8 and the interquartile range is 4.5.

For the distribution of species 2, the mean is 33.6, the median is 33.5, the standard deviation is 8.2 and the interquartile range is 12.

The centre of the distributions is very different: it is much higher for species 2. The spread of the ages of species 1 is very small but very large for species 2.

11. a. Key: 1 | 2 = 12 points

Kindergarten	Stem	Prep.
3	0	5
4 3	1	2 7
8 5	2	5 7
6 2	3	2 5
7 1	4	4 6
0	5	2

b. For the distribution of scores of the kindergarten children, the mean is 28.9, the median is 30, the standard deviation is 15.4 and the interquartile range is 27.

For the distribution of scores for the prep. children, the mean is 29.5, the median is 29.5, the standard deviation is 15.3 and the interquartile range is 27.

The distributions are very similar. There is not a lot of difference between the way the kindergarten children and the prep. children scored.

12. B

13. C

14. Key: 7 | 2 = 72

Year 12	Stem	Year 11
2	7	0 2
6	7*	
2 0	8	0 1 3 3 4
8 6	8*	8
4 4 2 0 0	9*	0 2 4
	9*	6
0	10	

15. D

16. a. Key: 3 | 1 = 31

Mathematical Methods	Stem	General Mathematics
9 8	2*	8
4 3 1	3	0 1
9 9 6 6	3*	5 7 7 8
4 4 2 1 0	4	1 3 3 4
	4*	6 8 8

	Mathematical Methods	General Mathematics
Mean	36.86	39.21
Median	37.5	39.5
IQR	41 − 33 = 88	44 − 35 = 9
Standard deviation	5.29	6.58

b. Mathematical Methods has a slightly lower IQR and standard deviation. General Mathematics had a greater mean (39.21) compared to Mathematical Methods (39.5), as well as a greater median: 39.5 compared to 36.86. This suggests that students do better in General Mathematics as compared to Mathematical Methods by an average of two marks.

2.4 Exam questions

Note: Mark allocations are available with the fully worked solutions online.

1. E

2. D

3. a. mean = 20, median = 18, mode = 21, standard deviation = 4.7

   b. Min X = 17, Q_1 = 17.5, Med = 18, Q_3 = 21, Max X = 37, Range = 20, IQR = 3.5

   c. mean = 25.4, median = 21, mode = 18, standard deviation = 10.5

   d. Min X = 18, Q_1 = 18, Med = 21, Q_3 = 33, Max X = 45, Range = 27, IQR = 15

   e. Please see the worked solution for the sample response.

2.5 Parallel boxplots and dot plots

2.5 Exercise

Note: When comparing and contrasting data sets, answers will naturally vary. It is good practice to discuss your conclusions in a group to consider different viewpoints.

1. 6A: 13.8, 15, 16.1, 17.6, 18.8
   6B: 14.8, 16.4, 17.3, 19.5, 20.3

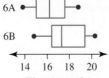

The boxplots show that 6A has a significantly lower median. The 6A median is lower than Q_1 of the 6B times; that is, the lowest 25% of times for 6B are greater than the lowest 50% of times for 6A.

2. 10A: 42, 59, 70.5, 87, 98
10D: 40, 62, 70.5, 84, 100

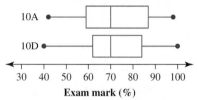

Exam mark (%)

The boxplots show that the medians are the same, but 10D has a higher mean. 10D also has the highest score of 100%, but 10D also has the lowest score. Since Q_1 and Q_3 are closer together for 10D, their results are more consistent around the median. The parallel dot plot confirms this but doesn't give any further information.*

3. a.

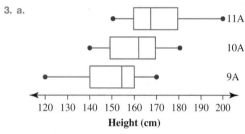

Height (cm)

b. Clearly, the median height increases from Year 9 to Year 11. There is greater variation in 9A's distribution than in 10A's. There is a wide range of heights in the lower 25% of 9A's distribution. There is a greater variation in 11A's distribution than in 10A's, with a wide range of heights in the top 25% of the 11A distribution.

4. a.

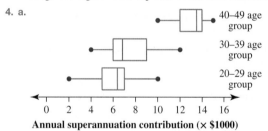

Annual superannuation contribution ($\times$ $1000)

b. Clearly, there is a great jump in contributions to superannuation for people in their 40s. The spread of contributions for that age group is smaller than for people in their 20s or 30s, suggesting that a high proportion of people in their 40s are conscious of superannuation. For people in their 20s and 30s, the range is greater, indicating a range of interest in contributing to super.

5. a. True   **b.** True   **c.** False   **d.** True

6. a. *The Pearlfishers*, which had a significantly higher median number of A-reserve tickets sold, as well as a higher minimum and maximum number of A-reserve tickets sold

b. *Orlando*, which had both a larger range and IQR of A-reserve tickets sold

7. D

8. C

9.

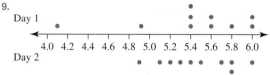

Diving score

The dives on day 1 were more consistent than the dives on day 2 with most of the dives between 5.4 and 6.0 (inclusive), despite two lower scoring dives. Day 2 was more spread with dives from 4.9 to 6.0 (inclusive). It must be noted that there were no very low scoring dives on the second day.

10. B

11. B

12. C

13.

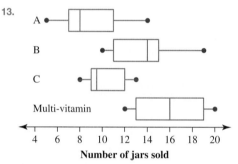

Number of jars sold

Overall, the biggest sales were of multi-vitamins, followed by vitamin B, then vitamin C and finally vitamin A.

14. For all four days, the median is the 6th score.
For all four days, Q_1 is the 3rd score. For all four days, Q_3 is the 9th score.

Day	Min.	Max.	Range	Median
Thursday	70	90	20	81
Friday	77	89	12	83
Saturday	81	89	8	86
Sunday	70	94	24	89

---

*2. 10A

40   45   50   55   60   65   70   75   80   85   90   95   100

10D

Exam mark (%)

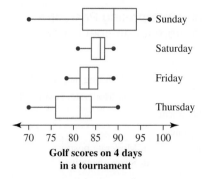

**Golf scores on 4 days
in a tournament**

2.5 Exam questions

Note: Mark allocations are available with the fully worked
solutions online.

1. E
2. a. Negatively skewed
   b. The median life expectancy increases as the years
      increase. There is a positive correlation.
3. C

2.6 Scatterplots

2.6 Exercise

1. Moderate positive linear relationship
2. Moderate positive linear relationship
3. a. Yes — positive association
   b. Yes — positive association
   c. Yes — positive association
   d. Yes — negative association
   e. Yes — positive association
   f. Yes — negative association
   g. No — no association
4. a. Weak, negative association of linear form
   b. Moderate, negative association of linear form
   c. Moderate, positive association of linear form
   d. Strong, positive association of linear form
   e. No association
   f. Non-linear association
5. B
6.

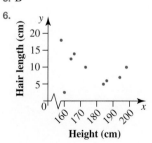

No relationship
7.

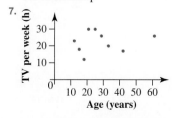

No relationship

8.

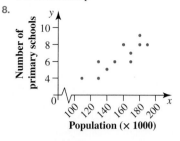

Moderate positive association of linear form, no outliers

9.

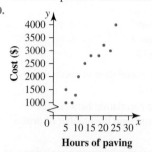

Strong positive association of linear form, no outliers

10.

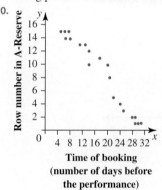

**Time of booking
(number of days before
the performance)**

Strong negative association of linear form, no outliers

11. B
12. a, b

13. a, b

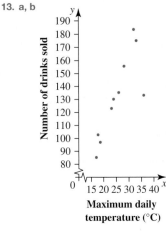

Maximum daily
temperature (°C)

14. a. There is a strong positive correlation between the dry cleaning items and the cost.

   b. There is a strong positive correlation between the maximum daily temperature and the number of drinks sold.

15. There appears to be no correlation between population and area of the various states and territories.

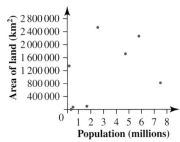

16. a.

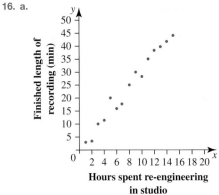

Hours spent re-engineering
in studio

   b. There is a relationship. It is strong, positive and linear.

2.6 Exam questions

Note: Mark allocations are available with the fully worked solutions online.

1. D

2. B

3. D

2.7 Estimating and interpreting Pearson's product–moment correlation coefficient

2.7 Exercise

1. a.i. $r = -0.9$ **ii.** Strong, negative, linear
   b.i. $r = 0.7$ **ii.** Moderate, positive, linear

2. a. Strong, positive, linear
   b. Weak, negative, linear

3. a. No association **b.** Moderate positive
   c. Strong negative **d.** Strong negative

4. a. Strong positive **b.** Strong positive
   c. Weak negative **d.** No association

5. a. i. $r \approx -0.8$
      ii. Strong, negative, linear association
   b. i. $r \approx 0.6$
      ii. Moderate, positive, linear association
   c. i. $r \approx 0.2$
      ii. No linear association
   d. i. $r \approx -0.2$
      ii. No linear association

6. a. i. $r \approx 1$
      ii. Perfect, positive, linear association
   b. i. $r \approx 0.8$
      ii. Strong, positive, linear association
   c. i. $r \approx 0$
      ii. No linear association
   d. i. $r \approx -0.7$
      ii. Moderate, negative, linear association

7. B

8. E

9. a. $r = 0.1$: no linear relationship
   b. $r = 0.2$: no linear relationship
   c. $r = 0.95$: strong and positive

10. D

11. C

12.

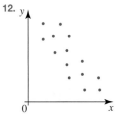

13. E

14. C

2.7 Exam questions

Note: Mark allocations are available with the fully worked solutions online.

1. D

2. A

3. A

2.8 Calculating r and the coefficient of determination, r^2

2.8 Exercise

1. a.

Height (cm)

b. No linear relationship, x^2

c. $r \approx 0.36$, weak, positive linear relationship

2. a.

Gestation time (weeks)

b. Strong, positive, linear relationship. $r \approx 0.95$

c. $r \approx 0.99$, very strong linear relationship

3. a.

Yearly salary ($\times$ $1000)

b. There is moderate, negative linear association. r is approximately -0.6.

c. $r = -0.66$. There is a moderate negative linear association between the yearly salary and the number of Brownlow votes. That is, the larger the yearly salary of the player, the fewer votes we might expect to see.

4. Coefficient of determination $= 0.59$
We can then conclude that 59% of the variation in the number of people found to have a blood alcohol reading over 0.05 can be explained by the variation in the number of booze buses in use. Thus the number of booze buses in use is a factor in predicting the number of drivers with a reading over 0.05.

5. Coefficient of determination is 0.7569. The portion of variation in the number of visits to the doctor that can be explained by the variation in the number of cigarettes smoked is about 76%.

6. a.

Annual advertising budget ($\times$ $1000)

b. There is strong, positive linear association. r is approximately 0.8.

c. $r = 0.98$

d. Coefficient of determination is $r = 0.96$.

e. The proportion of the variation in the yearly profit increase that can be explained by the variation in the advertising budget is 96%.

7. a.

Number of people in household

b. There is strong, positive association of a linear form and r is approximately 0.9.

c. $r = 0.98$

d. Coefficient of determination is 0.96.

e. The proportion of the variation in the weekly grocery costs that can be explained by the variation in the number of people in the household is 96%.

8. a.

Number of weeks on the diet

b. The scatterplot shows a strong, positive association of linear form.

c. It is appropriate since the scatterplot indicates an association showing linear form and there are no outliers.

d. $r \approx 0.9$

e. $r = 0.96$

f. We cannot say whether total weight loss is affected by the number of weeks people stayed on the Certain Slim diet. We can only note the degree of correlation.

g. $r^2 = 0.92$

h. The coefficient of determination tells us that 92% of the variation in total weight loss can be explained by the variation in the number of weeks on the Certain Slim diet.

2.8 Exam questions

Note: Mark allocations are available with the fully worked solutions online.

1. E
2. E
3. D

2.9 Cause and effect

2.9 Exercise

1. a.

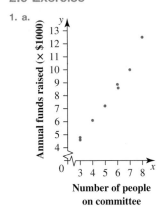

b. There is almost perfect positive correlation of a linear form and r is nearly 1.

c. $r = 0.99$

d. Causation cannot be established on a strong association alone.

2. There is a strong positive association between the number of books read and the spelling levels for this group of students.

3. E

4. B

5. C

6. It cannot be concluded that diabetes causes deterioration of the hip joint. The common cause could be age.

7. It cannot be concluded that holidays cause weight gain. The common cause could be any or a combination of the following: amount of exercise, alcohol consumption, diet.

8. Confounding variables could include time away from family, amount of exercise, diet.

2.9 Exam questions

Note: Mark allocations are available with the fully worked solutions online.

1. D
2. C
3. B

2.10 Review

2.10 Exercise

Multiple choice

1. A
2. A
3. E
4. B
5. E
6. D
7. D
8. E
9. A
10. E
11. D
12. C

Short answer

13. a. Key: $0 \mid 3 = 3$ hours

Full-time	Stem	Volunteer
1	0	
2 2	0	
4 4 3 3	0	
6 5	0	
	0	8
	1	0 1 1
	1	2 3 3
	1	4 5
	1	
	1	

b. Both distributions are symmetric with the same spread. The centre of the volunteers' distribution is much higher than that of the full-time firefighters' distribution. Clearly, the volunteers needed more counselling.

14.

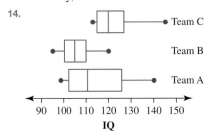

15. a.

		Party preferences	
		Liberal	**Labor**
Attitude	For	66.7%	18.9%
	Against	33.3%	81.1%
	Total	100.0%	100.0%

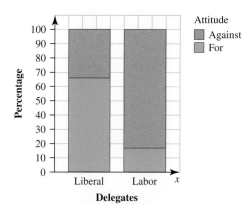

Delegates

b. Clearly, the reaction to uranium mining is affected by political affiliation.

16. a.

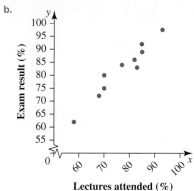

b. There appears to be an extremely weak association or no association between the variables.

17. r is approximately -0.7. There is a moderate, negative linear association between the variables x and y.

18. a. Lectures attended: explanatory variable; exam result: response variable

b.

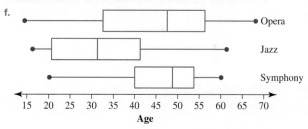

c. There is strong, positive correlation of a linear form between the variables and r is approximately equal to 0.8.

d. $r = 0.96$

e. The coefficient of determination is 0.93.

f. The proportion of the variation in the exam results that can be explained by the variation in the number of lectures attended is 93%.

Extended response

19. a. Key 1 | 6 = 16 years old

Jazz concert	Stem	Symphony concert
9 9 8 6	1	
9 7 4 3 0	2	0 3
9 8 4 3 0	3	0 5 9
6 5 3 2 0	4	2 5 5 7 8 8 9 9
	5	0 3 4 6 8 8
2	6	0

b.

	Summary	Symphony concert	Jazz concert
i.	X_{min}	20	16
ii.	Q_1	40.5	21.5
iii.	Median	48	31.5
iv.	Q_3	53.5	41
v.	X_{max}	60	62
vi.	Mean	45.45	32.35
vii.	IQR	13	19.5
viii.	Standard deviation	11.20	12.04

c. Overall, it appears that people who attended the symphony concert were older than those who attended the jazz concert. The spread of ages is nearly the same (slightly higher for the jazz audience).

d. Back-to-back stem plots can be used only for data with two categories. Since there are three events, parallel boxplots should be used.

e. $X_{min} = 12$, $Q_1 = 34$, median = 47.5, $Q_3 = 56.5$, $X_{max} = 68$, mean = 44.95, IQR = 22.5, standard deviation = 15.55

f.

g. Overall, the people who went to the symphony concert and to the opera were of similar ages and older than those who went to the jazz concert. The ages of people who went to the opera are the most spread out, while the ages of people who attended the symphony concert are the least spread out.

20. a. 188

b. 158

c.

	Engaged in sport activity outside school		
Level of stress	Regularly	Sometimes	Never
Low	47.1%	25.8%	16.2%
Medium	35.3%	32.3%	25.2%
High	17.6%	41.9%	58.6%
Total	100%	100%	100%

d.

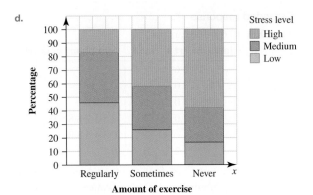

e. Overall, it appears that for this group of students, stress levels are related to the amount of physical activity they are engaged in outside of school.

2.10 Exam questions

Note: Mark allocations are available with the fully worked solutions online.

1. B

2. A

3. a. Place of capture

   b. 20 mm

   c. See the table at the bottom of the page.*

   d. As 52 mm is greater than this upper-fence value of 50 mm, it is an outlier.

   e. Please see the worked solution.

4. D

5. a. 19%

   b. 29 440 000

   c. Please see the sample response in the worked solution.

*2. c.

Place of capture	Wingspan (mm)				
	minimum	Q_1	median (M)	Q_3	maximum
forest	16	20	21	32	52
grassland	18	24	30	36	45

3 Investigating and modelling linear associations

Fully worked solutions for this topic are available online.

3.1 Overview

3.1.1 Introduction

A mathematical model uses mathematical concepts and language to demonstrate how something in the real world behaves. These models can be used to investigate an aspect of the natural world or society without being present in the situation. A very simple mathematical model is the use of the formula $V = l \times w \times h$, which describes the volume of a rectangular box. Rather than creating many boxes and measuring them to determine their volume, this formula can be used for a rectangular box of any dimensions.

A more complicated model uses more complex concepts, for example, a model required to send a rocket into space needs information about the energy needed to break through gravity, which is dependent on the size of the rocket, the starting and ending points of the journey and the energy provided by the preferred fuel. A trial-and-error model to determine the variables would compromise safety and waste valuable resources. Scientists set up mathematical models first and use these models to ensure the greatest chance of success.

A statistical model is a particular type of mathematical model. It is created by collecting data from a sample of the population that is under investigation. It uses this data to create an idealised way of predicting how any future data will behave. The manufacturer of a new running shoe wants to determine the best tread thickness for maximum support. Rather than spending years measuring tread thicknesses and the extent of running injuries across the world, researchers take a sample and use the data from the sample to make predictions.

KEY CONCEPTS

This topic covers the following key concepts from the VCE Mathematics Study Design:
- least squares line of best fit $y = a + bx$, where x represents the explanatory variable, and y represents the response variable; the determination of the coefficients a and b using technology, and the formulas $b = r\dfrac{s_y}{s_x}$ and $a = \bar{y} - b\bar{x}$
- modelling linear association between two numerical variables, including the:
  - identification of the explanatory and response variables
  - use of the least squares method to fit a linear model to the data
- interpretation of the slope and intercepts of the least squares line in the context of the situation being modelled, including:
  - use of the rule of the fitted line to make predictions being aware of the limitations of extrapolation
  - use of the coefficient of determination, r^2, to assess the strength of the association in terms of explained variation
  - use of residual analysis to check quality of fit
- data transformation and its use in transforming some forms of non-linear data to linearity using a square, logarithmic (base 10) or reciprocal transformation (applied to one axis only)
- interpretation and use of the equation of the least squares line fitted to the transformed data to make predictions.

Source: VCE Mathematics Study Design (2023–2027) extracts © VCAA; reproduced by permission.

3.2 Least squares line of best fit

LEARNING INTENTION

At the end of this subtopic you should be able to:
- calculate the equation of the least squares line of best fit using CAS
- calculate the least squares line of best fit by hand.

3.2.1 Calculating the least squares line of best fit using CAS

When data show a linear relationship and have no obvious outliers, a least squares line of best fit $y = a + bx$ can be found to model the data: x represents the explanatory variable and y represents the response variable.

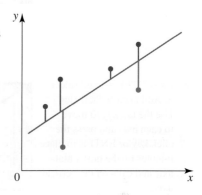

To understand the underlying theory behind least squares, consider the least squares line shown.

We wish to minimise the total of the vertical lines, or 'errors', in some way, for example balancing the errors above (positive) and below the line (negative). This is reasonable, but it is preferable to minimise the sum of the *squares* of each of these errors. This is the essential mathematics of least squares regression.

The calculation of the equation of a least squares line of best fit is simple using CAS.

tlvd-3569

WORKED EXAMPLE 1 Calculating the least squares line of best fit using CAS

A study shows the more calls a teenager makes on their mobile phone, the less time they spend on each call. Determine the equation of the linear least squares line for the number of calls made plotted against call time in minutes using the least squares method on CAS.

Number of minutes	1	3	4	7	10	12	14	15
Number of calls	11	9	10	6	8	4	3	1

Express coefficients correct to 2 decimal places and calculate the coefficient of determination to assess the strength of the association.

THINK	WRITE
1. Identify the explanatory and response variables.	Number of minutes: explanatory Number of calls: response
2. Enter the data into CAS to find the equation of the least squares line of best fit.	$y = 11.7327 - 0.634\,271x$
3. Write the equation with coefficients expressed to 2 decimal places.	$y = 11.73 - 0.63x$
4. Write the equation in terms of the variable names. Replace x with number of minutes and y with number of calls.	Number of calls $= 11.73 - 0.63 \times$ no. of minutes
5. Read the r^2 value from your calculator.	$r^2 = 0.87$, so we can conclude that 87% of the variation in y can be explained by the variation in x. Therefore, the strength of the linear association between y and x is strong.

**TI	THINK**	**DISPLAY/WRITE**	**CASIO	THINK**	**DISPLAY/WRITE**
1. Identify the explanatory and response variables.	Number of minutes: explanatory Number of calls: response	1. Identify the explanatory and response variables.	Number of minutes: explanatory Number of calls: response
2. On a Lists & Spreadsheet page, enter the data into the lists, named *minutes* and *calls*.	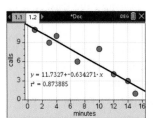	2. On a Statistics screen, enter the data into the lists, named *minutes* and *calls*.	
3. Press CTRL + I and select: 5: Add Data & Statistics Use the tab key to move to each axis and press the click key or ENTER to place minutes on the horizontal axis and calls on the vertical axis. Press MENU and select: 4: Analyze 6: Regression 2: Show Linear $(a + bx)$ *Note:* If the value of r^2 does not appear, tick Diagnostics in Settings.		3. Tap: Calc Regression Linear reg Complete the Set Calculation fields: Linear Reg XList: main\minutes YList: main\calls Freq: 1 Copy Formula y1 Copy Residual Off Tap OK.	
4. The answer appears on the screen.	$y = 11.7327 - 0.634\,271x$	4. The answer appears on the screen.	$y = ax + b$ $y = 11.7327 - 0.634\,271x$
5. Write the equation with the coefficients expressed to 2 decimal places and replace x with the number of minutes and y with the number of calls.	No. of calls $= 11.73 - 0.63 \times$ no. of minutes	5. Write the equation with the coefficients expressed to 2 decimal places and replace x with the number of minutes and y with the number of calls.	No. of calls $= 11.73 - 0.63 \times$ no. of minutes
6. The answer appears on the screen.	$r^2 = 0.87$ (correct to 2 decimal places), so we can conclude that 87% of the variation in y can be explained by the variation in x. Therefore, the strength of the linear association between y and x is strong.	6. The answer appears on the screen.	$r^2 = 0.87$ (correct to 2 decimal places), so we can conclude that 87% of the variation in y can be explained by the variation in x. Therefore, the strength of the linear association between y and x is strong.

3.2.2 Calculating the least squares line of best fit by hand

The least squares line of best fit equation minimises the average deviation of the points in the data set from the line of best fit. This can be shown using the following summary data and formulas to arithmetically determine the least squares line of best fit equation.

$\bar{x}$ the mean of the explanatory variable (x-variable)

$\bar{y}$ the mean of the response variable (y-variable)

s_x the standard deviation of the explanatory variable

s_y the standard deviation of the response variable

r Pearson's product–moment correlation coefficient

Formula to use for least squares line of best fit

The general form of the least squares line of best fit is

$$y = a + bx$$

where the slope of the least squares line is $b = r\dfrac{s_y}{s_x}$

the y-intercept of the least squares line is $a = \bar{y} - b\bar{x}$.

Alternatively, if the general form is given as $y = mx + c$, then $m = r\,\dfrac{s_y}{s_x}$ and $c = \bar{y} - m\bar{x}$.

tlvd-3570

WORKED EXAMPLE 2 Calculating the least squares line of best fit by hand

A study to find a relationship between the height of husbands and the height of their wives revealed the following details.

Mean height of husbands: 180 cm

Mean height of wives: 169 cm

Standard deviation of the height of husbands: 5.3 cm

Standard deviation of the height of wives: 4.8 cm

Correlation coefficient, $r = 0.85$

The form of the least squares line of best fit is to be:

height of wife $= a + b \times$ height of husband

a. State the response variable.

b. Calculate the value of b (correct to 2 significant figures).

c. Calculate the value of a (correct to 4 significant figures).

d. Use the equation of the least squares line to predict the height of a wife whose husband is 195 cm tall (correct to the nearest cm).

THINK	WRITE
a. Recall that the response variable is the subject of the equation in $y = a + bx$ form, that is, y.	a. The response variable is the height of the wife.
b. 1. The value of b is the gradient of the least squares line. Write the formula and state the required values.	b. $r = 0.85$, $S_y = 4.8$ and $S_x = 5.3$
2. Substitute the values into the formula and evaluate b.	$\begin{aligned} b &= 0.85 \times \dfrac{4.8}{5.3} \\ &= 0.7698 \\ &\approx 0.77 \end{aligned}$

c. 1. The value of a is the y-intercept of the least squares line. Write the formula and state the required values.

c. $a = \bar{y} - b\bar{x}$
$\bar{y} = 169$, $\bar{x} = 180$ and $b = 0.7698$ (from part **b**)

2. Substitute the values into the formula and evaluate a.

$a = 169 - 0.7698 \times 180$
$= 30.436$
≈ 30.44

d. 1. State the equation of the least squares line of best fit, using the values calculated from parts **b** and **c**. In this equation, y represents the height of the wife and x represents the height of the husband.

d. $y = 30.44 + 0.77x$ or
height of wife $= 30.44 + 0.77 \times$ height of husband

2. The height of the husband is 195 cm, so substitute $x = 195$ into the equation and evaluate.

$= 30.44 + 0.77 \times 195$
$= 180.59$

3. Write the answer in a sentence, rounding your answer correct to the nearest cm.

Using the equation of the least squares line of best fit found, the wife's height would be 181 cm.

 Resources

 Interactivity Fitting a straight line using least squares regression (int-6254)

3.2 Exercise

Students, these questions are even better in jacPLUS

 Receive immediate feedback and access sample responses

 Access additional questions

 Track your results and progress

Find all this and MORE in jacPLUS

1. **WE1** A study shows that as the temperature increases, the sales of air conditioners increase.

Temperature (°C)	21	23	25	28	30	32	35	38
Air conditioner sales	3	7	8	14	17	23	25	37

Determine the equation of the linear least squares line for the number of air conditioners sold per week plotted against the temperature in °C using the least squares method on CAS technology.
Also calculate the coefficient of determination.
Express the values correct to 2 decimal places and comment on the association between temperature and air conditioner sales.

2. Consider the following data set: x represents the month, y represents the number of dialysis patients treated.

x	1	2	3	4	5	6	7	8
y	5	9	7	14	14	19	21	23

Use CAS to determine the equation of the linear least squares line and the coefficient of determination, with values correct to 2 decimal places.

3. Determine the equation of the linear least squares line for the following data set using the least squares method. Comment on the strength of the association.

x	4	6	7	9	10	12	15	17
y	10	8	13	15	14	18	19	23

4. Determine the equation of the linear least squares line for the following data set using the least squares method.

x	1	2	3	4	5	6	7	8	9
y	35	28	22	16	19	14	9	7	2

5. Determine the equation of the linear least squares line for the following data set using the least squares method.

x	−4	−2	−1	0	1	2	4	5	5	7
y	6	7	3	10	16	9	12	16	11	21

6. **WE2** A study was conducted to find the relationship between the height of basketball players and the height of netball players. The following details were found.

Mean height of the basketball players: 182 cm
Mean height of the netball players: 166 cm
Standard deviation of the height of the basketball players: 6.1 cm
Standard deviation of the height of the netball players: 5.2 cm
Correlation coefficient, $r = 0.82$

The form of the least squares line of best fit is to be:

$height\ of\ basketball\ player = a + b \times height\ of\ netball\ player$

a. Name the explanatory variable.
b. Calculate the value of b (correct to 2 significant figures).
c. Calculate the value of a (correct to 2 decimal places).

7. Given the summary details $\bar{x} = 4.4$, $s_x = 1.2$, $\bar{y} = 10.5$, $s_y = 1.4$ and $r = -0.67$, determine the value of b and a for the equation of the least squares line of best fit $y = a + bx$.

8. Determine the least squares line of best fit equations, given the following summary data.

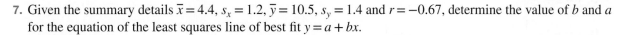

a. $\bar{x} = 5.6$ $s_x = 1.2$ $\bar{y} = 110.4$ $s_y = 5.7$ $r = 0.7$
b. $\bar{x} = 110.4$ $s_x = 5.7$ $\bar{y} = 5.6$ $s_y = 1.2$ $r = -0.7$
c. $\bar{x} = 25$ $s_x = 4.2$ $\bar{y} = 10\,200$ $s_y = 250$ $r = 0.88$
d. $\bar{x} = 10$ $s_x = 1$ $\bar{y} = 20$ $s_y = 2$ $r = -0.5$

9. The following summary details were calculated from a study to find a relationship between Mathematics exam marks and English exam marks from the results of 120 Year 12 students.

Mean Mathematics exam mark = 64%
Mean English exam mark = 74%
Standard deviation of Mathematics exam mark = 14.5%
Standard deviation of English exam mark = 9.8%
Correlation coefficient, $r = 0.64$

The form of the least squares line of best fit is to be:

$$Mathematics\ exam\ mark = a + b \times English\ exam\ mark$$

a. Name the response variable (y-variable).
b. Calculate the value of b for the least squares line of best fit (correct to 2 decimal places).
c. Calculate the value of a for the least squares line of best fit (correct to 2 decimal places).
d. Use the least squares line to predict the expected Mathematics exam mark if a student scores 85% in the English exam (correct to the nearest percentage).

10. Repeat questions **5**, **6** and **7**, collecting the values for $\bar{x}$, s_x, $\bar{y}$, s_y and r from CAS. Use the data to calculate the least squares regression equation. Compare your answers to the ones you obtained earlier. Comment on what you notice.

11. A mathematician is interested in the behaviour patterns of their kitten and collects the following data on two variables. Help them manipulate the data.

x	1	2	3	4	5	6	7	8	9	10
y	20	18	16	14	12	10	8	6	4	2

a. Fit a least squares line of best fit.
b. Comment on any interesting features of this line.
c. Fit the 'opposite least squares line' using the following data.

x	20	18	16	14	12	10	8	6	4	2
y	1	2	3	4	5	6	7	8	9	10

12. The best estimate of the least squares line of best fit for the scatterplot is:

A. $y = 2x$

B. $y = \dfrac{1}{2}x$

C. $y = 2 + \dfrac{1}{2}x$

D. $y = -2 + \dfrac{1}{2}x$

E. $y = -1 + \dfrac{1}{2}x$

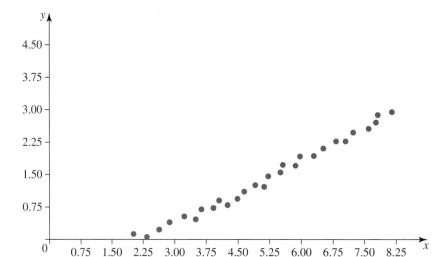

13. The life span of adults in a certain country over the last 220 years has been recorded.

Year	1800	1820	1840	1860	1880	1900	1920	1940	1960	1980	2000	2020
Life span (years)	51.2	52.4	51.7	53.2	53.1	54.7	59.9	62.7	63.2	66.8	72.7	79.2

a. Fit a least squares line of best fit to the data.
b. Plot the data and the least squares line on a scatterplot.
c. Discuss whether you think the data really look linear.

14. The price of an international telephone call changes as the duration of the call increases. The cost of a sample of calls from Melbourne to Slovenia are summarised in the table.

Cost of call ($)	1.25	1.85	2.25	2.50	3.25	3.70	4.30	4.90	5.80
Duration of call (seconds)	30	110	250	260	300	350	420	500	600

Cost of call ($)	7.50	8.00	9.25	10.00	12.00	13.00	14.00	16.00	18.00
Duration of call (seconds)	840	1000	1140	1200	1500	1860	2400	3600	7200

a. Name the explanatory variable.
b. Fit a least squares line of best fit to the data.
c. View the data on a scatterplot and comment on the reliability of the least squares line in predicting the cost of telephone calls. (i.e consider whether the least squares line you found proves that costs of calls and duration of calls are related).
d. Calculate the coefficient of determination and comment on the linear relationship between duration and cost of call.

15. **MC** In a study to find a relationship between the height of plants and the hours of daylight they were exposed to, the following summary details were obtained.

Mean height of plants $= 40$ cm
Mean hours of daylight $= 8$ hours
Standard deviation of plant height $= 5$ cm
Standard deviation of daylight hours $= 3$ hours
Pearson's correlation coefficient $= 0.9$

The most appropriate regression equation is:
A. height of plant (cm) $= -13.6 + 0.54 \times$ hours of daylight
B. height of plant (cm) $= -8.5 + 0.34 \times$ hours of daylight
C. height of plant (cm) $= 2.1 + 0.18 \times$ hours of daylight
D. height of plant (cm) $= 28.0 + 1.50 \times$ hours of daylight
E. height of plant (cm) $= 35.68 + 0.54 \times$ hours of daylight

16. Consider the following data set.

x	1	2	3	4	5	6
y	12	16	17	21	25	29

a. Perform a least squares regression on the first two points only.
b. Now add the 3rd point and repeat.
c. Repeat for the 4th, 5th and 6th points.
d. Comment on your results.

▶ **Question 1 (1 mark)**

Source: VCE 2020, Further Mathematics Exam 1, Section A, Q13; © VCAA.

MC A least squares line of the form $y = a + bx$ is fitted to a scatterplot.

Which one of the following is always true?
- **A.** As many of the data points in the scatterplot as possible will lie on the line.
- **B.** The data points in the scatterplot will be divided so that there are as many data points above the line as there are below the line.
- **C.** The sum of the squares of the shortest distances from the line to each data point will be a minimum.
- **D.** The sum of the squares of the horizontal distances from the line to each data point will be a minimum.
- **E.** The sum of the squares of the vertical distances from the line to each data point will be a minimum.

▶ **Question 2 (1 mark)**

Source: VCE 2019, Further Mathematics Exam 1, Section A, Q9; © VCAA.

MC A least squares line is used to model the relationship between the monthly *average temperature and latitude* recorded at seven different weather stations. The equation of the least squares line is found to be

$$average\ temperature = 42.9842 - 0.877447 \times latitude$$

When the numbers in this equation are correctly rounded to three significant figures, the equation will be
- **A.** *average temperature* $= 42.984 - 0.877 \times latitude$
- **B.** *average temperature* $= 42.984 - 0.878 \times latitude$
- **C.** *average temperature* $= 43.0 - 0.878 \times latitude$
- **D.** *average temperature* $= 42.9 - 0.878 \times latitude$
- **E.** *average temperature* $= 43.0 - 0.877 \times latitude$

▶ **Question 3 (1 mark)**

Source: VCE 2018, Further Mathematics Exam 1, Section A, Q8; © VCAA.

MC The scatterplot below displays the *resting pulse rate*, in beats per minute, and the *time spent exercising*, in hours per week, of 16 students. A least squares line has been fitted to the data.

The equation of this least squares line is closest to
- **A.** *resting pulse rate* $= 67.2 - 0.91 \times$ *time spent exercising*
- **B.** *resting pulse rate* $= 67.2 - 1.10 \times$ *time spent exercising*
- **C.** *resting pulse rate* $= 68.3 - 0.91 \times$ *time spent exercising*
- **D.** *resting pulse rate* $= 68.3 - 1.10 \times$ *time spent exercising*
- **E.** *resting pulse rate* $= 67.2 + 1.10 \times$ *time spent exercising*

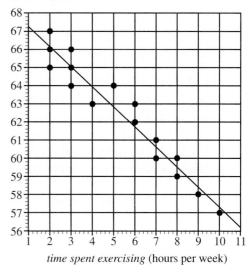

More exam questions are available online.

3.3 Interpretation, interpolation and extrapolation

LEARNING INTENTION

At the end of this subtopic you should be able to:
- use the model to make predictions, being aware of the problem of extrapolation
- model linear trends using the least squares line of best fit, interpret the model in the context of the trend being modelled, use the model to make forecasts with consideration of the limitations of extending forecasts too far into the future.

3.3.1 Interpreting slope and intercept (*b* and *a*)

Once you have a linear least squares line, the **slope** and **intercept** can give important information about the data set.

The slope (*b*) indicates the change in the response variable as the explanatory variable increases by 1 unit. The *y*-intercept indicates the value of the response variable when the explanatory variable = 0.

tlvd-3571

WORKED EXAMPLE 3 Interpreting the least squares equation

In a study of the growth in the number of a species of bacterium, it was assumed that the growth is linear. However, it is very expensive to measure the number of bacteria in a sample.

Day of experiment	1	4	5	9	11
Number of bacteria	500	1000	1100	2100	2500

Given the data listed, determine:
a. the equation, describing the relationship between the two variables
b. the rate at which the number of bacteria is growing
c. the number of bacteria at the start of the experiment.

THINK	WRITE
a. 1. Determine the equation of the least squares line of best fit using CAS.	a. $y = a + bx$
2. Replace *x* and *y* with the variables in question.	Number of bacteria = 202.5 + 206.25 × day of experiment
b. The rate at which the number of bacteria is growing is given by the gradient of the least squares line of best fit.	b. *b* is 206.25, hence on average the number of bacteria increases by approximately 206 per day.
c. The number of bacteria at the start of the experiment is given by the *y*-intercept of the least squares line of best fit.	c. The *y*-intercept is 202.5, hence the initial number of bacteria present was approximately 203.

3.3.2 Interpolation and extrapolation

As we have already observed, any linear regression method produces a linear equation in the form:

$$y = a + bx$$

where b is the gradient or slope and a is the y-intercept.

This equation can be used to 'predict' the y-value for a given value of x. These are only approximations, since the least squares line itself is only an estimate of the 'true' relationship between the two variables. However, the equation and least squares line can still be used, in some cases, to provide additional information about the data set (i.e. make predictions).

There are two types of prediction: **interpolation** and **extrapolation**.

Interpolation

Interpolation is the use of the least squares line to predict values within the range of data in a set, that is, the values are *in between* the values already in the data set. If the data are highly linear (r near $+1$ or -1), then we can be confident that our interpolated value is quite accurate. If the data are not highly linear (r near 0), then our confidence is duly reduced.

For example, medical information collected from a patient every third day would establish data for day 3, 6, 9, ... and so on. After performing regression analysis, it is likely that an interpolation for day 4 would be accurate, given a high r value.

Extrapolation

Extrapolation is the use of the least squares line to predict values outside the range of data in a set, that is, values that are *smaller than the smallest value* already in the data set or *larger than the largest value*.

Two problems may arise in attempting to extrapolate from a data set.

Firstly, it may not be reasonable to extrapolate too far away from the given data values.

For example, suppose there is a weather data set for five days. Even if it is highly linear (r near $+1$ or -1), a least squares line used to predict the same data 15 days in the future is highly risky. Weather has a habit of randomly fluctuating and patterns rarely stay stable for very long.

Secondly, the data may be highly linear in a narrow band of the given data set.

For example, there may be data on stopping distances for a train at speeds of between 30 and 60 km/h. Even if they are highly linear in this range, it is unlikely that things are similar at very low speeds (0–15 km/h) or high speeds (over 100 km/h).

Generally, one should feel *more confident about the accuracy of a prediction derived from interpolation* than one derived from extrapolation. Of course, it still depends upon the correlation coefficient (r). The closer to linearity the data are, the more confident our predictions in all cases.

Interpolation and extrapolation

Interpolation is the use of the least squares line to predict values *within* the range of data in a set.

Extrapolation is the use of the least squares line to predict values *outside* the range of data in a set.

WORKED EXAMPLE 4 Using interpolation to predict the value of an unknown

Using interpolation and the following data set, predict the height of an 8-year-old student.

Age (years)	1	3	5	7	9	11
Height (cm)	60	76	115	126	141	148

THINK	WRITE
1. Determine the equation of the least squares line of best fit using your calculator. (Age is the explanatory variable, and height is the response variable.)	$y = 55.63 + 9.23x$
2. Replace x and y with the variables in question.	Height $= 55.63 + 9.23 \times$ age
3. Substitute 8 for age into the equation and evaluate.	When age $= 8$, Height $= 55.63 + 9.23 \times 8$ $= 129.5\,(cm)$
4. Write the answer.	At age 8, the predicted height is 129.5 cm.

WORKED EXAMPLE 5 Using extrapolation to predict the value of an unknown

Use extrapolation and the data from Worked example 4 to predict the height of the student when they turn 15. Discuss the reliability of this prediction.

THINK	WRITE
1. Use the regression equation to calculate the student's height at age 15.	Height $= 55.63 + 9.23 \times$ age $= 55.63 + 9.23 \times 15$ $= 194.08$ cm
2. Analyse and comment on the result.	Since we have extrapolated the result (i.e. since the greatest age in our data set is 11 and we are predicting outside the data set), we cannot claim that the prediction is reliable.

1. **WE3** A study on the growth in height of a monkey in its first six months is assumed to be linear and is shown in the data table.

Month from birth	1	2	3	4	5	6
Height (cm)	15	19	23	27	30	32

Determine:

  a. the equation, describing the relationship between the two variables
  b. the rate at which the monkey is growing
  c. the height of the monkey at birth.

2. The outside temperature is assumed to increase linearly with time after 6 am as shown in the data table.

Hours after 6 am	0.5	1.5	3	3.5	5
Temperature (°C)	15	18	22	23	28

Determine:

  a. the equation, describing the relationship between the two variables
  b. the rate at which the temperature is increasing
  c. the temperature at 6 am.

3. A drug company wishes to test the effectiveness of a drug to increase red blood cell counts in people who have a low count. The following data are collected.

Day of experiment	4	5	6	7	8	9
Red blood cell count	210	240	230	260	260	290

Determine:

  a. the equation, describing the relationship between the variables in the form $y = a + bx$
  b. the rate at which the red blood cell count was changing
  c. the red blood cell count at the beginning of the experiment (i.e. on day 0).

4. A wildlife exhibition is held over six weekends and features still and live displays. The number of live animals being exhibited varies each weekend.
The number of animals participating, together with the number of visitors to the exhibition each weekend, is shown.

Number of animals	6	4	8	5	7	6
Number of visitors	311	220	413	280	379	334

Determine:

  a. the rate of increase of visitors as the number of live animals is increased by one
  b. the predicted number of visitors if there are no live animals.

5. An electrical goods warehouse produces the following data showing the selling price of electrical goods to retailers and the volume of those sales.

Selling price ($)	60	80	100	120	140	160	200	220	240	260
Sales volume (× 1000)	400	300	275	250	210	190	150	100	50	0

Perform a least squares regression analysis and discuss the meaning of the gradient and *y*-intercept.

6. a. **WE4** Using interpolation and the following data set, predict the height of a 10-year-old student.

Age (years)	1	3	4	8	11	12
Height (cm)	65	82	92	140	157	165

b. **WE5** Use extrapolation and the above data to predict the height of the student when they turn 16.
c. Discuss the reliability of this prediction.

7. a. Using interpolation and the following data set, predict the length of Matt's pet snake when it is 15 months old.

Age (months)	1	3	5	8	12	18
Length (cm)	48	60	71	93	117	159

b. Use extrapolation and the above data to predict the length of Matt's pet snake when it is 2 years old.
c. Discuss the reliability of the prediction.

8. A study of the dining-out habits of various income groups in a particular suburb produces the results shown.

Weekly income ($)	100	200	300	400	500	600	700	800	900	1000
Number of restaurant visits per year	5.8	2.6	1.4	1.2	6	4.8	11.6	4.4	12.2	9

Use the data to predict:

a. the number of visits per year by a person on a weekly income of $680
b. the number of visits per year by a person on a weekly income of $2000.

9. Fit a least squares line of best fit to the following data.

x	0	1	2	4	5	6	8	10
y	2	3	7	12	17	21	27	35

Determine:

a. the regression equation
b. *y* when *x* = 3
c. *y* when *x* = 12
d. *x* when *y* = 7
e. *x* when *y* = 25
f. which of b to e above are extrapolations.

10. The following table represents the costs for shipping a consignment of shoes from Melbourne factories. The cost is given in terms of distance from Melbourne. There are two factories that can be used.

Distance from Melbourne (km)	10	20	30	40	50	60	70	80
Factory 1 cost ($)	70	70	90	100	110	120	150	180
Factory 2 cost ($)	70	75	80	100	100	115	125	135

a. Determine the least squares line of best fit equation for each factory.
b. Determine which factory is likely to have the lowest cost to ship to a shop in Melbourne (i.e. distance from Melbourne = 0 km).

c. Determine which factory is likely to have the lowest cost to ship to Mytown, 115 kilometres from Melbourne.

d. State which factory has the most 'linear' shipping rates.

11. A factory produces calculators. The least squares line of best fit for cost of production (C) as a function of numbers of calculators (n) produced is given by:

$$C = 600 + 7.76\,n$$

Furthermore, this function is deemed accurate when producing between 100 and 1000 calculators.

a. Calculate the cost to produce 200 calculators.

b. Determine the number of calculators that could be produced for $2000.

c. Calculate the cost to produce 10 000 calculators.

d. Calculate the 'fixed' costs for this production.

e. State which of **a** to **c** above is an interpolation.

12. A study of the relationship between IQ and results in a mathematics exam produced the following results.

IQ	80		92	102	105		107	111	115	121
Test result(%)	56	60	68	65		74	71	73		92

Unfortunately, some of the data were lost. Copy and complete the table by using the least squares equation with the data that were supplied.

Note: Only use (x, y) pairs if both are in the table.

13. The least squares line of best fit for a starting salary (s) as a function of number of years of schooling (n) is given by the rule:

$$s = 37\,000 + 1800\,n$$

a. Calculate the salary for a person who completed 10 years of schooling.

b. Calculate the salary for a person who completed 12 years of schooling.

c. Calculate the salary for a person who completed 15 years of schooling.

d. Mary earned $60 800. Calculate her likely schooling experience.

e. Discuss the reasonableness of predicting salary on the basis of years of schooling.

14. Fit a least squares line of best fit to the following data.

q	0	1	3	7	10	15
r	12	18	27	49	64	93

Determine:

a. the regression equation

b. r when $q = 4$

c. r when $q = 18$

d. q when $r = 100$

e. which of **b** to **d** is extrapolation.

15. A plumbing company's charges follow the least squares line of best fit:

$$C = 180 + 80\,n$$

where C is the total cost and n is the number of hours of work. This function is accurate for a single 8-hour day.

a. Calculate the total cost if the plumber worked for 3 hours.

b. If the total charge was $1250, calculate how long the plumber worked, correct to 2 decimal places.

c. Calculate the total cost if the plumber worked 9 hours and 30 minutes.

d. Determine the call-out fee (the cost to come out before they start any work).

e. State which of **a** to **c** is an extrapolation.

16. A comparison between AFL memberships sold and the amount of money spent on advertising by the club was investigated.

Advertising (in millions $)	2.3	1.8	1.2	0.8	1.6	0.6	1.0
Members	81 363	67 947	58 846	55 597	62 295	54 946	57 295

a. Determine the least squares line of best fit equation. Round the coefficients to the nearest whole number.
b. Using the least squares line of best fit equation, calculate how many members you would expect to have if $2 million was spent on advertising. Discuss whether this is extrapolation or interpolation.
c. If you wanted 70 000 members, calculate how much you would expect to have to pay on advertising.
d. Calculate the coefficient of determination and use it to explain the association between membership numbers and the amount of money spent on advertising.

3.3 Exam questions

Question 1 (1 mark)
Source: VCE 2016, Further Mathematics Exam 1, Section A, Q9; © VCAA.

MC The scatterplot below shows life expectancy in years plotted against the Human Development Index (HDI) for a large number of countries in 2011.

A least squares line has been fitted to the data and the resulting residual plot is also shown.

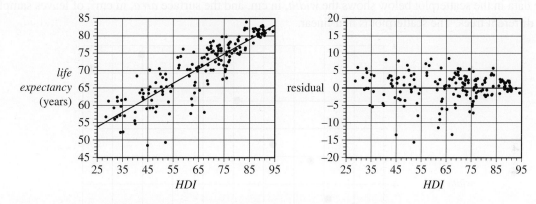

The equation of this least squares line is:

$$\text{life expectancy} = 43.0 + 0.422\ \text{HDI}$$

The coefficient of determination is $r^2 = 0.875$.

Given the information above, determine which one of the following statements is **not** true.
- **A.** The value of the correlation coefficient is close to 0.94.
- **B.** 12.5% of the variation in life expectancy is not explained by the variation in the Human Development Index.
- **C.** On average, life expectancy increases by 43.0 years for each 10-point increase in the Human Development Index.
- **D.** Ignoring any outliers, the association between life expectancy and the Human Development Index can be described as strong, positive and linear.
- **E.** Using the least squares line to predict the life expectancy in a country with a Human Development Index of 75 is an example of interpolation.

Question 2 (3 marks)

Source: VCE 2015, Further Mathematics Exam 2, Section A, Q3; © VCAA.

The scatterplot below plots male life expectancy (*male*) against female life expectancy (*female*) in 1950 for a number of countries. A least squares regression line has been fitted to the scatterplot as shown.

The slope of this least squares regression line is 0.88

a. Interpret the slope in terms of the variables *male* life expectancy and *female* life expectancy.

The equation of this least squares regression line is
$$male = 3.6 + 0.88 \times female$$

b. In a particular country in 1950, *female* life expectancy was 35 years.
Use the equation to predict *male* life expectancy for that country.

c. The coefficient of determination is 0.95
Interpret the coefficient of determination in terms of male life expectancy and female life expectancy.

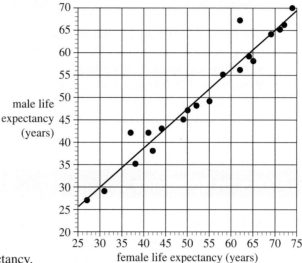

Question 3 (1 mark)

Source: VCE 2013, Further Mathematics Exam 1, Section A, Q10; © VCAA.

MC The data in the scatterplot below shows the *width*, in cm, and the surface *area*, in cm^2, of leaves sampled from 10 different trees. The scatterplot is non-linear.

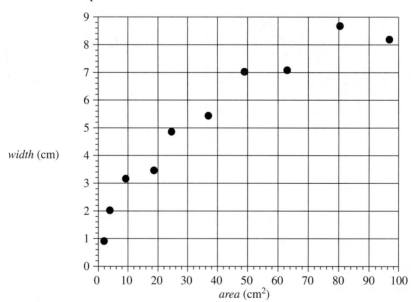

To linearise the scatterplot, $(width)^2$ is plotted against *area* and a least squares regression line is then fitted to the linearised plot.

The equation of this least squares regression line is

$$(width)^2 = 1.8 + 0.8 \times area$$

Using this equation, a leaf with a surface area of 120 cm^2 is predicted to have a width, in cm, closest to

A. 9.2   **B.** 9.9   **C.** 10.6   **D.** 84.6   **E.** 97.8

More exam questions are available online.

3.4 Residual analysis

3.4.1 Calculating residuals

A residual plot is a way to test the assumption that there is a linear relationship between two variables. A residual is the vertical difference between each data point and the least squares line.

Scientists gather data on the age of eggs and size for a particular moth. They enter the records in the table shown.

x	2	3	5	8	9	9
y	3	7	12	10	12	16

They then plot each point, and fit a least squares line as shown in Figure 1. They then decide to calculate the residuals.

The *residuals* are simply the vertical distances from the line to each point. These lines are shown as green and pink bars in Figure 2.

Figure 1

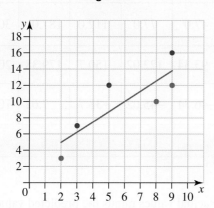

Figure 2

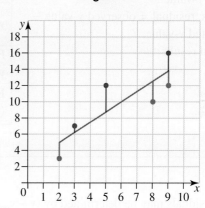

Finally, they calculate the residuals for each data point. This is done in two steps.

Step 1. Calculate the *predicted* value of y using the regression equation.

Step 2. Calculate the *difference* between this predicted value and the original value.

Residual value

Residual value = actual y-value − predicted y-value

WORKED EXAMPLE 6 Calculating residuals

Consider the data set shown. Determine the equation of the least squares line of best fit and calculate the residuals.

x	1	2	3	4	5	6	7	8	9	10
y	5	6	8	15	24	47	77	112	187	309

THINK

1. Determine the equation of a least squares line of best fit using a calculator.

2. Use the equation of the least squares line of best fit to calculate the predicted y-values (these are labelled as y_{pred}) for every x-value.
That is, substitute each x-value into the equation, evaluate and record results.

WRITE

$y = -78.7 + 28.7x$

x-value	1	2	3	4	5
y-value	5.0	6.0	8.0	15.0	24.0
Predicted y-value	−50.05	−21.38	7.3	35.98	64.66
Residual $(y - y_{pred})$	55.05	27.38	0.7	−20.98	−40.66

3. Calculate residuals for each point by subtracting predicted y-values from the actual y-value.
That is, residual = observed y-value − predicted y-value.
Record results.

x-value	6	7	8	9	10
y-value	47.0	77.0	112.0	187.0	309.0
Predicted y-value	93.34	122.02	150.7	179.38	208.06
Residual $(y - y_{pred})$	−46.34	−45.02	−38.7	7.62	100.94

Notes

1. The residuals may be determined by $(y - y_{pred})$, that is, the actual values minus the predicted values.
2. The *sum* of all the residuals *always* adds to 0 (or very close to 0 after rounding) when a least squares line of best fit is used. This can act as a check for our calculations.

3.4.2 Introduction to residual analysis

Calculating the residuals as in Worked example 6 is only the first step. We need to determine if there is a pattern in the residuals. We do this by plotting the *residuals* against the *original x-values*. If there is a pattern, it should become clearer after they are plotted.

Types of residual plots

There are three basic types of **residual plots**. Each type indicates whether or not a linear relationship exists between the two variables under investigation.

Note: The points are joined together to show the patterns more clearly.

Type	Description	Diagram
1.	The points of the residuals are randomly scattered above and below the *x*-axis. The original data probably have a *linear* relationship	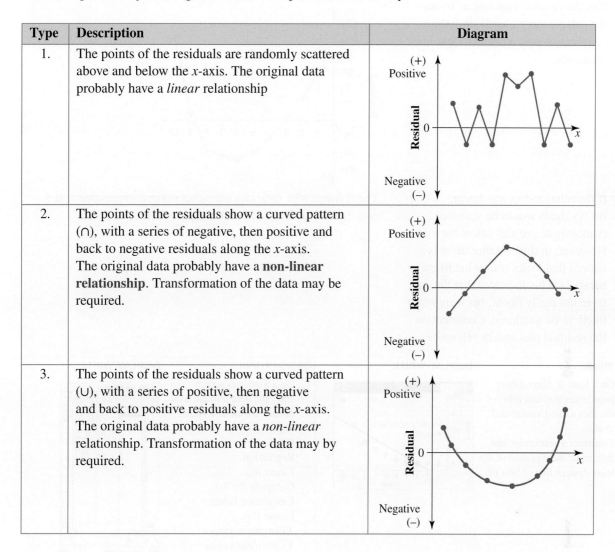
2.	The points of the residuals show a curved pattern (∩), with a series of negative, then positive and back to negative residuals along the *x*-axis. The original data probably have a **non-linear relationship**. Transformation of the data may be required.	
3.	The points of the residuals show a curved pattern (∪), with a series of positive, then negative and back to positive residuals along the *x*-axis. The original data probably have a *non-linear* relationship. Transformation of the data may by required.	

The transformation of data suggested in the last two residual plots will be studied in more detail in the next section.

Using the same data as in Worked example 6, plot the residuals and discuss the features of the residual plot.

THINK

1. Generate a table of values of residuals against *x*.

WRITE

x-value	1	2	3	4	5
Residual $(y - y_{\text{pred}})$	55.05	27.38	0.7	−20.98	−40.66
x-value	6	7	8	9	10
Residual $(y - y_{\text{pred}})$	−46.34	−45.02	−38.7	7.62	100.94

2. Plot the residuals against x. To see the pattern more clearly, join the consecutive points with straight line segments.

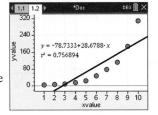

3. If the relationship was linear, the residuals would be scattered randomly above and below the x-axis. However, in this instance there is a pattern that looks somewhat like a parabola. This indicates that the data were not really linear, but were more likely to be quadratic. Comment on the residual plot and its relevance.

The residual plot indicates a distinct pattern suggesting that a non-linear model could be more appropriate.

TI	THINK	DISPLAY/WRITE

1. On a Lists & Spreadsheet page, enter the data into the lists named *xvalue* and *yvalue*.
Construct a scatterplot and determine the equation of the least squares line of best fit.

2. Return to the Lists & Spreadsheets page. Name Column C predicted and Column D residual.
In the formula cell under the heading predicted, complete the entry
line as:
$= -78.7333 + 28.6788 \times$ *xvalue*
Press ENTER.
In the formula cell under the heading residual, complete the entry
line as:
$= yvalue - predicted$
Press ENTER.

CASIO	THINK	DISPLAY/WRITE

1. On a Statistics screen, enter the data into the lists named *xvalue* and *yvalue*.
Tap:
Calc
Regression
Linear reg
Complete the Set Calculation fields:
Linear Reg
XList: main*xvalue*
YList: main*yvalue*
Freq: 1
Copy Formula y1
Copy Residual list 3
Tap OK.

2. Tap:
SetGraph
Setting
On the Set StatGraphs screen,
select the following:
XList: main\xvalue
YList: list 3
Freq: 1
Mark: square
Tap Set.

3. Return to the Data and Statistics page.
Press:
MENU
4: Analyze
7: Residuals
2: Show residual plot

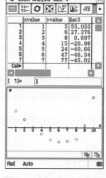

3. Tap the graph icon.

4. If the relationship was linear, the residuals would be scattered randomly above and below the *x*-axis. However, in this instance, there is a pattern that looks somewhat like a parabola. This indicates that the data were not really linear, but were more likely to be quadratic. Comment on the residual plot and its relevance.

The residual plot indicates a distinct pattern suggesting that a non-linear model could be more appropriate.

4. If the relationship was linear, the residuals would be scattered randomly above and below the *x*-axis. However, in this instance, there is a pattern that looks somewhat like a parabola. This indicates that the data were not really linear, but were more likely to be quadratic. Comment on the residual plot and its relevance.

The residual plot indicates a distinct pattern suggesting that a non-linear model could be more appropriate.

3.4 Exercise

Students, these questions are even better in jacPLUS

 Receive immediate feedback and access sample responses

 Access additional questions

 Track your results and progress

Find all this and MORE in jacPLUS ▶

1. **WE6+7** Consider the data set shown.

x	1	2	3	4	5	6	7	8	9
y	12	20	35	40	50	67	83	88	93

a. Determine the equation of the least squares line of best fit and calculate the residuals.
b. Plot the residuals and discuss the features of the residual plot.
c. Comment on whether your result is consistent with the coefficient of determination.

2. Consider the following data.

x	5	7	10	12	15	18	25	30	40
y	45	61	89	122	161	177	243	333	366

a. Determine the equation of the least squares line of best fit and calculate the residuals.
b. Plot the residuals and discuss the features of the residual plot.
c. Comment on whether your result is consistent with the coefficient of determination.

3. Consider the following data.

x	1	2	3	4	5	6
y	1	9.7	12.7	13.7	14.4	14.5

a. Calculate the residuals for the given data.
b. Plot the residuals and discuss whether the relationship between **x** and **y** is linear.

4. **MC** Select which of the following scatterplots shows a linear relationship between the variables.

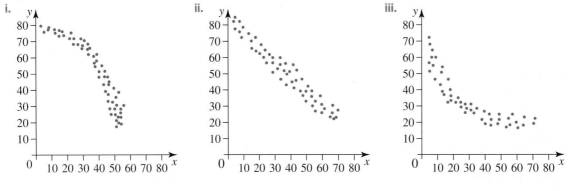

A. All of them   B. None of them   C. i and iii only   D. ii only   E. ii and iii only

5. Consider the following table from a survey conducted at a new computer manufacturing factory. It shows the percentage of defective computers produced on eight different days after the opening of the factory.

Day	2	4	5	7	8	9	10	11
Defective rate (%)	15	10	12	4	9	7	3	4

a. The results of least squares regression were: $b = -1.19$, $a = 16.34$, $r = -0.87$. Given $y = a + bx$, use the above information to calculate the predicted defective rates (y_{pred}).
b. Calculate the residuals ($y - y_{pred}$).
c. Plot the residuals and comment on the likely linearity of the data.
d. Estimate the defective rate after the first day of the factory's operation.
e. Estimate when the defective rate will be at zero. Comment on this result.

6. The following data represent the number of tourists booked into a hotel in Central Queensland during the first week of a drought. (Assume Monday = 1.)

Day	Mon.	Tues.	Wed.	Thurs.	Fri.	Sat.	Sun.
Bookings in hotel	158	124	74	56	31	35	22

The results of the least squares line of best fit were:
$b = -22.5$, $a = 161.3$, $r = -0.94$, where $y = a + bx$.

a. Determine the predicted hotel bookings (y_{pred}) for each day of the week.
b. Determine the residuals ($y - y_{pred}$).
c. Plot the residuals and comment on the likely linearity of the data.
d. Comment on whether this least squares line would be typical for this hotel.

7. **MC** A least squares line of best fit is fitted to the points shown in the scatterplot.

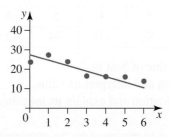

Select which of the following looks most similar to the residual plot for the data.

A.
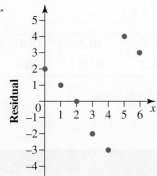

B.
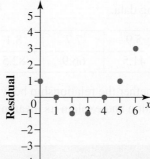

C.
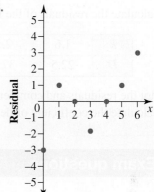

D.
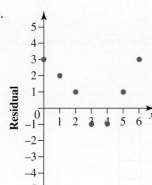

E.
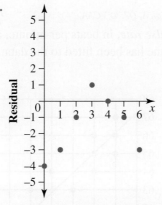

8. From each table of residuals, decide whether or not the relationship between the variables is likely to be linear.

a.

x	y	Residual
1	2	−1.34
2	4	−0.3
3	7	−0.1
4	11	0.2
5	21	0.97
6	20	2.3
7	19	1.2
8	15	−0.15
9	12	−0.9
10	6	−2.8

b.

x	y	Residual
23	56	0.12
21	50	−0.56
19	43	1.30
16	41	0.20
14	37	−1.45
11	31	2.16
9	28	−0.22
6	22	−3.56
4	19	2.19
3	17	−1.05

c.

x	y	Residual
1.2	23	0.045
1.6	25	0.003
1.8	24	−0.023
2.0	26	−0.089
2.2	28	−0.15
2.6	29	−0.98
2.7	34	−0.34
2.9	42	−0.01
3.0	56	0.45
3.1	64	1.23

9. Consider the following data set.

x	0	1	2	3	4	5	6	7	8	9	10
y	1	4	15	33	60	94	134	180	240	300	390

a. Plot the data and fit a least squares line of best fit.
b. Determine the correlation coefficient and interpret its value.
c. Calculate the coefficient of determination and explain its meaning.
d. Determine the residuals.
e. Construct the residual plot and use it to comment on the appropriateness of the assumption that the relationship between the variables is linear.

10. a. Calculate the residuals of the following data.

k	1.6	2.5	5.9	7.7	8.1	9.7	10.3	15.4
D	22.5	37.8	41.5	66.9	82.5	88.7	91.6	120.4

b. Plot the residuals and comment on whether the relationship between x and y is linear
c. Calculate the coefficient of determination and explain its meaning.

3.4 Exam questions

Question 1 (1 mark)

Source: VCE 2018, Further Mathematics Exam 1, Section A, Q7; © VCAA.

MC The scatterplot displays the *resting pulse rate*, in beats per minute, and the *time spent exercising*, in hours per week, of 16 students. A least squares line has been fitted to the data.

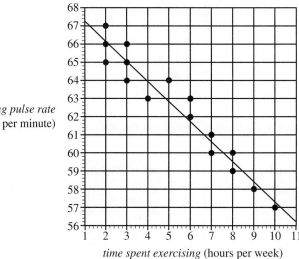

resting pulse rate (beats per minute)

time spent exercising (hours per week)

Using this least squares line to model the association between *resting pulse rate* and *time spent exercising*, the residual for the student who spent four hours per week exercising is closest to
A. –2.0 beats per minute.
B. –1.0 beats per minute.
C. –0.3 beats per minute.
D. 1.0 beats per minute.
E. 2.0 beats per minute.

▶ **Question 2 (1 mark)**

Source: VCE 2017, Further Mathematics Exam 1, Section A, Q9; © VCAA

MC The scatterplot below shows the *wrist* circumference and *ankle* circumference, both in centimetres, of 13 people. A least squares line has been fitted to the scatterplot with ankle circumference as the explanatory variable.

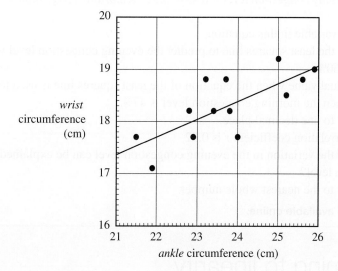

ankle circumference (cm)

When the least squares line on the scatterplot is used to predict the wrist circumference of the person with an ankle circumference of 24 cm, the residual will be closest to

A. −0.7   B. −0.4   C. −0.1   D. 0.4   E. 0.7

▶ **Question 3 (7 marks)**

Source: VCE 2018, Further Mathematics Exam 2, Section A, Q2; © VCAA.

The congestion level in a city can also be recorded as the percentage increase in travel time due to traffic congestion in peak periods (compared to non-peak periods).

This is called the percentage congestion level.

The percentage congestion levels for the morning and evening peak periods for 19 large cities are plotted on the scatterplot below.

a. Determine the median percentage congestion level for the morning peak period and the evening peak period.
Write your answers in the appropriate boxes provided below. **(2 marks)**

Median percentage congestion level for morning peak period ☐ %

Median percentage congestion level for evening peak period ☐ %

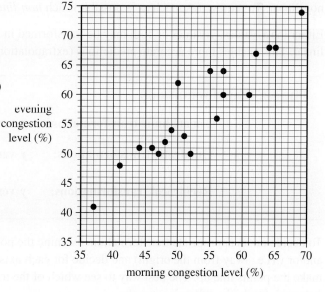

morning congestion level (%)

A least squares line is to be fitted to the data with the aim of predicting evening congestion level from morning congestion level.

The equation of this line is

$$evening\ congestion\ level = 8.48 + 0.922 \times morning\ congestion\ level$$

b. Name the response variable in this equation. **(1 mark)**

c. Use the equation of the least squares line to predict the evening congestion level when the morning congestion level is 60%. **(1 mark)**

d. Determine the residual value when the equation of the least squares line is used to predict the evening congestion level when the morning congestion level is 47%.
Round your answer to one decimal place. **(2 marks)**

e. The value of the correlation coefficient r is 0.92
What percentage of the variation in the evening congestion level can be explained by the variation in the morning congestion level?
Round your answer to the nearest whole number. **(1 mark)**

More exam questions are available online.

3.5 Transforming to linearity

LEARNING INTENTION

At the end of this subtopic you should be able to:
- transform data to achieve linearity in the case of clear non-linearity and repeat the modelling process using the transformed data.

3.5.1 Six transformations

Although linear regression might produce a 'good' fit (high r value) to a set of data, the data set may still be non-linear. To remove (as much as is possible) such *non-linearity*, the data can be transformed.

Either the x-values, y-values or both may be transformed in some way so that the transformed data are more linear. This enables more accurate predictions (extrapolations and interpolations) from the regression equation.

The six transformations		
Logarithmic transformations:	y **versus** $\log_{10}(x)$	$\log_{10}(y)$ **versus** x
Quadratic transformations:	y **versus** x^2	y^2 **versus** x
Reciprocal transformations:	y **versus** $\dfrac{1}{x}$	$\dfrac{1}{y}$ **versus** x

To decide on an appropriate transformation, examine the points on a scatterplot with high values of x and/or y (i.e. away from the origin) and decide for each axis whether it needs to be stretched or compressed to make the points line up. The best way to see which of the transformations to use is to look at a number of 'data patterns'. Study the following graphs.

Quadratic transformations

1. Use a y versus x^2 transformation.

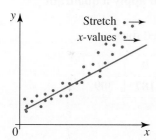

2. Use a y versus x^2 transformation.

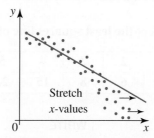

3. Use a y^2 versus x transformation.

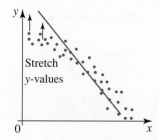

4. Use a y^2 versus x transformation.

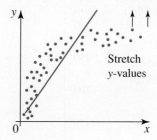

Logarithmic and reciprocal transformations

1. Use a y versus $\log_{10}(x)$ or y versus $\dfrac{1}{x}$ transformation.

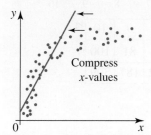

2. Use a y versus $\log_{10}(x)$ or y versus $\dfrac{1}{x}$ transformation.

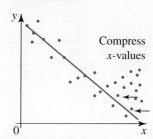

3. Use a $\log_{10}(y)$ versus x or $\dfrac{1}{y}$ versus x transformation.

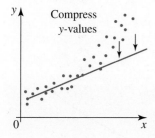

4. Use a $\log_{10}(y)$ versus x or $\dfrac{1}{y}$ versus x transformation.

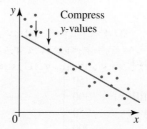

3.5.2 Testing transformations

As there are at least two possible transformations for any given non-linear scatterplot, the decision as to which is best comes from the coefficient of correlation. The least squares regression equation that has a Pearson correlation coefficient closest to 1 or -1 should be considered as the most appropriate. It is sometimes more useful to use a linear function rather than one of the six non-linear functions.

WORKED EXAMPLE 8 Testing quadratic transformations

a. **Plot the following data on a scatterplot, consider the shape of the graph and apply a quadratic transformation.**

b. **Calculate the equation of the least squares line of best fit for the transformed data.**

x	1	2	3	4	5	6	7	8	9	10
y	5	6	8	15	24	47	77	112	187	309

THINK

a. 1. Plot the data to check that a quadratic transformation is suitable. Looking at the shape of the graph, the best option is to stretch the x-values. This requires an x^2 transformation.

WRITE

a.

2. Square the x-values to give a transformed data set by using CAS.

x	1	2	3	4	5	6	7	8	9	10
x^2	1	4	9	16	25	36	49	64	81	100
y	5	6	8	15	24	47	77	112	187	309

b. 1. Determine the equation of the least squares line of best fit for the transformed data. Using CAS: y-intercept $(a) = -28.0$ gradient $(b) = 2.78$ correlation $(r) = 0.95$.

b. $y = a + bx$
$y = -28.0 + 2.78x_T$ where $x_T = x^2$; that is,
$y = -28.0 + 2.78x^2$

2. Plot the new transformed data.
Note: These data are still not truly linear, but are 'less' parabolic. Perhaps another transformation would improve things even further. This could involve transforming the y-values, such as $\log_{10}(y)$, and applying another linear regression.

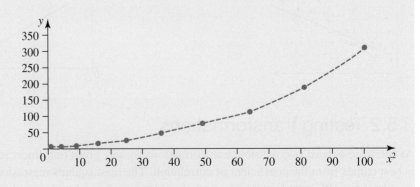

TI | THINK	DISPLAY/WRITE	CASIO | THINK	DISPLAY/WRITE

a. 1. On a Lists & Spreadsheet page, enter the data into the lists named *xvalue* and *yvalue*. Construct a scatterplot.

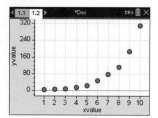

a. 1. On a Statistics screen, enter the data into the lists named *xvalue* and *yvalue*.

2. Looking at the shape of the graph, the best option is to stretch the *x*-axis. This requires an x^2 transformation.

Stretching the *x*-axis requires an x^2 transformation.

3. Return to the Lists & Spreadsheets page and in Column C, square the values on the horizontal axis.

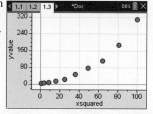

2. Looking at the shape of the graph, the best option is to stretch the *x*-axis. This requires an x^2 transformation.

Stretching the *x*-axis requires an x^2 transformation.

3. Rename list 3 as 'xsquare'. Place the cursor in the calculation cell at the bottom of list 3.
Type:
xvalue^2
Press EXE.

4. Construct a scatterplot with the *xsquared* values on the horizontal axis and *yvalues* on the vertical axis.

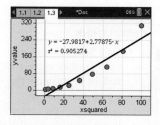

Construct a scatterplot with the *xsquared* values on the horizontal axis and *yvalues* on the vertical axis.

b. 1. Press MENU and select:
4: Analyze
6: Regression
2: Show Linear (a + bx)

b. 1. Tap:
Calc
RegressionLinear reg
Complete the Set Calculation fields:
Linear Reg
XList: main\xsquared
YList: main\yvalue
Freq: 1
Copy Formula: y1
Copy Residual: Off
Tap OK.

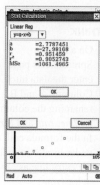

2. The answer appears on the screen.
$y = -28.98 + 2.78x^2$
(correct to 2 decimal places)

Note: These data are still not truly linear, but are 'less' parabolic. Perhaps another transformation would improve things even further. This could involve transforming the y-values, such as $\log_{10}(y)$, and applying another linear regression.

2. The answer appears on the screen.
$y = -28.98 + 2.78x^2$
(correct to 2 decimal places)

Note: These data are still not truly linear, but are 'less' parabolic. Perhaps another transformation would improve things even further. This could involve transforming the y-values, such as $\log_{10}(y)$, and applying another linear regression.

WORKED EXAMPLE 9 Applying logarithmic transformations

a. **Transform the data by applying a logarithmic transformation to the y-variable.**

Time after operation (h)	x	1	2	3	4	5	6	7	8
Heart rate (beats/min)	y	100	80	65	55	50	51	48	46

b. **Calculate the equation of the least squares line of best fit for the transformed data.**
c. **Comment on the value of r.**

THINK

a. Transform the y-value data by calculating the log of y-values or, in this problem, the log of heart rate.

WRITE

a.

Time	x	1	2	3	4	5	6	7	8
$\log_{10}$ (heart rate)	$\log_{10}(x)$	2	1.903	1.813	1.740	1.694	1.708	1.681	1.663

b. 1. Use a calculator to find the equation of the least squares line of best fit x and log (y).

b. $\log_{10}(y) = 1.98 - 0.05x$

2. Rewrite the equation in terms of the variables in question.

$\log_{10}(\text{heart rate}) = 1.98 - 0.05 \times \text{time}$
(i.e. time = number of hours after the operation)

c. State the value of r and comment on the result.

c. $r = -0.93$
Given that the value of r is -0.93, the correlation is strong and negative.

TI | THINK

a. 1. On a Lists & Spreadsheets page, enter the data into Columns A and B, renaming the columns as *time*, *rate* and *lgrate*. In the formula cell under lgrate, complete the entry line as:
$= \log(rate)$
Press ENTER.

DISPLAY/WRITE

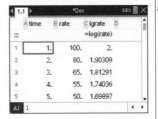

CASIO | THINK

a. 1. On a Statistics screen, enter the data into list 1 and list 2, renaming the lists as *time*, *rate* and *lgrate*. Place the cursor in the calculation cell at the bottom of list 3 (lgrate). Type:
log (rate)
Press EXE.

DISPLAY/WRITE

2. Press CTRL + I and select: 5: Add Data & Statistics
Select time for the horizontal axis and lgrate for the vertical axis.

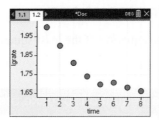

2. Construct a scatterplot with the time values on the horizontal axis and lgrate on the vertical axis.

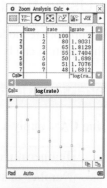

b. 1. Press MENU and then select:
4: Analyze
6: Regression
2: Show Linear $(a + bx)$

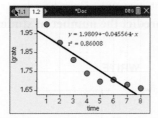

b. 1. Tap:
Calc
Regression
Linear Reg
Complete the Set Calculation fields:
Linear Reg
XList: main\time
YList: main\lgrate
Freq: 1
Copy Formula: y 1
Copy Residual: Off
Tap OK.

2. The answer is on the screen.

$\log_{10}$ (heart rate)
$= 1.98 - 0.05 \times$ time
(correct to 2 decimal places)
Time = number of hours after the operation

2. The answer is on the screen.

$\log_{10}$ (heart rate)
$= 1.98 - 0.05 \times$ time
(correct to 2 decimal places)
Time = number of hours after the operation

c. The value of r^2 is on the screen.
Note: The value of r is negative as the least squares line has a negative gradient.

$r^2 = 0.860\,08$
$r = -0.93$
There is a slight improvement of the correlation coefficient that resulted from applying a logarithmic transformation.

c. The value of r is on the screen.

$r = -0.93$
There is a slight improvement of the correlation coefficient that resulted from applying a logarithmic transformation.

Further investigation

Often all appropriate transformations need to be performed to choose the best one. Extend Worked example 9 by compressing the y data using the reciprocals of the y data or even compress the x data. Go back to the steps for transforming the data. Did you get a better r value and thus a more reliable line of best fit? (*Hint:* The best transformation gives $r = -0.98$.)

3.5.3 Using the transformed line for predictions

Once the appropriate model has been established and the equation of the least squares line of best fit has been found, the equation can be used for predictions.

WORKED EXAMPLE 10 Using the transformed equation to make predictions

a. **Using CAS, apply a reciprocal transformation to the following data.**

Temperature (°C)	x	5	10	15	20	25	30	35
Number of students in a class wearing jumpers	y	18	10	6	5	3	2	2

b. **Use the transformed least squares equation to predict the number of students wearing jumpers when the temperature is 12°C.**

THINK

a. 1. Construct the scatterplot. Temperature is the explanatory variable, while the number of students wearing jumpers is the response variable.
   Therefore, put *temperature* on the horizontal axis and *students* on the vertical axis.

WRITE

a.

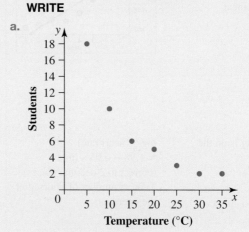

2. The *x*-values should be compressed, so it may be appropriate to transform the *x*-data by calculating the reciprocal of temperature. Reciprocate each *x*-value $\left(\dfrac{1}{x}\right)$.

$\dfrac{1}{\text{Temperature}}$	$\dfrac{1}{x}$	$\dfrac{1}{5}$	$\dfrac{1}{10}$	$\dfrac{1}{15}$	$\dfrac{1}{20}$	$\dfrac{1}{25}$	$\dfrac{1}{30}$	$\dfrac{1}{35}$
Number of students wearing jumpers	y	18	10	6	5	3	2	2

3. Use CAS to find the equation of the least squares line of best fit for $\dfrac{1}{x}$ and y.

$y = -0.4354 + 94.583x_T$, where $x_T = \dfrac{1}{x}$ or

$y = -0.4354 + \dfrac{94.583}{x}$

4. Replace *x* and *y* with the variables in question.

The number of students in class wearing

jumpers $= -0.4354 + \dfrac{94.583}{\text{Temperature}}$

b. 1. Substitute 12 for *x* into the equation of the least squares line and evaluate.

b. Number of students wearing jumpers

$= -0.4354 + \dfrac{94.583}{\text{Temperature}}$

$= -0.4354 + \dfrac{94.583}{12}$

$= 7.447$

2. Write your answer to the nearest whole number.

7 students are predicted to wear jumpers when the temperature is 12 °C.

TI	THINK	DISPLAY/WRITE	CASIO	THINK	DISPLAY/WRITE

TI | THINK

a. 1. Construct the scatterplot. Temperature is the explanatory variable, while the number of students wearing jumpers is the response variable. Therefore, put *temperature* on the horizontal axis and *number of students* on the vertical axis.

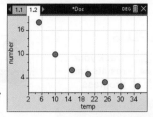

2. The values should be compressed, so transform the data on the horizontal axis by calculating the reciprocal of temperature. Return to the Lists & Spreadsheets page and label Column C as reciptemp.In the formula cell of Column C, complete the entry line as:
$= 1 \div \text{temp}$
Press ENTER.

3. Press CTRL + I and select:
5: Add Data & Statistics
Select reciptemp for the horizontal axis and number for the vertical axis. Press MENU and then select:
4: Analyze
6: Regression
2: Show Linear (a + bx)

4. The answer is on the screen.

$$y = -0.44 + 94.58 \times \frac{1}{x}$$

(correct to 2 decimal places)

5. Replace x and y with the variables in question.

No. of students
$$= -0.44 + \frac{94.58}{\text{Temperature}}$$

DISPLAY/WRITE

CASIO | THINK

a. 1. Construct the scatterplot. Temperature is the explanatory variable, while the number of students wearing jumpers is the response variable. Therefore, put *temperature* on the horizontal axis and *number of students* on the vertical axis.

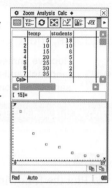

2. The values should be compressed, so transform the data on the horizontal axis by calculating the reciprocal of the temperature.
Rename list 3 as *reciptem*. Place the cursor in the calculation cell at the bottom of list 3 (reciptem).
Type:
$1 \div \text{temp}$
Press EXE.

3. Construct a scatterplot with the reciptem values on the horizontal axis and students on the vertical axis.
Tap:
Calc
Regression
Linear Reg
Complete the Set Calculation fields:
Linear Reg
XList: main\reciptem
YList: main\students
Freq: 1
Copy Formula: y1
Copy Residual: Off
Tap OK.

4. The answer is on the screen.

$$y = -0.44 + 94.58 \times \frac{1}{x}$$

(correct to 2 decimal places)

5. Replace x and y with the variables in question.

No. of students
$$= -0.44 + \frac{94.58}{\text{Temperature}}$$

 Resources

Interactivities Linearising data (int-6491)
Transforming to linearity (int-6253)

3.5 Exercise

1. **WE8**
   a. Plot the following data on a scatterplot, consider the shape of the graph and apply a quadratic transformation.

x	1	2	3	4	5	6	7	8	9
y	12	19	29	47	63	85	114	144	178

   b. Calculate the equation of the least squares line of best fit for the transformed data.

2. a. Plot the following data on a scatterplot, consider the shape of the graph and apply a quadratic transformation.

x	3	5	9	12	16	21	24	33
y	5	12	38	75	132	209	291	578

   b. Calculate the equation of the least squares line of best fit for the transformed data.

3. Apply a quadratic (x^2) transformation to the following data set. The least squares line has been determined as $y = 186 - 27.7x$ with $r = -0.91$.

x	2	3	4	5	7	9
y	96	95	92	90	14	-100

4. **WE9** Apply a logarithmic transformation to the following data, which represent the speed of a car as a function of time, by transforming the y-variable.

Time (s)	1	2	3	4	5	6	7	8
Speed $(\mathrm{m\,s^{-1}})$	90	71	55	45	39	35	32	30

5. Apply a logarithmic transformation to the following data by transforming the y-variable.

x	5	10	15	20	25	30	35	40
y	1000	500	225	147	99	70	59	56

6. The *average* heights of 50 students of various ages were measured as follows.

Age group (years)	9	10	11	12	13	14	15	16	17	18
Average height (cm)	128	144	148	154	158	161	165	164	166	167

   The original linear regression yielded:
   $$height = 104.7 + 3.76 \times age, \text{ with } r = 0.92$$

   a. Plot the original data and least squares line.
   b. Apply a $\log_{10}(x)$ transformation.
   c. Perform a least squares analysis on the transformed data and comment on your results.

7. **a.** Use the transformed data from question **6** to predict the heights of students of the following ages:

    i. 7 years old   **ii.** 10.5 years old   **iii.** 20 years old.

  b. State which of the predictions in part **a** were obtained by interpolating.

8. Comment on the suitability of transforming the data of question **6** to improve predictions of heights for students under 8 years old or over 18.

9. **a.** **WE10** Using CAS, apply a reciprocal transformation to the *x*-variable of the following data.

Time after 6 pm (h)	1	2	3	4	5	6	7	8
Temperature (°C)	32	22	16	11	9	8	7	7

  b. Use the transformed least squares equation to predict the temperature at 10.30 pm.

10. **a.** Using CAS, apply a reciprocal transformation to the *x*-variable of the following data.

x	2	5	7	9	10	13	15	18
y	120	50	33	15	9	5	2	1

  b. Use the transformed least squares equation to predict *y* when $x = 12$.
  c. Use the transformed least squares equation to predict *x* when $y = 20$.

11. **a.** Apply a reciprocal transformation to the following data obtained by a Physics student studying light intensity.

Distance from light source (metres)	1	2	3	4	5	10
Intensity (candlepower)	90	60	28	22	20	12

  b. Use the transformed least squares equation to predict the intensity at a distance of 20 metres.

12. For each of the following scatterplots, suggest an appropriate transformation(s).

  a.
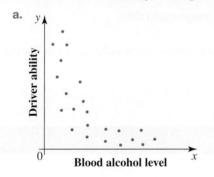

  b.
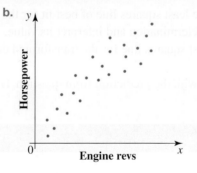

  c.
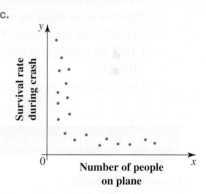

13. Use the equation $y = -12.5 + 0.2x^2$, found after transformation, to predict values of *y* for the given *x*-value (correct to 2 decimal places):

  a. $x = 2.5$ **b.** $x = -2.5$.

14. Use the equation $y = -25 + 1.12 \log_{10}(x)$, found after transformation, to predict values of *y* for the given *x*-value (correct to 2 decimal places):

  a. $x = 2.5$ **b.** $x = -2.5$

15. Use the equation $\log_{10}(y) = 0.03 + 0.2x$, found after transformation, to predict values of *y* for the given *x*-value (correct to 2 decimal places):

  a. $x = 2.5$ **b.** $x = -2.5$.

16. Use the equation $\dfrac{1}{y} = 12.5 + 0.2x$, found after transformation, to predict values of y for the given x-value (correct to 2 decimal places):

    a. $x = 2.5$ b. $x = -2.5$.

17. The seeds in a sunflower are arranged in spirals for a compact head. Counting the number of seeds in the successive circles starting from the centre and moving outwards, the following number of seeds were counted.

Circle	1st	2nd	3rd	4th	5th	6th	7th	8th	9th	10th
Number of seeds	3	5	8	13	21	34	55	89	144	233

a. Plot the data and fit a least squares line of best fit.
b. Calculate the correlation coefficient and interpret its value.
c. Using the equation of the least squares line, predict the number of seeds in the 11th circle.
d. Calculate the residuals.
e. Construct the residual plot. Discuss whether the relation between the number of the circle and the number of seeds is linear.
f. Determine what type of transformation could be applied to:

    i. the x-values. Explain why.
    ii. the y-values. Explain why.

18. Apply a $\log_{10}(y)$ transformation to the data used in question **17**.

a. Fit a least squares line of best fit to the transformed data and plot it with the data.
b. Calculate the correlation coefficient. Determine whether there was an improvement. Comment on the reasons for this outcome.
c. Determine the equation of the least squares line of best fit for the transformed data.
d. Calculate the coefficient of determination and interpret its value.
e. Using the equation of the least squares line for the transformed data, predict the number of seeds for the 11th circle.
f. Describe how this compares with the prediction from question **17**.

3.5 Exam questions

▶ Question 1 (1 mark)
Source: VCAA 2020, Further Mathematics Exam 1, Section A, Q14; © VCAA.

MC In a study, the association between the *number of tasks* completed on a test and the time allowed for the test, in hours, was found to be non-linear.

The data can be linearised using a $\log_{10}$ transformation applied to the variable number of tasks.

The equation of the least squares line for the transformed data is

$$\log_{10}(number\ of\ tasks) = 1.160 + 0.03617 \times time$$

This equation predicts that the *number of tasks* completed when the time allowed for the test is three hours is closest to

  A. 13   **B.** 16   **C.** 19   **D.** 25   **E.** 26

Source: VCE 2018, Further Mathematics Exam 1, Section A, Q11; © VCAA.

MC Freya uses the following data to generate the scatterplot below.

x	1	2	3	4	5	6	7	8	9	10
y	105	48	35	23	18	16	12	12	9	9

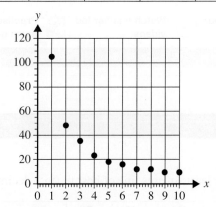

The scatterplot shows that the data is non-linear.

To linearise the data, Freya applies a reciprocal transformation to the variable y.

She then fits a least squares line to the transformed data.

With x as the explanatory variable, the equation of this least squares line is closest to

A. $\dfrac{1}{y} = -0.0039 + 0.012x$ **B.** $\dfrac{1}{y} = -0.025 + 1.1x$ **C.** $\dfrac{1}{y} = 7.8 - 0.082x$

D. $y = 45.3 + 59.7 \times \dfrac{1}{x}$ **E.** $y = 59.7 + 45.3 \times \dfrac{1}{x}$

Question 3 (1 mark)

Source: VCE 2018, Further Mathematics Exam 1, Section A, Q12; © VCAA.

MC A $\log_{10}(y)$ transformation was used to linearise a set of non-linear bivariate data.

A least squares line was then fitted to the transformed data.

The equation of this least squares line is

$$\log_{10}(y) = 3.1 - 2.3x$$

This equation is used to predict the value of y when $x = 1.1$

The value of y is closest to

A. −0.24   **B.** 0.57   **C.** 0.91   **D.** 1.6   **E.** 3.7

More exam questions are available online.

3.6 Review

3.6.1 Summary

doc-38034

Hey students! Now that it's time to revise this topic, go online to:

 Access the topic summary

 Review your results

 Watch teacher-led videos

 Practise VCAA exam questions

Find all this and MORE in jacPLUS

3.6 Exercise

Multiple choice

1. **MC** For the following data set, the least squares line of best fit (to 2 decimal places) is:

x	25	36	45	78	89	99	110
y	78	153	267	456	891	1020	1410

A. $y = 14.42 + 381.97x$ 　　　**B.** $y = 14.42 - 381.97x$ 　　　**C.** $y = -381.97 + 14.42x$

D. $y = -38 + 14x$ 　　　**E.** $y = 38 - 14x$

2. **MC** Given the following summary statistics:

$$\bar{x} = 154.4 \quad s_x = 5.8 \quad \bar{y} = 172.5 \quad s_y = 7.4 \quad r = 0.9$$

the values of b and a, respectively, for the equation of the least squares line $y = a + bx$ are:

A. 0.71 and 32.72 　　　**B.** 1.15 and -5.06 　　　**C.** 0.44 and 10.1

D. 0.04 and -0.16 　　　**E.** -1.32 and 3.8

Use the following information to answer questions 3 to 5.

A least squares line of best fit is fitted to the 7 points as shown.

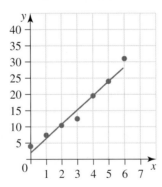

3. **MC** The equation of the least squares line is closest to:

A. $y = x$ 　　　**B.** $y = \dfrac{1}{5}x$ 　　　**C.** $y = 2x$ 　　　**D.** $y = 5x$ 　　　**E.** $y = 20x$

4. **MC** From the data points $x = 6$, $y = 33$, if the least squares line is used to predict the y-value when $x = 6$, the residual will be:

A. 0 　　　**B.** 1 　　　**C.** -1 　　　**D.** 2 　　　**E.** 3

5. **MC** The residual plot would look most similar to:

A.

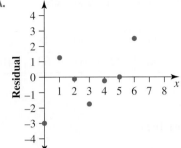

B.

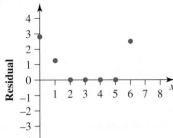

C.

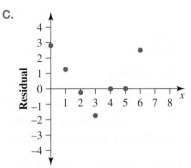

D.

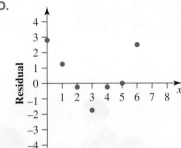

E.

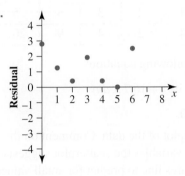

6. **MC** After a transformation, a relationship was found to be $y = 0.4x^2 + 12.1$. The predicted value for y given that $x = 2.5$ is:

A. 6.25   B. 2.5   C. 14.6   D. 13.1   E. 12.5

Short answer

7. Use the following summary statistics to calculate:
   a. the slope, b, of the least squares line of best fit
   b. the y-intercept, a, of the least squares line of best fit where $y = a + bx$ is the equation of the least squares line.

$$\bar{x} = 15 \qquad s_x = 5 \qquad \bar{y} = 10 \qquad s_y = 2.5 \qquad r = -0.9$$

8. a. Determine the equation for the least squares line of best fit for the following data.

x	1	2	4	8	9	10	12	15
y	23	21	20	14	16	9	12	5

   b. Using the equation you determined in part **a**, copy and complete the following table of predicted values.

x	3	5	7	9	11	13	15	17	20
y_{pred}									

   c. For the least squares line of best fit from **a**, determine the residuals.

Extended response

9. The data in the table show the number of hours spent by students learning to touch-type and their corresponding speed in words per minute (wpm).

Time (h)	20	33	22	39	40	37	46	44	24	36	50	48	29
Speed (wpm)	34	46	38	53	52	49	60	58	36	42	65	63	40

a. State which variable is explanatory and which is response.
b. Represent the data on a scatterplot.
c. Use the scatterplot to comment on the relationship between the two variables.

10. Consider this data set, which measures the sales figures for a new salesperson.

Day	1	2	3	4	5	6	7	8
Units sold	1	2	4	9	20	44	84	124

The least squares line yielded the following equation:

$$Units\ sold = -39.1 + 16.7 \times day$$

The correlation coefficient was 0.90.

a. Use CAS to construct the scatterplot of the data. Comment on the kind of relationship between the variables the scatterplot suggests.
b. Comment on using the least squares line to predict for small values of the explanatory variable.
c. Use the equation of the least squares line to predict the sales figures for the 10th day. Comment on the reliability of this prediction.
d. Transform the data from part a using a quadratic (x^2) transformation.
e. Determine a least squares line of best fit on the transformed data from part d.
f. Use the least squares line for the transformed data to predict the sales figures for the 10th day. Comment on whether this is a better prediction than the one found in part c.

11. A mining company wishes to predict its gold production output. It collected the following data over a 9-month period.

Month (1 = January)	Jan.	Feb.	Mar.	Apr.	May	June	July	Aug.	Sept.
Production (tonnes)	3	8	10.8	12	11.6	14	15.5	15	18.1

a. Determine the equation of the least squares line of best fit.
b. Using the line from part a, predict the production after 12 months.
c. Comment on the accuracy, usefulness and simplicity of the methods.
d. Looking at the original data set, discuss whether linearity is a reasonable assertion.
e. Research into gold mines indicates that after about 10 months, production tends not to increase as rapidly as in earlier months. Given this information, a logarithmic transformation is suggested. Transform the original data using this method.

f. Fit a straight line to this transformed data using the least squares line of best fit.
g. Discuss whether or not this transformation has removed any non-linearity.
h. Predict the level of production of gold after 12 months using the equation obtained in part f. Compare the prediction from part b with the one obtained using the logarithmic transformation.

12. A number of students were asked to record the number of hours they spent studying for a test and the mark obtained on that test. The time spent studying varied between 1.5 and 8 hours, while test results ranged from 28% to 92%. The equation of the least squares line of best fit was found to be:

$$test\ mark\ (\%) = 21.6 + 8.6 \times number\ of\ hours\ spent\ studying$$

a. Interpret the value of the gradient in the context of this problem.
Interpret the value of the y-intercept.

b. Use the equation of the least squares line of best fit to predict the test result for Nathan who spent 6 hours studying for the test. Give your answer correct to the nearest whole number. Is this an example of interpolation or extrapolation? Explain.

c. Rachel was sick on the day of the test and will have to do it on another day. She claims spending 9 hours studying for the test. Use the equation of the least squares line of best fit to predict the test result for Rachel. Comment on reliability of this prediction.

3.6 Exam questions

Question 1 (5 marks)

Source: VCE 2020, Further Mathematics Exam 2, Section A, Q4; © VCAA.

The *age*, in years, *body density*, in kilograms per litre, and *weight*, in kilograms, of a sample of 12 men aged 23 to 25 years are shown in the table below.

Age (years)	Body density (kg/litre)	Weight (kg)
23	1.07	70.1
23	1.07	90.4
23	1.08	73.2
23	1.08	85.0
24	1.03	84.3
24	1.05	95.6
24	1.07	71.7
24	1.06	95.0
25	1.07	80.2
25	1.09	87.4
25	1.02	94.9
25	1.09	65.3

a. For these 12 men, determine

  i. their median *age*, in years **(1 mark)**

  ii. the mean of their *body density*, in kilogram per litre. **(1 mark)**

b. A least squares line is to be fitted to the data with the aim of predicting *body density* from *weight*.

  i. Name the explanatory variable for this least squares line. **(1 mark)**

  ii. Determine the slope of this least squares line.
    Round your answer to three significant figures. **(1 mark)**

c. What percentage of the variation in *body density* can be explained by the variation in *weight*?
Round your answer to the nearest percentage. **(1 mark)**

Question 2 (7 marks)

Source: VCAA 2020, Further Mathematics Exam 2, Section A, Q5; © VCAA.

The scatterplot below shows *body density*, in kilograms per litre, against waist *measurement*, in centimetres, for 250 men.

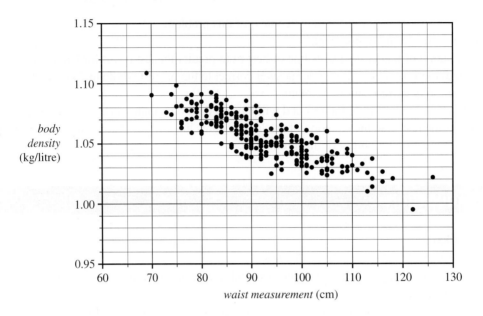

When a least squares line is fitted to the scatterplot, the equation of this line is

$$body\,density = 1.195 - 0.001512 \times waist\,measurement$$

a. Draw the graph of this least squares line on the **scatterplot above**. **(1 mark)**
(*Answer on the scatterplot above*)

b. Use the equation of this least squares line to predict the *body density* of a man whose waist *measurement* is 65 cm.
Round your answer to two decimal places. **(1 mark)**

c. When using the equation of this least squares line to make the prediction in **part b.**, are you extrapolating or interpolating? **(1 mark)**

d. Interpret the slope of this least squares line in terms of a man's *body density* and waist *measurement*. **(1 mark)**

e. In this study, the body density of the man with a waist measurement of 122 cm was 0.995 kg/litre.
Show that, when this least squares line is fitted to the scatterplot, the residual, rounded to two decimal places, is −0.02 **(1 mark)**

f. The coefficient of determination for this data is 0.6783.
Write down the value of the correlation coefficient *r*.
Round your answer to three decimal places. **(1 mark)**

g. The residual plot associated with fitting a least squares line to this data is shown below.

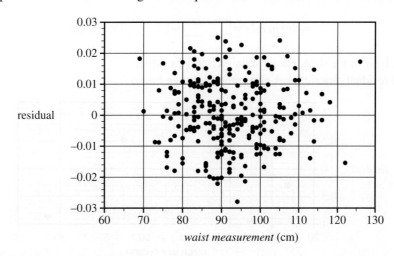

waist measurement (cm)

Does this residual plot support the assumption of linearity that was made when fitting this line to this data? Briefly explain your answer. **(1 mark)**

Question 3 (5 marks)

Source: VCAA 2020, Further Mathematics Exam 2, Section A, Q6; © VCAA.

The table below shows the mean age, in years, and the mean height, in centimetres, of 648 women from seven different age groups.

	Age group						
	Twenties	**Thirties**	**Forties**	**Fifties**	**Sixties**	**Seventies**	**Eighties**
Mean age (years)	26.3	35.2	45.3	55.3	65.1	74.8	83.1
Mean height (cm)	167.1	164.9	164.8	163.4	161.2	158.4	156.7

Data: J Sorkin et al., 'Longitudinal change in height of men and women: Implications for interpretation of the body mass index', *American Journal of Epidemiology*, vol. 150, no. 9, 1999, p. 971

a. What was the difference, in centimetres, between the *mean height* of the women in their twenties and the *mean height* of the women in their eighties? **(1 mark)**

A scatterplot displaying this data shows an association between the *mean height* and the *mean age* of these women. In an initial analysis of the data, a line is fitted to the data by eye, as shown.

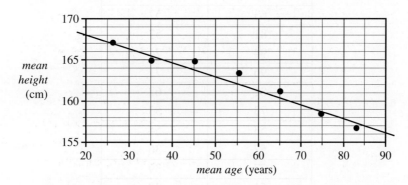

mean age (years)

b. Describe this association in terms of strength and direction. **(1 mark)**

c. The line on the scatterplot passes through the points (20, 168) and (85, 157).
Using these two points, determine the equation of this line. Write the values of the intercept and the slope in the appropriate boxes below.

Round your answers to three significant figures.   **(1 mark)**

$$mean\ height = \boxed{} + \boxed{} \times mean\ age$$

d. In a further analysis of the data, a least squares line was fitted.
The associated residual plot that was generated is shown below.

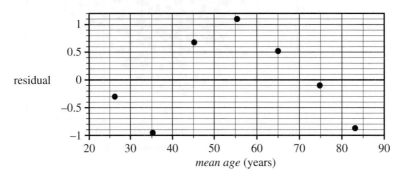

The residual plot indicates that the association between the *mean height* and the *mean age* of women is non-linear.
The data presented in the table on page 12 is repeated below. It can be linearised by applying an appropriate transformation to the variable *mean age*.

Mean age (years)	26.3	35.2	45.2	55.3	65.1	74.8	83.1
Mean height (cm)	167.1	164.9	164.8	163.4	161.2	158.4	156.7

Apply an appropriate transformation to the variable *mean age* to linearise the data. Fit a least squares line to the transformed data and write its equation below.
Round the values of the intercept and the slope to four significant figures.   **(2 marks)**

Question 4 (3 marks)

Source: VCAA 2019, Further Mathematics Exam 2, Section A, Q6; © VCAA.

The relative humidity (%) at 9 am and 3 pm on 14 days in November 2017 is shown in the table below.

Relative humidity (%)	
9 am	**3 pm**
100	87
99	75
95	67
63	57
81	57
94	74
96	71
81	62
73	53
53	54
57	36
77	39
51	30
41	32

Data: Australian Government, Bureau of Meteorology, < www.bom.gov.au/>

A least squares line is to be fitted to the data with the aim of predicting the relative humidity at 3 pm (*humidity 3 pm*) from the relative humidity at 9 am (*humidity 9 am*).

   a. Name the explanatory variable. **(1 mark)**

   b. Determine the values of the intercept and the slope of this least squares line.
     Round both values to three significant figures and write them in the appropriate boxes provided. **(1 mark)**

$$humidity\ 3\,pm = \boxed{} + \boxed{} \times humidity\ 9\,am$$

   c. Determine the value of the correlation coefficient for this data set.
     Round your answer to three decimal places. **(1 mark)**

▶ **Question 5 (8 marks)**

Source: VCAA 2019, Further Mathematics Exam 2, Section A, Q5; © VCAA.

The scatterplot below shows the atmospheric pressure, in hectopascals (hPa), at 3 pm (*pressure 3 pm*) plotted against the atmospheric pressure, in hectopascals, at 9 am (*pressure 9 am*) for 23 days in November 2017 at a particular weather station.

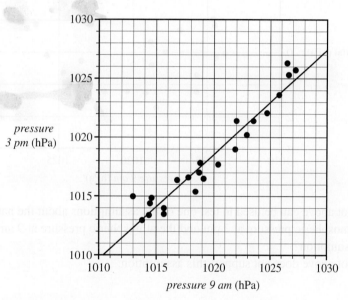

Data: Australian Government, Bureau of Meteorology,

A least squares line has been fitted to the scatterplot as shown.

The equation of this line is

$$pressure\ 3\,pm = 111.4 + 0.8894 \times pressure\ 9\,pm$$

   a. Interpret the slope of this least squares line in terms of the atmospheric pressure at this weather station at 9 am and at 3 pm. **(1 mark)**

   b. Use the equation of the least squares line to predict the atmospheric pressure at 3 pm when the atmospheric pressure at 9 am is 1025 hPa.
     Round your answer to the nearest whole number. **(1 mark)**

   c. Is the prediction made in **part b**. an example of extrapolation or interpolation? **(1 mark)**

   d. Determine the residual when the atmospheric pressure at 9 am is 1013 hPa.
     Round your answer to the nearest whole number. **(1 mark)**

e. The mean and the standard deviation of pressure 9 am and pressure 3 pm for these 23 days are shown in Table 4 below.

Table 4

	Pressure 9 am	Pressure 3 pm
Mean	1019.7	1018.3
Standard deviation	4.5477	4.1884

    i. Use the equation of the least squares line and the information in Table 4 to show that the correlation coefficient for this data, rounded to three decimal places, is $r = 0.966$ **(1 mark)**

    ii. What percentage of the variation in *pressure 3 pm* is explained by the variation in *pressure 9 am*? Round your answer to one decimal place. **(1 mark)**

f. The residual plot associated with the least squares line is shown below.

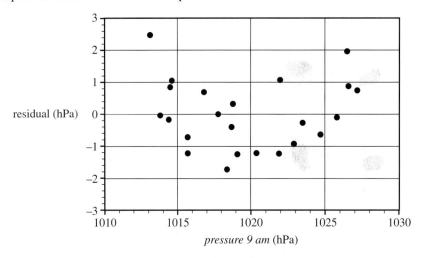

pressure 9 am (hPa)

    i. The residual plot above can be used to test one of the assumptions about the nature of the association between the atmospheric pressure at 3 pm and the atmospheric pressure at 9 am. What is this assumption? **(1 mark)**

    ii. The residual plot above does not support this assumption. Explain why. **(1 mark)**

More exam questions are available online.

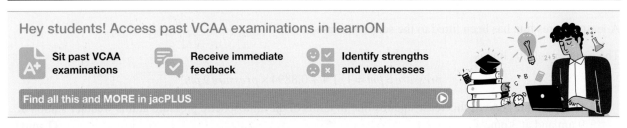

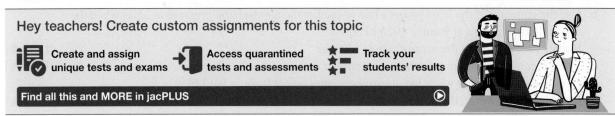

Answers

Topic 3 Investigating and modelling linear associations

3.2 Least squares line of best fit

3.2 Exercise

1. $y = -37.57 + 1.87x$ or air conditioner sales
   $= -37.57 + 1.87 \times$ (temperature), $r^2 = 0.97$, strong linear association

2. $y = 2.11 + 2.64x$ or number of dialysis patients
   $= 2.11 + 2.64 \times$ (month), $r^2 = 0.95$, strong linear association

3. $y = 4.57 + 1.04x$

4. $y = -1.691 - 3.72x$

5. $y = 9.06 + 1.20x$

6. a. Explanatory variable = height of netball players
   b. $b = 0.96$
   c. $a = 22.64$

7. $b = -0.781\,67$, $a = 13.94$, $y = 13.94 - 0.78x$

8. a. $y = 91.78 + 3.33x$ b. $y = 21.87 - 0.15x$
   c. $y = 8890.48 + 52.38x$ d. $y = 30 - x$

9. a. The Mathematics exam mark
   b. 0.95
   c. -6.07
   d. 75%

10. The least-squares regression equations are exactly the same as obtained in questions 3, 4 and 5.

11. a. $y = 22 - 2x$ b. A 'perfect' fit   c. $y = 11 - 0.5x$

12. E

13. a. $y = -164.7 + 0.119x$
    b.

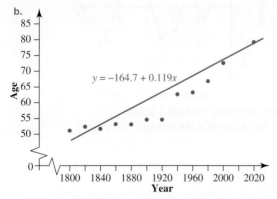

    c. The data definitely are not linear; there are big increases between 1900 and 1940 and between 1960 and 2020.

14. a. Duration of call is the explanatory variable.
    b. Cost of call $(\$) = \$4.27 + \$0.002\,57 \times$ duration of call (sec)

c.

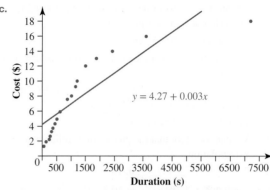

The line does not fit closely for all data points. The equation is not reliable due to outliers. If you eliminate the last two calls, then there is a direct relationship.

d. $r^2 = 0.73$. Therefore, the linear association is only moderate.

15. D

16. a. $y = 8 + 4x$, perfect fit, but meaningless
    b. $y = 10 + 2.5x$, good fit, but almost meaningless
    c. $y = 9.5 + 2.8x$, $y = 8.9 + 3.1x$, $y = 8.4 + 3.3x$, good fit
    d. The answers appear to be converging towards a 'correct' line.

3.2 Exam questions

Note: Mark allocations are available with the fully worked solutions online.

1. E
2. E
3. D

3.3 Interpretation, interpolation and extrapolation

3.3 Exercise

1. a. $y = 12.13 + 3.49x$ or monkey height (cm)
      $= 12.13 + 3.49 \times$ (month from birth)
   b. 3.49 cm/month
   c. 12.13 cm

2. a. $y = 13.56 + 2.83x$ or temperature (°C)
      $= 13.56 + 2.83 \times$ (time after 6 a.m.)
   b. 2.83°C/hr
   c. 13.56°C

3. a. Red blood cell count $= 157.3 + 14 \times$ day of experiment
   b. 14 cells per day
   c. 157

4. a. 48.5, or 49 people per extra animal
   b. 31.8, or 32 visitors

5. Sale volume $= 464 - 1.72 \times$ selling price, $r = -0.98$. Gradient shows a drop of 1720 sales for every \$1 increase in the price of the item. Clearly, the y-intercept is nonsensical in this case since an item is not going to be sold for \$0! This is a case where extrapolation of the line makes no sense.

6. a. $y = 56.03 + 9.35x$ or height (cm) $= 56.03 + 9.35 \times$
   (age) Height (cm) $= 149.53$ cm

   b. Height (cm) $= 205.63$ cm

   c. We cannot claim that the prediction is reliable, as it uses extrapolation.

7. a. $y = 40.10 + 6.54x$ or length (cm) $= 40.10 + 6.54 \times$
   (months)Length (cm) $= 138.20$ cm

   b. Length (cm) $= 197.06$ cm

   c. We cannot claim that the prediction is reliable, as it uses extrapolation.

8. a. 7   b. 18

9. a. $y = 0.286 + 3.381x$ b. 10.4
   c. 40.9   d. 1.99
   e. 7.31   f. c

10. a. Factory 1: cost $= 43.21 + 1.51 \times$ distance
    Factory 2: cost $= 56.61 + 0.964 \times$ distance

    b. Factory 1 is cheaper at \$43.21 (compared to Factory 2 at \$56.61).

    c. Factory 2 is cheaper at \$167.47 (compared to Factory 1 at \$216.86).

    d. Factory 2 is marginally more linear (Factory 1: $r^2 = 0.9332$; Factory 2: $r^2 = 0.9763$).

11. a. \$2152   b. 180   c. \$78 200
    d. \$600   e. **a, b** only

12.

IQ	Test result (%)
80	56
87	60
92	68
102	65
105	73
106	74
107	71
111	73
115	80
121	92

13. a. \$55 000   b. \$58 600
    c. \$64 000   d. About 13 years
    e. Please see the Worked solutions.

14. a. $r = 11.73 + 5.35q$ b. $r = 33.13$
    c. $r = 108.03$ d. $q = 16.50$
    e. **c** and **d**

15. a. $C = \$420$ b. $n = 13.38$ hours
    c. $C = \$940$ d. Call out fee $= \$180$
    e. **b** and **c**

16. a. $y = 42\,956 + 14\,795 \times$ (millions spent on advertising)
    Members $= 42\,956 + 14\,795 \times$ (millions spent on advertising)

    b. 72 546 members
    Interpolation

    c. \$1.83 million

    d. $r^2 = 0.90$, strong linear relationship

3.3 Exam questions

Note: Mark allocations are available with the fully worked solutions online.

1. C

2. a. The male life expectancy increases by 0.88 years with each year of increase in the female life expectancy.

   b. 34.4 years

   c. About 95% of the variability in the male life expectancy is explained by the variability of the female life expectancy using the linear least squares model.

3. B

3.4 Residual analysis

3.4 Exercise

1. a. Using CAS: $y = -0.03 + 10.85x$
   See the table at the bottom of the page.*

   b.

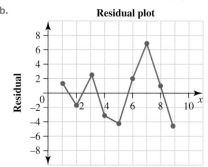

Residual plot

   c. $r^2 = 0.98$, consistent with a linear relationship

2. a. Using CAS: $y = 0.69 + 9.82x$
   See the table at the bottom of the page.*

---

*1. a.

Predicted y	10.82	21.67	32.52	43.37	54.22	65.07	75.92	86.77	97.62
Residual ($y - y_{predicted}$)	1.18	−1.67	32.52	−3.37	−4.22	1.93	7.08	1.23	−4.62
x	1	2	3	4	5	6	7	8	9
y	12	20	35	40	50	67	83	88	93

*2. a.

Predicted y	49.79	69.43	98.89	118.53	147.99	177.45	246.19	295.29	393.49
Residual ($y - y_{predicted}$)	−4.79	−8.43	−9.39	3.47	13.01	−0.45	−3.19	37.71	−27.49
x	5	7	10	12	15	18	25	30	40
y	45	61	89	122	161	177	243	333	366

b.

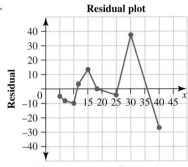

Residual plot

c. $r^2 = 0.98$, consistent with a linear relationship

3. a. $r^2 = 0.98$, consistent with a linear relationship

x	y	y_{pred}	Residual
1	1	5.1	-4.1
2	9.7	7.46	2.24
3	12.7	9.82	2.88
4	13.7	12.18	1.52
5	14.4	14.54	-0.14
6	14.5	16.9	-2.4

b. By examining the original scatterplot and residual plot, data are clearly not linear.

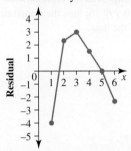

4. D

5. a, b

Day	Defective rate (%)	y_{pred}	Residual
2	15	13.96	1.04
4	10	11.58	-1.58
5	12	10.39	1.61
7	4	8.01	-4.01
8	9	6.82	2.18
9	7	5.63	1.37
10	3	4.44	-1.44
11	4	3.25	0.75

c.

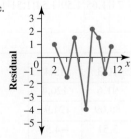

No apparent pattern in the residuals — likely to be linear

d. 15.15

e. 13.7 days. Unlikely that extrapolation that far from data points is accurate. Unlikely that there would be 0% defectives.

6. a, b

Day	Bookings in hotel	y_{pred}	Residual
1	158	138.8	19.2
2	124	116.3	7.7
3	74	93.8	-19.8
4	56	71.3	-15.3
5	31	48.8	-17.8
6	35	26.3	8.7
7	22	3.8	18.2

c.

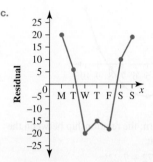

Slight pattern in residuals — may not be linear

d. Decline in occupancy likely due to drought — an atypical event

7. C

8. a. Non-linear   **b.** Linear   **c.** Non-linear

9. a. $y = -57.73 + 37.93x$

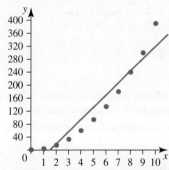

b. $r = 0.958$. This means that there is a strong positive relationship between variables x and y.

c. 0.9177; therefore, 91.8% of the variation in y can be explained by the variation in x.

d.

x	Residual
0	58.7
1	23.8
2	−3.1
3	−23.1
4	−34.0
5	−37.9
6	−35.8
7	−27.8
8	−5.7
9	16.4
10	68.5

e.

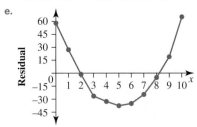

There is a clear pattern; the relationship between the variables is non-linear.

10. a. Using CAS: $D = 13.72 + 7.22k$

See the table at the bottom of the page.*

b.

Residual plot

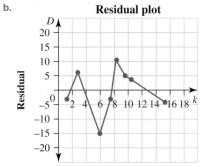

Since the data randomly jump from above to below the k-axis, the data probably have a linear relationship.

c. $r^2 = 0.94$; 94% of variation in D can be explained by variation in k.

3.4 Exam questions

Note: Mark allocations are available with the fully worked solutions online.

1. B

2. B

3. a. Morning median 52%, Evening median 56%

b. Evening congestion level

c. 63.8%

d. −1.8%

e. 85%

3.5 Transforming to linearity

3.5 Exercise

1. a.

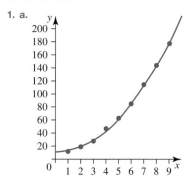

Apply an x^2 transformation to stretch the x-axis.

b.

x^2	1	4	9	16	25	36	49	64	81
y	12	19	29	47	63	85	114	144	178

$y = 11.14 + 2.07x^2$ with $r = 0.999\,76$

This transformation has improved the correlation coefficient from 0.97 to 0.999 76; thus the transformed equation is a better fit of the data.

2. a.

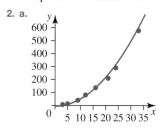

Apply an x^2 transformation to stretch the x-axis.

b.

x^2	9	25	81	144	256	441	576	1089
y	5	12	38	75	132	209	291	578

$y = -4.86 + 0.53x^2$ with $r = 0.9990$

This transformation has improved the correlation coefficient from 0.96 to 0.9990, thus the transformed equation is a better fit of the data.

3. $y = 128.15 - 2.62x_T$ where $x_T = x^2$, $r = -0.97$, which shows some improvement.

4.

Time (sec)	1	2	3	4	5	6	7	8
Log_{10} (speed) (m s^{-1})	1.95	1.85	1.74	1.65	1.59	1.54	1.51	1.48

---

*10. a.

k	1.6	5.9	178	7.7	8.1	9.7	10.3	15.4
D	22.5	37.8	41.5	66.9	82.5	88.7	91.6	120.4
$D_{\text{predicted}}$	25.27	31.77	56.32	69.31	72.20	83.75	88.09	124.91
Residual ($D - D_{\text{predicted}}$)	−2.77	6.03	−14.82	−2.41	10.30	4.95	3.51	−4.51

$\log_{10}(\text{speed}) = 1.97 - 0.07 \times \text{time with } r = -0.97$
Therefore, applying a logarithmic transformation improved the correlation coefficient.

5.

x	5	10	15	20	25	30	35	40
$\log_{10}(y)$	3	2.70	2.35	2.17	2.00	1.85	1.77	1.75

$\log_{10} y = 3.01 - 0.04x$ with $r = -0.96$
Therefore, applying a logarithmic transformation resulted in a significant improvement of the correlation coefficient.

6. a.

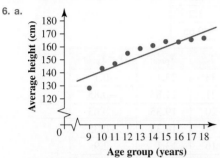

b.

log (age group)	Average height (cm)
0.954	128
1	144
1.041	148
1.079	154
1.114	158
1.146	161
1.176	165
1.204	164
1.230	166
1.255	167

c. $y = 24.21 + 117.2x_T$ where $x_T = \log_{10}(x)$, $r = 0.95$, most non-linearity removed.

7. a. i. 123.3 cm   ii. 143.9 cm   iii. 176.7 cm

b. a ii

8. Normal growth is linear only within given range; eventually the student stops growing. Thus, logarithmic transformation is a big improvement over the original regression.

9. a. See the table at the bottom of the page.*

$y = 3.85 + 29.87x_T$, where $x_T = \dfrac{1}{x}$

or

$\text{Temperature} = 3.85 + \dfrac{29.87}{\text{Time after 6 pm}}$

b. Temperature = 10.49 °C

10. a.

$\dfrac{1}{x}$	$\dfrac{1}{2}$	$\dfrac{1}{5}$	$\dfrac{1}{7}$	$\dfrac{1}{9}$	$\dfrac{1}{10}$	$\dfrac{1}{13}$	$\dfrac{1}{15}$	$\dfrac{1}{18}$
y	120	50	33	15	9	5	2	1

$y = -13.51 + 273.78x_T$, where $x_T = \dfrac{1}{x}$

or

$y = -13.51 + \dfrac{273.78}{x}$

b. $y = 9.31$

c. $x = 8.17$

11. a. $y = 2.572 + 90.867x_T$, where $x_T = \dfrac{1}{x}$ $(r = 0.9788)$

b. The intensity is 7.1 candlepower.

12. a. Compress the y- or x-values using logs or reciprocals.

b. Stretch the y-values using y^2 or compress the x-values using logs or reciprocals.

c. Compress the y- or x-values using logs or reciprocals.

13. a. -11.25 b. -11.25

14. a. -24.55

b. Cannot take the log of a negative number.

15. a. 3.39   b. 0.34

16. a. -0.08 b. -0.08

17. a.

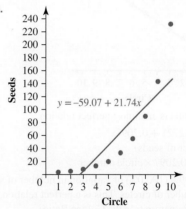

b. $r = 0.87$, which means it is a strong and positive relationship.

c. 180

d.

Circle	Seeds	Residual
1	3	40.33
2	5	20.59
3	8	1.85
4	13	-14.89
5	21	-28.63
6	34	-37.37
7	55	-38.11
8	89	-25.84
9	144	7.41
10	233	74.67

*9. a.

$\dfrac{1}{\text{Time after 6 pm (h)}}$	1	$\dfrac{1}{2}$	$\dfrac{1}{3}$	$\dfrac{1}{4}$	$\dfrac{1}{5}$	$\dfrac{1}{6}$	$\dfrac{1}{7}$	$\dfrac{1}{8}$
Temperature (°C)	32	22	16	11	9	8	7	7

e.

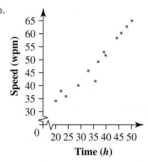

Circle

No, the relationship is not linear.

f. i. We can stretch the x-values towards linearity by using an x^2 transformation.

ii. We can compress the y-values towards linearity by using either a $\log_{10}(y)$ or a $\frac{1}{y}$ transformation.

18. a.

(graph: Log seeds vs Circle, y-axis from 0.6 to 2.4, x-axis 1 to 10)

Circle

b. $r = 0.9999$, this is an almost perfect relation.

c. $\log_{10}(y) = 0.2721 + 0.2097x$
$\log_{10}$ (number of seeds)
$= 0.2721 + 0.2097 \times$ circle number

d. 0.9999, 99.99% (100.0%) of variation in number of seeds is due to number of circles. This is a perfect relation, often found in nature (see the Golden Ratio).

e. 378

f. This is a much better prediction as it follows a steep upward trend.

3.5 Exam questions

Note: Mark allocations are available with the fully worked solutions online.

1. C

2. A

3. E

3.6. Review

3.6 Exercise

Multiple choice

1. C

2. B

3. D

4. E

5. C

6. C

Short answer

7. a. -0.45 b. 16.75

8. a. $y = 24.35 - 1.25x$, $r = -0.96$

b.

x	y_{pred}
3	20.6
5	18.1
7	15.6
9	13.1
11	10.6
13	8.1
15	5.6
17	3.2
20	-0.65

c.

x	y	y_{pred}	$y - y_{pred}$
1	23	23.1	-0.1
2	21	21.85	-0.85
4	20	19.35	0.65
8	14	14.35	-0.35
9	16	13.1	2.9
10	9	11.85	-2.85
11	12	10.6	1.4
15	5	5.6	-0.6

Extended response

9. a. Hours spent touch-typing — explanatory, speed of touch-typing — response.

b.

(graph: Speed (wpm) vs Time (h), y-axis 30 to 65, x-axis 20 to 50)

Time (h)

c. Strong, positive, linear relationship between the two variables

10. a. Likely to be a y versus x^2 relationship

b. A poor predictor for most values of x

c. 128; caution should always be taken when predicting outside the data set.

d.

Day	1	4	9	16	25	36	49	64
Units sold	1	2	4	9	20	44	84	124

e. $y = -13.86 + 1.96x_T$, where $x_T = x^2$

f. 182

11. a. $y = 4.27 + 1.55x$

b. 22.87

c. Simplicity of eye fitting versus accuracy in this case is quite good. Little difference in the sum of squared errors. Least squares regression gives quite a different answer from the other two methods, with consequent change in errors. (The three-median method is subject to errors due to outliers and computational errors.)

d. Not very linear, logarithmic transformation suggested

e.

$\log_{10}$(month)	Production (tonnes)
0	3
0.301	8
0.477	10.8
0.602	12
0.699	11.6
0.778	14
0.845	15.5
0.903	15
0.954	18.1

f. $y = 3.30 + 14.08x_T$, where $x_T = \log_{10}(x)$

g. Square of residual error is reduced, correlation is closer to 1, graph looks more linear

h. Prediction of 18.49 against 22.87 using untransformed data. Given the nature of the data, likely to be more accurate.

12. a. On average, the test mark increases by 8.6% with every extra hour spent on studying for the test.
   Any student who did not study for the test can expect to get 21.6%.

   b. 73%; this prediction is an example of interpolation, as 6 hours is within the original set of data.

   c. 99%; this prediction is an example of extrapolation, as 9 hours is outside the original set of data. Therefore, it is probably not very reliable.

3.6 Exam questions

Note: Mark allocations are available with the fully worked solutions online.

1. a.i. 24   ii. 1.065
   b.i. Weight   ii. Slope = −0.001 12
   c. 29%

2. a. The line needs to go through approximately (60, 1.104) and(130, 0.998).
   See the figure at the bottom of the page.*

   b. 1.10 kg/L

   c. Extrapolating

d. The body density decreases by 0.001 512 kg/L for every cm increase in waist measurement.

e. −0.02

f. $r = -0.824$

g. Yes, as the residual plot is random and centred at residual equal to zero. It has no defined pattern.

3. a. 10.4 cm

   b. Strong and negative

   c. *mean height* = 171 + (−0.169) × *mean age*

   d. *mean height* = 167.9 + (−0.001 621) × (*mean age*)²

4. a. *humidity 9 am*

   b. *humidity 3 pm* = −26.1 + 0.765 × *humidity 9 am*

   c. $r = 0.871$

5. a. Pressure at 3 pm increased by 0.8894 hPa for every 1 hPa increase in pressure at 9 am.

   b. 1023

   c. Interpolation

   d. 3 hPa

   e.i. 0.966   ii. 93.3%

   f. i. Assumption: There is a linear relationship between the atmospheric pressure at 3 pm and the atmospheric pressure at 9 am.

      ii. The residual plot is not random enough to support the assumption.

*2. a.

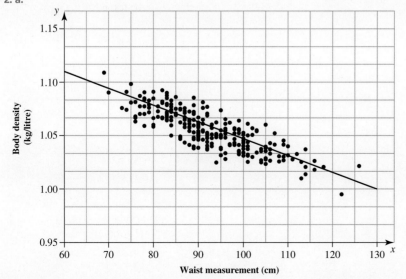

Waist measurement (cm)

4 Investigating and modelling time series data

4.1 Overview

4.1.1 Introduction

Those who study time series data are interested in trends and patterns in data. These may be short- or long-term trends and patterns. Climate scientists at NASA have used time series data and other technologies to determine that the Earth's average surface temperature has increased by 1.1 degrees Celsius since the late nineteenth century with the 10 warmest years on record occurring since 2005.

Modelling using time series data can be used to forecast and monitor predicted sales figures, population growth, mortality rates and long-term health outcomes, endangered species, weather patterns, food consumption and production, economic growth, interest rate fluctuations, and future consumer and voting trends. This information can be used to improve people's lives and ensure that our planet continues to thrive. The more information that governments, political lobby groups, business and other organisations can obtain, the more the information can be used in both positive and negative ways. It is important that individuals are aware and informed regarding the mining of their personal data and the ways in which it can be used to manipulate choices and influence patterns of behaviour. Every action that connects one to a website leaves a digital footprint. This footprint can be used to track an individual's political ideology, friends, consumer habits, credit rating, personal hobbies, location and interests. This may have a significant impact on an individual's ability to obtain employment, buy a house or a car, have a relationship or travel overseas.

KEY CONCEPTS

This topic covers the following key concepts from the VCE Mathematics Study Design:
- qualitative features of time series plots; recognition of features such as trend (long-term direction), seasonality (systematic, calendar related movements) and irregular fluctuations (unsystematic, short-term fluctuations); possible outliers and their sources, including one-off real-world events, and signs of structural change such as a discontinuity in the time series
- numerical smoothing of time series data using moving means with consideration of the number of terms required (using centring when appropriate) to help identify trends in time series plots with large fluctuations
- graphical smoothing of time series plots using moving medians (involving an odd number of points only) to help identify long-term trends in time series with large fluctuations
- seasonal adjustment including the use and interpretation of seasonal indices and their calculation using seasonal and yearly means
- line to a time series with time as the explanatory variable (data de-seasonalised where necessary), and the use of the model to make forecasts (with re-seasonalisation where necessary) including consideration of the possible limitations of fitting a linear model and the limitations of extending into the future.

Source: VCE Mathematics Study Design (2023–2027) extracts © VCAA; reproduced by permission.

4.2 Time series plots and trends

4.2.1 Types of time series

In this topic, we consider cases where the *x*-variable is time. Time goes up in even increments such as hours, days, weeks or years. In these cases, we have what is called a **time series**. The main purpose of a time series is to see how some quantity varies with time.

For example, a company may wish to record its daily sales figures over a 10-day period.

Time	Day 1	Day 2	Day 3	Day 4	Day 5	Day 6	Day 7	Day 8	Day 9	Day 10
Sales ($)	5200	5600	6100	6200	7000	7100	7500	7700	7700	8000

We could also make a graph of this time series as shown.

As can be seen from this graph, there seems to be a trend upwards — clearly, this company is increasing its revenues.

In time series data, trend, seasonal, cyclic and irregular fluctuations are important features to be observed.

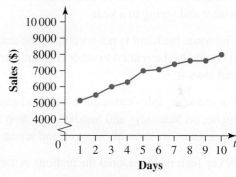

Trend

When there is long-term increase or decrease, it is said to be a **trend**. Trend is not always linear.

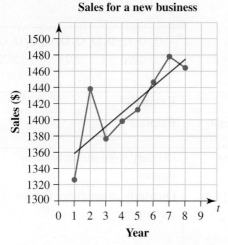

The graph shows an increasing trend.

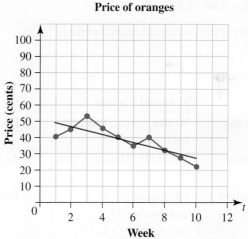

The graph shows a decreasing trend.

Share price of a new company

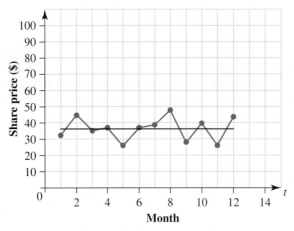

Month

The graph shows no trend.

Seasonal fluctuations

Certain data seem to fluctuate during the year, as the seasons change; this is termed a **seasonal time series**. The most obvious example of this would be total rainfall during summer, autumn, winter and spring in a year.

The name *seasonal* is not specific to the seasons of a year. It could also be related to other constant periods of highs and lows.

For example, sales figures at a fast-food store could be consistently higher on Saturdays and Sundays and drop off on weekdays. Here the seasons are days of the week and repeat once every week.

A key feature of seasonal fluctuations is that the seasons occur at the same time each cycle.

Here are some common seasonal periods.

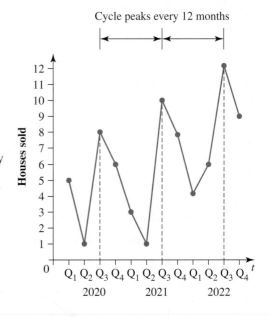

	Seasonal periods	**Cycle**	**Example**
Seasons	Winter, spring, summer, autumn	Four seasons in a *year*	Rainfall
Months	Jan., Feb., Mar., …, Nov., Dec.	12 months in a *year*	Grocery store monthly sales figures
Quarters	1st quarter (Q_1), 2nd quarter (Q_2), 3rd quarter (Q_3), 4th quarter (Q_4)	Four quarters in a *year*	Quarterly expenditure figures of a company
Week	Monday to Friday	Five days in a *week*	Daily sales for a store open from Monday to Friday only
Days	Monday, Tuesday, Wednesday, Thursday, Friday, Saturday, Sunday	Seven days in a *week*	Number of hamburgers sold at a takeaway store daily

Cyclic fluctuations

Like seasonal time series, **cyclic time series** show fluctuations upwards and downwards, but not according to season. Businesses often have cycles where at times profits increase, then decline, then increase again.

For example, consider the sales of a new major software product. At first, sales are slow, then they pick up as the product becomes popular. When enough people have bought the product, sales may fall off until a new version of the product comes on the market, causing sales to increase again.

This cycle can be repeated many times, which is why there are many versions of some software products.

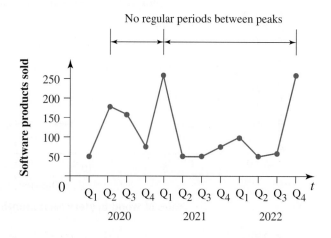

Irregular fluctuations

Irregular time series show fluctuations that occur at random. This can be caused by external events such as floods, wars, new technologies or inventions, or anything that results from random causes.

There is no obvious way to predict the direction of the time series or even when it changes direction.

In the graph shown, there are a couple of minor fluctuations at $t = 4$ and $t = 8$, and a major one at $t = 12$. The major fluctuation could have been caused by a change in government that positively affected profits.

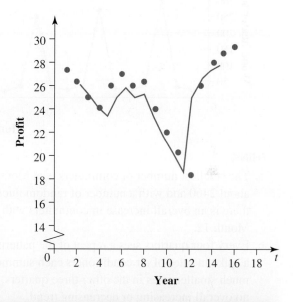

WORKED EXAMPLE 1 Describing the pattern in the time series

Describe the pattern in each of the following time series graphs.

a.

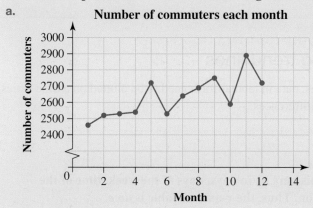

b.

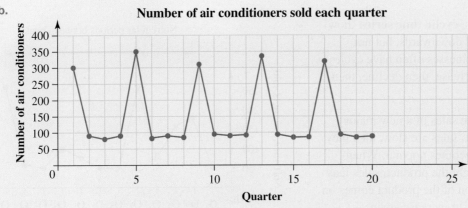

Number of air conditioners sold each quarter

c.

Sales of wood heaters each month from January 2018

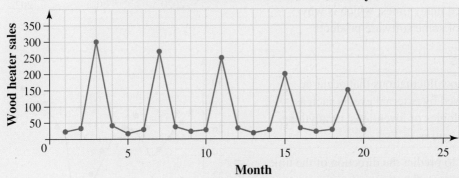

THINK	WRITE
a. The smallest number of commuters is at Month 1 at about 2460 and with a number of random fluctuations there is an overall increase in commuters with 2710 for Month 12.	The time series graph displays an increasing trend with random fluctuations.
b. Every four quarters sees a repeat of the pattern, with a spike in sales of air conditioners each summer and much smaller sales in the other three quarters. There is no overall increasing or decreasing trend.	The time series graph displays seasonality only with no obvious trend.
c. Every four quarters sees a repeat of the pattern, with a spike in sales of wood heaters each winter but the spike is declining in each consecutive year.	The time series graph displays seasonality with an overall downward trend.

4.2.2 Plotting time series and fitting trend lines

To make better judgements about the type of time series, data in tabular form need to be plotted on a time series plot. This is similar to a scatterplot with some notable differences.

Plot data on a time series plot

1. **The explanatory variable is always *time*. This may be in days, days of the week, time of the day, weeks, months, quarters, years and so on. Thus, the *x*-axis variable is *time*.**
2. **As the periods are often labels and not numerical, the *x*-axis may be scaled using these period labels.**
3. **The points are connected. As they occur in chronological order, joining the points assists in identifying the type of time series and if a trend exists.**

To enter the data into CAS, time periods that are labels (and not numerical) need to be converted to numerals. For this, an **association table** is needed. An association table summarises how the time periods are converted to numerical values. The first point is converted to 1, the second to 2 and so on, until the series is fully converted.

Here are two examples.

Example 1

Week 1 Mon.	Week 1 Tues.	Week 1 Wed.	Week 1 Thurs.	Week 1 Fri.	Week 1 Sat.	Week 1 Sun.	Week 2 Mon.	Week 2 Tues.
1	2	3	4	5	6	7	8	9

Example 2

Jan. 2022	Feb. 2022	Mar. 2022	Apr. 2022	May 2022	June 2022	July 2022	Aug. 2022
1	2	3	4	5	6	7	8

After we have plotted a time series graph, if there is a noticeable trend (upward, downward or flat) we can add a trend line, in the form of a straight line, to the data.

The trend lines we plot in this topic will be made using the least squares method we covered in Topic 3. While these trend lines can be calculated manually, it is expected that you will use CAS to hasten this process.

tlvd-3574

WORKED EXAMPLE 2 Plotting time series data

The following table displays the school fees collected over a 10-week period.

Week beginning	8 Jan.	15 Jan.	22 Jan.	29 Jan.	5 Feb.	22 Feb.	19 Feb.	26 Feb.	5 Mar.	12 Mar.
$ × 1000	1.5	2.5	14.0	4.5	13.0	4.5	8.5	0.5	5.0	1.0

Plot the data and decide on the type of time series pattern. If there is a trend, fit a straight line.

THINK

1. Set up an association table. Add another row and enter the numerical time code for each of the 10 weeks starting at 1, 2, …, through to 10.

2. Construct a plot of the data. Place weeks on the horizontal axis and school fees on the vertical axis.

WRITE

Week beginning	8 Jan.	15 Jan.	22 Jan.	29 Jan.	5 Feb.	12 Feb.	19 Feb.	26 Feb.	5 Mar.	12 Mar.
Time code	1	2	3	4	5	6	7	8	9	10
$ × 1000	1.5	2.5	14.0	4.5	13.0	4.5	8.5	0.5	5.0	1.0

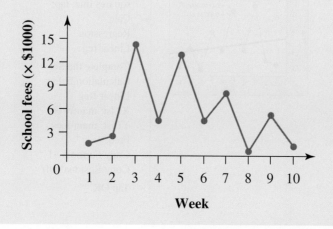

3. Identify the pattern as either seasonal, cyclical or irregular. If there is a trend, determine whether it is upwards or downwards.	The school fees can be classified as cyclical or irregular with a downward trend. This is evident by the reducing totals in school fees collected.
4. Use CAS to fit a least squares line.	$y = 7.2 - 0.31x$, where y represents school fees in thousands of dollars and $x = 1$ corresponds to the week beginning on 8 January.

TI	THINK	DISPLAY/WRITE	CASIO	THINK	DISPLAY/WRITE
1. On a Lists & Spreadsheet page, enter the data into the lists, named *week* and *fees*.	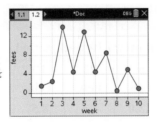	1. On a Statistics screen, enter the data into the lists, named *week* and *fees*.	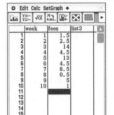
2. Press CTRL + I and select: 5: Add Data & Statistics Use the tab key to move to each axis and press the click key or ENTER to place *week* on the horizontal axis and *fees* on the vertical axis. To join the dots, press MENU and then select: 2: Plot Properties 1: Connect Data Points	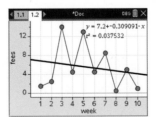	2. Tap: SetGraph Settings Complete the Set StatGraphs fields as: Draw: On Type: *xy* Line XList: main\week YList: main\fees Freq: 1 Mark: Square Tap Set and then the graph icon.	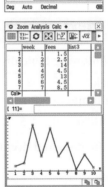
3. The answer appears on the screen.	The school fees can be classified as cyclical or irregular with a downward trend. This is evident by the reducing totals in school fees collected.	3. The answer appears on the screen.	The school fees can be classified as cyclical or irregular with a downward trend. This is evident by the reducing totals in school fees collected.
4. To fit a least squares line, press MENU and select: 4: Analyze 6: Regression 2: Show Linear $(a + bx)$	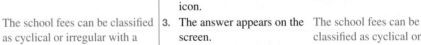	4. To determine the least squares line, tap: Calc Regression Linear reg Complete the Set Calculation fields: Linear Reg XList: main\week YList: main\fees Freq: 1 Copy Formula y1 Copy Residual Off Tap OK.	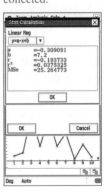

5. The answer appears on the screen.	$y = 7.2 - 0.31x$, where y represents school fees in thousands of dollars and $x = 1$ corresponds to the week beginning on 8 January.	5. The answer appears on the screen.	$y = 7.2 - 0.31x$, where y represents school fees in thousands of dollars and $x = 1$ corresponds to the week beginning on 8 January.

4.2 Exercise

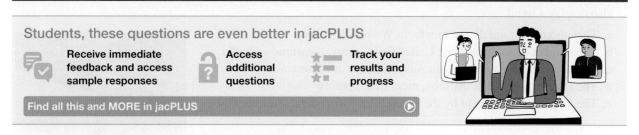
1. **WE1** Describe the pattern in each of the following time series graphs.

a.

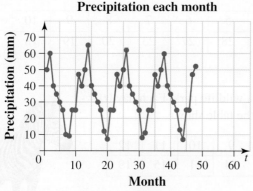

b.

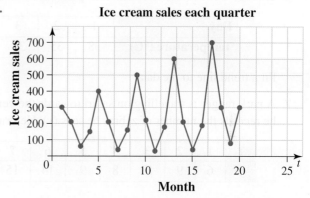

c.

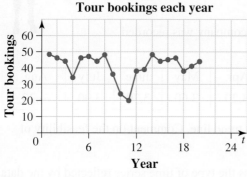

2. **MC** Select which of the following statements is false.

   A. For a time series plot with a seasonal trend, the seasons occur at the same time each cycle.
   B. A cyclic trend may not have regular periods between cycles.
   C. A time series plot with an upward trend cannot have irregular fluctuations.
   D. A cyclic trend is different from a seasonal trend.
   E. A time series plot must show an upward or downward trend.

3. **WE2** The following table shows the sales for the first eight years of a new business.

Year	1	2	3	4	5	6	7	8
Sales ($)	1326	1438	1376	1398	1412	1445	1477	1464

Plot the data and decide on the type of time series pattern. If there is a trend, fit a straight line.

4. Data were recorded about the number of families who moved from Melbourne to Ballarat over the last 10 years.

Year	2013	2014	2015	2016	2017	2018	2019	2020	2021	2022
Number moved	97	118	125	106	144	155	162	140	158	170

Plot the data and decide on the type of time series pattern. If there is a trend, fit a straight line.

5. For the following time series, identify whether they are likely to be seasonal, cyclic or irregular and if they also display a trend.

a. The amount of rainfall, per month, in Western Victoria
b. The number of soldiers in the United States army, measured annually
c. The number of people living in Australia, measured annually
d. The share price of BHP Billiton, measured monthly
e. The number of seats held by the Liberal Party in Federal Parliament

6. This table shows the maximum temperature in Victoria over a 10-day period.

Day	1	2	3	4	5	6	7	8	9	10
Temperature (°C)	38	35	34	30	28	27	23	20	19	18

Fit a trend line to the data.

7. A park ranger is travelling on safari towards the centre of a wildlife park. Each day (t), he records the number of zebras he sees (y). He draws up the table shown.

t	1	2	3	4	5	6	7	8	9	10	11	12
y	6	9	13	8	9	14	15	17	14	11	15	19

Fit a trend line to the data.

8. The monthly share price of a recently privatised telephone company was recorded as follows.

Month	Jan.	Feb.	Mar.	Apr.	May	June	July	Aug.
Price ($)	2.50	2.70	3.00	3.20	3.60	3.70	3.90	4.20

Graph the data (let 1 = Jan., 2 = Feb., ... and so on) and fit a trend line. Comment on the feasibility of predicting share prices for the following year.

9. Plot the following monthly sales data for umbrellas. Discuss the type of time series reflected by the data and the limitations of a trend line.

Month	Jan.	Feb.	Mar.	Apr.	May	June	July	Aug.	Sept.	Oct.	Nov.	Dec.
Sales	5	10	15	40	70	95	100	90	60	35	20	10

10. Consider the data in the table shown, which represent the price of oranges over a 19-week period.

Week	1	2	3	4	5	6	7	8	9	10	11	12	13	14	15	16	17	18	19
Price (cents)	40	45	53	46	40	45	62	58	67	60	72	60	64	78	74	66	78	81	80

a. Fit a straight trend line to the data.
b. Predict the price in week 25.

11. The following table represents the quarterly sales figures (in $000s) of a popular software product.

Quarter	Q_1-20	Q_2-20	Q_3-20	Q_4-20	Q_1-21	Q_2-21	Q_3-21	Q_4-21	Q_1-22	Q_2-22	Q_3-22	Q_4-22
Sales	120	135	150	145	140	120	100	110	120	140	190	220

Plot the data and fit a trend line. Discuss the type of time series best reflected by these data.

12. The number of employees at the Comnatpac Bank was recorded over a 10-month period.

Month	Mar.	Apr.	May	June	July	Aug.	Sept.	Oct.	Nov.	Dec.
Employees	6100	5700	5400	5200	4800	4400	4200	4000	3700	3300

Plot and fit a trend line to the data. Comment on the data trend.

4.2 Exam questions

▶ **Question 1 (1 mark)**

Source: VCE 2021, Further Mathematics Exam 1, Section A, Q12; © VCAA.

MC The time series plot below shows the quarterly *sales*, in thousands of dollars, of a small business for the years 2010 to 2020.

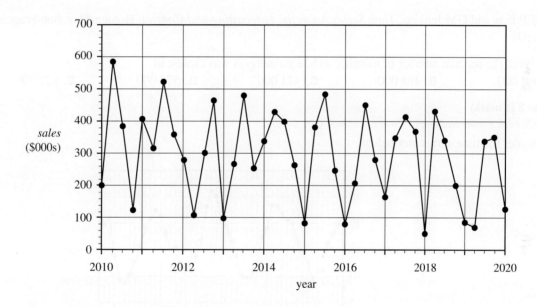

The time series plot is best described as having
  A. seasonality only.
  B. irregular fluctuations only.
  C. seasonality with irregular fluctuations.
  D. a decreasing trend with irregular fluctuations.
  E. a decreasing trend with seasonality and irregular fluctuations.

▷ **Question 2 (1 mark)**
Source: VCE 2020, *Further Mathematics Exam 1, Section A, Q19;* © VCAA.

`MC` The time series plot below displays the number of airline passengers, in thousands, each month during the period January to December 1960.

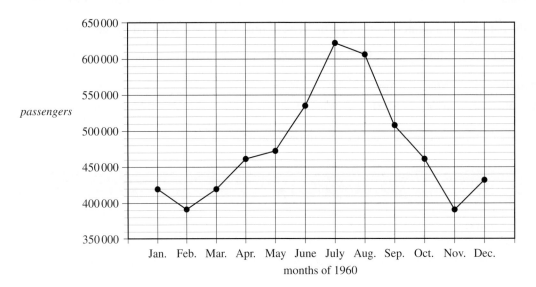

Data: GEP Box and GM Jenkins, *Time Series Analysis: Forecasting and Control*, Holden-Day, San Francisco, 1970, p. 531

During 1960, the median number of monthly airline *passengers* was closest to
    A. 461 000   **B.** 465 000   **C.** 471 000   **D.** 573 000   **E.** 621 000

▷ **Question 3 (1 mark)**
Source: VCE 2016, *Further Mathematics Exam 1, Section A, Q13;* © VCAA.

`MC` Consider the time series plot below.

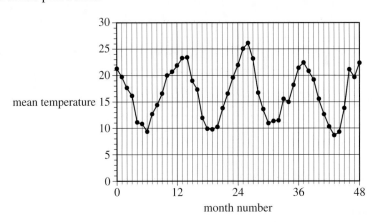

The pattern in the time series plot shown above is best described as having
    A. irregular fluctuations only.
    B. an increasing trend with irregular fluctuations.
    C. seasonality with irregular fluctuations.
    D. seasonality with an increasing trend and irregular fluctuations.
    E. seasonality with a decreasing trend and irregular fluctuations.

More exam questions are available online.

4.3 Fitting the least squares line and forecasting

4.3.1 Association tables and forecasting

As seen in section 4.2.2, an association table is often required to convert period labels to a numerical value so that a straight-line equation can be calculated. It is best to set up an extra row if data are in tabular form, or to change the labels shown on the axis of a time series plot to numerical values.

For **forecasting**, use the association table to devise a time code for any period in the future. This time code will then be used in the straight-line equation.

From the three examples, we can calculate the time codes as follows:

Example 1

Year	2019	2020	2021	2022
Time code	1	2	3	4

2026 would have a time code of 8.

Example 2

1st quarter 2021	1
2nd quarter 2021	2
3rd quarter 2021	3
4th quarter 2021	4
1st quarter 2022	5
2nd quarter 2022	6

1st quarter 2023 would have a time code of 9.

Example 3

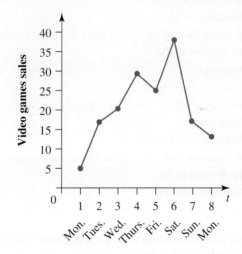

Monday week 4 would have a time code of 22.

WORKED EXAMPLE 3 Fitting a least squares line to time series data

A new hair salon has opened in a shopping centre. Customer numbers for its first days are shown in the following table.

Period	Week 1							Week 2		
	Mon.	Tues.	Wed.	Thurs.	Fri.	Sat.	Sun.	Mon.	Tues.	Wed.
Number of customers	9	9	11	13	16	18	19	20	23	27

Fit a straight line to the data set using the least squares line method.
Use the equation of the straight line to predict the number of customers for:

a. Monday week 4

b. Thursday week 2.

THINK	WRITE
1. Complete an association table, where Monday week 1 is 1, Tuesday week 1 is 2, Wednesday week 2 is 10.	(see table below)

	Week 1							Week 2		
Period	Mon.	Tues.	Wed.	Thurs.	Fri.	Sat.	Sun.	Mon.	Tues.	Wed.
Time code	1	2	3	4	5	6	7	8	9	10
Number of customers	9	9	11	13	16	18	19	20	23	27

2. Use a calculator to determine the equation of the least squares line of best fit.

$y = 5.67 + 1.97x$

Number of customers $= 5.67 + 1.97 \times$ time code, where time code 1 corresponds to Monday of week 1.

3. **a.** For Monday week 4, the time code is 22. Substitute $t = 22$ into the equation and evaluate. Round to the nearest integer.

a. Number of customers $= 5.67 + 1.97 \times$ time code
$= 5.67 + 1.97 \times 22$
$= 49.01$
Number of customers $= 49$

b. For Thursday week 2, the time code is 11. Substitute $t = 11$ into the equation and evaluate. Round to the nearest whole number.

b. Number of customers $= 5.67 + 1.97 \times$ time code
$= 5.67 + 1.97 \times 11$
$= 27.34$
Number of customers $= 27$

Note: Remember that forecasting is an extrapolation and if you go too far into the future, the prediction is not reliable, as the trend may change.

Once an equation has been determined for a time series, it can be used to analyse the situation.

For the period given in Worked example 3, the equation is:

$$Number\ of\ customers = 5.67 + 1.97 \times time\ code.$$

The y-intercept (5.67) has no real meaning, as it represents the time code of 0, which is the day before the opening of the salon. The gradient or rate of change is of more importance. It indicates that the number of customers is changing, in this instance growing by approximately 2 customers per day (gradient of 1.97).

tlvd-4976

WORKED EXAMPLE 4 Making predictions with time series data

The forecast equation for calculating the share price, *y*, of a sugar company was obtained from data of the share price over the past 5 years. The equation is $y = 1.56 + 0.42t$, where $t = 1$ represents the year 2021, $t = 2$ represents the year 2022 and so on.
a. Rewrite the equation putting it in the context of the question.
b. Interpret the values of the gradient and *y*-intercept.
c. Predict the share price in 2033.

THINK	WRITE
a. The *x*-variable represents the time codes and the *y*-variable represents the share price in dollars.	a. Share price = $1.56 + $0.42 × time code Time code $t = 1$ is 2021, $t = 2$ is 2022 and so on.
b. The *y*-intercept of 1.56 represents the starting value; that is, when $t = 0$. The gradient of 0.42 represents the rate of change in share price with respect to time. That is, it will grow as it has a positive gradient.	b. The *y*-intercept of $1.56 represents the approximate value of the shares in 2020. The gradient of $0.42 means that on average the share price will grow by $0.42 (42 cents) each year.
c. If $t = 1$ is 2021, then for 2033, the time code will be $t = 13$. Substitute into the equation given to predict the share price in 2033.	c. Share price = $1.56 + $0.42 × time code = $1.56 + $0.42 × 13 = $1.56 + $5.46 = $7.02

Before we fit a straight line to a data set using a least squares line of best fit, it is useful to draw a scatterplot.

Benefits of drawing a scatterplot before fitting a straight line

This is beneficial since it can:
1. demonstrate how close the points are to a straight line or if a curve is a better fit for the data
2. demonstrate if there are outliers in the data set that could affect the least squares line.

If there is an outlier and we are using the equation to make a prediction, then this can bias our prediction. Removing the outlier from the data would be more likely to give a better prediction since the least squares line will fit the data more closely.

WORKED EXAMPLE 5 Plotting the data and identifying outliers

The following table shows the sales for the first 8 years of a new business.

Year	1	2	3	4	5	6	7	8
Sales ($)	1326	1438	1376	1398	1412	1445	1477	1464

a. Plot the data and decide if there are any outliers.
b. Re-plot the data after removing any outliers. Explain a possible reason for the outlier.
c. If there is a trend, fit a straight line to the data using a least squares line method.
d. Compare the gradient and *y*-intercept to those of the following trend line:
sales = 1342.96 + 16.45 × year.

THINK	**WRITE**
a. Plot all the points on a scatterplot.	**a.** 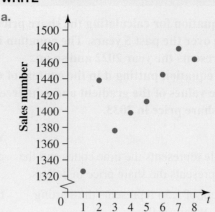
b. Looking at the scatterplot it appears that the second point is an outlier from a least squares line of best fit. Therefore, we remove this point and plot another scatterplot.	**b.** 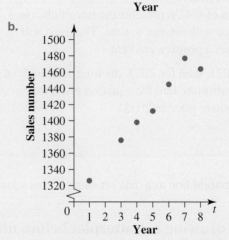

Since we don't know the year in the second year of business or the type of business, it is difficult to pinpoint a reason; however, they may have released a new line or developed something new to get this jump in sales. There could be many reasons for this outlier.

THINK	**WRITE**
c. This looks like an upward trend, so use CAS to complete a least squares line of best fit on the reduced data set.	**c.** The linear trend line is $y = 1309.35 + 21.55x$, where y is the sales in \$ and $x = 1$ corresponds to 1 year after the new business opening.
d. Compare the gradient and y-intercept to the trend line given, where the gradient is 16.45 and the y-intercept is 1342.96.	**d.** With the removal of the outlier, the gradient has increased and the y-intercept has decreased. This would now be an improved trend line to make predictions from.

1. **WE2** The number of people who watched the Channel 9 news over a fortnight is shown in the following table.

Day	Mon.	Tues.	Wed.	Thurs.	Fri.	Sat.	Sun.
Viewers (1 000 000s)	1.20	1.18	1.16	1.18	0.9	0.75	1.0

Day	Mon.	Tues.	Wed.	Thurs.	Fri.	Sat.	Sun.
Viewers (1 000 000s)	1.21	1.23	1.19	1.16	0.95	0.68	0.98

Fit a straight line to the data set using a least squares line method.
Use the equation of the straight line to predict the number of viewers for:

a. Wednesday week 3
b. Monday week 4.

2. Data were recorded on the number of road fatalities for a two-year period in Australia.

Month in year 1	01	02	03	04	05	06
Driver fatalities	115	117	132	153	130	131

Month in year 1	07	08	09	10	11	12
Driver fatalities	109	117	113	143	107	124

Month in year 2	01	02	03	04	05	06
Driver fatalities	126	98	104	115	134	111

Month in year 2	07	08	09	10	11	12
Driver fatalities	107	94	104	121	119	120

Fit a straight line to the data set using the least squares method and use the straight line to predict the number of fatalities in:

a. June year 7
b. April year 10.

3. The following table represents the number of cars remaining to be completed on an assembly line.

Time (hours)	1	2	3	4	5	6	7	8	9
Cars remaining	32	26	27	23	16	17	13	10	9

Fit a straight line to the data using the least squares line method.

a. Predict the number of cars remaining to be completed after 11 hours.
b. Calculate the rate at which the number of cars on the assembly line is being reduced.
c. From the equation of the trend line, it should be possible to predict when there are no cars left on the assembly line. This is done by determining the value of t that makes $y = 0$. Calculate the time when there will be no cars left on the assembly line.

4. When the MicroHard Company first started, it employed only one person. The company has consistently grown, so that after 12 months there are 14 employees. The time series data are shown by the graph.

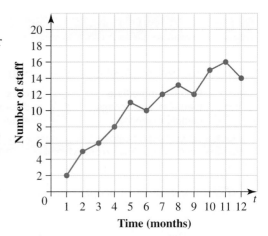

   a. Fit a straight line to the data using the least squares method.
   b. Predict the number of employees after a *further* 12 months.

5. The table below shows the share price of MicroHard during a volatile period in the stock market.

Day	1	2	3	4	5	6	7	8	9	10	11	12	13	14	15	16
Price ($)	2.75	3.30	3.15	2.25	2.10	1.80	1.50	2.70	4.10	4.20	3.55	1.65	2.60	2.95	3.25	3.70

   a. Using CAS, fit a least squares line of best fit.
   b. State the type of time series shown in the table.

6. The following time series shows the number of visitors to a website over a 9-month period.

Time (months)	1	2	3	4	5	6	7	8	9
Visitors (millions)	2.00	2.20	2.50	3.10	3.60	4.70	6.10	7.20	8.50

Plot the data and fit a line using the least squares line method. Comment on this line as a predictor of further growth.

7. **WE3** The forecast equation for calculating the share price, y, of a bank company was obtained from data of the share price over the past seven years. The equation is $y = 22.74 + 0.28t$, where $t = 1$ represents the year 2020.

   a. Rewrite the equation, putting it in the context of the question.
   b. Interpret the values of the gradient and y-intercept.
   c. Predict the share price in 2030.

8. The forecast equation for calculating the share price, y, of a mining company was obtained from data of the share price over the past four years. The equation is $y = 18.57 - 0.1t$, where $t = 1$ represents the year 2020.

   a. Rewrite the equation, putting it in the context of the question.
   b. Interpret the values of the gradient and y-intercept.
   c. Predict the share price in 2029.

9. The forecast equation for calculating price, y, of shares in a steel company was obtained from data of the share prices over the past six years. The equation is:

$$y = 2.56 + 0.72t$$

where $t = 1$ represents the year 2020, $t = 2$ represents the year 2021 and so on.

   a. Rewrite the equation, putting it in the context of the question.
   b. Interpret the values of the gradient and the y-intercept.
   c. Predict the share price in 2030.

10. The Teeny-Tiny-Tot Company has started to make prams. Its sales figures for the
    first eight months are given in the table shown.

Date	Jan.	Feb.	Mar.	Apr.	May	June	July	Aug.
Sales	65	95	130	115	145	170	190	220

    a. Using the sequence Jan. = 1, Feb. = 2, …, calculate the equation of the trend line
    using the least squares line of best fit method.

    b. Plot the data points and the trend line on the same set of axes.

    c. Use the trend line equation to predict the company's sales for December.

    d. Comment on the suitability of the trend line as a predictor of future sales,
    supporting your arguments with mathematical statements.

11. The sales figures of Harold Courtenay's latest novel (in thousands of units) are
    given in this table. The book was released a week before the first figures were
    collected.

Time (weeks)	1	2	3	4	5	6	7	8	9
Sales (× 1000)	1	3	5	17	21	25	28	27	26

    a. Calculate the equation of the trend line for the data using the least squares line of best fit method.

    b. Plot the data points and the trend line on the same set of axes.

    c. Use the trend line equation to predict the sales for weeks 10, 12 and 14.

    d. Comment on the suitability of the trend line as a predictor of future sales, supporting your arguments with
    mathematical statements.

12. The average quarterly price of coffee (per 100 kg) has been recorded for 3 years.

Quarter	Q_1-20	Q_2-20	Q_3-20	Q_4-20	Q_1-21	Q_2-21	Q_3-21	Q_4-21	Q_1-22	Q_2-22	Q_3-22	Q_4-22
Price ($)	358	323	316	336	369	333	328	351	389	387	393	402

    a. Calculate the equation of the trend line for the data using the least squares
    line of best fit method.

    b. Plot the data points and the trend line on the same set of axes.

    c. Use the trend line equation to predict the price for the next quarter.

    d. Comment on the suitability of the trend line as a predictor of future prices,
    supporting your arguments with mathematical statements.

13. A mathematics teacher gives her students a test each month for 10 months, and the class average is recorded.
    The tests are carefully designed to be of similar difficulty.

Test	Feb.	Mar	Apr.	May	June	July	Aug.	Sept.	Oct.	Nov
Mark (%)	57	63	62	67	65	68	70	72	74	77

    a. Calculate the equation of the trend line for the data using the least squares line of best fit method.

    b. Plot the data points and the trend line on the same set of axes.

    c. Use the trend line equation to predict the results for the last exam in December.

    d. Comment on the suitability of the trend line as a predictor of future marks, supporting your arguments
    with mathematical statements.

14. **WE4** The following table shows the All Ordinaries values from the Australian stock market over eight years. The All Ordinaries consists of the 500 largest eligible companies.

Year	1995	1996	1997	1998	1999	2000	2001	2002
All Ords	1925	2140	2435	2500	2820	3200	2950	3050

a. Plot the data and decide if there are any outliers. Complete a least squares line of best fit including all data points.
b. Re-plot the data after removing any outliers.
c. If there is a trend, fit a straight line to the data using a least squares method.
d. Compare the gradient and y-intercept of the two least squares lines.

15. The following table shows the stock price for a company in 2022.

Month 2022	Jan.	Feb.	Mar.	Apr.	May	June	July	Aug.	Sept.
Stock price ($)	192.06	204.62	235.00	261.09	256.88	251.53	257.25	243.10	283.75

a. Plot the data and decide if there are any outliers. Complete a least squares line of best fit including all data points.
b. Re-plot the data after removing any outliers.
c. If there is a trend, fit a straight line to the data using a least squares line method.
d. Compare the gradient and y-intercept of the two least squares lines.

16. The average cost of a hotel room in Sydney in 2019 is shown in the table.

Month 2019	Jan.	Feb.	Mar.	Apr.	May	June	July	Aug.	Sept.
Hotel price ($)	250	240	235	237	239	230	228	237	332

a. Plot the data and decide if there are any outliers. Complete a least squares line of best fit including all data points.
b. Re-plot the data after removing any outliers.
c. If there is a trend, fit a straight line to the data using a least squares line of best fit method.
d. Compare the gradient and y-intercept of the two least squares lines and give a possible explanation for the outlier.

4.3 Exam questions

Question 1 (1 mark)
Source: VCE 2021, Further Mathematics Exam 1, Section A, Q10; © VCAA.

MC Oscar walked for nine consecutive days. The time, in minutes, that Oscar spent walking on each day is shown in the table below.

Day	1	2	3	4	5	6	7	8	9
Time	46	40	45	34	36	38	39	40	33

A least squares line is fitted to the data.

The equation of this line predicts that on day 10 the time Oscar spends walking will be the same as time he spent walking on

A. day 3   **B.** day 4   **C.** day 6   **D.** day 8   **E.** day 9

▶ **Question 2 (1 mark)**

Source: VCE 2021, Further Mathematics Exam 1, Section A, Q11; © VCAA.

MC The table below shows the *weight*, in kilograms, and the *height*, in centimetres, of 10 adults.

Weight (kg)	Height (cm)
59	173
67	180
69	184
84	195
64	173
74	180
76	192
56	169
58	164
66	180

A least squares line is fitted to the data.

The least squares line enables an adult's *weight* to be predicted from their *height*.

The number of times that the predicted value of an adult's *weight* is greater than the actual value of their *weight* is

    A. 3   **B.** 4   **C.** 5   **D.** 6   **E.** 7

▶ **Question 3 (1 mark)**

Source: VCE 2019, Further Mathematics Exam 1, Section A, Q14; © VCAA.

MC The *time*, in minutes, that Liv ran each day was recorded for nine days.

These times are shown in the table below.

Day number	1	2	3	4	5	6	7	8	9
Time (minutes)	22	40	28	51	19	60	33	37	46

The time series plot below was generated from this data.

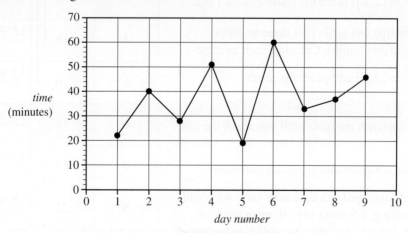

A least squares line is to be fitted to the time series plot shown above.

The equation of this least squares line, with *day number* as the explanatory variable, is closest to

    A. *day number* = 23.8 + 2.29 × *time*   **B.** *day number* = 28.5 + 1.77 × *time*
    C. *time* = 23.8 + 1.77 × *day number*   **D.** *time* = 23.8 + 2.29 × *day number*
    E. *time* = 28.5 + 1.77 × *day number*

More exam questions are available online.

4.4 Smoothing using the moving mean with an odd number of points

4.4.1 Moving-mean smoothing

When the data fluctuates a lot, it is often hard to see the underlying trend. In order to reveal the trend, we remove some of these fluctuations before attempting to fit the trend line. This process is referred to as smoothing.

Moving-mean smoothing

Moving-mean smoothing is an option that is preferred for data sets with few random fluctuations.

Moving-mean method using three points

Consider the following example:

Notice how the third column in the table is computed from the second column.

1. Take the first three y-values (i.e. first, second and third) and calculate their average $\left(\dfrac{12 + 10 + 15}{3} = 12.3\right)$; place the result against $t = 2$.

2. Move down one line and again take three y-values (i.e. second, third and fourth). Calculate their average $\left(\dfrac{10 + 15 + 13}{3} = 12.7\right)$ and place the result against $t = 3$.

3. Continue moving down the table until you reach the last three points.

This process is called a *3-point moving-mean smoothing*.

The number of points averaged at a time may vary: we could have a 4-point smoothing, a 5-point smoothing or even an 11-point smoothing. Later in the topic, we will discuss how to choose the number of points for smoothing.

Time (t)	Data (y)	Moving mean
1	12	
2	10	$\dfrac{12 + 10 + 15}{3} = 12.3$
3	15	$\dfrac{10 + 15 + 13}{3} = 12.7$
4	13	$\dfrac{15 + 13 + 16}{3} = 14.7$
5	16	$\dfrac{13 + 16 + 13}{3} = 14.0$
6	13	$\dfrac{16 + 13 + 18}{3} = 15.7$
7	18	$\dfrac{13 + 18 + 21}{3} = 17.3$
8	21	$\dfrac{18 + 21 + 19}{3} = 19.3$
9	19	

Note: There are fewer smoothed points than original points.

For a 3-point smooth, 1 point at either end is 'lost'; while for a 5-point smooth, 2 points at either end are 'lost'.

WORKED EXAMPLE 6 Smoothing with a 3-point moving mean.

The temperature of a sick patient was measured every two hours and the results recorded.

Time (hours)	2	4	6	8	10	12	14	16
Temp. (°C)	36.5	37.2	36.9	37.1	37.3	37.2	37.5	37.8

a. Use a 3-point moving-mean technique to smooth the data.
b. Plot both original and smoothed data on the same set of axes.
c. Predict the temperature for 18 hours using the last smoothed value.

THINK

a. 1. Put the data in a table.
   2. Calculate a 3-point moving mean for each data point.
   Note: The 'lost' values are at $t = 2$ and $t = 16$. Therefore, the first point plotted is (4, 36.87).

WRITE

a.

Time (hr)	Temp. (°C)	Smoothed temp. (°C)
2	36.5	
4	37.2	$\frac{1}{3}(36.5 + 37.2 + 36.9) = 36.87$
6	36.9	$\frac{1}{3}(37.2 + 36.9 + 37.1) = 37.07$
8	37.1	$\frac{1}{3}(36.9 + 37.1 + 37.3) = 37.1$
10	37.3	$\frac{1}{3}(37.1 + 37.3 + 37.2) = 37.2$
12	37.2	$\frac{1}{3}(37.3 + 37.2 + 37.5) = 37.33$
14	37.5	$\frac{1}{3}(37.2 + 37.5 + 37.8) = 37.5$
16	37.8	

b. 1. Plot the data. The smoothed line is the pink one.
   Note: The smoothed data start at the 2nd time point and finish at the 7th point.

b.

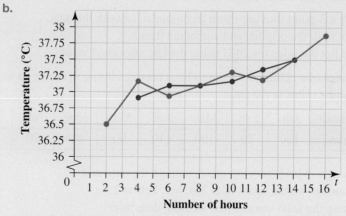

2. Comment on the result.

The smoothed line has removed much of the fluctuation of the original time series and, in fact, clearly exposes the upwards trend in temperature.

c. The last smoothed data point is 37.5.

c. The temperature at 18 hours is predicted to be approximately 37.78 °C.

on Resources

Interactivity Smoothing: moving mean method (int-6255)

4.4 Exercise

1. **WE5** The temperature of a pool was measured every hour over a summer's day and the results recorded.

Time (hr)	1	2	3	4	5	6	7	8
Temp (°C)	20.3	21.8	20.9	22.0	23.4	24.9	24.3	25.3

a. Use a 3-point moving-mean technique to smooth the data.
b. Plot both the original and smoothed data on the same set of axes.
c. Predict the temperature for 9 hours using the last smoothed value.

2. The membership numbers of a football club over an eight-year period are shown in the table.

Year (t)	2015	2016	2017	2018	2019	2020	2021	2022
Members (m)	38 587	42 498	45 972	57 408	71 271	72 688	78 427	72 170

a. Use a 3-point moving-mean technique to smooth the data.
b. Plot both the original and smoothed data on the same set of axes.
c. Predict the number of members in 2024 using the last smoothed value.

3. The following table represents sales of a textbook.

Year (t)	2015	2016	2017	2018	2019	2020	2021	2022
Sales (y)	2250	2600	2400	2750	2900	2450	3100	3400

a. Use a 3-point moving-mean technique to smooth the data.
b. Plot both the original and smoothed data.
c. Predict the sales for 2023 using the last smoothed value.

4. The sales of a certain car seem to have been declining in recent months. The management wishes to determine if this is the case.

Month	Jan.	Feb.	Mar.	Apr.	May	June	July	Aug.	Sept.	Oct.	Nov.	Dec.
Sales	120	70	100	110	90	80	70	90	80	100	60	60

a. Using a 3-point moving mean, smooth the data and comment on the result. Use Jan. = 1, Feb. = 2, ...
b. Using the least squares line method, determine the equation of the trend line for the smoothed data.
c. Use the equation to predict the number of sales for March next year. Comment on the prediction.
d. Perform a 5-point moving-mean smoothing on the data and discuss the result.

5. Consider the quarterly rainfall data shown. Rainfall has been measured over a three-year period.

Time (*t*)	Spring 2018	Summer 2018	Autumn 2019	Winter 2019	Spring 2019	Summer 2019
Rainfall (mm)	100	50	65	120	90	50

Time (*t*)	Autumn 2020	Winter 2020	Spring 2020	Summer 2020	Autumn 2021	Winter 2021
Rainfall (mm)	60	110	85	40	50	100

Perform a 3-point moving-mean smoothing and comment on whether there is an underlying trend.

6. The attendance at a country football league's games was recorded over 10 years. Management wishes to see if there is a trend.

Year	2013	2014	2015	2016	2017	2018	2019	2020	2021	2022
Attendance (× 1000)	75	72	69	74	66	72	61	64	69	65

a. Perform a 3-point moving-mean smoothing on the data and comment on the result.
b. Using a least squares line of best fit on the smoothed data, calculate the equation of the trend line.
c. Use the equation from part b to predict the attendance in 2024. Comment on the prediction.

7. a. Use technology to complete a 3-point moving mean smoothing on the following data, which represent sales figures for a 21-week period.

Week	Sales	Smoothed data
1	34	
2	27	
3	31	
4	37	
5	41	
6	29	
7	32	
8	37	
9	47	
10	38	
11	41	

Week	Sales	Smoothed data
12	44	
13	47	
14	49	
15	41	
16	52	
17	48	
18	44	
19	49	
20	56	
21	54	

b. Use a spreadsheet or CAS to perform a 5-point moving-mean smoothing on the data in part a and discuss your result.

8. The sales of a new car can vary due to the effect of advertising and promotion. The sales figures for Nassin Motor Company's new sedan are shown in the table.

Month	Feb.	Mar.	Apr.	May	June	July	Aug.	Sept.	Oct.	Nov.	Dec.
Sales	141	270	234	357	267	387	288	303	367	465	398

Use 5-point moving means to smooth the data. Plot the data, and use the last smoothed value to predict sales for the next month.

9. a. Coffee price data are shown in the table.

Quarter	Q_1-20	Q_2-20	Q_3-20	Q_4-20	Q_1-21	Q_2-21
Price ($)	358	323	316	336	369	333

Quarter	Q_3-21	Q_4-21	Q_1-22	Q_2-22	Q_3-22	Q_4-22
Price ($)	328	351	389	387	393	402

Perform a 3-point moving average to smooth the data. Plot the smoothed and original data and comment on your result.

b. Use a spreadsheet or CAS to perform a 5-point moving-mean smoothing on the data in part **a** and discuss your result.

10. A large building site requires varying numbers of workers. The weekly employment figures over the last seven weeks have been recorded.

Week	Number of employees
1	67
2	78
3	54
4	82
5	69
6	88
7	94

By performing a 3-point moving-mean smoothing, predict the number of people required for the next week.

4.4 Exam questions

Question 1 (1 mark)

Source: VCE 2021, Further Mathematics Exam 1, Section A, Q14; © VCAA.

MC A garden centre sells garden soil.

The table below shows the daily *quantity* of garden soil sold, in cubic metres, over a one-week period.

Day	Monday	Tuesday	Wednesday	Thursday	Friday	Saturday	Sunday
Quantity (m^3)	234	186				346	346

The *quantity* of garden soil sold on Wednesday, Thursday and Friday is not shown.

The five-mean smoothed *quantity* of garden soil sold on Thursday is $206 \, m^3$.

The three-mean smoothed *quantity* of garden soil sold on Thursday, in cubic metres, is

A. 143
B. 166
C. 206
D. 239
E. 403

▶ **Question 2 (1 mark)**

Source: VCE 2020, Further Mathematics Exam 1, Section A, Q20; © VCAA.

MC The time series plot below displays the number of airline passengers, in thousands, each month during the period January to December 1960.

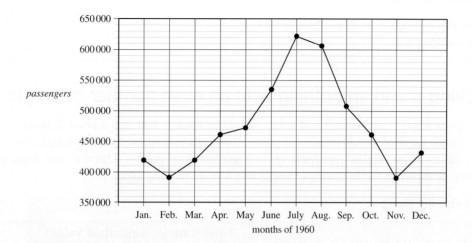

Data: GEP Box and GM Jenkins, *Time Series Analysis: Forecasting and Control*, Holden-Day, San Francisco, 1970, p. 531

During the period January to May 1960, the total number of airline *passengers* was 2 160 000.

The five-mean smoothed number of passengers for March 1960 is

A. 419 000 **B.** 424 000 **C.** 430 000 **D.** 432 000 **E.** 434 000

▶ **Question 3 (1 mark)**

Source: VCE 2019, Further Mathematics Exam 1, Section A, Q16; © VCAA.

MC The time series plot below shows the *monthly rainfall* at a weather station, in millimetres, for each *month* in 2017.

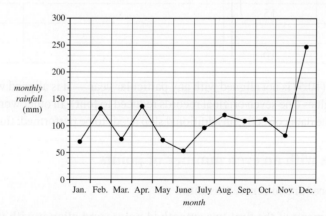

If seven-mean smoothing is used to smooth this time series plot, the number of smoothed data points would be

A. 3 **B.** 5 **C.** 6 **D.** 8 **E.** 10

More exam questions are available online.

4.5 Smoothing using the moving mean with an even number of points

4.5.1 Moving-mean smoothing with an even number of points

There are situations when an even number of points should be used — that is, a 4-point, 6-point or even 12-point moving mean. When we used an odd number of points, the result was automatically centred; that is, the y-data had the same t-values as the original (except at the first and last 'lost' points). This does not occur with an even-point smoothing, as shown in the following example of a 4-point moving mean.

Consider the following example.

| Time | y-value | 4-point mean (smoothed value) | |
		Calculation	Result
2016	6		
2017	10		
		$(6 + 10 + 14 + 12) \div 4$	10.5
2018	14		
		$(10 + 14 + 12 + 11) \div 4$	11.75
2019	12		
		$(14 + 12 + 11 + 15) \div 4$	13
2020	11		
		$(12 + 11 + 15 + 16) \div 4$	13.5
2021	15		
2022	16		

Observe that the first mean (10.5) is not aligned with any particular year — it is aligned with 2017.5! Also note that there are now three 'lost' values (the seven original records reduced to four). In other words, the moving mean is not centred properly. To align the data correctly, an additional step needs to be performed; this is called *centring*.

Centring the data with an even number of points

Use the following procedure to centre the data:

Step 1: Calculate the mean of the first two smoothed points and align it with the 3rd time point.

Step 2: Calculate the mean of the next two smoothed points and align it with the 4th time point.

Step 3: Repeat, leaving two blank entries at both top and bottom of the table.

This is demonstrated in the following table, using the data from the previous table.

The first mean (11.125) is now aligned with 2018, the second (12.375) aligned with 2019 and so on. This process not only introduces an extra step, but an *extra averaging* (or smoothing) as well.

Time	y-value	4-point mean (smoothed value)		4-point mean after centring	
		Calculation	Result	Calculation	Result
2016	6				
2017	10				
		(6 + 10 + 14 + 12) ÷ 4	10.5		
2018	14			(10.5 + 11.75) ÷ 2	11.125
		(10 + 14 + 12 + 11) ÷ 4	11.75		
2019	12			(11.75 + 13) ÷ 2	12.375
		(14 + 12 + 11 + 15) ÷ 4	13		
2020	11			(13 + 13.5) ÷ 2	13.25
		(12 + 11 + 15 + 16) ÷ 4	13.5		
2021	15				
2022	16				

tlvd-4978

WORKED EXAMPLE 7 Smoothing with a 4-point moving mean

The quarterly sales figures for a dress shop (in thousands of dollars) were recorded over a two-year period.

Time	Summer	Autumn	Winter	Spring	Summer	Autumn	Winter	Spring
Sales (× $1000)	27	22	19	25	31	25	22	29

Perform a centred 4-point moving-mean smoothing and plot the result. Comment on any trends revealed.

THINK

1. Arrange the data in a table.
Note: Code the time column.

2. Calculate a 4-point moving mean in column 3.
3. Average the pairs of averages to determine the 4-point centred data. This is done in column 4.

WRITE

Time	Sales	4-point moving mean	4-point centred moving mean
1	27		
2	22		
		(27 + 22 + 19 + 25) ÷ 4 = 23.25	
3	19		(23.25 + 24.25) ÷ 2 = 23.75
		(22 + 19 + 25 + 31) ÷ 4 = 24.25	
4	25		(24.25 + 25.00) ÷ 2 = 24.63
		(19 + 25 + 31 + 25) ÷ 4 = 25.00	
5	31		(25.00 + 25.75) ÷ 2 = 25.38
		(25 + 31 + 25 + 22) ÷ 4 = 25.75	
6	25		(25.75 + 26.75) ÷ 2 = 26.25
		(31 + 25 + 22 + 29) ÷ 4 = 26.75	
7	22		
8	29		

4. Plot the data. The smoothed line is the pink one. *Note*: The smoothed data start at the 3rd time point and finish at the 6th point.

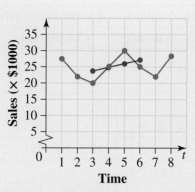

5. Interpret the results.

Observe the steadily increasing trend (even with only four smoothed points) that was not obvious from the original data.

Determining the number of points for a moving mean

When smoothing data, it is important to decide on the number of points to be used. The data size for different data sets are summarised in the following table.

Data type	Number of points to use
Small data sets	Number of points should be small
Large data sets	Number of points should be large, producing a smoother trend line
Cyclic variation	Length of the cycle
Seasonal variation	Number of seasons

So, for example, for monthly figures use 12-point smoothing, for quarterly data use 4-point smoothing and for weekly data use 7-point smoothing.

4.5 Exercise

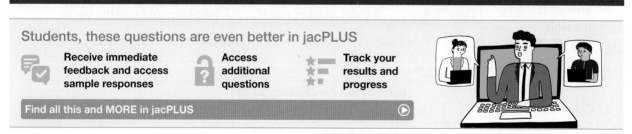

Students, these questions are even better in jacPLUS

Receive immediate feedback and access sample responses

Access additional questions

Track your results and progress

Find all this and MORE in jacPLUS

1. WE6 The quarterly sales figures for a shoe shop (in thousands of dollars) were recorded over a two-year period.

Time	Summer	Autumn	Winter	Spring	Summer	Autumn	Winter	Spring
Sales ($\times \$1000$)	59	48	43	50	63	52	47	61

a. Perform a centred 4-point moving-mean smoothing and plot the results.
b. Comment on any trend revealed.

2. The quarterly electricity cost figures for a shop were recorded over a two-year period.

Time	Q_1-21	Q_2-21	Q_3-21	Q_4-21	Q_1-22	Q_2-22	Q_3-22	Q_4-22
Cost ($)	554	503	467	587	636	533	493	684

a. Perform a centred 4-point moving-mean smoothing and plot the results.
b. Comment on any trend revealed.

3. Consider the following data table.

t	1	2	3	4	5	6	7	8	9	10
y	75	54	62	60	70	45	54	59	62	64

a. Perform a 4-point centred moving mean to smooth the following data and plot the result.
b. Comment on any trends that you observe.

4. The price of oranges fluctuates from season to season. Data have been recorded for three years.

t	Autumn 2020	Winter 2020	Spring 2020	Summer 2020	Autumn 2021	Winter 2021
Price	45	67	51	44	52	76

t	Spring 2021	Summer 2021	Autumn 2022	Winter 2022	Spring 2022	Summer 2022
Price	63	48	58	80	66	52

Perform a 4-point centred moving-mean smoothing, and plot the data and comment on any trends.

5. a. The time series represents the temperature of a hospital patient over 15 days. Use technology to complete the following table.

Day	Temperature	4-point moving mean	4-point centred moving mean
1	36.6		
2	36.4		
		36.75	
3	36.8		
		36.825	
4	37.2		
		36.85	
5	36.9		
		36.95	
6	36.5		
		37	
7	37.2		
		37.05	
8	37.4		
		37.275	
9	37.1		
		37.375	
10	37.4		
		37.25	
11	37.6		
		37.275	
12	36.9		
		37.325	
13	37.2		
		37.15	
14	37.6		
15	36.9		

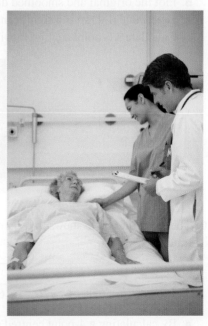

b. Using the smoothed data, determine the equation of the least squares line of best fit.
c. Use the trend line to predict the temperature of the patient on day 16.

6. **MC** If you start with 12 points of data and carry out a 4-point centred moving average, the number of points you end up with is:

   A. 8   B. 7   C. 6   D. 5   E. 4

7. **MC** You have data points for each day in the month of April and complete a 6-point centred moving average. The number of points you end up with is:

   A. 30   B. 28   C. 26   D. 24   E. 22

8. The sales of summer clothing vary according to the season. The following table gives seasonal sales data (in thousands of dollars) for three years at a Darryl Jones department store.

Season	Q_3-19	Q_4-19	Q_1-20	Q_2-20	Q_3-20	Q_4-20
Sales	78	92	90	73	62	85

Season	Q_1-21	Q_2-21	Q_3-21	Q_4-21	Q_1-22	Q_2-22
Sales	83	70	61	78	74	59

   a. Calculate a 4-point centred moving average.
   b. Plot the original and smoothed data on the same set of axes.
   c. Determine if there is an underlying upward or downward trend.
   d. Calculate a 6-point centred moving mean on the data.

9. An athlete wishes to measure their performance in running a 1 km race. They record their times over the last 10 days.

Day	1	2	3	4	5	6	7	8	9	10
Time (s)	188	179	183	180	173	171	182	168	171	166

   a. Perform a 4-point centred moving-mean smoothing.
   b. Plot the original and smoothed data on the same set of axes.
   c. Determine if there is a significant improvement in their times.
   d. Calculate a 6-point centred moving mean on the data.

10. The following table shows the share price index of Industrial Companies during an unstable fortnight's trading.

Day	1	2	3	4	5	6	7	8	9	10
Index	678	726	692	714	689	687	772	685	688	712

   a. By calculating a 4-point centred moving mean, determine if there is an upward or downward trend.
   b. Use the smoothed data in part a to draw the least squares line of best fit.
   c. Use the trend line to predict the price index of the Industrial Companies after 15 days.

Question 1 (1 mark)
Source: VCE 2020, Further Mathematics Exam 1, Section A, Q18; © VCAA.

MC The table below shows the monthly rainfall for 2019, in millimetres, recorded at a weather station, and the associated long-term seasonal indices for each month of the year.

Table 4

	Jan.	Feb.	Mar.	Apr.	May	June	July	Aug.	Sep.	Oct.	Nov.	Dec.
Monthly rainfall (mm)	18.4	17.6	46.8	23.6	92.6	77.2	80.0	86.8	93.8	55.2	97.3	69.4
Seasonal index	0.728	0.734	0.741	0.934	1.222	0.973	1.024	1.121	1.159	1.156	1.138	1.072

Data: adapted from © Commonwealth of Australia 2020, Bureau of Meteorology

The six-mean smoothed monthly rainfall with centring for August 2019 is closest to
    A. 67.8 mm   **B.** 75.9 mm   **C.** 81.3 mm   **D.** 83.4 mm   **E.** 86.4 mm

Question 2 (1 mark)
Source: VCE 2018, Further Mathematics Exam 1, Section A, Q15; © VCAA.

MC The table below shows the monthly profit, in dollars, of a new coffee shop for the first nine months of 2018.

Month	Jan.	Feb.	Mar.	Apr.	May	June	July	Aug.	Sept.
Profit ($)	2890	1978	2402	2456	4651	3456	2823	2678	2345

Using four-mean smoothing with centring, the smoothed profit for May is closest to
    A. $2502   **B.** $3294   **C.** $3503   **D.** $3804   **E.** $4651

Question 3 (1 mark)
Source: VCE 2013, Further Mathematics Exam 1, Section A, Q13; © VCAA.

MC The time series plot below displays the number of guests staying at a holiday resort during summer, autumn, winter and spring for the years 2007 to 2012 inclusive.

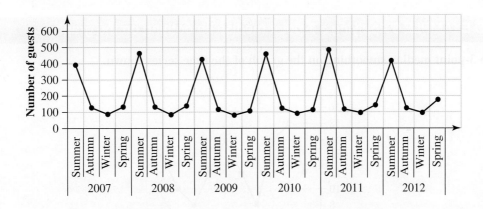

The table below shows the data from the times series plot for the years 2007 and 2008.

Year	Season	Number of guests
2007	summer	390
	autumn	126
	winter	85
	spring	130
2008	summer	460
	autumn	136
	winter	86

Using four-mean smoothing with centring, the smoothed number of guests for winter 2007 is closest to

A. 85   **B.** 107   **C.** 183   **D.** 192   **E.** 200

More exam questions are available online.

4.6 Median smoothing from a graph

> **LEARNING INTENTION**
>
> At the end of this subtopic you should be able to:
> - identify key qualitative features of a time series plot including trend (using smoothing if necessary).

4.6.1 Median smoothing

An alternative to moving-mean (average) smoothing is to replace the averaging of a group of points with the median of each group. This is a technique that requires no calculations and, in the scope of this course, will only be addressed graphically.

Generally, the effect of median smoothing is to remove some random fluctuations. It performs poorly on cyclical or seasonal fluctuations — unless the size of the range being used (3, 5, 7, ... points) is chosen carefully. Provided the graph has clearly marked data points, it is possible to calculate a median smooth directly from it.

tlvd-4979

> **WORKED EXAMPLE 8 Applying a 3-point median smoothing on the graph**
>
> **Perform a 3-point median smoothing on the graph of the time series shown.**
>
>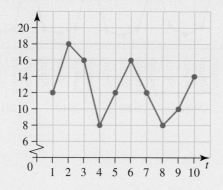

THINK

1. Read the data values and compute the median.

WRITE

The 1st group of 3 points is: 12, 18, 16 — median = 16.
The 2nd group of 3 points is: 18, 16, 8 — median = 16.
The 3rd group of 3 points is: 16, 8, 12 — median = 12.
The 4th group of 3 points is: 8, 12, 16 — median = 12.
The 5th group of 3 points is: 12, 16, 12 — median = 12.
The 6th group of 3 points is: 16, 12, 8 — median = 12.
The 7th group of 3 points is: 12, 8, 10 — median = 10.
The 8th group of 3 points is: 8, 10, 14 — median = 10.

2. Plot the medians on the graph.
Note: Median smoothing has indicated a downward trend that is probably not in the real time series. This indicates that moving-average smoothing would be the preferred option.

 Resources

 Interactivity Smoothing: median method (int-6256)

4.6 Exercise

Students, these questions are even better in jacPLUS

 Receive immediate feedback and access sample responses

 Access additional questions

★= **Track your results and progress**

Find all this and **MORE in jacPLUS** ▶

1. **WE7** Perform a 3-point median smoothing on the graph of the time series shown.

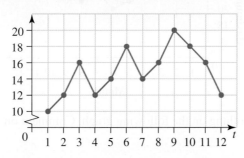

2. Perform a 3-point median smoothing on the graph of the time series shown.

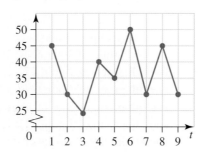

3. Determine the number of data points left if:
   a. you have a 12-point time series and perform a 3-point median smoothing on the data
   b. you have an 11-point time series and perform a 5-point median smoothing on the data.

4. Perform a 3-point median smoothing on the graph below and plot the result.

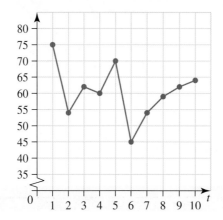

5. The maximum daily temperatures for a year were recorded as a monthly average.

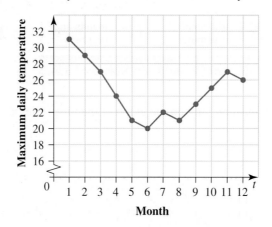

   Perform a 3-point median smoothing on the following graph. Comment on your result.

6. Consider the graphical time series shown below.

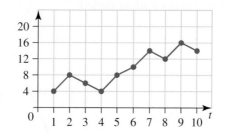

   a. Perform a 3-point median smoothing on the graph.
   b. Comment on the effectiveness of the 3-point median smoothing.

7. Consider the graphical time series shown.

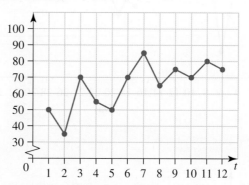

   a. Perform a 3-point median smoothing on the graph.
   b. Comment on the effectiveness of the 3-point median smoothing.

8. Perform a 3-point median smoothing on the graph shown, which represents the share price of the HAL computer company over the last 15 days.

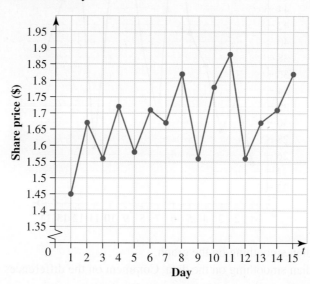

9. Perform a 5-point median smoothing on the graph shown, which represents the share price of the Pear-Shaped Computer Company over an 8-week trading period.

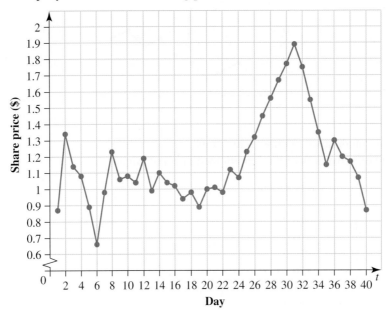

10. Consider the following data.

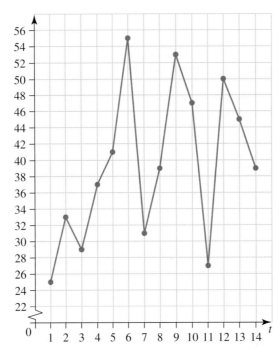

a. Perform a 3-point median smoothing on the graph.
b. Perform a 5-point median smoothing on the data. Comment on the differences of the two smoothing techniques.

4.6 Exam questions

▶ **Question 1 (1 mark)**

Source: VCE 2021, Further Mathematics Exam 1, Section A, Q13; © VCAA.

MC The time series plot below shows the *points scored* by a basketball team over 40 games.

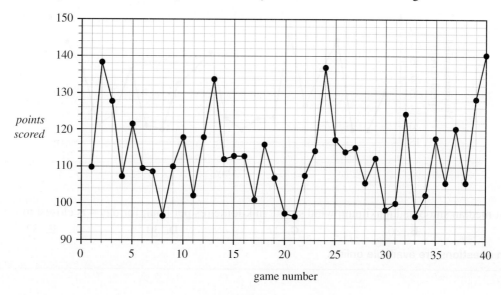

The nine-median smoothed *points scored* for game number 10 is closest to

A. 102   **B.** 108   **C.** 110   **D.** 112   **E.** 117

▶ **Question 2 (1 mark)**

Source: VCE 2019, Further Mathematics Exam 1, Section A, Q13; © VCAA.

MC The *time*, in minutes, that Liv ran each day was recorded for nine days.

These times are shown in the table below.

Day number	1	2	3	4	5	6	7	8	9
Time (minutes)	22	40	28	51	19	60	33	37	46

The time series plot below was generated from this data.

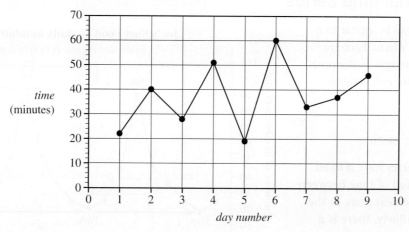

Both three-median smoothing and five-median smoothing are being considered for this data.

Both of these methods result in the same smoothed value on *day number*

A. 3   **B.** 4   **C.** 5   **D.** 6   **E.** 7

Source: VCE 2017, Further Mathematics Exam 1, Section A, Q14; © VCAA.

MC The wind speed at a city location is measured throughout the day.

The time series plot below shows the daily maximum wind speed, in kilometres per hour, over a three-week period.

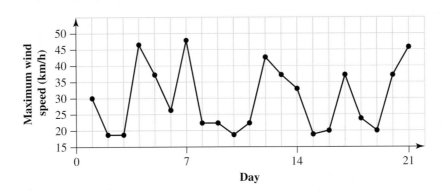

The 7-point median smoothed maximum wind speed, in kilometres per hour, for day 4 is closest to

   A. 22   **B.** 26   **C.** 27   **D.** 30   **E.** 32

More exam questions are available online.

4.7 Seasonal adjustment

> **LEARNING INTENTION**
>
> At the end of this subtopic you should be able to:
> * calculate, interpret and apply seasonal indices
> * model linear trends using the least squares line of best fit, interpret the model in the context of the trend being modelled, use the model to make forecasts with consideration of the limitations of extending forecasts too far into the future.

4.7.1 Seasonal time series

A seasonal time series is similar to a cyclical time series where there are defined peaks and troughs in the time series data, except for one notable difference.

Seasonal time series

Seasonal time series have a fixed and regular period of time between one peak and the next peak in the data values. Similarly, there is a fixed and regular period of time between one trough and the next trough.

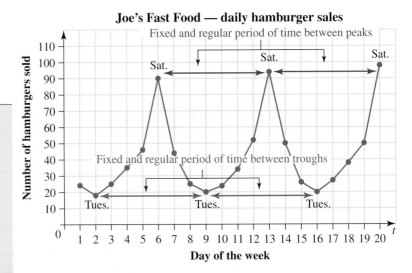

Joe's Fast Food — daily hamburger sales

As we have seen in the sections on fitting a straight line to a time series, it is difficult to determine an effective linear equation for such data. As well, the sections on smoothing indicated that seasonal data may not lend themselves to the techniques of moving-mean or median smoothing. We may have to accept that the data vary from season to season and treat each record individually.

For example, the unemployment rate in Australia is often quoted as '6.8% — seasonally adjusted'. The government has accepted that each season has its own time series, more or less independent of the other seasons. How can we remove the effect of the season on our time series? The technique of seasonally adjusting, or **deseasonalising**, will modify the original time series, hopefully removing the seasonal variation, and exposing any other fluctuations (cyclic or irregular) that may be 'hidden' by seasonal variation.

4.7.2 Deseasonalising time series

The process of deseasonalising time series data involves calculating **seasonal indices**. A seasonal index compares a particular season to the average season. That is, the seasonal index measures by what factor a particular season is above or below the average of all seasons for the cycle.

For example:

Seasonal index = 1.3 means that season is 1.3 times the average season (that is, the figures for this season are 30% above the seasonal average). It is a peak or high season.

Seasonal index = 0.7 means that season is 0.7 times the average season (that is, the figures for this season are 30% below the seasonal average). It is a trough or low season.

Seasonal index = 1.0 means that season is the same as the average season. It is neither a peak nor a trough.

To deseasonalise the data, we divide each value by the corresponding seasonal index.

To calculate the deseasonalised value

$$\text{Deseasonalised figure or value} = \frac{\text{actual original figure or value}}{\text{Seasonal index}}$$

The method of deseasonalising time series is best demonstrated with an example. Observe carefully the various steps, which must be performed in the order shown.

WORKED EXAMPLE 9 Deseasonalising data

Unemployment figures have been collected over a five-year period and presented in this table. It is difficult to see any trends.

Season	2018	2019	2020	2021	2022
Summer	6.2	6.5	6.4	6.7	6.9
Autumn	8.1	7.9	8.3	8.5	8.1
Winter	8.0	8.2	7.9	8.2	8.3
Spring	7.2	7.7	7.5	7.7	7.6

a. **Calculate the seasonal indices.**
b. **Deseasonalise the data using the seasonal indices.**
c. **Plot the original and deseasonalised data.**
d. **Comment on your results, supporting your statements with mathematical evidence.**

THINK	WRITE

a. 1. Calculate the yearly averages over the four seasons for each year and put them in a table.

a. 2018: $(6.2 + 8.1 + 8.0 + 7.2) \div 4 = 7.375$
2019: $(6.5 + 7.9 + 8.2 + 7.7) \div 4 = 7.575$
2020: $(6.4 + 8.3 + 7.9 + 7.5) \div 4 = 7.525$
2021: $(6.7 + 8.5 + 8.2 + 7.7) \div 4 = 7.775$
2022: $(6.9 + 8.1 + 8.3 + 7.6) \div 4 = 7.725$

Year	2018	2019	2020	2021	2022
Average	7.375	7.575	7.525	7.775	7.725

2. Divide each term in the original time series by its yearly average. That is, divide each value for 2018 by the yearly average for 2018 (i.e. by 7.375); next divide each value for 2019 by the yearly average for 2019 (i.e. by 7.575), etc.

Summer 2018: $6.2 \div 7.375 = 0.8407$
Autumn 2018: $8.1 \div 7.375 = 1.0983$
Winter 2018: $8.0 \div 7.375 = 1.0847$
Spring 2018: $7.2 \div 7.375 = 0.9763$
Summer 2019: $6.5 \div 7.575 = 0.8581$
$\vdots$
Spring 2022: $7.6 \div 7.725 = 0.9838$

Season	2018	2019	2020	2021	2022
Summer	0.8407	0.8581	0.8505	0.8617	0.8932
Autumn	1.0983	1.0429	1.1030	1.0932	0.8932
Winter	1.0847	1.0825	1.0498	1.0547	1.0744
Spring	0.9763	1.0165	0.9967	0.9904	0.9838

3. Determine the seasonal averages from this second table. That is, calculate the average of all five values for summer; next calculate the average of all values for autumn and so on. These are called *seasonal indices*.

Summer: $(0.8427 + 0.8581 + 0.8505 + 0.8617 + 0.8932) \div 5 = 0.8612.$
Autumn: $(1.0983 + 1.0429 + 1.1030 + 1.0932 + 1.0485) \div 5 = 1.0772.$
Winter: $(1.0847 + 1.0825 + 1.0498 + 1.0547 + 1.0744) \div 5 = 1.0692.$
Spring: $(0.9763 + 1.0165 + 0.9967 + 0.9904 + 0.9838) \div 5 = 0.9927.$

Season	Summer	Autumn	Winter	Spring
Seasonal index	0.8612	1.0772	1.0692	0.9927

b. Divide each term in the original series by its seasonal index. That is, divide all summer figures by the summer seasonal index (0.8612), divide all autumn figures by the autumn seasonal index (1.0772) and so on. This gives the seasonally adjusted or deseasonalised time series.
Note: Your answers may vary a little, depending upon how and when you rounded your calculations.

b. Summer 18: $6.2 \div 0.8612 = 7.1992$
Autumn 18: $8.1 \div 1.0772 = 7.5195$
$\vdots$
Spring 22: $7.6 \div 0.9927 = 7.6559$

	2018	2019	2020	2021	2022
Summer	7.199	7.548	7.431	7.780	8.012
Autumn	7.520	7.334	7.705	7.891	7.520
Winter	7.482	7.669	7.389	7.669	7.763
Spring	7.253	7.757	7.555	7.757	7.656

c. Graph the original and the seasonally adjusted (deseasonalised) time series.

c.

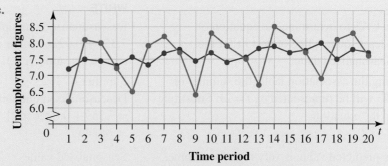

d. Note that most, but not all, of the seasonal variation has been removed. However, by using least squares, we could more confidently fit a straight line to the deseasonalised data.

d. There appears to be a slight upward trend in unemployment figures.

4.7.3 Forecasting with seasonal time series

In the previous section, we smoothed the seasonal variation and are now able to see any trends more clearly. If there is an upward or downward secular trend, then a straight-line equation can be calculated and used for making predictions into the future. Using the least squares line of best fit method on the *deseasonalised* data is always preferred.

Once the equation of the least squares line for deseasonalised data has been obtained, it can be used for forecasting.

However, the prediction obtained using such an equation will also be deseasonalised or smoothed out to the average season. But as we have the relevant seasonal indices, we should be able to use it to remove the smoothing, that is, to **re-seasonalise** the predicted value.

Formula for re-seasonalising

Re-seasonalised figure or value = deseasonalised figure or value × seasonal index

---

tlvd-4980

WORKED EXAMPLE 10 Determining the least squares line from deseasonalised data

Use the deseasonalised data from Worked example 9 to determine the equation of the straight line for the deseasonalised data using the least squares line of best fit method.

	2018	2019	2020	2021	2022
Summer	7.199	7.548	7.431	7.780	8.012
Autumn	7.520	7.334	7.705	7.891	7.519
Winter	7.482	7.699	7.389	7.669	7.763
Spring	7.253	7.757	7.555	7.757	7.656

Predict the unemployment figure for summer in 2023. The deseasonalised data are reproduced in the table. (The seasonal index for summer is 0.8608.)

THINK	WRITE
1. Use CAS to determine the equation of the least squares line of best fit for the deseasonalised data.	Deseasonalised unemployment % $7.359 + 0.0226 \times$ time code, where time code 1 represents summer 2018.
2. Using the association table, the summer of 2023 will be represented by $t = 21$. Substitute into the equation.	Summer of 2023: Deseasonalised unemployment % $= 7.359 + 0.0226 \times 21$ $= 7.834\%$
3. The predicted value is very high for summer. Re-seasonalise by using the seasonal index for summer, which was 0.8612. That is, this was a season or period of low unemployment.	Seasonalised value $=$ deseasonalised value $\times$ seasonal index $= 7.834 \times 0.8612$ $= 6.75\%$

TI | THINK	DISPLAY/WRITE
1. On a Lists & Spreadsheet page, enter the data into the lists named *time* and *unemploy*.	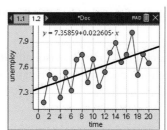

2. Press CTRL + I and select:
5: Add Data & Statistics
Use the tab key to move to each axis and press the click key or ENTER to place time on the horizontal axis and unemploy on the vertical axis.

To join the dots, press MENU and then select:
2: Plot Properties
1: Connect Data Points
To fit a least squares line of best fit, press MENU and select:
4: Analyze
6: Regression
2: Show Linear
$(a + bx)$

3. The answer appears on the screen.

Deseasonalised unemployment (%) $7.359 + 0.0026 \times$ time code, where time code 1 represents summer 2018.

CASIO | THINK	DISPLAY/WRITE
1. On a Statistics screen, enter the data into the lists, named *time* and *unemploy*.	

2. To determine the least squares line of best fit, tap:
Calc
Regression
Linear reg

Complete the Set Calculation fields:
Linear Reg
X List: main\time
Y List: main\unemploy
Freq: 1
Copy Formula y1
Copy Residual Off

Tap OK.

3. The answer appears on the screen.

Deseasonalised unemployment (%) $7.359 + 0.0226 \times$ time code, where time code 1 represents summer 2018.

Note: Forecasting comes with limitations and cannot be used for an infinite time period. It should be used only for a small time period after your most recent recorded data.

WORKED EXAMPLE 11 Using seasonal indices

Quarterly sales figures for a pool chemical supplier between 2017 and 2022 were used to determine the following seasonal indices.

Season	1st quarter	2nd quarter	3rd quarter	4th quarter
Summer	1.8	1.2	0.2	0.8

Using the seasonal indices provided in the table, determine:
a. the deseasonalised figure if the actual sales figure for the second quarter in 2021 was $4680
b. the deseasonalised figure if the actual sales figure for the third quarter in 2021 was $800
c. the predicted value if the deseasonalised predicted value for the first quarter in 2023 is $4000.

THINK	WRITE
Use the formula for deseasonalising.	
a. Use the 2nd quarter seasonal index.	a. Deseasonalised figure $= \dfrac{\text{actual figure}}{\text{seasonal index}}$ $= \dfrac{4680}{1.2}$ $= \$3900$
b. Use the 3rd quarter seasonal index to obtain the deseasonalised figure.	b. Deseasonalised figure $= \dfrac{\text{actual figure}}{\text{seasonal index}}$ $= \dfrac{800}{0.2}$ $= \$4000$
c. Use the re-seasonalising formula and select the 1st quarter seasonal index.	c. Re-seasonalised figure $=$ deseasonalised figure $\times$ seasonal index $= 4000 \times 1.8$ $= \$7200$ The forecast sales figure for the first quarter in 2023 is $7200.

4.7.4 Seasonal indices

Finally, note that the sum of all the seasonal indices gives a specific result, which can be used to answer certain types of queries.

Sum of the seasonal indices

The sum of the seasonal indices is equal to the number of seasons.

This can be summarised as follows.

Type of data	Number of seasons	Cycle	Sum of all the seasonal indices
Monthly figures	12	A year	12
Quarterly figures	4	A year	4
Fortnightly figures	26	A year	26
Daily figures for data from Monday to Friday only	5	A week	5
Daily figures for data from Monday to Sunday	7	A week	7

WORKED EXAMPLE 12 Calculating missing seasonal indices

A fast food store that is open seven days a week has the following seasonal indices.

Season	Monday	Tuesday	Wednesday	Thursday	Friday	Saturday	Sunday
Index	0.5	0.2	0.5	0.6		2.2	1.1

The index for Friday has not been recorded. Calculate the missing index.

THINK	WRITE
1. The sum of the seasonal indices is equal to the number of seasons.	There are 7 seasons (Monday to Sunday); therefore, the sum of indices is 7.
2. The missing index is the sum of all the other seasons subtracted from the total.	Friday index
	$= 7 - (\text{sum of all the other indices})$
	$= 7 - (0.5 + 0.2 + 0.5 + 0.6 + 2.2 + 1.1)$
	$= 7 - 5.1$
	$= 1.9$

 Resources

 interactivity Seasonal adjustment (int-6257)

Note: Your answers may vary slightly, depending upon rounding. Try to round correct to 4 decimal places for all intermediate calculations.

1. **WE8,9** The price of 1 litre of milk has been collected over a five-year period and is presented in the table. It is difficult to see any trends, other than seasonal ones.

Season	2018	2019	2020	2021	2022
Summer	1.55	1.60	1.50	1.60	1.45
Autumn	1.50	1.55	1.50	1.45	1.40
Winter	1.65	1.75	1.70	1.70	1.60
Spring	1.70	1.80	1.75	1.80	1.65

a. Calculate the seasonal indices.
b. Deseasonalise the data using the seasonal indices.
c. Plot the original and deseasonalised data.
d. Comment on your results, supporting your statements with mathematical evidence.
e. Use the deseasonalised data from part **b** to determine the equation of the straight line for the deseasonalised data using the least squares line of best fit method.
f. Predict the price of 1 litre of milk for summer 2023. (The seasonal index for summer is 0.96.)

2. The average price of 1 litre of petrol has been collected over a five-year period and is presented in the table. It is difficult to see any trends, other than seasonal ones.

Season	2018	2019	2020	2021	2022
Summer	1.60	1.60	1.57	1.63	1.58
Autumn	1.48	1.50	1.46	1.31	1.27
Winter	1.40	1.35	1.39	1.28	1.19
Spring	1.65	1.60	1.63	1.69	1.59

a. Calculate the seasonal indices.
b. Deseasonalise the data using the seasonal indices.
c. Plot the original and deseasonalised data.
d. Comment on your results, supporting your statements with mathematical evidence.
e. Use the deseasonalised data from part **b** to determine the equation of the straight line for the deseasonalised data using the least squares line of best fit method.
f. Predict the price of 1 litre of petrol for summer 2023. (The seasonal index for summer is 1.0731.)

3. The price of sugar ($/kg) has been recorded over three years on a seasonal basis.

Season	2020	2021	2022
Summer	1.03	0.98	0.95
Autumn	1.26	1.25	1.21
Winter	1.36	1.34	1.29
Spring	1.14	1.07	1.04

a. Compute the seasonal indices.
b. Deseasonalise the data using the seasonal indices.
c. Plot the original and deseasonalised data.
d. Comment on your results, supporting your statements with mathematical evidence.

4. Data on the total seasonal rainfall (in mm) have been accumulated over a six-year period.

Season	2017	2018	2019	2020	2021	2022
Summer	103	97	95	117	118	120
Autumn	93	84	82	100	99	98
Winter	143	124	121	156	155	151
Spring	123	109	107	125	122	124

a. Compute the seasonal indices.
b. Deseasonalise the original time series.
c. Plot the original and deseasonalised time series.
d. Comment on your result, supporting your statements with mathematical evidence.

5. It is known that young people (18–25) have problems in finding work; these problems are different from those facing older people. The youth unemployment statistics are recorded separately from the overall data. Using the youth unemployment figures for the five years shown:

Season	2018	2019	2020	2021	2022
Summer	7.6	7.7	7.8	7.7	7.9
Autumn	10.9	11.3	11.9	12.6	13.1
Winter	11.7	12.4	12.8	13.5	13.9
Spring	9.9	10.5	10.8	11.4	11.9

a. compute the seasonal indices
b. deseasonalise the time series
c. plot the original and deseasonalised time series
d. comment on your result, supporting your statements with mathematical evidence.

6. The unemployment rate in a successful European economy is given in the table as a percentage.

Quarter	1	2	3	4
2020	5.8	4.9	3.5	6.7
2021	6.1	5.1	3.2	6.5
2022	5.7	4.5	4.1	7.1

a. Compute the seasonal indices.
b. Deseasonalise the time series.
c. Plot the original and deseasonalised time series.
d. Determine the equation of the line of best fit for the deseasonalised data using the least squares method.
e. Use the equation of the line from part d to predict the unemployment rate for:

   i. quarter 1 in 2023   ii. quarter 3 in 2027.

7. It is possible to seasonally adjust time series for periods other than the usual four seasons. Consider an expensive restaurant that wishes to study its customer patterns on a daily basis. In this case a 'season' is a single day and there are seven seasons in a weekly cycle.

Data are total revenue each day, shown in the table that follows.

Season	Week 1	Week 2	Week 3	Week 4	Week 5
Monday	1036	1089	1064	1134	1042
Tuesday	1103	1046	1085	1207	1156
Wednesday	1450	1324	1487	1378	1408
Thursday	1645	1734	1790	1804	1789
Friday	2078	2204	2215	2184	2167
Saturday	2467	2478	2504	2526	2589
Sunday	1895	1786	1824	1784	1755

Modify the spreadsheet solution to allow for these seven seasons and deseasonalise the following data over a five-week period. Comment on your result, supporting your statements with mathematical evidence.

8. **MC** A line of best fit for deseasonalised data was given as:

$$Deseasonalised\ monthly\ sales = 10\,000 + 1500 \times time\ code$$

where June 2022 represents $t = 1$.

The predicted actual expected sales figure for June 2023, if the June seasonal index is 0.8, would be:

A. $23 600   B. $29 500   C. $19 500   D. $36 875   E. $35 000

Questions 9 and 10 relate to the following table, which contains the seasonal indices for the monthly sales of spring water in a particular supermarket.

Season	Jan.	Feb.	Mar.	Apr.	May	June	July	Aug.	Sept.	Oct.	Nov.	Dec.
Index	1.05		1.0	1.0	0.95	0.85	0.8	0.9	0.95	1.05	1.10	1.15

9. **MC** The seasonal index missing from the table is:

A. 1.0   B. 1.05   C. 1.10   D. 1.15   E. 1.20

10. **MC** If the actual sales figure for June was $102 000, then the deseasonalised figure would be:

A. $96 900   B. $86 700   C. $107 368.42   D. $120 000   E. $102 000

11. Complete the following table of seasonal indices.

Season	Summer	Autumn	Winter	Spring
Index	1.23	0.89		1.45

12. The following table gives the deseasonalised figures and corresponding seasonal indices for umbrella sales.

Season	Jan.	Feb.	Mar.	Apr.	May	June	July	Aug.	Sept.	Oct.	Nov.	Dec.
Number of umbrellas sold (deseasonalised)	24	24	25	26	25	27	27	28	30	31	33	34
Index	1.15	0.90	0.20	0.20	0.35	0.45	3.0	2.10	2.15	0.95	0.40	0.15

a. Determine the equation of the straight line for the deseasonalised data using the least squares line of best fit method.
b. Predict the umbrella sales for January the following year.

13. **WE10** Quarterly sales figures of a pet supplier between 2017 and 2022 were used to determine the following seasonal indices.

Season	1st quarter	2nd quarter	3rd quarter	4th quarter
Seasonal index	1.4	1.2	0.6	0.8

Using the seasonal indices provided in the table, calculate the following.

a. Calculate the deseasonalised figure if the actual sales figure for the second quarter in 2021 was $4345.
b. Calculate the deseasonalised figure if the actual sales figure for the third quarter in 2021 was $950.
c. Calculate the predicted value if the deseasonalised predicted value for the first quarter in 2023 is $5890.

14. Quarterly sales figures of a surf shop between 2017 and 2022 were used to determine the following seasonal indices.

Season	1st quarter	2nd quarter	3rd quarter	4th quarter
Seasonal index	1.7	1.3	0.4	0.6

Using the seasonal indices provided in the table, calculate the following.

a. Calculate the deseasonalised figure if the actual sales figure for the second quarter in 2021 was $8945.
b. Calculate the deseasonalised figure if the actual sales figure for the third quarter in 2021 was $3250.
c. Calculate the predicted value if the deseasonalised predicted value for the first quarter in 2023 is $7950.

15. Quarterly sales figures for an ice-cream parlour between 2020 and 2022 were used to determine the following seasonal indices.

Season	1st quarter	2nd quarter	3rd quarter	4th quarter
Seasonal index	1.50	1.00	0.25	1.25

Using the seasonal indices provided in the table, calculate the following.

a. Calculate the deseasonalised figure if the actual sales figure for the second quarter in 2021 was $3000.
b. Calculate the deseasonalised figure if the actual sales figure for the third quarter in 2021 was $800.
c. Calculate the predicted value if the deseasonalised predicted value for the first quarter in 2023 is $3200.

16. **WE11** A pizza shop that is open seven days a week has the following seasonal indices. The index for Friday has not been recorded.

Season	Monday	Tuesday	Wednesday	Thursday	Friday	Saturday	Sunday
Index	0.4	0.3	0.6	0.8		1.8	1.3

Calculate the missing index.

17. A cinema open seven days a week has the following seasonal indices.

Season	Monday	Tuesday	Wednesday	Thursday	Friday	Saturday	Sunday
Index	0.4	1.1	0.5	0.6	1.3		1.6

Calculate the missing index.

18. A newsagency that is open seven days a week has the following seasonal indices.

Season	Monday	Tuesday	Wednesday	Thursday	Friday	Saturday	Sunday
Index	0.5	0.2		0.6	1.5	2.2	1.1

Determine the value of the missing index.

4.7 Exam questions

Question 1 (1 mark)
Source: VCE 2021, Further Mathematics Exam 1, Section A, Q15; © VCAA.

MC The table below shows the number of visitors to an art gallery during the summer, autumn, winter and spring quarters for the years 2017 to 2019.

The quarterly average is also shown for each of these years.

Season	2017	2018	2019
summer quarter	29 685	25 420	31 496
autumn quarter	27 462	23 320	29 874
winter quarter	25 564	21 097	27 453
spring quarter	26 065	22 897	28 149
Quarterly average	27 194.0	23 183.5	29 243.0

The seasonal index for summer is closest to

A. 1.077   **B.** 1.081   **C.** 1.088   **D.** 1.092   **E.** 1.096

Question 2 (1 mark)
Source: VCE 2021, Further Mathematics Exam 1, Section A, Q16; © VCAA.

MC The number of visitors to a regional animal park is seasonal.

Data is collected and deseasonalised before a least squares line is fitted.

The equation of the least squares line is

$$deseasonalised\ number\ of\ visitors = 2349 - 198.5 \times month\ number$$

where *month number* 1 is January 2020.

The seasonal indices for the 12 months of 2020 are shown in the table below.

Month number	1	2	3	4	5	6	7	8	9	10	11	12
Seasonal index	1.10	1.25	1.15	0.95	0.85	0.75	0.80	0.85	0.95	1.10	1.15	1.10

The actual number of visitors predicted for February 2020 was closest to

A. 1562   **B.** 1697   **C.** 1952   **D.** 2245   **E.** 2440

Question 3 (1 mark)

Source: VCE 2020, Further Mathematics Exam 1, Section A, Q17; © VCAA.

MC The table below shows the monthly rainfall for 2019, in millimetres, recorded at a weather station, and the associated long-term seasonal indices for each month of the year.

	Jan.	Feb.	Mar	Apr.	May	June	July	Aug.	Sep.	Oct.	Nov.	Dec.
Monthly rainfall (mm)	18.4	17.6	46.8	23.6	92.6	77.2	80.0	86.8	93.8	55.2	97.3	69.4
Seasonal index	0.728	0.734	0.741	0.934	1.222	0.973	1.024	1.121	1.159	1.156	1.138	1.072

Data: adapted from © Commonwealth of Australia 2020, Bureau of Meteorology

The deseasonalised rainfall for May 2019 is closest to

- **A.** 71.3 mm
- **B.** 75.8 mm
- **C.** 86.1 mm
- **D.** 88.1 mm
- **E.** 113.0 mm

More exam questions are available online.

4.8 Review

4.8.1 Summary

doc-38035

Hey students! Now that it's time to revise this topic, go online to:

 Access the topic summary

 Review your results

 Watch teacher-led videos

 Practise VCAA exam questions

Find all this and MORE in jacPLUS

4.8 Exercise

Multiple choice

1. **MC** The price of oranges over a 16-month period is recorded as shown. The trend can be described as:

   A. cyclic.
   B. seasonal.
   C. irregular.
   D. secular.
   E. non-existent.

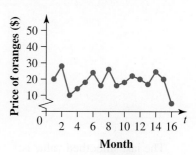

2. **MC** From another 16-month time series, for the price of apples, it was found that the least squares trend line was: $price = 8.45 + 0.415 \times month$. A prediction for the price of apples in month 18 is:

   A. 8.45
   B. 0.42
   C. 6.64
   D. 15.92
   E. unable to be determined with the above information.

3. **MC** A least squares trend line has been fitted to the time series in the figure shown.
   Its equation is most likely to be:

   A. $y = 10t$
   B. $y = -8t + 10$
   C. $y = 8t$
   D. $y = 8t - 10$
   E. $y = 8t + 10$

4. **MC** Consider the following data.

Time	2016	2017	2018	2019	2020	2021	2022
y-value	12	13	16	16	17	19	22

The value, after a 4-point moving average smoothing after centring, plotted against the year 2019 is:

A. 16.25   B. 14.25   C. 15.5   D. 17   E. 14.875

5. **a.** **MC** The following data represent the number of employees in a car manufacturing plant. The data are smoothed using a 3-point moving average.

Year	2015	2016	2017	2018	2019	2020	2021	2022
Number	350	320	300	310	270	240	200	160

The first two points in the smoothed trend line are:

A. 320 and 300   **B.** 320 and 310   **C.** 323 and 310

D. 335 and 310   **E.** 323 and 273

b. **MC** Determine how many points the smoothed trend in part **a** will contain.

A. 8   **B.** 7   **C.** 6   **D.** 5   **E.** 4

6. **MC** A 3-point median smoothing is performed on the data in the figure shown.

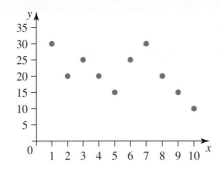

The last smoothed value is:

A. 25   **B.** 21.7   **C.** 20   **D.** 15   **E.** 9

7. **MC** Seasonal indices and adjustment can be used when:

A. there are irregular variations in the data.

B. there are seasonal variations along with a secular trend.

C. there are seasonal variations only.

D. there are seasonal or cyclic variations.

E. there are at least four seasons' worth of data.

8. **MC** The seasonal indices in the table were obtained from a time series.

Season	Spring	Summer	Autumn	Winter
Index	1.12	0.78	0.92	

a. The value of the winter's seasonal index is:

A. 1.18   **B.** 0.94   **C.** 1.08

D. 1.06   **E.** None of the above.

b. **MC** Using the data, a seasonally adjusted value for summer when the original value was 520, is closest to:

A. 406   **B.** 667   **C.** 464

D. 614   **E.** None of the above.

9. **MC** If you were to smooth a quarterly gas bill data over a number of years, it would be best to use:

A. 3-point smoothing.

B. 4-point (centred) smoothing.

C. 5-point smoothing.

D. 6-point (centred) smoothing.

E. 7-point smoothing.

10. **MC** The time series plot shows the revenue from sales (in dollars) each month made by a Queensland souvenir shop over a 3-year period.

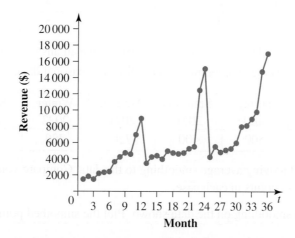

This time series plot indicates that, over the 3-year period, revenue from sales each month showed:

A. no overall trend.

B. no correlation.

C. positive skew.

D. an increasing trend only.

E. an increasing trend with seasonal variation.

Short answer

11. The number of uniforms sold in a school uniform shop is reported in the table.

Month	January	February	March	April	May	June
Number of uniforms sold	118	92	53	20	47	102

Month	July	August	September	October	November	December
Number of uniforms sold	90	42	35	26	12	58

Fit a trend line to the data. State the type of trend that is best reflected by the data. Explain these trends.

12. Fit a least squares line for the following data, which represent the sales at a snack bar during the recent Melbourne show.

Day	1	2	3	4	5	6	7	8
Sales	2300	2200	2600	3100	2900	3200	3300	3500

State the gradient and y-intercept.

13. A hotel records the number of rooms booked over an 11-day period. Fit a trend line using the least squares method.

Day	1	2	3	4	5	6	7	8	9	10	11
Rooms	12	18	15	20	22	20	25	24	26	28	30

a. State the gradient and y-intercept, rounded to 2 decimal places.

b. Predict the number of rooms booked for days 12 and 13.

14. Perform a 3-point moving average smoothing on the following rainfall data.

Day	1	2	3	4	5	6	7	8
Rainfall (mm)	2	5	4	6	3	7	6	9

Plot the original and smoothed data on the same set of axes. Give all answers rounded to 1 decimal place.

15. **a.** Apply a 5-point moving average smoothing to the following seasonal data of coat sales.

Season	Winter 2021	Spring 2021	Summer 2021	Autumn 2021	Winter 2022	Spring 2022	Summer 2022	Autumn 2022
Sales ($)	690	500	400	720	780	660	550	440

   b. Apply a 4-point centred moving average smoothing to the data. Compare your results. Comment on the number of smoothed data points in each case.

16. Perform a 3-point median smoothing on the data shown. Plot the smoothed points and join them with straight-line segments.

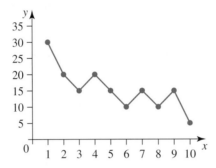

17. The seasonal indices for the price of shares in CSP fruit canneries are shown in the table.

Season	Winter	Spring	Summer	Autumn
Index	1.7	0.6	0.5	1.2

Use the seasonal indices to deseasonalise the following data.

Season	Share price	
	Seasonalised	Deseasonalised
Spring	150	
Summer	100	
Autumn	300	
Winter	400	

Extended response

18. Jazza's online store has been open for the past three weeks. The sales figures for the store were recorded and tabulated as follows.

Jazza's online store daily sales figures — number of video games sold							
	Mon.	Tues.	Wed.	Thurs.	Fri.	Sat.	Average daily sales for the week
Week 1	10	8	12	15	24	45	
Week 2	12	9	14	18	26	53	
Week 3	15	10	16	21	33	58	

a. Plot the data as a time series plot and comment on the type of trend that exists. Justify your choice.
b. State the number of seasons.
c. Calculate the average daily sales for each of the weeks.
d. Complete the table of seasonal indices for each day.

Jazza's online store daily sales figures — number of video games sold						
	Mon.	**Tues.**	**Wed.**	**Thurs.**	**Fri.**	**Sat.**
Week 1	$\frac{10}{19} = 0.5263$	0.4211	$\frac{12}{19} =$	0.7895	1.2632	
Week 2	$\frac{12}{} = 0.5455$	0.4091	$\frac{14}{} = 0.6364$	0.8182		2.4091
Week 3	0.5882	0.3922			1.2941	2.2745

e. Complete the following table of seasonal indices.

	Mon.	**Tues.**	**Wed.**	**Thurs.**	**Fri.**	**Sat.**
Seasonal indices	0.5533	$\frac{1.2224}{3} = 0.4075$	0.6318	$\frac{2.4312}{} = 0.8104$	1.2462	

f. Interpret what an index of 2.3507 means.

19. The following table shows deseasonalised video games sales figures.

Jazza's online store daily sales figures — number of video games sold						
	Mon.	**Tues.**	**Wed.**	**Thurs.**	**Fri.**	**Sat.**
Week 1	18.07	19.63	18.99	18.51	19.26	19.14
Week 2	21.69	22.09	22.16	22.21	20.86	22.55
Week 3	27.11	24.54	25.32	25.91	26.48	24.69

a. Determine the equation of the trend line using the least squares method, and interpret the values of the gradient and the y-intercept.
b. Using the trend line, predict the deseasonalised sales figures for:
  i. Monday week 4   ii. Saturday week 4   iii. Saturday week 6.
c. Using the deseasonalised values from part b, calculate the actual expected future sales for each. Comment on the reliability of the predictions.

20. This question relates to the following data, which represent seasonal rainfall (mm) in an Australian city.

Season	1	2	3	4	5	6	7	8	9	10	11	12
Rainfall (mm)	43	75	41	13	47	78	50	19	51	83	55	25

a. Plot the data points and try to fit a trend line by eye. Comment on the ease of fitting the line to this plot.
b. Fit a trend line using the least squares technique. Compare your result with the previous one.
c. To smooth out the seasonal variation, 3-point and 5-point moving average smoothings are tried. Compare the results of these two methods with the results from questions a and b by plotting the smoothed data.
d. Upon observing the results with the 5-point smoothing, a trend appears. Take the data from the 5-point moving average smoothing and fit a straight line using the least squares method.
Put the first smoothed point at $t = 3$ and then centre the time data. State the y-intercept and gradient. Compare this trend line with that from question b.

e. Given the seasonal nature of the data, a 4-point moving average smoothing is tried. After calculating the 4-point moving average, fit a least squares line, following the method of question **d**.
Compare the results obtained with those from question **d**.

f. Finally, try seasonal adjustment. Take $t = 1$ to be summer and calculate the seasonal indices. Then, seasonally adjust the data.

g. Take the seasonally adjusted data from question **f** and fit a trend line using the least squares method. Comment on this result.

4.8 Exam questions

Question 1 (5 marks)

Source: VCE 2021, Further Mathematics Exam 2, Section A, Q4; © VCAA.

The time series plot below shows that the winning time for both men and women in the 100 m freestyle swim in the Olympic Games has been decreasing during the period 1912 to 2016.

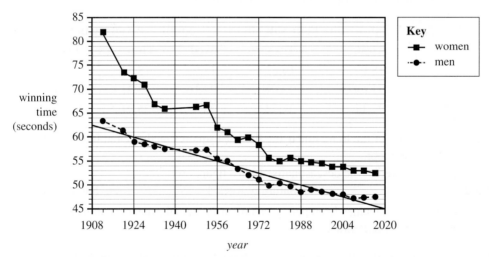

Data: International Olympic Committee, <https://olympics.com/en/olympic-games/olympic-results>

Least squares lines are used to model the trend for both men and women.

The least squares line for the men's winning time has been drawn on the time series plot above.

The equation of the least squares line for men is

$$winning\ time\ men = 356.9 - 0.1544 \times year$$

The equation of the least squares line for women is

$$winning\ time\ women = 538.9 - 0.2430 \times year$$

a. Draw the least squares line for *winning time women* on the **time series plot above**. **(1 mark)**
(*Answer on the time series plot above.*)

b. The difference between the women's predicted winning time and the men's predicted winning time can be calculated using the formula

$$difference = winning\ time\ women - winning\ time\ men$$

Use the equations of the least squares lines and the formula above to calculate the *difference* predicted for the 2024 Olympic Games.

Round your answer to one decimal place. **(2 marks)**

c. The Olympic Games are held every four years. The next Olympic Games will be held in 2024, then 2028, 2032 and so on.

In which **Olympic year** do the two least squares lines predict that the winning time for women will first be faster than the winning time for men in the 100 m freestyle? **(2 marks)**

Question 2 (3 marks)

Source: VCE 2019, Further Mathematics Exam 2, Section A, Q6; © VCAA.

The total rainfall, in millimetres, for each of the four seasons in 2015 and 2016 is shown in the table below.

	Total rainfall (mm)			
Year	Summer	Autumn	Winter	Spring
2015	142	156	222	120
2016	135	153	216	96

a. The seasonal index for winter is shown in the next table below.
Use the values in the previous table to find the seasonal indices for summer, autumn and spring.
Write your answers in the next table, rounded to two decimal places. **(2 marks)**

	Summer	Autumn	Winter	Spring
Seasonal index			1.41	

b. The total rainfall for each of the four seasons in 2017 is shown in the table below. **(1 mark)**

	Total rainfall (mm)			
Year	Summer	Autumn	Winter	Spring
2017	141	156	262	120

Use the appropriate seasonal index from the previous table to deseasonalise the total rainfall for winter in 2017.

Round your answer to the nearest whole number.

Question 3 (9 marks)

Source: VCE 2018, Further Mathematics Exam 2, Section A, Q3; © VCAA.

The table below shows the yearly average traffic congestion levels in two cities, Melbourne and Sydney, during the period 2008 to 2016. Also shown is a time series plot of the same data.

The time series plot for Melbourne is incomplete.

	Congestion level (%)								
Year	2008	2009	2010	2011	2012	2013	2014	2015	2016
Melbourne	25	26	26	27	28	28	28	29	33
Sydney	28	30	32	33	34	34	35	36	39

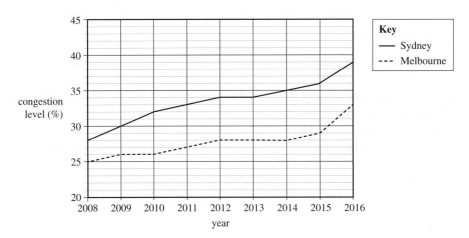

a. Use the data in the table above to complete the **time series plot above** for Melbourne. **(1 mark)**

b. A least squares line is used to model the trend in the time series plot for Sydney. The equation is

$$congestion\,level = -2280 + 1.15 \times year$$

   i. Draw this least squares line on the **time series plot**. **(1 mark)**

   ii. Use the equation of the least squares line to determine the average rate of increase in percentage congestion level for the period 2008 to 2016 in Sydney. **(1 mark)**
   Write your answer in the box provided below.

                                            % per year

   iii. Use the least squares line to predict when the percentage congestion level in Sydney will be 43%. **(1 mark)**

c. The yearly average traffic congestion level data for Melbourne is repeated in the following table.

	Congestion level (%)								
Year	2008	2009	2010	2011	2012	2013	2014	2015	2016
Melbourne	25	26	26	27	28	28	28	29	33

When a least squares line is used to model the trend in the data for Melbourne, the intercept of this line is approximately -1514.75556
Round this value to four significant figures. **(1 mark)**

d. Use the data in the table in part **c** to determine the equation of the least squares line that can be used to model the trend in the data for Melbourne. The variable *year* is the explanatory variable.
Write the values of the intercept and the slope of this least squares line in the appropriate boxes provided below.
Round both values to four significant figures. **(2 marks)**

              congestion level =   –   × year

e. Since 2008, the equations of the least squares lines for Sydney and Melbourne have predicted that future traffic congestion levels in Sydney will always exceed future traffic congestion levels in Melbourne.
Explain why, quoting the values of appropriate statistics. **(2 marks)**

▶ **Question 4 (1 mark)**

Source: VCE 2017, Further Mathematics Exam 1, Section A, Q15; © VCAA.

MC The wind speed at a city location is measured throughout the day.

The time series plot below shows the daily *maximum wind speed*, in kilometres per hour, over a three-week period.

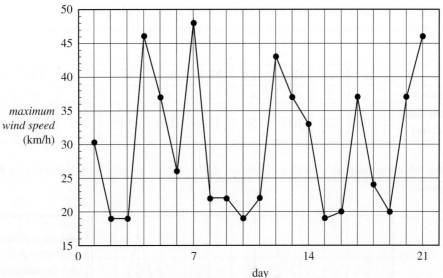

The table below shows the daily *maximum wind speed*, in kilometres per hour, for the days in week 2.

Day	8	9	10	11	12	13	14
Maximum wind speed (km/h)	22	22	19	22	43	37	33

A four-point moving mean with centring is used to smooth the time series data above.

The smoothed *maximum wind speed*, in kilometres per hour, for day 11 is closest to

A. 22   **B.** 24   **C.** 26   **D.** 28   **E.** 30

▶ **Question 5 (4 marks)**

Source: VCE 2016, Further Mathematics Exam 2, Q4; © VCAA.

The time series plot below shows the *minimum rainfall* recorded at the weather station each month plotted against the *month number* (1 = January, 2 = February, and so on).

Rainfall is recorded in millimetres.

The data was collected over a period of one year.

Key
- —●— unsmoothed
- —✕— five-median smoothed

minimum rainfall (mm)

month number

a. Five-point median smoothing has been used to smooth the time series plot above.

The first four smoothed points are shown as crosses (×).

Complete the five-median smoothing by marking smoothed values with crosses (×) on the **time series plot above.** **(2 marks)**

The maximum daily rainfall each month was also recorded at the weather station.

The table below shows the *maximum daily rainfall* each month for a period of one year.

Month	Jan.	Feb.	Mar.	Apr.	May	June	July	Aug.	Sep.	Oct.	Nov.	Dec.
Month number	1	2	3	4	5	6	7	8	9	10	11	12
Maximum daily rainfall (mm)	79	123	100	156	174	186	149	162	124	140	225	119

The data in the table has been used to plot *maximum daily rainfall* against *month number* in the time series plot below.

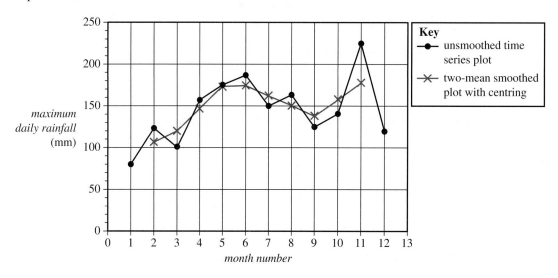

b. Two-mean smoothing with centring has been used to smooth the time series plot above.

The smoothed values are marked with crosses (×).

Using the data given in the table, show that the two-mean smoothed rainfall centred on October is 157.25 mm. **(2 marks)**

More exam questions are available online.

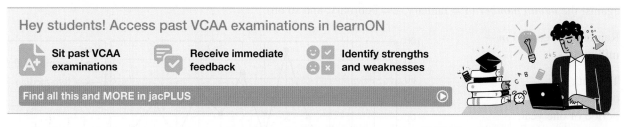
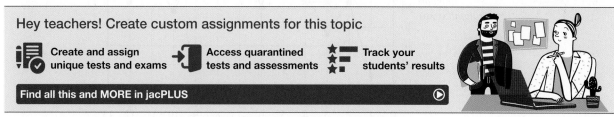

Answers

Topic 4 Investigating and modelling time series data

4.2 Time series plots and trends

4.2 Exercise

Note: Your answers may vary slightly due to using the 'by eye' method.

1. a. The time series plot displays seasonality only.
   b. The time series plot displays seasonality with an upward trend.
   c. The time series plots display irregular fluctuations with no obvious trend.

2. C

3.

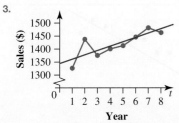

Upward trend with a possible outlier

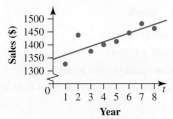

$Sales = 1342.96 + 16.45 \times year$

4.

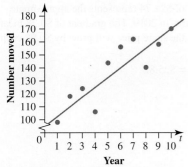

Irregular with an upward trend

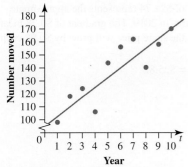

$Number\ moved = 97.8 + 7.22 \times year$

5. a. Seasonal
   b. Irregular
   c. Upward trend
   d. Irregular or cyclical with slight upward trend
   e. Cyclical

6. Definite downward trend

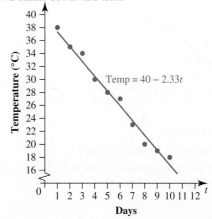

7. Although there are some random variations, the time series could also be cyclical, with an upward trend.

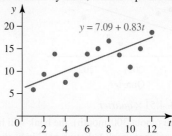

8. Predicting a share price in the following year is an extrapolated value (outside the plotted values) and can only be treated as an approximate value at best.

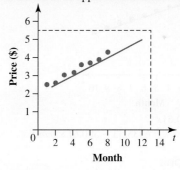

$Price = 2.26 + 0.24t$

9. The time series is cyclical, so it is difficult to fit an appropriate trend line. No noticeable trend to the data.

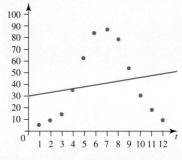

10. a.

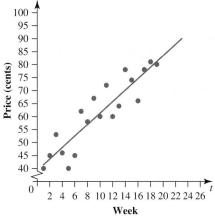

Price $= 39.42 + 2.21t$

b. Prediction for $t = 25$ is about 95 cents.

11. Difficult to fit an accurate trend line, due to likely cyclical nature of the software sales business.

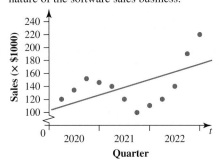

Sales $= 111.52 + 4.51 \times quarter$

12. At the current rate (about 300/month), the bank will have no employees in another year! Although not likely, there is a clear downward trend.

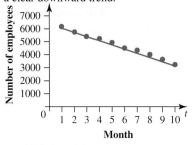

$y = 6333.33 - 300.61x$
Number of employees $= 6333.33 - 300.61 \times months$

4.2 Exam questions

Note: Mark allocations are available with the fully worked solutions online.

1. D

2. A

3. C

4.3 Fitting the least squares line and forecasting

4.3 Exercise

1. $y = 1.18 - 0.02x$ or millions of viewers $= 1.18 - 0.02 \times$ time code
  a. Number of viewers $= 840\ 000$
  b. Number of viewers $= 740\ 000$

2. $y = 128.1 - 0.77x$ or driver fatalities $= 128.1 - 0.77 \times$ time code
  a. Driver fatalities $= 68$ **b.** Driver fatalities $= 42$

3. $y = 33.72 - 2.9t$, where $y =$ cars remaining and $t =$ time
  a. 2   **b.** 2.9 per hour
  c. 11.63 hours

4. a. $y = 2.68 + 1.16x$, where $y =$ numbers of staff and $x =$ months, where $x = 1$ is the first month of business

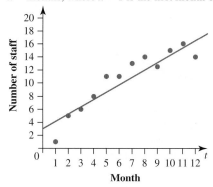

  b. $y = 30.52$, approximately 31 staff

5. a. Least squares: $y = 2.48 + 0.044t$
  b. Probably a random (or cyclical) time series.

6. $y = 0.283 + 0.83x$; given the exponential nature of data, it is a very poor predictor.

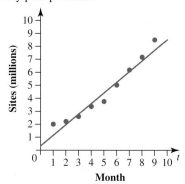

7. a. *Share price ($) = \$22.74 + \$0.28 \times time code*
    Time code of $t = 1$ represents 2020
  b. The y-intercept of \$22.74 represents the approximate value of the shares in 2019. The gradient of \$0.28 means that on average the share price will grow by \$0.28 each year.
  c. Share price ($) $= \$25.82$

8. a. *Share price ($) = $18.57 − $0.1 × time code*
   Time code of $t = 1$ represents 2020

b. The y-intercept of $18.57 represents the approximate value of the shares in 2019. The gradient of $−$0.1$ means that on average the share price will decline by $0.1 (10 cents) each year.

c. Share price ($) = $17.57

9. a. *Share price = $2.56 + $0.72 × time code*
   Time code $t = 1$ represents 2020, $t = 2$ represents 2021 and so on.

b. The y-intercept of $2.56 represents the approximate value of the shares in 2019. The gradient of $+$0.72$ means that the share value will grow by $0.72 (72 cents) each year.

c. $10.48

10. a. $y = 49.64 + 20.36t$, where $y =$ sales and $t =$ months

b.

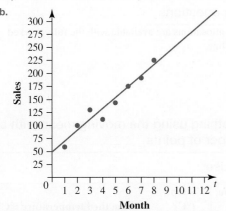

c. y(Dec.) = 293.93 (294)

d. The trend line fits the data well, so it is a good predictor of future trends.

11. a. $y = −1.83 + 3.77t$, where $y =$ sales and $t =$ weeks

b.

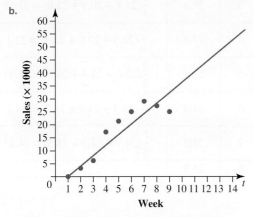

c. $y(10) = 35.83, y(12) = 43.37, y(14) = 50.90$

d. A poor predictor, given the nature of the data

12. a. $y = 315.8 + 6.35\ t$, where $y =$ price and $t =$ quarters

b.

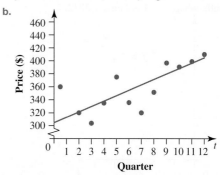

c. $398

d. The first two years displayed seasonal fluctuations, the last year an upward trend, so overall trend line would be a poor predictor.

13. a. $y = 56.9 + 1.93t$, where $y =$ marks and $t =$ months

b.

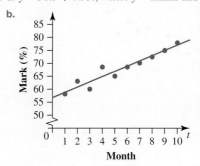

c. 78%

d. Given fairly even increase in averages, the trend line is an excellent predictor.

14. a. $y = 1848.57 + 173.1x$, where $y =$ All Ordinaries value and $x =$ time, where $x = 1$ is 1995.

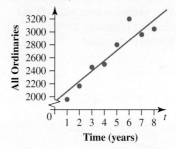

b. Taking out the 2000 data value:

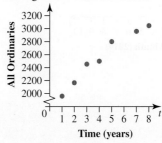

c. $y = 1777.14 + 192.14x$, where $y =$ All Ordinaries and $x =$ time

d. The gradient increased and the y-intercept is lower when the outlier is taken out.

15. a. $y = 199.71 + 8.62x$, where $y =$ stock price (\$) and $x =$ month, where $x = 1$ is January 2022.

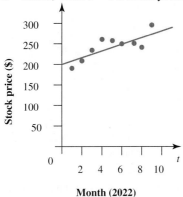

Month (2022)

b. The April value is a possible outlier.

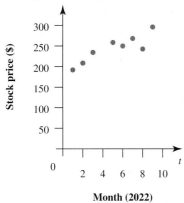

Month (2022)

c. $y = 193.72 + 9.13x$, where $y =$ stock price and $x =$ month

d. The gradient increased and the y-intercept was lower when the April value was taken out.

16. a. $y = 222.72 + 4.97x$, where $y =$ hotel price (\$) and $x =$ months, where $x = 1$ is January 2019.
See table at the bottom of the page*

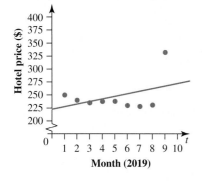

Month (2019)

b. Outlier = September value

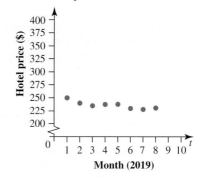

Month (2019)

c. Hotel price (\$) $= 245.79 - 1.95 \times$ month

d. Gradient (positive to negative), y-intercept (increased)

4.3 Exam questions

Note: Mark allocations are available with the fully worked solutions online.

1. B

2. D

3. E

4.4 Smoothing using the moving mean with an odd number of points

4.4 Exercise

1. a.

Time (h)	Temp (°C)	Smoothed temperature (°C)
1	20.3	
2	21.8	$\frac{1}{3}(20.3 + 21.8 + 20.9) = 21$
3	20.9	$\frac{1}{3}(21.8 + 20.9 + 22.0) = 21.57$
4	22.0	$\frac{1}{3}(20.9 + 22.0 + 23.4) = 22.1$
5	23.4	$\frac{1}{3}(22.0 + 23.4 + 24.9) = 23.43$
6	24.9	$\frac{1}{3}(23.4 + 24.9 + 24.3) = 24.2$
7	24.3	$\frac{1}{3}(24.9 + 24.3 + 25.3) = 24.83$
8	25.3	

---

*16. a.

Month (2019)	1	2	3	4	5	6	7	8	9
Hotel price (\$)	250	240	235	237	239	230	228	237	332

b.

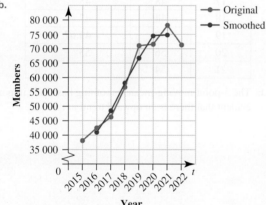

c. 24.83 °C

2. **a.** See the table at the bottom of the page.*

b.

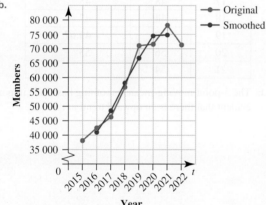

c. 74 428 members

3. **a.** Smoothed data:

2417	2583	2683	2700	2817	2983

b.

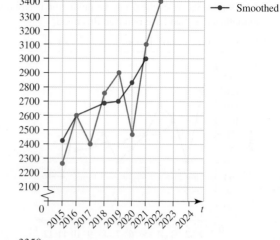

c. 3350

4. **a.** Smoothed data: possible downward trend, but still fluctuations.

96.7	93.3	100	93.3	80	80	80	90	80	73.3

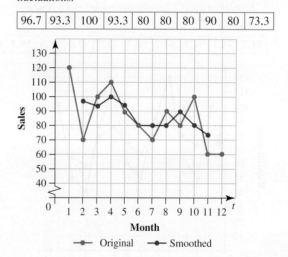

*2. a.

Year (t)	Members (m)	Smoothed members (m)
2015	38 587	
2016	42 498	$\frac{1}{3}(38\ 587 + 42\ 498 + 45\ 972) = 42\ 352$
2017	45 972	$\frac{1}{3}(42\ 498 + 45\ 972 + 57\ 408) = 48\ 626$
2018	57 408	$\frac{1}{3}(45\ 972 + 57\ 408 + 71\ 271) = 58\ 217$
2019	71 271	$\frac{1}{3}(57\ 408 + 71\ 271 + 72\ 688) = 67\ 122$
2020	72 688	$\frac{1}{3}(71\ 271 + 72\ 688 + 78\ 427) = 74\ 129$
2021	78 427	$\frac{1}{3}(72\ 688 + 78\ 427 + 72\ 170) = 74\ 428$
2022	72 170	

b. Price $= 99.78 - 2.39t$

c. 63.93, the data are cyclical and the prediction is based on smoothed data that have removed this trend.

d.

| 98 | 90 | 90 | 88 | 82 | 84 | 80 | 78 |

Month
— Original — Smoothed

Smoothed data: a definite downward trend is now apparent.

5.

| 71.7 | 78.3 | 91.7 | 86.7 | 66.7 | 73.3 | 85.0 | 78.3 | 58.3 | 63.3 |

Quarter
— Original — Smoothed

Smoothed data did not remove seasonal fluctuation; from the figure, there may be a slight trend downward.

6. a.

| 72 | 71.7 | 69.7 | 70.7 | 66.3 | 65.7 | 64.7 | 66 |

Smoothed data show a clear downward trend.

b. Attendance $= 74.47 - 1.11x$

c. 61, this is a reasonable prediction as long as the trend continues to decline as given by the negative gradient.

7. a.

Week	Sales	Smoothed data
1	34	
2	27	30.67
3	31	31.67
4	37	36.33
5	41	35.67
6	29	34
7	32	32.67
8	37	38.67
9	47	40.67
10	38	42
11	41	41
12	44	44
13	47	46.67
14	49	45.67
15	41	47.33
16	52	47
17	48	48
18	44	47
19	49	49.67
20	56	53
21	54	

b. The 5-point moving-mean smoothing has made it more evident that there is an upward trend in the data.

8. a. See the table and graph at the bottom of the page.*
Smoothed data: Some but not all seasonal fluctuation removed.

b. It has further smoothed out the seasonal fluctuations.

9.

253.8	303	306.6	320.4	322.4	362	364.2

See the graph at the bottom of the page.*
Smoothed data: most random variation smoothed, slight upward trend possible. Prediction for 12th month is 364.

10.

66.33	71.33	68.33	79.67	83.67

Smoothed data; prediction for week 8 is 84.

4.4 Exam questions

Note: Mark allocations are available with the fully worked solutions online.

1. B

2. D

3. C

*8. a.

332.3	325	340.3	346	343.3	337.3	356	375.7	389.7	394

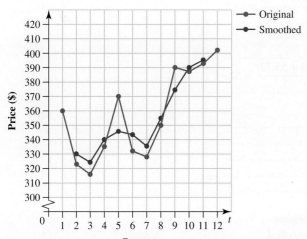

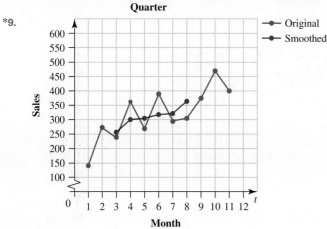

4.5 Smoothing using the moving mean with an even number of points

4.5 Exercise

1. **a.** See the table and graph at the bottom of the page.*
   b. The data shows a steady increasing trend.

Time	Sale	4-point moving mean	4-point centred moving mean
1	59		
2	48		
		$\dfrac{(59 + 48 + 43 + 50)}{4} = 50$	
3	43		$\dfrac{(50 + 51)}{2} = 50.5$
		$\dfrac{(48 + 43 + 50 + 63)}{4} = 51$	
4	50		$\dfrac{(51 + 52)}{2} = 51.5$
		$\dfrac{(43 + 50 + 63 + 52)}{4} = 52$	
5	63		$\dfrac{(52 + 53)}{2} = 52.5$
		$\dfrac{(50 + 63 + 52 + 47)}{4} = 53$	
6	52		$\dfrac{(53 + 55.75)}{2} = 54.38$
		$\dfrac{(63 + 52 + 47 + 61)}{4} = 55.75$	
7	47		
8	61		

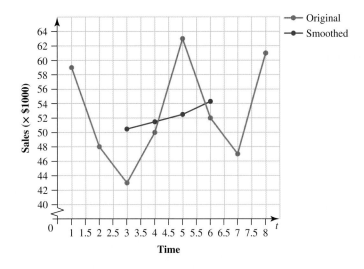

2. a. See the table at the bottom of the page.*

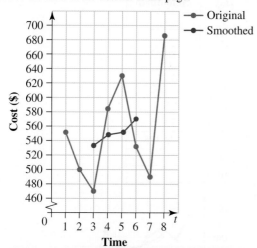

b. This data shows a steady increasing trend. This is not obvious with the original data.

3. a.

t	3	4	5	6	7	8
y	62.125	60.375	58.25	57.125	56	57.375

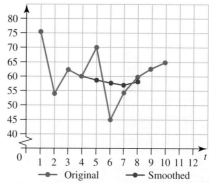

b. Smoothed data indicate a general downward trend, possibly with cyclic fluctuations in original data.

---

*2. a.

Time	Sale	4-point moving mean	4-point centred moving mean
1	554		
2	503		
		$\dfrac{(554 + 503 + 467 + 587)}{4} = 527.75$	
3	467		$\dfrac{(527.75 + 548.25)}{2} = 538$
		$\dfrac{(503 + 467 + 587 + 636)}{4} = 548.25$	
4	587		$\dfrac{(548.25 + 555.75)}{2} = 552$
		$\dfrac{(467 + 587 + 636 + 533)}{4} = 555.75$	
5	636		$\dfrac{(555.75 + 562.25)}{2} = 559$
		$\dfrac{(587 + 636 + 533 + 493)}{4} = 562.25$	
6	533		$\dfrac{(562.25 + 586.50)}{2} = 574.38$
		$\dfrac{(636 + 533 + 493 + 684)}{4} = 586.5$	
7	493		
8	684		

4.

t	Price ($)
Autumn '20	
Winter '20	
Spring '20	52.625
Summer '20	54.63
Autumn '21	57.25
Winter '21	59.25
Spring '21	60.50
Summer '21	61.75
Autumn '22	62.63
Winter '22	63.50
Spring '22	
Summer '22	

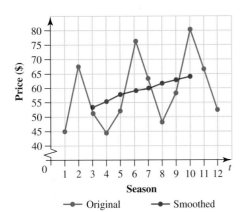

Smoothed data indicate a strong upward trend of almost 11 cents over 3 years.

5. a.

Day	Temperature	4-point moving mean	4-point centred moving mean
1	36.6		
2	36.4		
		36.75	
3	36.8		36.7875
		36.825	
4	37.2		36.8375
		36.85	
5	36.9		36.9
		36.95	
6	36.5		36.975
		37	
7	37.2		37.025
		37.05	
8	37.4		37.1625
		37.275	
9	37.1		37.325
		37.375	
10	37.4		37.3125
		37.25	
11	37.6		37.2625
		37.275	
12	36.9		37.3
		37.325	
13	37.2		37.2375
		37.15	
14	37.6		
15	36.9		

b. Temperature $= 36.766 + 0.056d$

c. 37.7 °C

6. A

7. D

8. a.

Day	Smoothed
1	
2	
3	180.625
4	177.75
5	176.625
6	175
7	173.25
8	172.375
9	
10	

b.

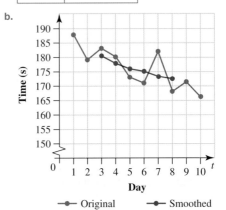

c. Yes, there is a significant improvement in times.

d.

Day	Time (s)	6-point moving average	6-point centred moving average
1	188		
2	179		
3	183		
		179	
4	180		178.5
		178	
5	173		177.09
		176.17	
6	171		175.17
		174.17	
7	182		173
		171.83	
8	168		
9	171		
10	166		

9. a.

Season	Smoothed
Q_3-19	
Q_4-19	
Q_1-20	81.25
Q_2-20	78.375
Q_3-20	76.625
Q_4-20	75.375
Q_1-21	74.875
Q_2-21	73.875
Q_3-21	71.875
Q_4-21	69.375
Q_1-22	
Q_2-22	

b.

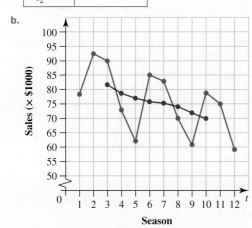

c. Smoothed data indicate a clear downward trend.

d.

Day	Smoothed
1	
2	
3	
4	36.78
5	36.92
6	37.03
7	37.07
8	37.14
9	37.24
10	37.27
11	37.29
12	37.29
13	
14	
15	

10. a.

Day	Index
1	678
2	726
3	692
4	714
5	689
6	687
7	772
8	685
9	688
10	712

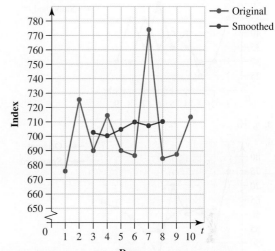

b.

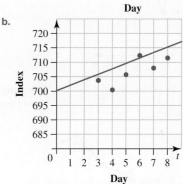

$y = 700.2 + 1.88x$, where y = price index and x = day number.

c. $y = 728.4$; therefore, the price index is approximately 728 after 15 days.

4.5 Exam questions

Note: Mark allocations are available with the fully worked solutions online.

1. C

2. B

3. D

4.6 Median smoothing from a graph

4.6 Exercise

1.

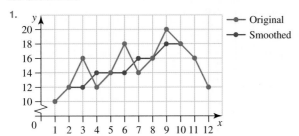

2.

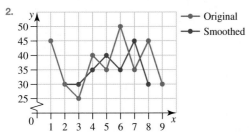

3. a. 10   **b.** 7

4.

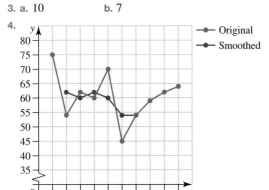

5.

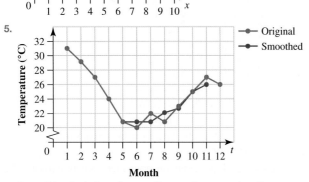

Smoothing had virtually no effect on data — only minor variations smoothed.

6. a.

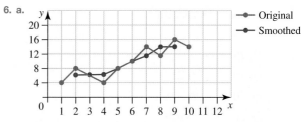

b. Effective at smoothing out small random variations.

7. a.

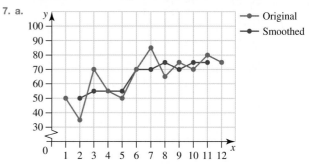

b. Smoothed out much of the variation; indicates a slight upward trend.

8.

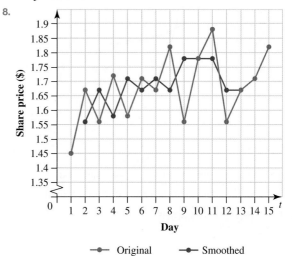

9. See the graph at the bottom of the page.*

10. a.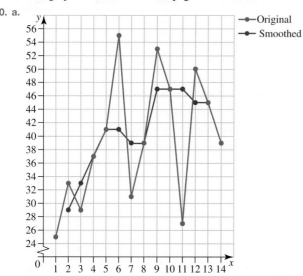

b.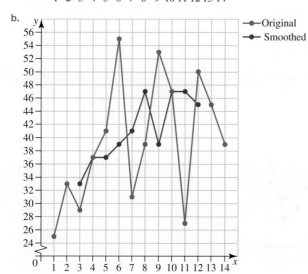

The 3-point median smoothing in part **a** reduces the fluctuations more than the 5-point median smoothing in part **b**.

4.6 Exam questions

Note: Mark allocations are available with the fully worked solutions online.

1. C
2. E
3. D

4.7 Seasonal adjustment

4.7 Exercise

1. a.

Season	Summer	Autumn	Winter	Spring
Seasonal index	0.96	0.92	1.04	1.08

b.

Season	2018	2019	2020	2021	2022
Summer	1.61	1.67	1.56	1.67	1.51
Autumn	1.63	1.68	1.63	1.58	1.52
Winter	1.59	1.68	1.63	1.63	1.54
Spring	1.57	1.67	1.62	1.67	1.53

c.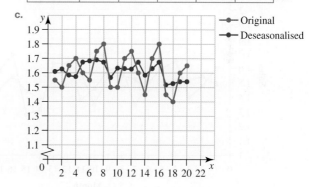

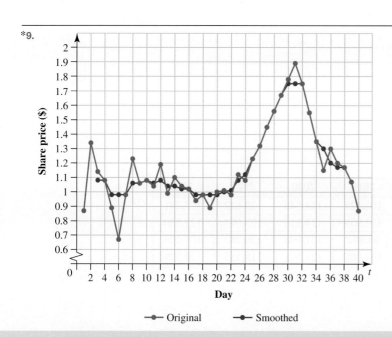

d. There seems to be a very slight downward trend in milk prices.

e. Deseasonalised 1 L of milk cost $= 1.6557 - 0.004\,40 \times$ time code, where time code 1 represents summer 2018.

f. Predicted summer 2023 price $= \$1.50$

2. a. Seasonal indices: 1.0731, 0.9423, 0.8874, 1.0972

b.

Season	2018	2019	2020	2021	2022
Summer	1.4910	1.4910	1.4630	1.5190	1.4724
Autumn	1.5706	1.5918	1.5494	1.3902	1.3478
Winter	1.5776	1.5213	1.5664	1.4424	1.3410
Spring	1.5038	1.4583	1.4856	1.5402	1.4491

c. See the graph at the bottom of the page.*

d. There appears to be a slight downward trend in the price of 1 litre of petrol.

e. Deseasonalised 1 L of petrol cost $= 1.5659 - 0.007\,38 \times$ time code, where time code 1 represents summer 2018.

f. Predicted summer 2023 price $= \$1.51$

3. a. Seasonal indices: 0.8504, 1.0692, 1.1467, 0.9336

b.

Season	2020	2021	2022
Summer	1.21	1.15	1.12
Autumn	1.18	1.17	1.13
Winter	1.19	1.17	1.12
Spring	1.22	1.15	1.11

c. See the graph at the bottom of the page.*

d. Slight trend downwards

4. a. Seasonal indices: 0.9394, 0.8044, 1.2274, 1.0288

b. See the table at the bottom of the page.*

*2. c.

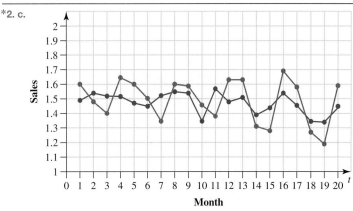

*3. c.

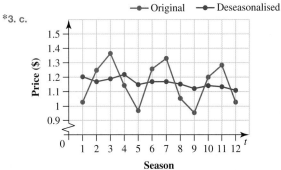

*4. b.

Season	2017	2018	2019	2020	2021	2022
Summer	109.644	103.257	101.128	124.548	125.612	127.741
Autumn	115.614	104.426	101.939	124.316	123.073	121.830
Winter	116.506	101.027	98.582	127.098	126.283	123.024
Spring	119.557	105.949	104.005	121.501	118.585	120.529

c.

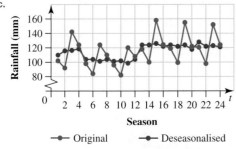

Season

Original ● Deseasonalised ●

d. Probable drought in 2018–2019

5. a. Seasonal indices: 0.7141, 1.100, 1.1832, 1.0027

b. See the table at the bottom of the page.*

c.

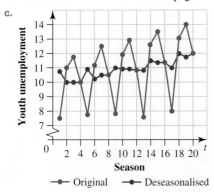

Season

Original ● Deseasonalised ●

d. Youth unemployment increases in all seasons except in summer.

6. a. Seasonal indices: 1.1143, 0.9183, 0.6829, 1.2845

b.

Quarter	1	2	3	4
2020	5.205	5.336	5.125	5.216
2021	5.474	5.554	4.686	5.060
2022	5.115	4.900	6.004	5.527

c.

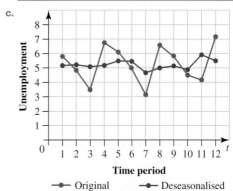

Time period

Original ● Deseasonalised ●

d. Deseasonalised unemployment rate
$= 5.1449 + 0.0188 \times t$

e. i. 6.0   **ii.** 3.9

7. Seasonal indices: 0.6340, 0.6613, 0.8328, 1.0353, 1.2820, 1.4848, 1.0691. Restaurant should probably close Mon.–Tues., steady sales over 5-week period — see the following table.
See the table and graph at the bottom of the page.*

8. A

9. E

---

*5. b.

Season	2018	2019	2020	2021	2022
Summer	10.6428	10.7828	10.9228	10.7828	11.0629
Autumn	9.9091	10.2727	10.8182	11.4545	11.9091
Winter	9.8884	10.4801	10.8181	11.4097	11.7478
Spring	9.8733	10.4717	10.7709	11.3693	11.8680

*7.

Season	Week 1	Week 2	Week 3	Week 4	Week 5
Monday	1634.07	1717.67	1678.23	1788.64	1643.53
Tuesday	1667.93	1581.73	1640.71	1825.19	1748.07
Wednesday	1741.11	1589.82	1785.54	1654.66	1690.68
Thursday	1588.91	1674.88	1728.97	1742.49	1728.00
Friday	1620.90	1719.19	1727.77	1703.59	1690.33
Saturday	1661.50	1668.91	1686.42	1701.24	1743.67
Sunday	1772.52	1670.56	1706.11	1668.69	1641.57

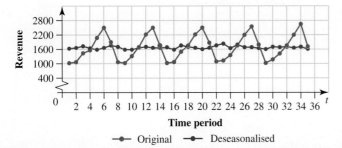

Time period

Original ● Deseasonalised ●

10. D

11. 0.43

12. a. Deseasonalised umbrella sales $= 21.8788 + 0.9161 \times t$

　b. 39

13. a. \$3620.83　　**b.** \$1583.33　　**c.** \$8246

14. a. \$6880.77　　**b.** \$8125　　**c.** \$13515

15. a. \$3000　　　**b.** \$3200　　**c.** \$4800

16. 1.8

17. 1.5

18. 0.9

4.7 Exam questions

Note: Mark allocations are available with the fully worked solutions online.

1. C

2. E

3. B

4.8 Review

4.8 Exercise

Multiple choice

1. C

2. D

3. E

4. A

5. a. C

　b. C

6. D

7. B

8. a. A　　　　　**b.** B

9. B

10. E

Short answer

11.

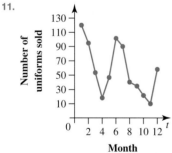

It's very difficult to fit a trend line as data is seasonal. Summer uniforms are mostly bought at the end of the year and then the beginning of the year. Winter uniforms are mostly bought near winter — these are peaks. The sales decrease throughout the other parts of the year.

12. Gradient $= 184.52$, y-intercept $= 2057.14$

13. a. Gradient $= 1.58$, y-intercept $= 12.33$

　b. $y(12) = 31$, $y(13) = 33$

14.

Day	2	3	4	5	6	7
Ave.	3.7	5.0	4.3	5.3	5.3	7.3

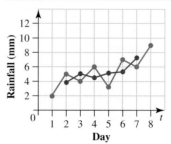

—●— Original　—●— Smoothed

15. a.

Season	Sales (\$)
Winter '21	
Spring '21	
Summer '21	618
Autumn '21	612
Winter '22	622
Spring '22	630
Summer '22	
Autumn '22	

b. Same number of smoothed data points. (In fact a 4-point centred is a special case of a 5-point smoothing.)

Season	Sales (\$)
Winter '17	
Spring '17	
Summer '17	588.75
Autumn '17	620
Winter '18	658.75
Spring '18	642.5
Summer '18	
Autumn '18	

16.

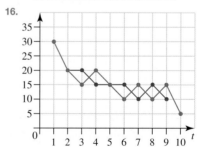

17. 250, 200, 250, 235

Extended response

18. a. The time series is seasonal. There are peaks and troughs occurring on the same days of the week. There is also an upward secular trend. This can also be seen from the table as each day, week after week, more video games are sold.

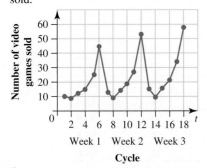

b. 6

c. Week 1: 19, week 2: 22, week 3: 25.5

d. See the table at the bottom of the page.*

e. See the table at the bottom of the page.*

f. An index of 2.3507 means that Saturdays have about 2.35 times the sales compared to the average daily sales.

19. a. Deseasonalised video games sales $= 17.43 + 0.4993t$
y-intercept of 17.4342 (17.43) means that these sales were expected the day before the data were calculated. The gradient of 0.4993 (0.5) means that as each day goes by, Jazza can expect an increase in sales by half a video game each day (or more logically by 1 video game every two days).

b. i. Monday week 4, $t = 19$. Deseasonalised video game sales $= 26.92$

ii. Saturday week 4, $t = 24$. Deseasonalised video game sales $= 29.41$

iii. Saturday week 6, $t = 36$. Deseasonalised video game sales $= 35.40$

c. i. Monday week 4: 15 video games approximately

ii. Saturday week 4: 69 video games approximately

iii. Saturday week 6: 83 video games approximately
The first two predictions are reliable as they are only one week into the future. The Saturday week 6 prediction of 83 video games is not so reliable as it is far into the future and the trend may change in the meantime, such as during holidays and so on.

20. a. Very difficult to fit an accurate trend line. However, there seems to be an upward trend.

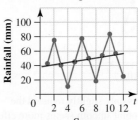

b. No improvement using this method.

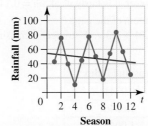

c. See the table at the bottom of the page.*

---

*18. d.

Week 1	$\frac{10}{19} = 0.5263$	0.4211	$\frac{12}{19} = 0.6316$	0.7895	1.2632	$\frac{45}{19} = 2.3684$
Week 2	$\frac{12}{22} = 0.5455$	0.4091	$\frac{14}{22} = 0.6364$	0.8182	$\frac{26}{22} = 1.1818$	2.4091
Week 3	0.5882	0.3922	$\frac{16}{25.5} = 0.6275$	$\frac{21}{25.5} = 0.8235$	1.2941	2.2745

*18. e.

	Mon.	Tues.	Wed.	Thurs.	Fri.	Sat.
Seasonal index	0.5533	$\frac{1.2224}{3} = 0.4075$	0.6318	$\frac{2.4312}{3} = 0.8104$	1.2464	$\frac{7.052}{3} = 2.3507$

*20. c.

Season	1	2	3	4	5	6	7	8	9	10	11	12
Rainfall	43	75	41	13	47	78	50	19	51	83	55	25
3-pt ave.		53	43	33.7	46	58.3	49	40	51	63	54.3	
5-pt ave.			43.8	50.8	45.8	41.4	49	56.2	51.6	46.6		

3-point smooth:

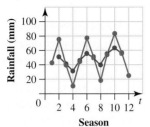

5-point smooth:

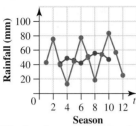

The 3-point average has only slightly reduced the variation, while the 5-point smooth seems more effective.

d. $y = 44.81 + 0.74t$

A positive trend. However, it is fairly weak ($m = 0.74$).

e. See the table at the bottom of the page.*

Least squares: $y = 42.01 + 1.36t$. A stronger increasing trend shown.

f.

Season	Seasonal index
Summer	0.9741
Autumn	1.6346
Winter	1.0041
Spring	0.3871

Seasonally adjusted data:

Season	Rainfall	Seas. adj.
1	43	44.1
2	75	45.9
3	41	40.8
4	13	33.6
5	47	48.2
6	78	47.7
7	50	49.8
8	19	49.1
9	51	52.4
10	83	50.8
11	55	54.8
12	25	64.6

g. $y = 37.70 + 1.66t$

Strongest trend yet ($m = 1.66$). Also, all 12 points used rather than 8 or 10 as in moving average smoothing.

4.8 Exam questions

Note: Mark allocations are available with the fully worked solutions online.

1. a. See the image at the bottom of the page.*

---

*20. e.

Season	1	2	3	4	5	6	7	8	9	10	11	12
Rainfall	43	75	41	13	47	78	50	19	51	83	55	25
4-pt ave.			43.5	44.4	45.9	47.8	49.0	50.1	51.4	52.8		

*1. a.

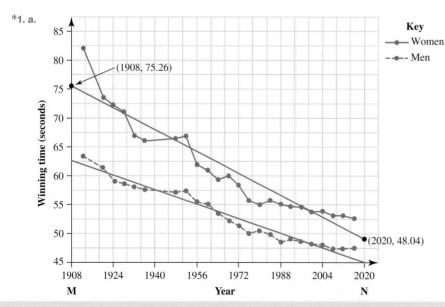

b. Winning time men $= 356.9 - 0.1544 \times 2024$
$$= 44.3944$$
Winning time women $= 538.9 - 0.2430 \times 2024$
$$= 47.068$$
Difference $= 47.068 - 44.3944 \simeq 2.7$ seconds.

c. The next Olympics will be in 2056.

2. a. The < seasonal indices are:

$$\text{Summer} = 0.89$$
$$\text{Autumn} = 1.00$$
$$\text{Spring} = 0.70$$

b. 186 mm

3. a. See the image at the bottom of the page.*

  b. **i.** See the image at the bottom of the page.**

    ii. The average rate of increase in percentage congestion level from 2008 to 2016 in Sydney is equal to the graph's gradient. This is 1.15% per year on average.

    iii. 2020

  c. −1515 (rounded to 4 s.f.)

  d. *congestion level* $= -1515 + 0.7667 \times year$

  e. Sydney's traffic congestion will increase faster than Melbourne's, meaning it will always exceed future traffic congestion levels in Melbourne.

4. D

5. a. Sample responses can be found in the worked solutions in the online resources.

  b. 157.25

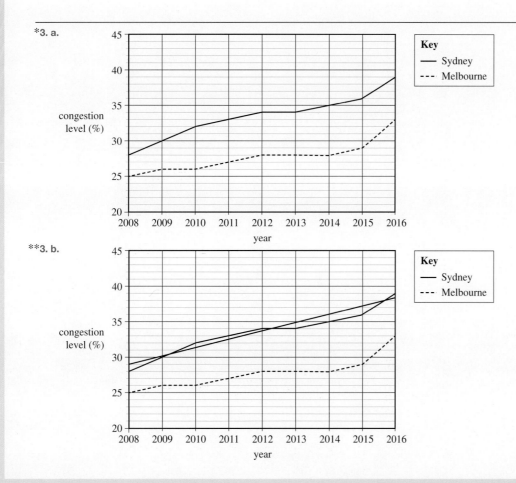

*3. a.

**3. b.

5 Modelling depreciation of assets using recursion

Fully worked solutions for this topic are available online.

5.1 Overview

5.1.1 Introduction

Many items such as antiques, jewellery or real estate increase in value (appreciate or have a capital gain) with time. On the other hand, items such as computers, vehicles or machinery decrease in value (depreciate) with time because of wear and tear, advances in technology or a lack of demand for those specific items.

The estimated loss in value of assets is called depreciation. Each financial year a business may set aside money equal to the depreciation of an item to cover the cost of the eventual replacement of that item. The estimated value of an item at any point in time is called its future value (or book value).

When the value becomes zero, the item is said to be 'written off'. At the end of an item's useful or effective life (as a contributor to a company's income), its future value is called its scrap value.

There are three methods to calculate depreciation. They are:
1. flat rate depreciation
2. reducing balance depreciation
3. unit cost depreciation

Currently, the Australian Taxation Office (ATO) allows depreciation of an asset as a tax deduction for businesses. This means that the annual depreciation reduces the amount of tax paid by a business in that year. The higher the depreciation, the greater the tax benefit. The ATO sets the percentage depreciation rates, either flat rate or reducing balance, but once a method is applied to an asset it cannot be changed for the life of the asset.

KEY CONCEPTS

This topic covers the following key concepts from the VCE Mathematics Study Design:
- use of a first-order linear recurrence relation of the form: $u_0 = a$, $u_{n+1} = Ru_n + d$ where a, R and d are constants to generate the terms of a sequence
- use of a recurrence relation to model and compare (numerically and graphically) flat rate, unit cost and reducing balance depreciation of the value of an asset with time, including the use of a recurrence relation to determine the depreciating value of an asset after n depreciation periods for the initial sequence
- use of the rules for the future value of an asset after n depreciation periods for flat rate, unit cost and reducing balance depreciation and their application.

Source: VCE Mathematics Study Design (2023–2027) extracts © VCAA; reproduced by permission.

5.2 A first-order linear recurrence relation

5.2.1 Generating the terms of a first-order recurrence relation

A sequence is a list of numbers where each successive number relates in a certain way to the numbers preceding it.

A **first-order recurrence relation** relates a term in a **sequence** to the previous term in the same sequence, which means that we only need an **initial value** to be able to generate all remaining terms of a sequence.

In a recurrence relation, the nth term is represented by u_n, with the next term being represented by u_{n+1} and the term directly before u_n being represented by u_{n-1}. The initial value of the sequence is represented by u_0.

We can define the sequence 1, 5, 9, 13, 17, ... with the following equation:

$$u_{n+1} = u_n + 4 \quad u_0 = 1$$

This expression is read as 'the next term is the previous term plus 4, starting at 1'.

Or, transposing the above equation, we get:

$$u_{n+1} - u_n = 4 \quad u_0 = 1$$

A first-order recurrence relation

A first-order recurrence relation defines a relationship between two successive terms of a sequence, for example, between:

u_n, the previous term $\qquad$ u_{n+1}, the next term.

The first term is represented by u_0.

A first-order linear recurrence relation can be written as $u_0 = a$, $u_{n+1} = Ru_n + d$.

Where a, R and d are constants.

---

tlvd-3914

WORKED EXAMPLE 1 Generating the values of a sequence

Write the first five terms of the sequence that starts at 1 and each successive term increases by 2.

THINK	WRITE
1. This is a first-order recurrence relation.	1. This is a first-order recurrence relation. It has the pattern $u_{n+1} = u_n + 2$, $u_0 = 1$.
2. Write out the sequence of five numbers.	2. 1, 3, 5, 7, 9

5.2.2 Starting term

Earlier, it was stated that a starting term was required to fully define a sequence. As can be seen below, the same pattern with a different starting point gives a different set of numbers.

$$u_{n+1} = u_n + 2 \qquad u_0 = 3 \qquad \text{gives } 3, 5, 7, 9, 11, \ldots$$
$$u_{n+1} = u_n + 2 \qquad u_0 = 2 \qquad \text{gives } 2, 4, 6, 8, 10, \ldots$$

WORKED EXAMPLE 2 Generating the values of a sequence using a calculator

Write the first five terms of the sequence defined by the first-order recurrence relation:

$$u_{n+1} = 3u_n + 5, \ u_0 = 2$$

THINK	WRITE
1. Since we know the u_0 or starting term, we can generate the next term, u_1, using the pattern: The next term is $3 \times$ the previous term $+ 5$. The five terms can be generated on the calculator (see instructions below).	$u_0 = 2$ $u_{n+1} = 3u_n + 5$ $\begin{array}{ll} 2 & 2 \\ 2 \times 3 + 5 & 11 \\ 11 \times 3 + 5 & 38 \\ 38 \times 3 + 5 & 119 \\ 119 \times 3 + 5 & 362 \end{array}$
2. Write the answer.	The sequence is 2, 11, 38, 119, 362.

TI | THINK	DISPLAY/WRITE	CASIO | THINK	DISPLAY/WRITE
1. On a Calculator page, press 2 and then ENTER. Complete the entry line as: $\times 3 + 5$ Then continue to press ENTER until the first five terms are displayed.		1. On the Main screen, press 2 and then EXE. Complete the entry line as: $\times 3 + 5$ Then continue to press EXE until the first five terms are displayed.	
2. The answer appears on the screen.	The sequence is 2, 11, 38, 119, 362.	2. The answer appears on the screen.	The sequence is 2, 11, 38, 119, 362.

 Resources

 Interactivity Initial values and first-order recurrence relations (int-6262)

1. **WE1** Write down the first five terms of a sequence that starts at -2 and each successive term decreases by 3.

2. Write down the first five terms of a sequence that starts at 3 and each successive term is double the previous term.

3. **WE2** Write the first five terms of the sequence defined by the first-order recurrence relation:

$$u_n = 4u_{n-1} + 3 \quad u_0 = 5$$

4. Write the first five terms of the sequence defined by the first-order recurrence relation:

$$f_{n+1} = 5f_n - 6 \quad f_0 = -2$$

5. Identify which of the following equations are complete first-order recurrence relations.
   a. $u_n = 2 + n$
   b. $u_n = u_{n-1} - 1 \quad u_0 = 2$
   c. $u_n = 1 - 3u_{n-1} \quad u_0 = 2$
   d. $u_n - 4u_{n-1} = 5$
   e. $u_n = -u_{n-1}$
   f. $u_n = n + 1 \quad u_1 = 2$
   g. $u_n = 1 - u_{n-1} \quad u_0 = 21$
   h. $u_n = a^{n-1} \quad u_2 = 2$
   i. $f_{n+1} = 3f_n - 1$
   j. $p_n = p_{n-1} + 7 \quad p_0 = 7$

6. Write the first five terms of each of the following sequences.
   a. $u_n = u_{n-1} + 2 \quad u_0 = 6$
   b. $u_n = u_{n-1} - 3 \quad u_0 = 5$
   c. $u_n = 1 + u_{n-1} \quad u_0 = 23$
   d. $u_{n+1} = u_n - 10 \quad u_0 = 7$

7. Write the first five terms of each of the following sequences.
   a. $u_n = 3u_{n-1} \quad u_0 = 1$
   b. $u_n = 5u_{n-1} \quad u_0 = -2$
   c. $u_n = -4u_{n-1} \quad u_0 = 1$
   d. $u_{n+1} = 2u_n \quad u_0 = -1$

8. Write the first five terms of each of the following sequences.
   a. $u_n = 2u_{n-1} + 1 \quad u_0 = 1$
   b. $u_n = 3u_{n-1} - 2 \quad u_0 = 5$

9. Write the first seven terms of each of the following sequences.
   a. $u_n = -u_{n-1} + 1 \quad u_0 = 6$
   b. $u_{n+1} = 5u_n \quad u_0 = 1$

10. **MC** Select the sequence that is generated by the following first-order recurrence relation.

$$u_n = 3u_{n-1} + 4 \quad u_0 = 2$$

   A. 2, 3, 4, 5, 6, ...
   B. 2, 6, 10, 14, 18, ...
   C. 2, 10, 34, 106, 322, ...
   D. 2, 11, 47, 191, 767, ...
   E. 6, 10, 14, 18, 22, ...

11. **MC** Select the sequence that is generated by the following first-order recurrence relation.

$$u_{n+1} = 2u_n - 1 \quad u_0 = -3$$

   A. $-3, 5, 9, 17, 33, ...$
   B. $-3, -5, -9, -17, -33, ...$
   C. $-3, 5, -3, 5, -3, ...$
   D. $-3, -8, -14, -26, -54, ...$
   E. $-3, -7, -15, -31, -63, ...$

12. Write the first-order recurrence relations for the following descriptions and generate the first five terms of the sequence.

   a. The next term is 3 times the previous term, starting at $\frac{1}{4}$.

   b. Next year's attendance at a motor show is 2000 more than the previous year's attendance, with an initial year attendance of 200 000.

   c. The next term is the previous term less 7, starting at 100.

   d. The next day's total sum is double the previous day's sum less 50, with a first-day sum of $200.

5.2 Exam questions

Question 1 (1 mark)

Source: VCE 2019, Further Mathematics Exam 1, Section A, Q17; © VCAA.

MC Consider the recurrence relation shown below.

$$A_0 = 3, \quad A_{n+1} = 2A_n + 4$$

The value of A_3 in the sequence generated by this recurrence relation is given by

   A. $2 \times 3 + 4$
   B. $2 \times 4 + 4$
   C. $2 \times 10 + 4$
   D. $2 \times 24 + 4$
   E. $2 \times 52 + 4$

Question 2 (1 mark)

Source: VCE 2017, Further Mathematics Exam 1, Section A, Core, Q18; © VCAA.

MC The first five terms of a sequence are 2, 6, 22, 86, 342 ...

The recurrence relation that generates this sequence could be

   A. $P_0 = 2, \quad P_n + 1 = P_n + 4$
   B. $P_0 = 2, \quad P_n + 1 = 2P_n + 2$
   C. $P_0 = 2, \quad P_n + 1 = 3P_n$
   D. $P_0 = 2, \quad P_n + 1 = 4P_n - 2$
   E. $P_0 = 2, \quad P_n + 1 = 5P_n - 4$

Question 3 (1 mark)

Source: VCE 2016 Further Mathematics Exam 1, Section A, Q17; © VCAA.

MC Consider the recurrence relation below.

$$A_0 = 2, \quad A_{n+1} = 3A_n + 1$$

The first four terms of this recurrence relation are

   A. 0, 2, 7, 22...
   B. 1, 2, 7, 22...
   C. 2, 5, 16, 49 ...
   D. 2, 7, 18, 54 ...
   E. 2, 7, 22, 67 ...

More exam questions are available online.

5.3 Modelling flat rate depreciation with a recurrence relation

5.3.1 Flat rate (straight line) depreciation

If an item depreciates by the **flat rate method**, then its value decreases by a fixed amount each unit time interval, generally each year. This depreciation value may be expressed in dollars or as a percentage of the original cost price.

Since the depreciation is the same for each unit time interval, the flat rate method is an example of straight line (linear) decay.

Formula to calculate the flat rate depreciation

As a recurrence relation:

$$V_{n+1} = V_n - d$$

where V_n is the value of the asset after n depreciating periods and d is the depreciation each time period.

Formula for after n periods of depreciation:

$$V_n = V_0 - nd$$

where v_0 is the initial value of the asset.

We can use this relationship to generate a depreciation schedule (table) that can be used to draw a graph of future value against time. The schedule displays the future value after each unit time interval, that is:

Time, n	Depreciation, d	Future value, V_n

tlvd-3915

WORKED EXAMPLE 3 Graphing flat rate depreciation

Fast Word Printing Company bought a new printing press for $15 000 and chose to depreciate it by the flat rate method. The depreciation was 15% of the cost price each year and its useful life was 5 years.

a. Calculate the annual depreciation.
b. Set up a recurrence relation to represent the depreciation.
c. Write a depreciation schedule for the useful life of the press and use it to draw a graph of future value against time.
d. Identify the scrap value.

THINK

a. 1. State the cost price.

   2. Calculate the depreciation rate as 15% of the cost price.

   3. Write your answer.

b.   Write the recurrence relation for flat rate depreciation and substitute in the value for d.

c. 1. Write a depreciation schedule for 0–5 years, using depreciation of $2250 each year and a starting value of $15 000. Using your calculator, in the entry type 15 000, press ENTER/EXE, then –2250. Continue to press ENTER/EXE five times to complete the schedule.

   2. Draw a graph of the tabled values for future value against time.

WRITE/DRAW

a. $V_0 = 15\,000$

$d = V_0 \times \dfrac{r}{100}$

$= 15\,000 \times \dfrac{15}{100}$

$= 2250$

Annual depreciation is $2250.

b. $V_{n+1} = V_n - d$
$V_{n+1} = V_n - 2250$

c.

Time, n (years)	Depreciation, d ($)	Future value, V_n ($)
0	—	15 000
1	2250	12 750
2	2250	10 500
3	2250	8250
4	2250	6000
5	2250	3750

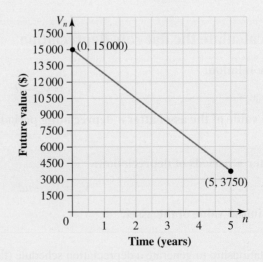

d. 1.	The press is scrapped after five years, so look at the schedule in part b.	d.	When $n = 5$
			$V_5 = 3750$
2.	Write the answer.		The scrap value is $3750.

The depreciation schedule gives the scrap value, as can be seen in Worked example 3. So too does a graph of future value against time, since it is only drawn for the item's useful life and its end point is the scrap value.

Businesses need to keep records of depreciation for tax purposes on a year-to-year basis. What if an individual wants to investigate the rate at which an item has depreciated over many years? An example is the rate at which a private car has depreciated. If a straight line depreciation model is chosen, then the following example demonstrates its application.

WORKED EXAMPLE 4 Calculating depreciation rate and future value

Jarrod bought his car five years ago for $15 000. Its current market value is $7500. Assuming straight line depreciation, calculate:
a. the car's annual depreciation rate
b. the relationship between the future value and time, and use your calculator to determine when the car will have a value of $3000.

THINK	WRITE
a. 1. Calculate the total depreciation over the five years and thus the rate of depreciation.	a. Total depreciation $=$ cost price $-$ current value $= \$15\,000 - \$7500$ $= \$7500$
	Rate of depreciation $= \dfrac{\text{total depreciation}}{\text{number of years}}$ $= \dfrac{\$7500}{5 \text{ years}}$ $= \$1500 \text{ per year}$
2. Write the answer.	The annual depreciation rate is $1500.
b. 1. Set up the recurrence relationship.	b. $V_{n+1} = V_n - 1500$ $V_0 = 15\,000$
2. Open a calculator screen and in the entry line type 15 000, then press ENTER/EXE. Type -1500 and press ENTER/EXE. Continue to press ENTER/EXE, counting the number of times that ENTER/EXE has been pressed until the value reaches 3000.	$V_n = 3000$ When $n = 8$
3. Write the answer.	The depreciation equation for the car is $V_n = 15\,000 - 1500n$. The future value will reach $3000 when the car is eight years old.

  Resources

Interactivity Depreciation: flat rate, reducing balance, unit cost (int-6266)

1. **WE3** A mining company bought a vehicle for $25 000 and chose to depreciate it by the flat rate method. The depreciation was 15% of the cost price each year and its useful life was five years.

   a. Calculate the annual depreciation.
   b. Set up a recurrence relation to represent the depreciation.
   c. Write a depreciation schedule for the item's useful life and draw a graph of future value against time.
   d. Identify the scrap value.

2. Write the recurrence relationships for the following items.

   a. A $30 000 car that is depreciated by 20% p.a. flat rate
   b. A $2000 display unit depreciated by 10% of its cost
   c. A $6000 piano depreciated by $500 p.a.

3. **WE4** For the situations described below, use a straight line depreciation model to determine:

   i. the annual rate of depreciation
   ii. the relationship between the future value and time, and use CAS to determine when the item is written off (has a value of $0).

   a. A car purchased for $50 000 with a current value of $25 000; it is now 5 years old.
   b. A stereo unit bought for $850 seven years ago; it has a current value of $150.
   c. A refrigerator with a current value of $285 that was bought 10 years ago for $1235.

4. A second-hand car is currently on sale for $22 500. Its value is expected to depreciate by $3200 per year.

   a. Write a recurrence relationship for this item.
   b. Use your calculator to work out the expected value of the car after 5 years.

5. All Clean carpet cleaners bought a new cleaner for $10 000 and chose to depreciate it by the flat rate method. The depreciation was 15% of the cost price each year and its useful life was five years.

   a. Calculate the annual depreciation.
   b. Use your calculator to create a depreciation schedule for the item's useful life and draw a graph of future value against time.

For the situations outlined in questions 6 and 7:

   a. write a depreciation schedule for the item's first five years and draw a graph of future value against time
   b. use your calculator to determine the scrap value.

6. A farming company chose to depreciate a tractor by the cost price method and the annual depreciation was $4000. The tractor was purchased for $45 000 and its useful life was 10 years.

7. A winery chose to depreciate a corking machine that cost $13 500 when new, by the cost price method. The annual depreciation was $2000 and its useful life was six years.

For the situations outlined in questions 8 and 9:

   a. calculate the annual depreciation
   b. set up a recurrence relation to represent the depreciation
   c. use your calculator to create a depreciation schedule for the item's useful life and draw a graph of future value against time
   d. calculate how long it will take for the item to reach its scrap value.

8. Machinery is bought for $7750 and depreciated by the flat rate method. The depreciation is 20% of the cost price each year and its scrap value is $1550.

9. An excavation company buys a digger for $92 000 and depreciates it by the flat rate method. The depreciation is 15% of the cost price per year and its scrap value is $9200.

10. Each of the following four graphs represents the flat rate depreciation of a different item. In each case determine:

   i. the cost price of the item
   ii. the annual depreciation
   iii. the time taken for the item to reach its scrap value or to be written off.

a.

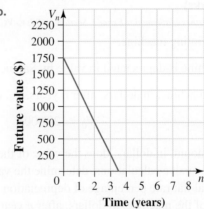

b.

c.

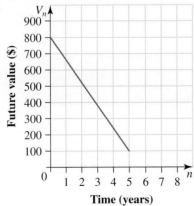

d.

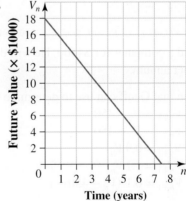

11. **MC** A 1-tonne truck, bought for $31 000, was depreciated using the flat rate method. If the scrap value of $5000 was reached after 5 years, the annual depreciation would be:

A. $31   B. $1000   C. $5200   D. $6200   E. $26 000

12. A business buys two different photocopiers at the same time. One costs $2200 and is to be depreciated by $225 per annum. It has a scrap value of $400. The other costs $3600 and is to be depreciated by $310 per annum. This one has a scrap value of $500.

a. State which machine would need to be replaced first.
b. Determine how much later the other machine would need to be replaced.

5.3 Exam questions

Question 1 (4 marks)
Source: VCE 2020, Further Mathematics Exam 2, Section A, Q7; © VCAA.

Samuel owns a printing machine.

The printing machine is depreciated in value by Samuel using flat rate depreciation.

The value of the machine, in dollars, after n years, V_n, can be modelled by the recurrence relation

$$V_0 = 120\,000, \qquad V_{n+1} = V_n - 15\,000$$

a. By what amount, in dollars, does the value of the machine decrease each year?   **(1 mark)**
b. Showing recursive calculations, determine the value of the machine, in dollars, after two years.   **(1 mark)**
c. What annual flat rate percentage of depreciation is used by Samuel?   **(1 mark)**
d. The value of the machine, in dollars, after n years, could also be determined using a rule of the form. Write down this rule for V_n.   **(1 mark)**

Question 2 (1 mark)
Source: VCE 2019, Further Mathematics Exam 1, Section A, Q19; © VCAA.

MC Geoff purchased a computer for $4500. He will depreciate the value of his computer by a flat rate of 10% of the purchase price per annum.

A recurrence relation that Geoff can use to determine the value of the computer after n years, V_n, is

A. $V_0 = 4500, \quad V_{n+1} = V_n - 450$
B. $V_0 = 4500, \quad V_{n+1} = V_n + 450$
C. $V_0 = 4500, \quad V_{n+1} = 0.9\,V_n$
D. $V_0 = 4500, \quad V_{n+1} = 1.1\,V_n$
E. $V_0 = 4500, \quad V_{n+1} = 0.1\,(V_n - 450)$

▶ **Question 3 (1 mark)**

Source: VCE 2017, Further Mathematics Exam 1, Section A, Core, Q22; © VCAA.

MC Consider the graph below.

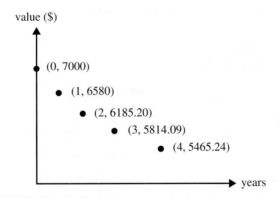

This graph could show the value of
- **A.** a piano depreciating at a flat rate of 6% per annum.
- **B.** a car depreciating with a reducing balance rate of 6% per annum.
- **C.** a compound interest investment earning interest at the rate of 6% per annum.
- **D.** a perpetuity earning interest at the rate of 6% per annum.
- **E.** an annuity investment with additional payments of 6% of the initial investment amount per annum.

More exam questions are available online.

5.4 Modelling reducing balance depreciation with a recurrence relation

LEARNING INTENTION

At the end of this subtopic you should be able to:
- apply first-order linear recurrence relations to model the reducing balance method for depreciating assets.

5.4.1 Reducing balance depreciation

If an item depreciates by the **reducing balance depreciation** method, then its value decreases by a fixed rate each unit time interval, generally each year. This rate is a fixed percentage of the previous value of the item. Reducing balance depreciation is an example of geometric decay.

Reducing balance depreciation is also known as *diminishing value depreciation*.

Reducing balance depreciation can be expressed by the recurrence relation shown below.

Formula for reducing balance depreciation as a recurrence relation

$$V_{n+1} = RV_n$$

where V_n is the value of the asset after n depreciating periods and $R = 1 - \dfrac{r}{100}$, where r is the depreciation rate.

Suppose the new $15 000 printing press considered in Worked example 3 was depreciated by the reducing balance method at a rate of 20% p.a. of the previous value.

a. Generate a depreciation schedule using a recurrence relation for the first five years of work for the press.
b. State the future value after five years.
c. Draw a graph of future value against time.

THINK

WRITE/DRAW

a. 1. Calculate the value of R.

a. $R = 1 - \dfrac{r}{100}$

$= 1 - \dfrac{20}{100}$

$= 0.8$

2. Write the recurrence relation for reducing balance depreciation and substitute in the known information.

$V_{n+1} = RV_n$

$V_{n+1} = 0.8V_n, \ V_0 = 15\,000$

3. Use the recurrence relation and your calculator to calculate the future value for the first five years (up to $n = 5$).
On a calculator screen type 15 000, press ENTER/EXE. Type $\times 0.8$, then press ENTER/EXE. Repeat this another four times.

15 000	15 000
15 000 × 0.8	12 000
12 000 × 0.8	9600
9600 × 0.8	7680
7680 × 0.8	6144
6144 × 0.8	4915.20

4. Draw the depreciation schedule.

Time, n (years)	Future value, V_n ($)
0	15 000
1	12 000
2	9600
3	7680
4	6144
5	4915.20

b. State the future value after five years from the depreciation schedule.

b. The future value of the press after five years will be $4915.20.

c. Draw a graph of the future value against time.

c.

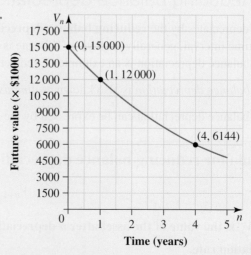

From Worked example 5, it is clear from the graph and the depreciation schedule that reducing balance depreciation results in greater depreciation during the early stages of the asset's life (the future value drops more quickly at the start since the annual depreciation falls from $3000 in year 1 to $1228.80 in year 4).

tlvd-3916

WORKED EXAMPLE 6 Generating depreciation schedules

A transport business bought a new bus for $60 000. The business has the choice of depreciating the bus by a flat rate of 20% of the cost price each year or by 30% of the previous value each year.
a. Generate depreciation schedules using both methods for a life of five years.
b. Draw graphs of the future value against time for both methods on the same set of axes.
c. Determine after how many years the reducing balance future value becomes greater than the flat rate future value.

THINK

a. 1. Calculate the flat rate depreciation per year.

2. Generate a flat rate depreciation schedule using your calculator for 0–5 years.

WRITE/DRAW

a. $d = 20\%$ of $60 000
   $= \$12\,000$ per year

Time, n (years)	Depreciation, d ($)	Future value, V_n ($)
0	—	60 000
1	12 000	48 000
2	12 000	36 000
3	12 000	24 000
4	12 000	12 000
5	12 000	0

3. Generate a reducing balance depreciation schedule. Annual depreciation is 30% of the previous value, so to calculate the future value multiply the previous value by 0.7. Using your calculator, continue to calculate the future value for a period of five years.

Time, n (years)	Future value, V_n ($)
0	60 000
1	$60\,000 \times 0.7 = 42\,000$
2	$42\,000 \times 0.7 = 29\,400$
3	$29\,400 \times 0.7 = 20\,580$
4	$20\,580 \times 0.7 = 14\,406$
5	$14\,406 \times 0.7 = 10\,084.20$

b. Draw graphs using values for V_n and n from the schedules. In this instance, the blue line is the flat rate value and the pink curve is the reducing balance value.

b.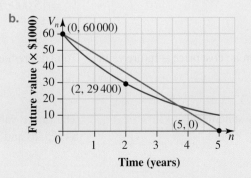

c. Look at the graph to see when the reducing balance curve lies above the flat rate line. State the first whole year after this point of intersection.

c. The future value for the reducing balance method is greater than that of the flat rate method after four years.

In the example above, for each consecutive year, the previous future value is multiplied by 0.7. So:

$$V_5 = 60\,000 \times 0.7 \times 0.7 \times 0.7 \times 0.7 \times 0.7$$
$$= 60\,000 \times 0.7^5$$

The reducing balance depreciation formula

$$V_n = V_0 R^n$$

where V_n is the future value after time

n R is the rate of depreciation $\left(1 - \dfrac{r}{100}\right)$

V_0 is the cost price
n is the time since purchase.

That is, given the cost price and depreciation rate, we can determine the future value (including scrap value) of an article at any time after purchase. Let us now see how we can use this formula.

WORKED EXAMPLE 7 Calculating future value and total depreciation

The printing press from Worked example 3 was depreciated by the reducing balance method at 20% p.a. Calculate the future value and total depreciation of the press after five years if it cost $15 000 new.

THINK	WRITE
1. State V_0, r and n.	$V_0 = 15\,000$, $r = 20$, $n = 5$
2. Calculate the value of R.	$R = 1 - \dfrac{20}{100}$ $= 0.8$
3. Substitute into the depreciation formula and evaluate.	$V_n = V_0 R$ $V_5 = 15\,000(0.8)^5$ $= 4915.2$
4. Total depreciation is: cost price − future value.	Total depreciation $= V_0 - V_5$ $= 15\,000 - 4915.2$ $= 10\,084.8$
5. Write a summary statement.	The future value of the press after five years will be $4915.20 and its total depreciation will be $10 084.80.

5.4.2 Effective life

The situation may arise where the scrap value is known and we want to know how long it will be before an item reaches this value, that is, its useful or **effective life**. So, in the reducing balance formula $V_n = V_0 R^n$, n is needed.

WORKED EXAMPLE 8 Calculating scrap value

A photocopier purchased for $8000 depreciates by 25% p.a. by the reducing balance method. If the photocopier has a scrap value of $1200, calculate how long it will be before this value is reached. Use the CAS calculator to solve for the number of years.

THINK	WRITE
1. State the values of V_n, V_0 and r.	$V_n = \$1200$, $V_0 = \$8000$ and $r = 25\%$
2. Calculate the value of R.	$R = 1 - \dfrac{25}{100}$ $= 0.75$
3. Substitute the values of the pronumerals into the formula and simplify.	$V_n = V_0 R^n$ $1200 = 8000 \times (0.75)^n$ Solving $1200 = 8000 \times (0.75)^n$ for n,
4. Use CAS to solve for n.	$n = 6.59$ years
5. Interest is compounded annually, so n represents years. Raise n to the next whole year.	As the depreciation is calculated once a year, $n = 7$ years.
6. Write the answer.	It will take 7 years for the photocopier to reach its scrap value.

TI	THINK	DISPLAY/WRITE	CASIO	THINK	DISPLAY/WRITE
4. On a Calculator page, press MENU and select: 3. Algebra 1: Solve Complete the entry line as: solve $(1200 = 8000 \times (0.75)^n, n)$ Then press ENTER.	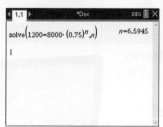	4. On the Main screen, complete the entry line as: Solve $(1200 = 8000 \times (0.75)^n, n)$ Then press EXE.	
The answer appears on the screen.	$n = 6.59$ years It will take 7 years for the photocopier to reach its scrap value.	The answer appears on the screen.	$n = 6.59$ years It will take 7 years for the photocopier to reach its scrap value.

1. **WE5** A laptop was bought new for $1500 and it depreciates by the reducing balance method at a rate of 17% p.a. of the previous book value.

   a. Generate a depreciation schedule using a recurrence relation for the first five years.
   b. State the future value after five years.
   c. Draw a graph of future value against time.

2. A road bike was bought new for $3300 and it depreciates by the reducing balance method at a rate of 13% p.a. of the previous value.

   a. Generate a depreciation schedule using a recurrence relation for the first five years.
   b. State the future value after five years.
   c. Draw a graph of future value against time.

3. A farming company chose to depreciate its new $60 000 bulldozer by the reducing balance method at a rate of 20% p.a. of the previous value.

   a. Write the recurrence relation that represents this depreciation.
   b. Draw a depreciation schedule for the first five years of the bulldozer's life.
   c. State its future value after five years.
   d. Draw a graph of future value against time.

4. A retail store chose to depreciate its new $4000 computer by the reducing balance method at a rate of 40% p.a. of the previous value.

   a. Write the recurrence relation that represents this depreciation.
   b. Draw a depreciation schedule for the first five years of the computer's life.
   c. State its future value after five years.
   d. Draw a graph of future value against time.

5. **WE6** A taxi company bought a new car for $29 990. The company has the choice of depreciating the car by a flat rate of 20% of the cost price each year or by 27% of the previous value each year.

   a. Generate depreciation schedules using both methods for a life of five years.
   b. Draw graphs of the future value against time for both methods on the same axes.
   c. Determine after how many whole years the reducing balance future value becomes greater than the flat rate future value.

6. A sailing rental company bought a new catamaran for $23 000. The company has the choice of depreciating the catamaran by a flat rate of 20% of the cost price each year or by 28% of the previous value each year.

a. Generate depreciation schedules using both methods for a life of five years.
b. Draw graphs of the future value against time for both methods on the same axes.
c. Determine after how many whole years the reducing balance future value becomes greater than the flat rate future value.

7. A café buys a cash register for $550. The owner has the choice of depreciating the register by the flat rate method (at 20% of the cost price each year) or the reducing balance method (at 30% of the previous value each year).

a. Draw depreciation schedules for both methods for a life of five years.
b. Draw graphs of future value against time for both methods on the same set of axes.
c. Determine after how many whole years the reducing balance future value becomes greater than the flat rate future value.

8. Speedy Cabs taxi service has bought a new taxi for $30 000. The company has the choice of depreciating the taxi by the flat rate method (at $33\frac{1}{3}$% of the cost price each year) or the diminishing value method (at 50% of the previous value each year).

a. Draw depreciation schedules for both methods for three years.
b. Draw graphs of future value against time for both methods on the same set of axes.
c. Determine after how many whole years the reducing balance future value becomes greater than the flat rate future value.

9. **WE7** Using the reducing balance formula, determine V_n (correct to 2 decimal places) given $V_0 = 45\,000$, $r = 15$, $n = 6$.

10. Using the reducing balance formula, determine V_n (correct to 2 decimal places) given $V_0 = 2675$, $r = 22.5$, $n = 5$.

11. Using the reducing balance formula, determine V_n (correct to 2 decimal places) given:
a. $V_0 = 20\,000$, $r = 20$, $n = 4$
b. $V_0 = 30\,000$, $r = 25$, $n = 4$.

12. **MC** A refrigerator costing $1200 is depreciated by the reducing balance method at 20% a year. After four years, its future value will be:
A. $240   B. $491.52   C. $960   D. $1105   E. $2488.32

13. The items below are depreciated by the reducing balance method at 25% p.a. Calculate the future value and total depreciation of:
a. a TV after eight years, if it cost $1150 new
b. a photocopier after four years, if it cost $3740 new
c. carpets after six years, if they cost $7320 new.

14. The items below are depreciated at 30% p.a. by the reducing balance method. Calculate the future value and total depreciation of:

   a. a lawn mower after five years, if it cost $685 new
   b. a truck after four years, if it cost $32 500 new
   c. a washing machine after three years, if it cost $1075 new.

15. **MC** After seven years, a new $3000 photocopier, which devalues by 25% of its value each year, will have depreciated by:

   A. $400.45 **B.** $750 **C.** $2250
   D. $2599.55 **E.** $2750

16. **MC** New office furniture valued at $17 500 is subjected to reducing balance depreciation of 20% p.a. and will reach its scrap value in 15 years. The scrap value will be:

   A. less than $300. **B.** between $300 and $400.
   C. between $400 and $500. **D.** between $500 and $600.
   E. between $600 and $700.

17. **WE8** Use your CAS calculator to determine how many years it takes an item of initial value $4500, depreciating at a reducing balance rate of 25% p.a., to reach a value of $900.

18. Use your CAS calculator to solve for n given reducing balance depreciation where $V_n = \$1500$, $V_0 = \$7600$, $r = 15\%$ p.a.

19. **MC** A new chainsaw bought for $1250 has a useful life of only three years. If it depreciates annually at a 60% reducing balance rate, its scrap value will be:

   A. $0 **B.** $60 **C.** $80 **D.** $250 **E.** $270

20. An item is depreciated by using reducing balance depreciation. Use your CAS calculator to solve for n (correct to 2 decimal places), given:

   a. $V_n = \$3000$, $V_0 = \$40\,000$, $r = 20$ **b.** $V_n = \$500$, $V_0 = \$3000$, $r = 30$.

5.4 Exam questions

Question 1 (1 mark)

Source: VCE 2020, Further Mathematics Exam 1, Section A, Q23; © VCAA.

MC Consider the following four recurrence relations representing the value of an asset after n years, V_n.

- $V_0 = 20\,000$, $V_{n+1} = V_n + 2500$
- $V_0 = 20\,000$, $V_{n+1} = V_n - 2500$
- $V_0 = 20\,000$, $V_{n+1} = 0.875\,V_n$
- $V_0 = 20\,000$, $V_{n+1} = 1.125\,V_n - 2500$

How many of these recurrence relations indicate that the value of an asset is depreciating?

   A. 0 **B.** 1 **C.** 2 **D.** 3 **E.** 4

Question 2 (5 marks)

Source: VCE 2017 Further Mathematics Exam 2, Section A, Core, Q5; © VCAA.

Alex is a mobile mechanic.

He uses a van to travel to his customers to repair their cars.

The value of Alex's van is depreciated using the flat rate method of depreciation.

The value of the van, in dollars, after n years, V_n, can be modelled by the recurrence relation shown below.

$$V_0 = 75\,000, \quad V_{n+1} = V_n - 3375$$

a. Recursion can be used to calculate the value of the van after two years.
Complete the calculations below by writing the appropriate numbers in the boxes provided. **(2 marks)**

$V_0 = 75\,000$

$V_1 = 75\,000 - \boxed{} = 71\,625$

$V_2 = \boxed{} - \boxed{} = \boxed{}$

b. i. By how many dollars is the value of the van depreciated each year? **(1 mark)**
ii. Calculate the annual flat rate of depreciation in the value of the van.
Write your answer as a percentage. **(1 mark)**
c. The value of Alex's van could also be depreciated using the reducing balance method of depreciation. The value of the van, in dollars, after n years, R_n, can be modelled by the recurrence relation shown below.

$$R_0 = 75\,000, \quad R_{n+1} = 0.943R_n$$

Calculate the annual percentage rate that the value of the van is depreciated by each year. **(1 mark)**

Question 3 (1 mark)
Source: VCE 2016 Further Mathematics Exam 1, Section A, Q19; © VCAA.
MC The purchase price of a car was $26\,000.

Using the reducing balance method, the value of the car is depreciated by 8% each year.

A recurrence relation that can be used to determine the value of the car after n years, C_n, is
A. $C_0 = 26\,000, \; C_{n+1} = 0.92\,C_n$
B. $C_0 = 26\,000, \; C_{n+1} = 1.08\,C_n$
C. $C_0 = 26\,000, \; C_{n+1} = C_{n+8}$
D. $C_0 = 26\,000, \; C_{n+1} = C_{n-8}$
E. $C_0 = 26\,000, \; C_{n+1} = 0.92\,C_n - 8$

More exam questions are available online.

5.5 Modelling unit cost depreciation with a recurrence relation

LEARNING INTENTION

At the end of this subtopic you should be able to:
• apply first-order linear recurrence relations to model the unit cost method for depreciating assets.

5.5.1 Unit cost method

The flat rate and reducing balance depreciations of an item are based on the age of the item. With the **unit cost method**, the depreciation is based on the possible maximum output (units) of the item. The useful life of a car could be expressed in terms of the distance travelled rather than a fixed number of years.

For example, 120 000 kilometres rather than six years. The actual depreciation of the car for the financial year would be a measure of the number of kilometres travelled. (The value of the car decreases by a certain amount for each kilometre travelled.)

Formula for the unit cost depreciation as a recurrence relation

$$V_{n+1} = V_n - d$$

where V_n is the value of the asset after n outputs and d is the depreciation per output.

---

tlvd-3917

WORKED EXAMPLE 9 Modelling unit cost with a recurrence relation

A motorbike purchased for **$12 000** depreciates at a rate of **$14 per 100 km driven.**
a. Set up a recurrence relation to represent the depreciation.
b. Use the recurrence relation to generate a depreciation schedule for the future value of the bike after it has been driven for 100 km, 200 km, 300 km, 400 km and 500 km.

THINK	WRITE
a. 1. Write the recurrence relation for unit cost depreciation as well as the known information.	a. $V_{n+1} = V_n - d$ $V_0 = 12\,000, d = 14$
2. Substitute the values into the recurrence relation.	$V_{n+1} = V_n - 14, V_0 = 12\,000$

b. 1. Use the recurrence relation and your calculator to calculate the future value for the first five outputs (up to 500 km). Open a calculator screen and type 12 000, press ENTER/EXE, then type −14 and continue to press ENTER/EXE five more times.

b.

12 000	12 000
12 000 − 14	11 986
11 986 − 14	11 972
11 972 − 14	11 958
11 958 − 14	11 944
11 944 − 14	11 930

2. Draw the depreciation schedule.

Distance driven (km)	Outputs (n)	Future value, V_n ($)
100	1	11 986
200	2	11 972
300	3	11 958
400	4	11 944
500	5	11 930

WORKED EXAMPLE 10 Calculating annual depreciation and scrap value for unit cost method

A taxi is bought for **$31 000** and it depreciates by **28.4 cents per kilometre driven**. In one year, the car is driven **15 614 km**. Calculate:
a. **the annual depreciation for this particular year**
b. **the taxi's useful life if its scrap value is $12 000.**

THINK	WRITE
a. 1. Depreciation amount = distance travelled × rate	a. Depreciation = $15\,614 \times \$0.284$ $= \$4434.38$
2. Write a summary statement.	Annual depreciation for the year is $4434.38.
b. 1. Total depreciation = cost price − scrap value $\text{Distance travelled} = \dfrac{\text{total depreciation}}{\text{rate of depreciation}}$ where rate of depreciation = 28.4 cents/km $= \$0.284$ per km	b. Total depreciation = $31\,000 - 12\,000$ $= \$19\,000$ $\text{Distance travelled} = \dfrac{19\,000}{0.284}$ $= 66\,901$ km
2. Write the answer.	The taxi has a useful life of 66 901 km.

WORKED EXAMPLE 11 Future value with unit cost method

A photocopier purchased for **$10 800** depreciates at a rate of **20 cents for every 100 copies** made. In its first year of use, **500 000** copies were made and in its second year, **550 000**. Calculate:
a. **the depreciation each year**
b. **the future value at the end of the second year.**

THINK	WRITE
a. To determine the depreciation, identify the rate and number of copies made. Express the rate of 20 cents per 100 copies in a simpler form of dollars per 100 copies, that is, $0.20 per 100 copies or $\dfrac{0.20}{100\,\text{copies}}$.	a. Depreciation = copies made × rate $\text{Depreciation 1st year} = 500\,000 \times \dfrac{0.20}{100\,\text{copies}}$ $= \$1000$ Depreciation in the first year is $1000. $\text{Depreciation 2nd year} = 550\,000 \times \dfrac{0.20}{100\,\text{copies}}$ $= \$1100$ Depreciation in the second year is $1100.
b. Future value = cost price − total depreciation	b. Total depreciation after 2 years $= 1000 + 1100$ $= \$2100$ Book value = $10\,800 - 2100$ $= \$8700$

Formula for the unit cost depreciation after n outputs

$$V_n = V_0 - nd$$

where V_n is the value of the asset after n outputs and d is the depreciation per output.

---

tlvd-3918

WORKED EXAMPLE 12 Calculating unit cost depreciation after n outputs

The initial cost of a vehicle was $27 850 and its scrap value is $5050. If the vehicle needs to be replaced after travelling 80 000 km (useful life):
a. calculate the depreciation rate (depreciation ($) per km)
b. calculate the amount of depreciation in a year when 16 497 km were travelled
c. set up an equation to determine the value of the vehicle after travelling n km
d. calculate the future value after the vehicle has been used for a total of 60 000 km
e. set up a depreciation schedule table that lists future value for every 20 000 km.

THINK	WRITE
a. 1. To calculate the depreciation rate, first determine the total depreciation. Total amount of depreciation = cost price − scrap value	a. Total amount of depreciation = 27 850 − 5050 = $22 800
2. Calculate the rate of depreciation. It is common to express rates in cents per use if less than a dollar.	Depreciation rate = $\dfrac{\text{total depreciation}}{\text{total distance travelled}}$ $= \dfrac{22\,800}{80\,000}$ = $0.285 per km = 28.5 cents per km
b. 1. Calculate the amount of depreciation using the rate calculated. Amount of depreciation is always expressed in dollars.	b. Amount of depreciation = amount of use × rate of depreciation = 16 497 × 28.5 = 470 165 cents = $4701.65
c. 1. Write the equation for unit cost depreciation after n outputs as well as the known information.	c. $V_n = V_0 - nd$ $V_0 = 27\,850,\ d = 0.285$
2. Substitute the values into the equation.	$V_n = 27\,850 - 0.285n$
d. 1. Use the equation from part c to determine the future value when $n = 60\,000$.	d. $V_{60\,000} = 27\,850 - 0.285 \times 60\,000$ $= 27\,850 - 17\,100$ $= 10\,750$
2. Write the answer.	The future value after the car has been used for 60 000 km is $10 750.

e. Calculate the future value for every 20 000 km of use and summarise in a table.

e.

Use, n (km)	Future value, V_n ($)
0	$27 850
20 000	$V_1 = 27\,850 - (20\,000 \times 0.285)$ $= \$22\,150$
40 000	$V_2 = 27\,850 - (20\,000 \times 0.285 \times 2)$ $= \$16\,450$
60 000	$V_3 = 27\,850 - (20\,000 \times 0.285 \times 3)$ $= \$10\,750$
80 000	$V_4 = 27\,850 - (20\,000 \times 0.285 \times 4)$ $= \$5\,050$

5.5 Exercise

1. **WE9** A washing machine purchased for $800 depreciates at a rate of 35 cents per wash.

   a. Set up a recurrence relation to represent the depreciation.
   b. Use the recurrence relation to generate a depreciation schedule for the future value of the washing machine after each of the first five washes.

2. An air conditioning unit purchased for $1400 depreciates at a rate of 65 cents per hour of use.

   a. Set up a recurrence relation to represent the depreciation.
   b. Use the recurrence relation to generate a depreciation schedule for the future value of the air conditioning unit at the end of each of the first five hours of use.

3. **WE10** Below are the depreciation details for a vehicle.
   Calculate:

   a. the annual depreciation in the first year
   b. the useful life (km).

Purchase price ($)	Scrap value ($)	Rate of depreciation (cents/km)	Distance travelled in first year (km)
29 600	12 000	28.5	14 000

4. A taxi was purchased for $42 000 and depreciates by 25 cents per km driven. During its first year, the taxi travelled 64 000 km; during its second year, it travelled 56 000 km. Calculate:

   a. the depreciation in each of the first two years
   b. the distance the car had travelled if its total depreciation was $20 000.

5. Consider the following table showing the depreciation details for two vehicles.

	Purchase price ($)	Scrap value ($)	Rate of depreciation (cents/km)	Distance travelled in first year (km)
a.	25 000	10 000	26	12 600
b.	21 400	8 000	21.6	13 700

   In each case, calculate:

   i. the annual depreciation in the first year
   ii. the useful life (km).

6. A company buys a $32 000 car that depreciates at a rate of 23 cents per km driven. It covers 15 340 km in the first year and has a scrap value of $9500.
   Calculate:

   a. the annual depreciation in the first year
   b. the car's useful life.

7. A new taxi is worth $29 500 and it depreciates at 27.2 cents per km travelled. In its first year of use, it travelled 28 461 km. Its scrap value was $8200.
   Calculate:

   a. the annual depreciation in the first year
   b. the taxi's useful life.

8. A photocopier purchased for $7200 depreciates at a rate of $1.50 per 1000 copies made. In its first year of use, 620 000 copies were made and in its second year, 540 000 were made. Calculate:

   a. the depreciation for each year
   b. the future value at the end of the second year.

9. A photocopier bought for $11 300 depreciates at a rate of 2.5 cents for every 10 copies made.

   a. Set up a recurrence relation to represent the depreciation.
   b. Copy and complete the table.

Time (years)	Outputs, n (10 copies)	Annual depreciation ($)	Future value at end of year, V_n ($)
1	35 000		
2	42 500		
3	37 620		
4	29 104		
5	38 562		

10. A corking machine bought for $14 750 depreciates at a rate of $2.50 for every 100 bottles corked.
   a. Set up a recurrence relation to represent the depreciation.
   b. Copy and complete the given table.

Time (years)	Outputs, n (100 bottles corked)	Depreciation ($)	Future value at end of year, V_n ($)
1	400		
2	425		
3	467		
4	382.5		
5	430.6		

11. **WE11** A photocopier is bought for $8600 and it depreciates at a rate of 22 cents for every 100 copies made. In its first year of use, 400 000 copies are made and in its second year, 480 000 copies are made. Calculate:
   a. the depreciation for each year
   b. the future value at the end of the second year.

12. A printing machine was purchased for $38 000 and depreciated at a rate of $1.50 per million pages printed. In its first year 385 million pages were printed and 496 million pages were printed in its second year. Calculate:
   a. the depreciation for each year
   b. the future value at the end of the second year.

13. **MC** A vehicle is bought for $25 900 and it depreciates at a rate of 21.6 cents per km driven. After its first year of use, in which it travels 13 690 km, the future value of the vehicle is closest to:
   A. $1000   B. $3000   C. $20 000   D. $23 000   E. $25 000

In each situation in questions 14 and 15, calculate:
   a. the depreciation for each year
   b. the future value at the end of the second year.

14. A company van is purchased for $32 600 and it depreciates at a rate of 24.8 cents per km driven. In its first year of use, the van travels 15 620 km; it travels 16 045 km in its second year.

15. A taxi is bought for $35 099 and depreciates at a rate of 29.2 cents per km driven. It travels 21 216 km in its first year of use and 19 950 km in its second year.

16. **WE12** A delivery service purchases a van for $30 000 and it is expected that the van will be written off after travelling 200 000 km. It is estimated that the van will travel 1600 km each week.
   a. Calculate the depreciation rate (charge per km).
   b. Calculate how long it will take for the van to be written off.
   c. Set up an equation to determine the value of the van after travelling n kilometres.
   d. Calculate the distance travelled for the van to depreciate by $13 800.
   e. Calculate the van's future value after it has travelled 160 000 kilometres.
   f. Set up a schedule table for the value of the van for every 20 000 kilometres.

17. A car bought for $28 395 depreciates at a rate of 23.6 cents for every km travelled. Copy and complete the given table.

Time (years)	Distance travelled, n (km)	Depreciation ($)	Future value at end of year, V_n ($)
1	13 290		
2	15 650		
3	14 175		
4	9 674		
5	16 588		

18. A car is bought for $35 000 and a scrap value of $10 000 is set for it. The following three options for depreciating the car are available:

  i. flat rate of 10% of the purchase price each year
  ii. 20% p.a. of the reducing balance
  iii. 25 cents per km driven (the car travels an average of 10 000 km per year).

  a. State which method will enable the car to reach its scrap value the earliest.
  b. If the car is used in a business, the annual depreciation can be claimed as a tax deduction. Calculate the tax deduction in the first year of use for each of the depreciation methods.
  c. Explain how your answers to part **b** would vary for the 5th year of use.

19. **MC** A machine that was bought for $8500 was depreciated at the rate of 2 cents per unit produced. By the time the book value had decreased to $2000, the number of units produced would be:

A. 75 000 **B.** 100 000 **C.** 125 000 **D.** 300 000 **E.** 325 000

20. **MC** An $8500 machine depreciates by 2 cents per unit. By the time the machine had depreciated by $5000, it would have produced:

A. 275 000 units.
B. 250 000 units.
C. 225 000 units.
D. 175 000 units.
E. 150 000 units.

5.5 Exam questions

Question 1 (3 marks)

Source: VCE 2021, Further Mathematics Exam 2, Section A Core, Q7; © VCAA.

Sienna owns a coffee shop.

A coffee machine, purchased for $12 000, is depreciated in value using the unit cost method.

The rate of depreciation is $0.05 per cup of coffee made.

The recurrence relation that models the year-to-year value, in dollars, of the coffee machine is

$$M_0 = 12\,000, \quad M_{n+1} = M_n - 1440$$

  a. Calculate the number of cups of coffee that the machine produces per year. **(1 mark)**
  b. The recurrence relation above could also represent the value of the coffee machine depreciating at a flat rate. What annual flat rate percentage of depreciation is represented? **(1 mark)**

c. Complete the rule below that gives the value of the coffee machine, M_n, in dollars, after n cups have been produced. Write your answers in the boxes provided. **(1 mark)**

$$M_n = \boxed{} + \boxed{} \times n$$

Question 2 (1 mark)

Source: VCE 2020, Further Mathematics Exam 1, Section A, Q29; © VCAA.

MC The value of a van purchased for $45\,000 is depreciated by $k\%$ per annum using the reducing balance method.

After three years of this depreciation, it is then depreciated in the fourth year under the unit cost method at the rate of 15 cents per kilometre.

The value of the van after it travels 30 000 km in this fourth year is $26 166.24.

The value of k is
 A. 9
 B. 12
 C. 14
 D. 16
 E. 18

Question 3 (1 mark)

Source: VCE 2019, Further Mathematics Exam 1, Section A, Q22; © VCAA.

MC A machine is purchased for $30\,000.

It produces 24 000 items each year.

The value of the machine is depreciated using a unit cost method of depreciation.

After three years, the value of the machine is $18 480.

A rule for the value of the machine after n units are produced, V_n, is
 A. $V_n = 0.872\,n$
 B. $V_n = 24\,000\,n - 3840$
 C. $V_n = 30\,000 - 24\,000\,n$
 D. $V_n = 30\,000 - 0.872\,n$
 E. $V_n = 30\,000 - 0.16\,n$

More exam questions are available online.

5.6 Review

5.6.1 Summary

doc-38036

5.6 Exercise

Multiple choice

1. **MC** Select the sequence that is generated by the first-order recurrence relation $u_{n+1} = u_n - 4, u_0 = -6$.

   A. $-4, -10, -16, -22, -28, \ldots$

   B. $-6, -10, -14, -18, -22, \ldots$

   C. $-6, -2, -6, -2, -6, \ldots$

   D. $-6, -2, 2, 6, 10, \ldots$

   E. $-6, 2, -10, 4, -14, \ldots$

2. **MC** A recurrence relation is defined as:

   $$V_{n+1} = 3V_n - 2 \quad V_0 = 2$$

   The first four terms of the relation are:

   A. 2, 6, 18, 54

   B. 1, 2, 3, 4

   C. 2, 4, 10, 28

   D. $-2, -8, -26, -80$

   E. 2, 4, 12, 34

3. **MC** A recurrence relation is defined as:

   $$V_{n+1} = 1.025V_n - 200 \quad V_0 = 10\,000$$

   The fifth term in this recurrence relation would be closest to:

   A. 11 038   B. 10 769   C. 10 506   D. 10 208   E. 10 110

4. **MC** The sequence 4, 2, 0, -2, -4, ... can be defined by the first-order recurrence relation:

   A. $u_{n+1} = 2u_n \qquad u_0 = 4$

   B. $u_{n+1} = -2u_n \qquad u_0 = 4$

   C. $u_{n+1} = u_n - 2 \qquad u_0 = 4$

   D. $u_{n+1} = 2u_n + 2 \qquad u_0 = 4$

   E. $u_{n+1} = 4u_n + 2 \qquad u_0 = 4$

5. **MC** The sequence -3, 12, -48, 192, -768, ... can be defined by the first-order recurrence relation:

   A. $A_{n+1} = -4A_n \qquad A_0 = -3$

   B. $A_{n+1} = -4A_n \qquad A_0 = 3$

   C. $A_{n+1} = -3A_n \qquad A_0 = -4$

   D. $A_{n+1} = -3A_n \qquad A_0 = 4$

   E. $A_{n+1} = 4A_n \qquad A_0 = -3$

6. **MC** A yacht was purchased for $54\,000$. It depreciated at a rate of 12% each year. The recurrence relation that could be used to determine the value of the yacht, Y_n, after n years is:

   A. $Y_{n+1} = -0.88Y_n \qquad Y_0 = 54\,000$

   B. $Y_{n+1} = Y_n - 6480 \qquad Y_0 = 54\,000$

   C. $Y_{n+1} = Y_n - 54\,000 \qquad Y_0 = 6480$

   D. $Y_{n+1} = 1.12Y_n \qquad Y_0 = 6480$

   E. $Y_{n+1} = 0.88Y_n \qquad Y_0 = 54\,000$

7. **MC** A library adds 300 new books to its collection each year. The collection began with 4000 books and it is claimed that no book has ever been removed.

A first-order recurrence relation that reflects this situation is:

A. $B_n = B_{n-1} + 300 \qquad B_0 = 300$

B. $B_n = B_{n-1} + 300 \qquad B_0 = 4000$

C. $B_n = 1.03B_{n-1} + 100 \qquad B_0 = 300$

D. $B_n = 1.03B_{n-1} + 4120 \qquad B_0 = 4000$

E. $B_n = 1.04B_{n-1} + 300 \qquad B_0 = 300$

8. **MC** Consider the first-order recurrence relations below.

$$V_{n+1} = V_n - 500 \qquad V_0 = 9500$$

$$P_{n+1} = 0.7P_n \qquad P_0 = 8000$$

V_{n+1} and P_{n+1} are examples of:

A. linear decay and geometric decay respectively.

B. geometric growth and linear decay respectively.

C. linear growth and geometric growth respectively.

D. geometric decay and linear decay respectively.

E. linear decay.

9. **MC** The reducing balance depreciation method is used to depreciate an air conditioner at a rate of 12.5% of the previous value. The purchase price of the air conditioner was $8400.

Select the recurrence relation that can be used to describe this situation.

A. $V_{n+1} = V_n - 1050, \ V_0 = 8400$

B. $V_{n+1} = 0.125V_n$

C. $V_{n+1} = 0.125V_n, \ V_0 = 8400$

D. $V_{n+1} = 0.875V_n, \ V_0 = 8400$

E. $V_{n+1} = 0.875V_n$

10. **MC** A coffee machine was purchased for $12 500. It has been depreciated using a flat rate model of $1000 per year. Select the number of years before it reaches a book value of $6500.

A. 2

B. 4

C. 6

D. 8

E. 10

Short answer

11. Write the first five terms of each of the following sequences.

a. $u_{n+1} = 4u_n - 3 \qquad u_0 = -1$

b. $u_{n+1} = 3 + 5u_n \qquad u_0 = 0$

12. A club loses 4% of its membership each year but adds 20 new members each year. The initial membership of the club was 300.

a. Write a first-order recurrence relation to describe this situation, stating clearly the terms you use.

b. State the number of members of the club at the end of four years.

13. A first-order recurrence relation is given as $C_{n+1} = 0.5C_n + 100$ and $C_0 = 400$. State the value of C_4.

14. A security firm purchased a van for $35 000. The firm decided to use the reducing balance method of depreciation at 8% per year. Let P_n be the value of the van after n years.

a. Write the recurrence relation for this situation.

b. Calculate the future value after five years.

15. Office equipment is bought for $18 000 and depreciated by the flat rate method. The depreciation is 15% of the cost price each year and its scrap value is $1800.
Calculate:

a. the annual depreciation
b. how long it will take for the equipment to reach its scrap value.

16. A 3D printer is purchased for $6500 and it depreciates by $5.20 for every 20 items printed. In its first year of use, 2000 items were printed and in its second year 2500 items were printed.
Calculate:

a. the depreciation each year
b. the future value at the end of the second year.

Extended response

17. A band has bought new electronic recording equipment for $150 000 and has a choice of depreciating the equipment by a flat rate of 15% of the cost price each year or by 25% of the previous value each year.

a. Using your calculator, generate depreciation schedules using both methods for a life of five years.
b. Determine the number of years it takes for the reducing balance future value to become greater than the flat rate future value.

18. Rachel bought a new limousine for her events management company. She paid $60 000 and is considering various methods of depreciating her new asset.

The limousine can be depreciated using the flat rate method. The depreciation rate is 12% of the cost price per year.

a. Calculate the annual depreciation of the limousine. Hence, set up a recurrence relation to represent the depreciation.
b. Generate the relationship between the book value of the limousine and time.
c. Calculate the book value of the limousine and total amount of depreciation after five years.

Another way to depreciate the limousine is to use unit cost depreciation method. It can be depreciated at a rate of 30 cents per kilometre driven.

d. Rachel estimates that her new limousine will be driven at least 25 000 km per year. According to this estimation, calculate the book value of the car after five years.

Rachel considers her final option: reducing balance depreciation method at 16% p.a.

e. Calculate the future value and the total depreciation of the limousine after five years.
f. State which of the three methods gives the greatest total depreciation over the five-year period.

5.6 Exam questions

Question 1 (1 mark)
Source: VCE 2020, Further Mathematics Exam 1, Section A, Q22; © VCAA.

MC An asset is purchased for $2480.

The value of this asset after n time periods, V_n, can be determined using the rule

$$V_n = 2480 + 45n$$

A recurrence relation that also models the value of this asset after n time periods is

A. $V_0 = 2480, \ V_{n+1} = V_n + 45n$
B. $V_n = 2480, \ V_{n+1} = V_n + 45n$
C. $V_0 = 2480, \ V_{n+1} = V_n + 45$
D. $V_1 = 2480, \ V_{n+1} = V_n + 45$
E. $V_n = 2480, \ V_{n+1} = V_n + 45$

Question 2 (4 marks)

Source: VCE 2019, Further Mathematics Exam 2, Section A, Q7; © VCAA.

Phil is a builder who has purchased a large set of tools.

The value of Phil's tools is depreciated using the reducing balance method.

The value of the tools, in dollars, after n years, V_n, can be modelled by the recurrence relation shown below.

$$V_0 = 60\,000, \qquad V_{n+1} = 0.9\,V_n$$

a. Use recursion to show that the value of the tools after two years, V_2, is \$48 600. **(1 mark)**
b. What is the annual percentage rate of depreciation used by Phil? **(1 mark)**
c. Phil plans to replace these tools when their value first falls below \$20 000.
   After how many years will Phil replace these tools? **(1 mark)**
d. Phil has another option for depreciation. He depreciates the value of the tools by a flat rate of 8% of the purchase price per annum.
   Let V_n be the value of the tools after n years, in dollars.
   Write down a recurrence relation, in terms of V_0, V_{n+1} and V_n, that could be used to model the value of the tools using this flat rate depreciation. **(1 mark)**

Question 3 (1 mark)

Source: VCE 2017, Further Mathematics Exam 1, Section A, Core, Q21; © VCAA.

MC A printer was purchased for \$680.

After four years the printer has a value of \$125.

On average, 1920 pages were printed every year during those four years.

The value of the printer was depreciated using a unit cost method of depreciation.

The depreciation in the value of the printer, per page printed, is closest to
   A. 3 cents.
   B. 4 cents.
   C. 5 cents.
   D. 6 cents.
   E. 7 cents.

Question 4 (1 mark)

Source: VCAA 2015, Further Mathematics Exam 1, Section B, Module 4, Q7; © VCAA.

MC The following graph shows the depreciating value of a van.

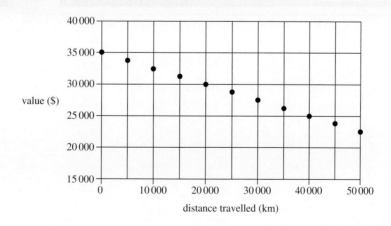

The graph could represent the van being depreciated using

    A. flat rate depreciation with an initial value of $35 000 and a depreciation rate of $25 per year.

    B. flat rate depreciation with an initial value of $35 000 and a depreciation rate of 25 cents per year.

    C. reducing balance depreciation with an initial value of $35 000 and a depreciation rate of 2.5% per annum.

    D. unit cost depreciation with an initial value of $35 000 and a depreciation rate of 25 cents per kilometre travelled.

    E. unit cost depreciation with an initial value of $35 000 and a depreciation rate of 25 cents per kilometer travelled.

Question 5 (1 mark)

Source: VCAA 2014, Further Mathematics Exam 1, Section B, Module 4, Q7; © VCAA.

MC New furniture was purchased for an office at a cost of $18 000.

Using flat rate depreciation, the furniture will be valued at $5000 after four years.

The expression that can be used to determine the value of the furniture, in dollars, after one year is

    A. $18\,000 - (4 \times 5000)$

    B. $18\,000 - \left(\dfrac{18\,000 - 5000}{4} \right)$

    C. $18\,000 - \dfrac{5000}{4}$

    D. $\dfrac{18\,000}{4} - 5000$

    E. $18\,000 \times 0.726$

More exam questions are available online.

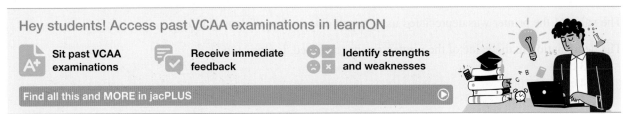

Hey students! Access past VCAA examinations in learnON

Sit past VCAA examinations   Receive immediate feedback   Identify strengths and weaknesses

Find all this and **MORE** in jacPLUS

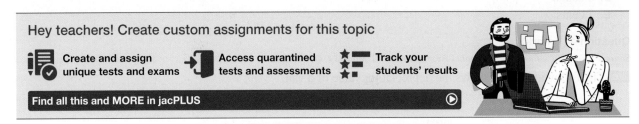

Hey teachers! Create custom assignments for this topic

Create and assign unique tests and exams   Access quarantined tests and assessments   Track your students' results

Find all this and **MORE** in jacPLUS

Answers

Topic 5 Modelling depreciation of assets using recursion

5.2 A first-order linear recurrence relation

5.2 Exercise

1. $-2, -5, -8, -11, -14$

2. $3, 6, 12, 24, 48$

3. $u_n = 4u_{n-1} + 3$, $u_0 = 5$
   The sequence is $5, 23, 95, 383, 1535$.

4. $f_{n+1} = 5f_n - 6$, $f_0 = -2$
   The sequence is $-2, -16, -86, -436, -2186$.

5. b, c, g, j

6. a. $6, 8, 10, 12, 14$ b. $5, 2, -1, -4, -7$
   c. $23, 24, 25, 26, 27$ d. $7, -3, -13, -23, -33$

7. a. $1, 3, 9, 27, 81$
   b. $-2, -10, -50, -250, -1250$
   c. $1, -4, 16, -64, 256$
   d. $-1, -2, -4, -8, -16$

8. a. $1, 3, 7, 15, 31$ b. $5, 13, 37, 109, 325$

9. a. The first seven terms are $6, -5, 6, -5, 6, -5, 6$.
   b. The first seven terms are
   $1, 5, 25, 125, 625, 3125, 15\,625$.

10. C

11. E

12. a. $u_{n+1} = 3u_n$, $u_0 = \dfrac{1}{4}$; $\dfrac{1}{4}, \dfrac{3}{4}, 2\dfrac{1}{4}, 6\dfrac{3}{4}, 20\dfrac{1}{4}$
    b. $u_{n+1} = u_n + 2000$, $u_0 = 200\,000$; $200\,000, 202\,000,$
    $204\,000, 206\,000, 208\,000$
    c. $u_{n+1} = u_n - 7$, $u_0 = 100$; $100, 93, 86, 79, 72$
    d. $u_{n+1} = 2u_n - 50$,
    $u_0 = \$200$; $\$200, \$350, \$650, \$1250, \$2450$

5.2 Exam questions

Note: Mark allocations are available with the fully worked solutions online.

1. D

2. D

3. E

5.3 Modelling flat rate depreciation with a recurrence relation

5.3 Exercise

1. a. $\$3750$
   b. $V_{n+1} = V_n - 3750$
   c.

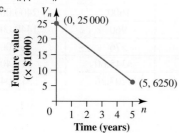

d. $\$6250$

2. a. $V_0 = 30\,000$, $V_{n+1} = V_n - 6000$
   b. $V_0 = 2000$, $V_{n+1} = V_n - 200$
   c. $V_0 = 6000$, $V_{n+1} = V_n - 500$

3. a. i. $\$5000$ per year   ii. 10 years
   b. i. $\$100$ per year   ii. 8.5 years
   c. i. $\$95$ per year   ii. 13 years

4. a. $V_0 = 22\,500$
   $V_{n+1} = V_n - 3200$
   b. $\$6500$

5. a. $\$1500$
   b.

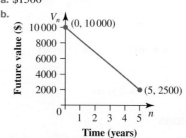

6. a.

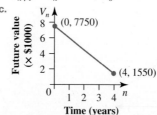

b. Scrap value is $\$5000$.

7. a.

V_n graph with axis *Future value ($)* from 2000 to 14 000, *Time (years)* 0 to 6, showing points $(0, 13\,500)$ and $(6, 1500)$.

b. Scrap value is $\$1500$.

8. a. $\$1550$
   b. $V_{n+1} = V_n - 1550$, $V_0 = 7750$
   c.

V_n graph with axis *Future value (× 1000)* from 2 to 8, *Time (years)* 0 to 4, showing points $(0, 7750)$ and $(4, 1550)$.

d. 4 years

9. a. $\$13\,800$
   b. $V_{n+1} = V_n - 13\,800$, $V_0 = 92\,000$

c.

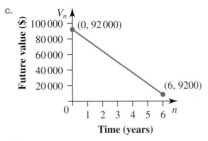

d. 6 years

10. a. i. $3000   ii. $400   iii. 5 years

b. i. $1750   ii. $500   iii. $3\frac{1}{2}$ years

c. i. $800   ii. $140   iii. 5 years

d. i. $18\,000   ii. $2400   iii. $7\frac{1}{2}$ years

11. C

12. a. The cheaper machine   b. 2 years

5.3 Exam questions

Note: Mark allocations are available with the fully worked solutions online.

1. a. $15\,000

   b. $90\,000

   c. 12.5%

   d. $V_n = 120\,000 - 15\,000n$, where $n = 0, 1, 2, ...$

2. A

3. B

5.4 Modelling reducing balance depreciation with a recurrence relation

5.4 Exercise

1. a.

Time, n (years)	Future value, V_n ($)
0	1500.00
1	1245.00
2	1033.35
3	857.68
4	711.87
5	590.86

b. $590.85

c.

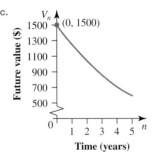

2. a.

Time, n (years)	Future value, V_n ($)
0	3300.00
1	2871.00
2	2497.77
3	2173.06
4	1890.56
5	1644.79

b. $1644.79

c.

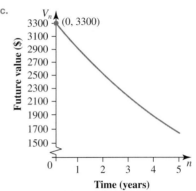

3. a. $V_{n+1} = 0.8V_n$, $V_0 = 60\,000$

b.

Time, n (years)	Future value, V_n ($)
0	60\,000.00
1	48\,000.00
2	38\,400.00
3	30\,720.00
4	24\,576.00
5	19\,660.80

c. $19\,660.80

d.

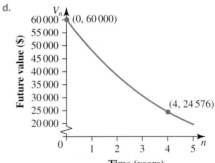

4. a. $V_{n+1} = 0.6V_n$, $V_0 = 4000$

b.

Time, n (years)	Depreciation ($)	Future value, V_n ($)
0	–	4000.00
1	1600	2400.00
2	960	1440.00
3	576	864.00
4	345.60	518.40
5	207.36	311.04

c. $311.04

d.

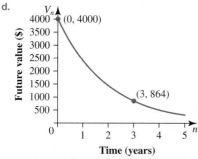

Time (years)

c. The future value for the reducing balance method is greater than the flat rate method after four years.

5. a.

Time, n (years)	Depreciation ($)	Future value, V_n ($)
0	–	29 900
1	5998	23 992
2	5998	17 994
3	5998	11 996
4	5998	5 998
5	5998	0

Time, n (years)	Future value, V_n ($)
0	29 900.00
1	21 892.70
2	15 981.67
3	11 666.62
4	8 516.63
5	6 217.14

b.

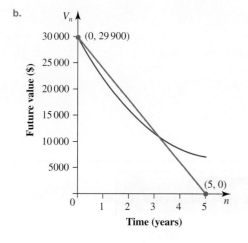

Time (years)

6. a.

Time, n (years)	Depreciation ($)	Future value, V_n ($)
0	–	23 000
1	4600	18 400
2	4600	13 800
3	4600	9 200
4	4600	4 600
5	4600	0

Time, n (years)	Future value, V_n ($)
0	23 000.00
1	16 560.00
2	11 923.20
3	8 584.70
4	6 180.99
5	4 450.31

b.

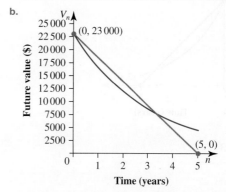

Time (years)

c. The future value for the reducing balance method is greater than the flat rate method after four years.

7. a. See table at bottom of the page*

b.
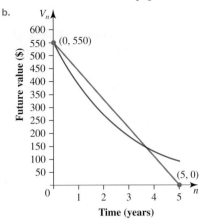

c. 4 years

8. a. See table at bottom of the page*

b.
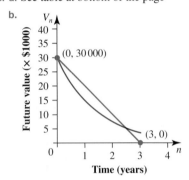

c. 3 years

9. $16971.73

10. $747.88

11. a. $8192   **b.** $9492.19

12. B

13. a. $115.13, $1034.87

  b. $1183.36, $2556.64

  c. $1302.80, $6017.20

14. a. $115.13, $569.87

  b. $7803.25, $24 696.75

  c. $368.73, $706.27

15. D

16. E

17. 6 years

18. 10 years

19. C

20. a. 11.61   **b.** 5.02

5.4 Exam questions

Note: Mark allocations are available with the fully worked solutions online.

1. C

2. a. $V_1 = 71\,625$, $V_2 = 68\,250$

  b. i. $3375   **ii.** 4.5% p.a

  c. 5.7% p.a.

3. A

---

*7

	Flat rate			Reducing balance	
Time (years)	Dep. ($)	Future value ($)	Time (years)	Dep. ($)	Future value ($)
0	–	550	0	–	550.00
1	110	440	1	165	385.00
2	110	330	2	115.50	269.50
3	110	220	3	80.85	188.65
4	110	110	4	56.60	132.06
5	110	0	5	39.62	92.44

*8

	Flat rate			Reducing balance	
Time (years)	Dep. ($)	Future value ($)	Time (years)	Dep. ($)	Future value ($)
0	–	30 000	0	–	30 000
1	10 000	20 000	1	15 000	15 000
2	10 000	10 000	2	7 500	7 500
3	10 000	0	3	3 750	3 750

5.5 Modelling unit cost depreciation with a recurrence relation

5.5 Exercise

1. a. $V_{n+1} = V_n - 0.35n$, $V_0 = 800$

b.

Number of washes (n)	Future value, V_n ($\$$)
1	799.65
2	799.30
3	798.95
4	798.60
5	798.25

2. a. $V_{n+1} = V_n - 0.65n$, $V_0 = 1400$

b.

Hours of use (n)	Future value, V_n ($\$$)
1	1399.35
2	1398.70
3	1398.05
4	1397.40
5	1396.75

3. a. $3990   **b.** 61754 km

4. a. $16000, $14000   **b.** 80000 km

5. a.i. $3276   **ii.** 57692 km

  b.i. $2959.20   **ii.** 62037 km

6. a. $3528.20   **b.** 97826 km

7. a. $7741.39   **b.** 78309 km

8. a. $930, $810   **b.** $5460

9. a. $V_{n+1} = V_n - 0.0025n$, $V_0 = 11300$

b.

Annual depreciation ($\$$)	Future value at end of year, V_n ($\$$)
875.00	10425.00
1062.50	9362.50
940.50	8422.00
727.60	7694.40
964.05	6730.35

10. a. $V_{n+1} = V_n - 0.025n$, $V_0 = 14750$

b.

Depreciation ($\$$)	Future value at end of year, V_n ($\$$)
1000.00	13750.00
1062.50	12687.50
1167.50	11520.00
956.25	10563.75
1076.50	9487.25

11. a. $880, $1056   **b.** $6664

12. a. $577.50, $744   **b.** $36678.50

13. D

14. a. $3873.76, $3979.16   **b.** $24747.08

15. a. $6195.07, $5825.40

  b. $23078.53

16. a. 15 cents per km

  b. 2 years, 21 weeks

  c. $V_n = 30000 - 0.15n$

  d. 92000 km

  e. $6000

  f.

Distance, n (km)	Future value ($\$$)
0	30000
20000	27000
40000	24000
60000	21000
80000	18000
100000	15000
120000	12000
140000	9000
160000	6000
180000	3000
200000	0

17.

Depreciation ($\$$)	Future value at end of year, V_n ($\$$)
3136.44	25258.56
3693.40	21565.16
3345.30	18219.86
2283.06	15936.80
3914.77	12022.03

18. a. i. 8 years   **ii.** 6 years   **iii.** 10 years

  Therefore, the reducing balance depreciation (6 years) enables the car to reach its scrap value the earliest.

  b. i. $3500   **ii.** $7000   **iii.** $2500

  c. i. $3500 (same)

    ii. $2867.20 ($4132.80 less)

    iii. $2500 (same)

19. E

20. B

5.5 Exam questions

Note: Mark allocations are available with the fully worked solutions online.

1. a. 28000

  b. 12%

  c. $M_n = 12000 - 0.05 \times n$

2. B

3. E

5.6 Review

5.6 Exercise

Multiple choice

1. B

2. C

3. D

4. C

5. A

6. E
7. B
8. A
9. D
10. C

Short answer

11. a. $-1, -7, -31, -127, -511$
    b. $0, 3, 18, 93, 468$

12. a. $M_{n+1} = 0.96M_n + 20, M_0 = 300$
    b. 330 members

13. 212.5

14. a. $P_{n+1} = 0.92P_n, P_0 = 35\,000$
    b. $23\,067.85

15. a. $2700   b. 6 years

16. a. $520, $650   b. $5330

Extended response

17. a.

Time (n)	Future value (flat rate)	Future value (reducing balance)
0	150 000	150 000
1	127 500	112 500
2	105 000	84 375
3	82 500	63 281.25
4	60 000	47 460.94
5	37 500	35 595.70
6	15 000	26 696.80

    b. 6 years

18. a. $7200; $V_{n+1} = V_n - 7200, V_0 = 60\,000$
    b. $V_n = 60\,000 - 7200n$
    c. Book value after five years is $24\,000; total depreciation is $36\,000.
    d. $22\,500
    e. Future value after five years is $25\,092.72; total depreciation is $34\,907.28.
    f. Unit cost depreciation

5.6 Exam questions

Note: Mark allocations are available with the fully worked solutions online.

1. C
2. a. Please see the worked solution.
   b. 10%
   c. 10 years
   d. $V_{(n+1)} = V_n - 0.08\,V_0$, where $V_0 = 60\,000$
3. E
4. D
5. B

6 Modelling compound interest investments and loans using recursion

Fully worked solutions for this topic are available online.

6.1 Overview

6.1.1 Introduction

Most people in the modern world will at some stage in their lives borrow and invest money. When a person enters the work force, they may want a car or a house or other goods and do not have the money to purchase these items outright. They then borrow the money believing that they have the means to service the loan and pay the money back over a specified time.

In Australia, every employer must contribute to their employees' Superannuation Guarantee Fund, which is set by the government at a percentage of their base salary. Many employees also make extra contributions to their superannuation fund. Self-employed people are encouraged to set up their own superannuation fund. The money in the superannuation funds is invested so that by the age of retirement, every person will have some money saved.

Given the likelihood that every individual will need to grapple with decisions pertaining to borrowing and investing money throughout their adult life, it is essential that everyone has a competent understanding of financial mathematics. This will ensure that they have thoroughly investigated all their options and that their decision-making is therefore informed.

KEY CONCEPTS

This topic covers the following key concepts from the VCE Mathematics Study Design:
- the concepts of simple and compound interest
- use of a recurrence relation to model and analyse (numerically and graphically) a compound interest investment or loan, including the use of a recurrence relation to determine the value of the compound interest loan or investment after *n* compounding periods for an initial sequence from first principles
- the difference between nominal and effective interest rates and the use of effective interest rates to compare investment returns and the cost of loans when interest is paid or charged, for example, daily, monthly, quarterly
- the future value of a compound interest investment or loan after *n* compounding periods and its use to solve practical problems.

Source: VCE Mathematics Study Design (2023–2027) extracts © VCAA; reproduced by permission.

6.2 Simple interest

LEARNING INTENTION

At the end of this subtopic you should be able to:
- represent simple interest as a recurrence relation to calculate the value of an investment
- calculate simple interest using a rule.

6.2.1 Simple interest as a first-order linear recurrence relation

Simple interest can be represented by a first-order linear recurrence relation.

> **Formula for simple interest as a linear recurrence relation**
>
> $$V_{n+1} = V_n + d, \ V_0 = \text{initial value};$$
>
> $$\text{where } d = \frac{V_0 \times r}{100}$$
>
> where V_n represents the value of the investment after n time periods
>
> V_{n+1} is the amount of the investment at the next time period after V_n
>
> d is the amount of interest earned per period
>
> V_0 is the initial (or principal) amount
>
> r is the interest rate per period.

tlvd-3919

WORKED EXAMPLE 1 Calculating simple interest as a recurrence relation

$325 is invested in a simple interest account for 5 years at 3% p.a. (*per annum* or *per year*).
a. **Set up a recurrence relation to calculate the value of the investment after n years.**
b. **Use the recurrence relation from part a to calculate the value of the investment at the end of each of the first 5 years.**

THINK	WRITE
a. 1. Write the formula to calculate the amount of interest earned per period (d).	a. $d = \dfrac{V_0 \times r}{100}$
2. List the values of V_0 and r.	$V_0 = 325, r = 3$
3. Substitute into the formula and evaluate the amount of interest earned per period.	$d = \dfrac{325 \times 3}{100}$ $= 9.75$
4. Use the values of d and V_0 to set up your recurrence relation.	$V_{n+1} = V_n + 9.75, V_0 = 325$

b. 1. Set up a table to determine the value of the investment for up to $n = 5$.

b.

$n+1$	V_n ($)	V_{n+1} ($)
1	325	$325 + 9.75 = 334.75$
2	334.75	$334.75 + 9.75 = 344.50$
3	344.50	$344.50 + 9.75 = 354.25$
4	354.25	$354.25 + 9.75 = 364$
5	364	$364 + 9.75 = 373.75$

2. Use the recurrence relation from part **a** to complete the table.

3. Write the answer.

The value of the investment at the end of each of the first 5 years is: $334.75, $344.50, $354.25, $364 and $373.75.

TI	THINK	DISPLAY/WRITE	CASIO	THINK	DISPLAY/WRITE
b. 1. On a Calculator page, type 325 and then press ENTER. Complete the entry line as: $+9.75$ Then continue to press ENTER until the first 5 terms are displayed on the screen.	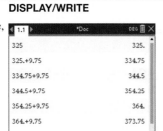	**1.** On the Main screen, type 325 and then press EXE. Complete the entry line as: $+9.75$ Then continue to press EXE until the first 5 terms are displayed on the screen.	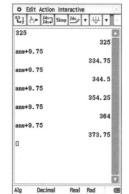
2. The answer appears on the screen.	The value of the investment at the end of each of the 5 years is: $334.75, $344.50, $354.25, $364, $373.75.	**2.** The answer appears on the screen.	The value of the investment at the end of each of the 5 years is: $334.75, $344.50, $354.25, $364, $373.75.

WORKED EXAMPLE 2 Calculating simple interest using a calculator

A bank offers 9% p.a. simple interest on an investment of $600.

a. Write the recurrence relation for this situation.

b. Using your calculator, complete a table to show the amount in the investment account at the end of 4 years.

c. Calculate the total interest earned after 4 years.

THINK

a. 1. Write the formula to calculate the amount of interest earned per period (d).

2. List the values of V_0 and r.

3. Substitute into the formula and evaluate the amount of interest earned per period.

WRITE

a. $d = \dfrac{V_0 \times r}{100}$

$V_0 = 600, r = 9$

$d = \dfrac{600 \times 9}{100}$

$= 54$

4. Use the values of d and V_0 to set up your recurrence relation.

$V_{n+1} = V_n + 54, V_0 = 600$

b. 1. On a calculator screen, type 600 and press ENTER/EXE; then, type $+54$ and press ENTER/EXE 4 times to obtain the amount in the investment account at the end of each of the 4 years.

b.

600	600
600 + 54	654
654 + 54	708
708 + 54	762
762 + 54	816

2. Write the answer.

At the end of 4 years there will be $816 in the account.

c. 1. The amount of interest earned is the total amount in the account at the end of the 4 years minus the principal (V_0, the initial amount of money in the account).

c. $I = 816 - 600$
$= 216$

2. Write the answer.

The interest earned in 4 years is $216.

6.2.2 Simple interest using a rule

You can also calculate the total value of a loan or investment by using the following rule.

Simple interest rule

$$\text{total value of loan or investment} = \text{initial amount (or principal)} + n \times d$$

$$V_n = V_0 + nd$$

where V_n represents the value of the investment after n time periods

d **is the amount of interest earned per period**

V_0 **is the initial amount (or principal).**

Simple interest is the percentage of the amount borrowed or invested multiplied by the number of time periods (usually years). The amount is added to the principal either as payment for the use of the money borrowed or as return on money invested.

In the case of simple interest, the total value of investment increases by the same amount per period. Therefore, if the values of the investment at the end of each time period are plotted, a straight-line graph is formed. Simple interest is an example of a linear model of growth.

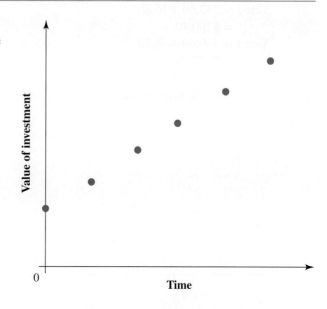

tlvd-3920

Jan invested \$210 with a building society in a savings account that paid 8% p.a. simple interest for 5 years.
a. **Use the simple interest rule to determine how much was in her account after 5 years.**
b. **Represent the account balance for each of the 5 years graphically.**

THINK	WRITE/DRAW
a. 1. Write the simple interest rule.	a. $V_n = V_0 + nd$
2. Write the formula to calculate the amount of interest earned per period.	$d = \dfrac{V_0 \times r}{100}$
3. List the values of V_0 and r, then substitute into the formula and evaluate.	$V_0 = 210,\ r = 8$ $d = \dfrac{210 \times 8}{100}$ $= \$16.80$
4. List the values of V_0, n and d to use in the simple interest rule.	$V_0 = 210,\ n = 5,\ d = \$16.80$
5. Substitute the values of the pronumerals into the rule and evaluate.	$V_n = 210 + 5 \times 16.80$ $= 294$
6. Write the answer.	The amount in her account after 5 years is \$294.
b. 1. As we are dealing with simple interest, the value of the investment increases each year by the same amount.	b. Increase per year $= \$16.80$
2. Draw a set of axes. Put time (in years) on the horizontal axis and the amount of investment (in \$) on the vertical axis. Plot the points: the initial value of investment is \$210 and it grows by \$16.80 each year. Year 1 $= 210 + 16.80$ $= \$226.80$ Year 2 $= 226.80 + 16.80$ $= \$243.60$ Year 3 $= 243.60 + 16.80$ $= \$260.40$ Year 4 $= 260.40 + 16.80$ $= \$277.20$ Year 5 $= 277.20 + 16.80$ $= \$294.00$ (as in part a).	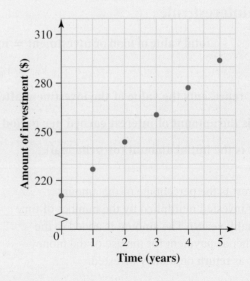

1. **WE1** $1020 is invested in a simple interest account for 5 years at 8.5% p.a.
   a. Set up a recurrence relation V_n to calculate the value of the investment after n years.
   b. Using your calculator and the recurrence relation from part **a**, determine the value of the investment at the end of each of the first 5 years.

2. $713 is invested in a simple interest account for 5 years at 6.75% p.a.
   a. Set up a recurrence relation B_n to calculate the value of the investment after n years.
   b. Using your calculator and the recurrence relation from part **a**, determine the value of the investment at the end of each of the first 5 years.

3. **WE2** A bank offers $12\frac{1}{2}$% p.a. simple interest on an investment of $1500. The amount in the bank is A_n after n years.
   a. Write the recurrence relation for this situation.
   b. Using your calculator, complete a table to show the amount in the investment account at the end of 2 years.
   c. Calculate the total interest earned after 2 years.

4. An amount of $2400 is invested at $5\frac{1}{2}$% p.a. for 18 months.
   a. Write the recurrence relation for this situation.
   b. Calculate the amount in the investment account at the end of 18 months.
   c. Calculate the total interest earned after 18 months.

5. Determine the value of the following investments at the end of each year of the investment by using a recurrence relation.
   a. $680 for 4 years at 5% p.a. simple interest
   b. $13 000 for 3 years at 7.5% p.a. simple interest

6. **WE3** $10 500 is invested in a simple interest account at 6.5% p.a.
   a. Use the simple interest rule to determine how much was in the account after 5 years.
   b. Represent the account balance for each of the 5 years graphically.

7. A building society offers an investment opportunity described using the following recurrence relation.

$$A_{n+1} = A_n + 565, \quad A_0 = 6200$$

   a. Explain in words the building society's investment offer.
   b. Calculate the interest rate per annum correct to 2 decimal places.

8. Silvio invested the $1500 he won playing the lottery with an insurance company bond that pays $12\frac{1}{4}$% p.a. simple interest provided he keeps the bond for 5 years.

   a. Calculate Silvio's total return from the bond at the end of the 5 years.

   b. Represent the balance at the end of each year graphically.

9. The value of a simple interest investment at the end of year 2 is $3377. At the end of year 3 the investment is worth $3530.50.
Use a recurrence relation to work out how much was invested.

10. Carol has $3000 to invest. Her aim is to earn $450 in interest at a rate of 5% p.a. Determine the number of years she should invest her money to earn this interest.

11. **MC** A loan of $1000 is taken over 5 years. The simple interest is calculated monthly. The total amount repaid for this loan is $1800. The simple interest rate per year on this loan is closest to:

   A. 8.9%   **B.** 16%   **C.** 36%   **D.** 5%   **E.** 11.1%

12. Calculate the interest charged or earned on the following loans and investments:

   a. $750 loaned at 12% p.a. simple interest for 4 years

   b. $7500 invested for 3 years at 1% per month simple interest

   c. $250 invested at $1\frac{3}{4}$% per month for $2\frac{1}{2}$ years

6.2 Exam questions

Question 1 (1 mark)

Source: VCE 2020, Further Mathematics Exam 1, Section A, Q24; © VCAA.

MC Manu invests $3000 in an account that pays interest compounding monthly.

The balance of his investment after n months, B_n, can be determined using the recurrence relation

$$B_0 = 3000, \; B_{n+1} = 1.0048 \times B_n$$

The total interest earned by Manu's investment after the first five months is closest to

   A. $57.60   **B.** $58.02   **C.** $72.00   **D.** $72.69   **E.** $87.44

Question 2 (1 mark)

Source: VCE 2013, Further Mathematics Exam 1, Section B, Module 4, Q7; © VCAA.

MC The graph shows the growth in value of a $1000 investment over a period of four years.

A different amount of money is invested under the same investment conditions for eight years.

In total, the amount of interest earned on this investment is $600.

The amount of money invested is

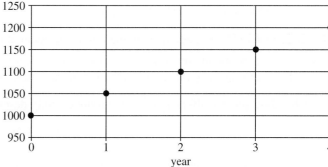

   A. $500   **B.** $600   **C.** $1500   **D.** $2000   **E.** $2400

Question 3 (1 mark)

MC Tayla wants to earn $1800 interest on a 5-year investment at 4.5% per annum simple interest. The amount she needs to invest is closest to

   A. $405   **B.** $1800   **C.** $7310   **D.** $8000   **E.** $9000

More exam questions are available online.

6.3 Compound interest as a geometric recurrence relation

LEARNING INTENTION

At the end of this subtopic you should be able to:
- represent compound interest as a geometric recurrence relation to calculate the value of an investment.

6.3.1 The geometric growth of compound interest

When interest is added to the initial amount (principal) at the end of an interest-bearing period, then both the interest and the principal earn further interest during the next period, which in turn is added to the balance. The time period in which the interest is added to the investment or loan is known as the **compounding period**. The interest is said to be compounded and is an example of geometric growth. The more often compounding occurs, the higher the total amount of interest received or paid.

This results in the increase of the account balance at regular intervals, as well as of the interest earned.

> **Formula for compound interest as a geometric recurrence relation**
>
> $$V_{n+1} = RV_n$$
>
> where V_n is the value of the investment after n compounding periods
>
> V_{n+1} is the amount of the investment at the next time period after V_n
>
> R is the growth or compounding factor, $R = 1 + \dfrac{r}{100}$
>
> r is the interest rate per compounding period
>
> V_0 is the initial value of the investment.

Consider $1000 invested for 4 years at an interest rate of 12% p.a. with interest compounded annually (added on each year). Use the recurrence relation $V_{n+1} = 1.12V_n$, $V_0 = 1000$ to calculate the amount in the account at the end of each year.

V_n ($)	Calculation	V_{n+1} ($)
1000	1.12×1000	1120.00
1120	1.12×1120	1254.40
1254.40	1.12×1254.40	1404.93
1404.93	1.12×1404.93	1573.52

So the balance after 4 years is $1573.52.

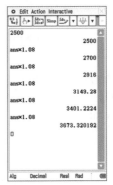

WORKED EXAMPLE 4 Calculating compound interest as a geometric recurrence relation

tlvd-3921

Laura invested \$2500 for 5 years at an interest rate of 8% p.a. with interest compounded annually.
a. Write Laura's investment as a recurrence relation.
b. Using your calculator, determine the amount invested at the end of 5 years.

THINK	WRITE
a. 1. Write the recurrence relation. Write the values of V_0 and r.	**a.** $V_{n+1} = RV_n$ $V_0 = 2500$ $r = 8$
2. Calculate R, the compounding factor.	$R = 1 + \dfrac{r}{100}$ $= 1 + \dfrac{8}{100}$ $= 1.08$
3. Substitute into the recurrence relation rule.	$V_{n+1} = 1.08\,V_n,\ V_0 = 2500$

b. 1. Using your calculator, type 2500, press ENTER/EXE, then type ×1.08.
Press ENTER/EXE 5 times to calculate the amount in the account for each of the 5 years.

b.

2500	2500
2500 × 1.08	2700
2700 × 1.08	2916
2916 × 1.08	3149.28
3149.28 × 1.08	3401.22
3401.22 × 1.08	3673.32

2. Write the answer.

At the end of 5 years, the amount invested is \$3673.32.

TI	THINK	DISPLAY/WRITE	CASIO	THINK	DISPLAY/WRITE

b. 1.	On a Calculator page, type 2500 and then press ENTER. Complete the entry line as: ×1.08 Then continue to press ENTER until the first 5 terms are displayed on the screen.	<table><tr><td>1.1</td><td>*Doc</td><td>DEG ☐ ✕</td></tr><tr><td>2500</td><td></td><td>2500.</td></tr><tr><td>2500.· 1.08</td><td></td><td>2700.</td></tr><tr><td>2700.· 1.08</td><td></td><td>2916.</td></tr><tr><td>2916.· 1.08</td><td></td><td>3149.28</td></tr><tr><td>3149.28· 1.08</td><td></td><td>3401.22</td></tr><tr><td>3401.2224· 1.08</td><td></td><td>3673.32</td></tr></table>	**1.**	On the Main screen, type 2500 and then press EXE. Complete the entry line as: ×1.08 Then continue to press EXE until the first 5 terms are displayed on the screen.	Edit Action Interactive
					2500
					    2500
					ans×1.08
					    2700
					ans×1.08
					    2916
					ans×1.08
					    3149.28
					ans×1.08
					    3401.2224
					ans×1.08
					    3673.320192
					☐
					Alg Decimal Real Rad

2.	The answer appears on the screen.	At the end of 5 years, the amount invested is \$3673.32.	**2.**	The answer appears on the screen.	At the end of 5 years, the amount invested is \$3673.32.

1. **WE4** Fraser invested $7500 for 4 years at an interest rate of 6% p.a. with interest compounded annually.

   a. Write Fraser's investment as a recurrence relation.
   b. Using your calculator, determine the amount invested at the end of 4 years.

2. Bobbi invested $3250 for 5 years at an interest rate of 7.5% p.a. with interest compounded annually.

   a. Using V_0 to represent the initial investment, write Bobbi's investment as a recurrence relation.
   b. Using your calculator, determine the amount invested at the end of 5 years.

3. Imani invested $600 in a short-term deposit for 6 months at an interest rate of 6% p.a. with interest compounded monthly.

   a. Using V_0 to represent the initial investment, write Imani's investment as a recurrence relation.
   b. Using your calculator, determine the amount invested at the end of the 6 months.

4. To join her friends on a holiday, Siobhan borrowed $2500 for 2 months at an interest rate of 12% p.a. with interest compounded monthly.

   a. Using V_0 to represent the initial investment, write Siobhan's loan as a recurrence relation.
   b. Calculate the amount of interest owing on the loan at the end of the two months.

The following information relates to questions 5–9.

Shannon invested money in a savings account. The amount in the account after n years, V_n, can be modelled by the recurrence relation below.

$$V_{n+1} = 1.05V_n, \quad V_0 = 3000$$

5. **MC** The annual percentage compound interest rate for this account is:
   A. 1.05%   **B.** 5%   **C.** 50%   **D.** 95%   **E.** 105%

6. **MC** The interest earned in the first year is:
   A. $50   **B.** $100   **C.** $150   **D.** $200   **E.** $300

7. **MC** The balance at the end of the first year is:
   A. $3000   **B.** $3050   **C.** $3100   **D.** $3150   **E.** $3200

8. **MC** An investment of $25 000 is invested in an account earning 4.7% interest with bi-annual compounding periods.
   The difference between the interest earned in the first and the second investment periods is:
   A. $13.81   **B.** $15.47   **C.** $587.50   **D.** $601.31   **E.** $25 587.50

9. **MC** The interest earned during the second year is:
   A. $100   **B.** $125   **C.** $150   **D.** $157.50   **E.** $172.50

10. **MC** If $4500 is invested for 10 years at 12% p.a. with interest compounding annually and the interest earned in the third year was $677.38, then the interest earned in the fourth year is closest to:

   A. $600   **B.** $625   **C.** $650   **D.** $677   **E.** $759

11. Declan invested $5750 for 4 years at an interest rate of 8% p.a. with interest compounded annually. Complete the table by calculating the values A, B, C, D, E and F.

V_n ($)	Interest ($)	V_{n+1} ($)
5750	A% of $5750 = 460$	6210
B	8% of $C = 496.80$	D
6706.80	8% of $6706.80 = 536.54$	7243.34
7243.34	8% of $7243.34 = 579.47$	E
F	8% of $7822.81 = 625.82$	8448.63

12. Alex invested $12 000 for 4 years at an interest rate of 7.5% p.a. with interest compounded annually. Complete the table by calculating the values A, B, C, D, E and F.

V_n ($)	Interest ($)	V_{n+1} ($)
12 000	7.5% of $12 000 = 900$	12 900
12 900	7.5% of $12 900 = A$	B
C	7.5% of $13 867.50 = 1040.06$	14 907.56
14 907.56	7.5% of $14 907.56 = 1118.07$	D
E	7.5% of $16 025.63 = 1201.92$	F

13. Sarina invested $5500 for 3 years at an interest rate of 6.5% p.a. with interest compounded monthly. Complete the table shown to determine the value of the investment at the end of 5 months.

V_n ($)	V_{n+1} ($)
5500	$1.005\,42 \times 5500 = 5529.81$
5529.81	

14. Wai Keat runs his own architecture business. A client has requested to pay off an account of $15 000 rather than pay the full amount by the due date. Wai Keat explains that he will need to charge interest on the money not paid by the due date at a rate of 1.2% per month.

   a. Using V_0, V_n and V_{n+1}, write a recurrence relation that models this situation, where n is the number of months after the due date.

   b. Determine how much interest the client will need to pay if she does not pay the full amount until 3 months after the due date.

15. **MC** Consider the following recurrence relations:

$$P_{n+1} = 0.95P_n, \quad P_0 = 3000$$

$$M_{n+1} = 1.15M_n, \quad M_0 = 3000$$

These are examples of:

A. linear decay and geometric growth respectively.
B. geometric decay and linear growth respectively.
C. linear decay and linear growth respectively.
D. geometric decay and geometric growth respectively.
E. geometric growth.

6.3 Exam questions

Question 1 (1 mark)

Source: VCE 2019, Further Mathematics Exam 1, Section A, Q18; © VCAA.

MC The value of a compound interest investment, in dollars, after n years, V_n, can be modelled by the recurrence relation shown below.

$$V_0 = 100\,000, \quad V_{n+1} = 1.01\,V_n$$

The interest rate, per annum, for this investment is

A. 0.01% **B.** 0.101% **C.** 1% **D.** 1.01% **E.** 101%

Question 2 (4 marks)

Source: VCE 2017, Further Mathematics Exam 2, Section A, Core, Q6; © VCAA.

Alex sends a bill to his customers after repairs are completed.

If a customer does not pay the bill by the due date, interest is charged.

Alex charges interest after the due date at the rate of 1.5% per month on the amount of an unpaid bill.

The interest on this amount will compound monthly.

   a. Alex sent Marcus a bill of $200 for repairs to his car.
      Marcus paid the full amount one month after the due date.
      How much did Marcus pay? **(1 mark)**
   b. Alex sent Lily a bill of $428 for repairs to her car.
      Lily did not pay the bill by the due date.
      Let A_n be the amount of this bill n months after the due date.
      Write down a recurrence relation, in terms of A_0, A_{n+1} and A_n, that models the amount of
      the bill. **(2 marks)**
   c. Lily paid the full amount of her bill four months after the due date.
      How much interest was Lily charged?
      Round your answer to the nearest cent. **(1 mark)**

Question 3 (2 marks)

Using the recurrence relation of $V_{n+1} = V_n \left(1 + \dfrac{4.8}{100} \right)$, $V_0 = \$25\,000$, calculate the value of V_3.

More exam questions are available online.

6.4 Compound interest using a rule

LEARNING INTENTION

At the end of this subtopic you should be able to:
- apply the annual compound interest rule
- calculate the total interest
- calculate non-annual compounding interest.

6.4.1 The value of the investment in terms of the initial investment

From the previous exercise we saw that we could write the value of a compounding investment as a recurrence relation.

This pattern can be expanded further to write the value of the investment in terms of the initial investment. This is known as the compound interest rule.

> **Formula for compound interest as a recurrence relation**
>
> $$V_n = V_0 R^n$$
>
> where V_n is the final or total amount (\$)
>
> V_0 is the principal (\$)
>
> R is the growth or compounding factor, $R = 1 + \dfrac{r}{100}$
>
> r is the interest rate per period
>
> n is the number of interest-bearing periods.

Note that the compound interest rule gives the *total amount* in an account, not just the interest earned as in the simple interest formula.

To calculate the total interest compounded, I, we can use the following rule.

> **Calculating the total interest**
>
> $$I = V_n - V_0$$
>
> where V_n is the final or total amount (\$)
>
> V_0 is the principal (\$).

If compound interest is used, the value of the investment at the end of each period grows by an increasing amount. Therefore, when plotted, the values of the investment at the end of each period form an exponential curve.

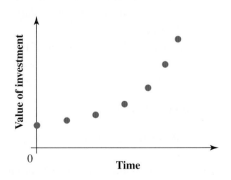

WORKED EXAMPLE 5 Calculating the annual compound interest

$5000 is invested for 4 years at 6.5% p.a. with interest compounded annually.
a. **Generate the compound interest rule for this investment.**
b. **Calculate the amount in the balance after 4 years and the interest earned over this period.**

THINK	WRITE
a. 1. Write the compound interest rule.	a. $V_n = V_0 \left(1 + \dfrac{r}{100}\right)^n$
2. List the values of n, r and P.	$n = 4, r = 6.5, V_0 = 5000$
3. Substitute into the rule.	$V_n = 5000 \left(1 + \dfrac{6.5}{100}\right)^n$
4. Simplify.	$V_n = 5000(1.065)^n$
b. 1. Substitute $n = 4$ into the rule.	b. $V_4 = 5000(1.065)^4$
2. Evaluate correct to 2 decimal places.	$= \$6432.33$
3. Subtract the principal from the balance.	$I = V_4 - V_0$ $= 6432.33 - 5000$ $= \$1432.33$
4. Write the answer.	The amount of interest earned is $1432.33 and the balance is $6432.33.

6.4.2 Non-annual compounding periods

In Worked example 5, interest was compounded annually. However, as shown in section 6.3, in many cases the interest is compounded more often than once a year — for example, quarterly (every 3 months), weekly or daily.

Calculating interest for non-annual compounding periods

number of interest periods, n = number of years × number of interest periods per year

$$\text{interest rate per period, } r = \dfrac{\text{interest rate per annum}}{\text{number of compounding periods per year}}$$

Note: Nominal interest rate per annum is the annual interest rate advertised by a financial institution.

WORKED EXAMPLE 6 Calculating quarterly compound interest

If $3200 is invested for 5 years at 6% p.a. with interest compounded quarterly:
a. **determine the number of compounding periods, n**
b. **determine the interest rate per compounding period, r**
c. **determine the balance of the account after 5 years**
d. **graphically represent the balance at the end of each quarter for 5 years. Describe the shape of the graph.**

THINK	WRITE/DRAW
a. Recall that the number of compounding periods, n, is the number of years multiplied by the number of interest periods per year.	a. $n = 5\,(\text{years}) \times 4\,(\text{quarters})$ $= 20$

▶

b. Convert % p.a. to % per quarter to match the time over which the interest is calculated. Divide r% p.a. by the number of compounding periods per year, namely 4. Write as a decimal.

b. $r\% = \dfrac{6\% \text{ p.a.}}{4}$

$= 1.5\%$ per quarter

$r = 1.5$

c. 1. Write the compound interest formula.

2. List the values of V_0, r and n.

3. Substitute into the formula.

4. Simplify.

5. Evaluate correct to 2 decimal places.

6. Write the answer.

c. $V_n = V_0 R^n$, where $R = \left(1 + \dfrac{r}{100}\right)$

$V_0 = 3200$, $r = 1.5$, $n = 20$

$V_{20} = 3200 \left(1 + \dfrac{1.5}{100}\right)^{20}$

$= 3200 \, (1.015)^{20}$

$= \$4309.94$

Balance of account after 5 years is $4309.94.

d. 1. Using CAS, calculate the balance at the end of each quarter and plot these values on the set of axes. (The first point is (0, 3200), which represents the principal.)

d.

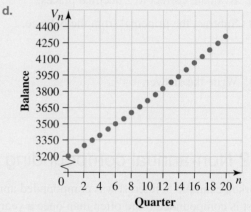

2. Comment on the shape of the graph.

The graph is exponential as the interest is added at the end of each quarter and the following interest is calculated on the *new* balance.

TI	THINK	DISPLAY/WRITE

d. 1. On a Lists & Spreadsheets page, enter the quarters as consecutive whole numbers from 0 to 20 in Column A. Rename Column A as quarter and Column B as balance. In the formula cell for the column named balance, complete the entry line as:
$= 3200 \times (1.015)^{\text{quarter}}$
Then press ENTER.

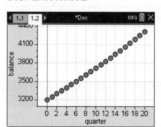

CASIO	THINK	DISPLAY/WRITE

1. On a Statistics screen, rename list1 as quarter and list2 as balance. Enter the quarters as consecutive whole numbers from 0 to 20 in list1. In the formula cell for the column named balance, complete the entry line as:
$= 3200 \times (1.015)^{\text{quarter}}$
Then press EXE.

2. Press CTRL + I and select: 5: Add Data & Statistics Construct a scatterplot with quarter on the horizontal axis and balance on the vertical axis.	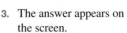	2. Tap: SetGraph Setting Complete the Set StatGraph fields as: Draw: On Type: Scatter XList: main\quarter YList: main\balance Freq: 1 Mark: Square Tap the graph icon.	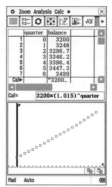
3. The answer appears on the screen.	The graph is exponential as the interest is added at the end of each quarter and the following interest is calculated on the new balance.	3. The answer appears on the screen.	The graph is exponential as the interest is added at the end of each quarter and the following interest is calculated on the new balance.

The situation often arises where we require a certain amount of money by a future date. It may be to pay for a holiday or to finance the purchase of a car. It is then necessary to know what principal should be invested now in order that it will increase in value to the desired final balance within the time available.

tlvd-3922

WORKED EXAMPLE 7 Calculating the principal

Calculate the principal that will grow to $4000 in 6 years, if interest is added quarterly at 6.5% p.a. (use CAS to solve the equation).

THINK	WRITE
1. Use CAS to solve the equation. Calculate n (there are 4 quarters in a year).	$n = 6 \times 4$ $= 24$
2. Calculate r, by converting % p.a. to % per quarter to match the time for which interest is calculated.	$r = \dfrac{6.5}{4}$ $= 1.625$
3. List the value of V_{24}.	$V_{24} = 4000$
4. Write the compound interest rule, substituting $R = 1 + \dfrac{r}{100}$. Use CAS to solve the equation.	$V_{24} = V_0 R^n$ $V_{24} = V_0 \left(1 + \dfrac{r}{100}\right)^n$ $4000 = V_0 \left(1 + \dfrac{1.625}{100}\right)^{24}$ Solving for V_0: solve $(4000 = V_0(1.016\,25)^{24}, V_0)$.
5. Evaluate correct to 2 decimal places.	$V_0 = 2716.73$
6. Write a summary statement.	$2716.73 would need to be invested.

TI | THINK	DISPLAY/WRITE	CASIO | THINK	DISPLAY/WRITE
4. On a Calculator page, complete the entry line as: solve $(4000 = v \times (1.016\,25)^{24}, v)$ Then press ENTER.		4. On the Main screen, complete the entry line as: solve $(4000 = v \times (1.016\,25)^{24}, v)$ Then press EXE.	
2. The answer appears on the screen.	$2716.73 would need to be invested.	2. The answer appears on the screen.	$2716.73 would need to be invested.

 Resources

  Interactivities Simple and compound interest (int-6265)

Non-annual compounding (int-6270)

6.4 Exercise

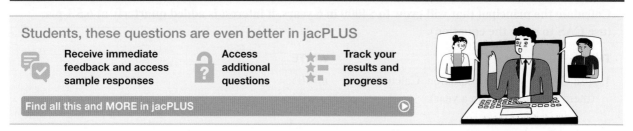

Students, these questions are even better in jacPLUS

Receive immediate feedback and access sample responses

Access additional questions

Track your results and progress

Find all this and MORE in jacPLUS

Note: Where it is the most efficient method, CAS should be used to solve the following problems.

1. **WE5** $2500 is invested for 5 years at 7.5% p.a. with interest compounded annually.
   a. Generate the compound interest rule for this investment.
   b. Determine the amount in the balance after 5 years and the interest earned over this period.

2. Determine the balance after $6750 is invested for 7 years at 5.25% p.a. with interest compounded annually.

3. **WE6** If $4200 is invested for 3 years at 7% p.a. with interest compounded quarterly:
   a. determine the number of compunding periods, n
   b. determine the interest rate per period, r
   c. determine the balance of the account after 3 years
   d. graphically represent the balance at the end of each quarter for 3 years. Describe the shape of the graph.

4. If $7500 is invested for 2 years at 5.5% p.a. with interest compounded monthly:
   a. determine the number of compounding periods, n
   b. determine the interest rate per period, r
   c. determine the balance of the account after 2 years
   d. graphically represent the balance at the end of each month for 2 years. Describe the shape of the graph.

5. **WE7** Determine the principal that will grow to $5000 in 5 years if interest is added quarterly at 7.5% p.a.

6. Determine the principal that will grow to $6300 in 7 years, if interest is added monthly at 5.5% p.a.

7. Use the compound interest rule to calculate the amount, V_n, when:
   a. $V_0 = \$500, n = 2, r = 8$
   b. $V_0 = \$1000, n = 4, r = 13$
   c. $V_0 = \$3600, n = 3, r = 7.5$
   d. $V_0 = \$2915, n = 5, r = 5.25$.

8. Using a recurrence relation, calculate: i the balance, and ii the interest earned (interest compounded annually) after:
   a. $2000 is invested for 1 year at 7.5% p.a.
   b. $2000 is invested for 2 years at 7.5% p.a.
   c. $2000 is invested for 6 years at 7.5% p.a.

9. Calculate the number of interest-bearing periods, n, if interest is compounded:
   a. annually for 5 years
   b. quarterly for 5 years
   c. semi-annually for 4 years
   d. monthly for 6 years
   e. 6-monthly for $4\frac{1}{2}$ years
   f. quarterly for 3 years and 9 months.

10. Calculate the interest rate per period, r, if the annual rate is:
    a. 6% and interest is compounded quarterly
    b. 4% and interest is compounded half-yearly
    c. 18% and interest is compounded monthly
    d. 7% and interest is compounded quarterly.

11. $1500 is invested for 2 years into an account paying 8% p.a.
    a. Determine the balance for interest that is compounded yearly.
    b. Determine the balance for interest that is compounded quarterly.
    c. Determine the balance for interest that is compounded monthly.
    d. Determine the balance for interest that is compounded weekly.
    e. Compare your answers to parts a, b, c and d.

12. Determine the amount that accrues in an account that pays compound interest at a nominal rate of:
    a. 7% p.a. if $2600 is invested for 3 years (compounded monthly)
    b. 8% p.a. if $3500 is invested for 4 years (compounded monthly)
    c. 11% p.a. if $960 is invested for $5\frac{1}{2}$ years (compounded fortnightly)
    d. 7.3% p.a. if $2370 is invested for 5 years (compounded weekly)
    e. 15.25% p.a. if $4605 is invested for 2 years (compounded daily).

13. **MC** The greatest return is likely to be made if interest is compounded:
    A. annually   B. semi-annually   C. quarterly   D. monthly   E. fortnightly.

14. **MC** If $12 000 is invested for $4\frac{1}{2}$ years at 6.75% p.a., compounded fortnightly, the amount of interest that would accrue would be closest to:
    A. $3600
    B. $4200
    C. $5000
    D. $12 100
    E. $16 300

15. Use the compound interest rule to determine the principal, V_0, when:
    a. $V_n = \$5000, r = 9, n = 4$
    b. $V_n = \$2600, r = 8.2, n = 3$
    c. $V_n = \$3550, r = 1.5, n = 12$
    d. $V_n = \$6661.15, r = 0.8, n = 36$

16. Calculate the principal that will grow to:

    a. $3000 in 4 years, if interest is compounded 6-monthly at 9.5% p.a.
    b. $2000 in 3 years, if interest is compounded quarterly at 9% p.a.
    c. $5600 in $5\frac{1}{4}$ years, if interest is compounded quarterly at 8.7% p.a.
    d. $10 000 in $4\frac{1}{4}$ years, if interest is compounded monthly at 15% p.a.
    e. Calculate the interest accrued in each part a–d.

6.4 Exam questions

Question 1 (1 mark)

Source: VCE 2015, Further Mathematics Exam 1, Section B, Module 4, Q4; © VCAA.

MC Mary invests $1200 for two years.

Interest is calculated at the rate of 3.35% per annum, compounding monthly.

The amount of interest she earns in two years is closest to

    A. $6.71   **B.** $40.82   **C.** $80.40   **D.** $81.75   **E.** $83.03

Question 2 (1 mark)

Source: VCE 2014, Further Mathematics Exam 1, Section B, Module 4, Q8; © VCAA.

MC Robert invested $6000 at 4.25% per annum with interest compounding quarterly.

Immediately after interest is paid at the end of each quarter, he adds $500 to his investment.

The value of Robert's investment at the end of the third quarter, after his $500 has been added, is closest to

    A. $6193   **B.** $7569   **C.** $7574   **D.** $7709   **E.** $8096

Question 3 (1 mark)

MC If an investment is worth $7035.04 after two months of compounding interest (paid monthly) at the rate of 3% per annum, determine the initial amount invested.

    A. $7017.50   **B.** $7000   **C.** $6990   **D.** $7010   **E.** $7200

More exam questions are available online.

6.5 Calculating rate or time for compound interest

> **LEARNING INTENTION**
>
> At the end of this subtopic you should be able to:
> • calculate the interest rate or time for compound interest for an investment
> • calculate the time period in compound interest for an investment.

6.5.1 Calculating the rate in compound interest

Sometimes we know how much we can afford to invest as well as the amount we want to have at a future date. Using the compound interest rule we can calculate the interest rate that is needed to increase the value of our investment to the amount we desire. This allows us to 'shop around' various financial institutions for an account that provides the interest rate we want.

We must first determine the interest rate per period, r, and convert this to the corresponding nominal rate per annum.

tlvd-3923

WORKED EXAMPLE 8 Calculating the interest rate for an investment

Calculate the interest rate per annum (correct to 2 decimal places) that would enable an investment of $3000 to grow to $4000 over 2 years if interest is compounded quarterly.

THINK	WRITE
1. List the values of V_n, V_0 and n. For this example, n needs to represent quarters of a year and therefore r will be evaluated in % per quarter.	$V_n = \$4000$ $V_0 = \$3000$ $n = 2 \times 4$ $\quad = 8$
2. Write the compound interest rule and substitute the known values.	$V_n = V_0 R^n$ $4000 = 3000 R^8$
3. Replace R with $1 + \dfrac{r}{100}$.	$4000 = 3000 \left(1 + \dfrac{r}{100}\right)^8$
4. Use CAS to solve the equation for r.	Solve $\left(4000 = 3000 \left(1 + \dfrac{r}{100}\right)^8\right)$ for r.
5. Write the solution for r.	$r = -203.66$ or $r = 3.661\,46$
6. $r > 1$	so $r = 3.661\,46$ $r\% = 3.661\,46\%$ per quarter
7. Multiply r by the number of interest periods per year to get the annual rate (correct to 2 decimal places).	Annual rate $= r\%$ per quarter $\times 4$ $\quad = 3.661\,46\%$ per quarter $\times 4$ $\quad = 14.65\%$ per annum
8. Write the answer.	Interest rate of 14.65% p.a. is required, correct to 2 decimal places.

TI | THINK	DISPLAY/WRITE	CASIO | THINK	DISPLAY/WRITE
4. On a Calculator page, complete the entry line as: solve $\left(4000 = 3000 \times \left(1 + \dfrac{r}{100}\right)^8, r\right)$ Then press ENTER.	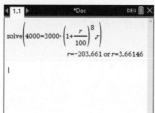	4. On the Main screen, complete the entry line as: solve $\left(4000 = 3000 \times \left(1 + \dfrac{r}{100}\right)^8, r\right)$ Then press EXE.	
5. The answer appears on the screen.	$r = -203.66$ or $r = 3.66$	5. The answer appears on the screen.	$r = -203.66$ or $r = 3.66$

6.5.2 Calculating time in compound interest

We have seen how the compound interest formula, $V_n = V_0 R^n$, where $R = 1 + \dfrac{r}{100}$ can be manipulated to solve situations where V_n, V_0 and r were unknown.

To determine n, the number of interest-bearing periods (that is, to determine the time period of an investment), use the compound interest rule and solve for n.

The value obtained for n may be a whole number, but it is more likely to be a decimal. That is, the time required will lie somewhere between two consecutive integers. The smaller of the two integers represents insufficient time for the investment to amount to the balance desired; the larger integer represents more than enough time.

WORKED EXAMPLE 9 Calculating the time period for an investment

Calculate how many years it will take $2000 to amount to $3500 at 8% p.a. with interest compounded annually.

THINK	WRITE
1. State the values of V_n, V_0 and r.	$V_n = \$3500$, $V_0 = \$2000$ and $r = 8\%$ p.a.
2. Write the rule for compound interest and substitute in the values.	$V_n = V_0 R^n$, $R = 1 + \dfrac{r}{100}$ $$3500 = 2000 \left(1 + \frac{8}{100} \right)^n$$
3. Use CAS to solve for n.	$\text{solve} \left(3500 = 2000 \left(1 + \dfrac{8}{100} \right)^n, n \right)$ $n = 7.27$ years
4. Interest is compounded annually, so n represents years. Raise n to the next whole year.	As the interest is compounded annually, $n = 7.27$ years.
5. Write the answer.	It will take 7.27 years for $2000 to amount to $3500.

---

tlvd-3924

WORKED EXAMPLE 10 Calculating the number of compounding periods

Calculate the number of compounding periods, n, required and hence the time it will take $3600 to amount to $5100 at a rate of 7% p.a., with interest compounded quarterly.

THINK	WRITE
1. State the values of V_n, V_0 and r.	$V_n = \$5100$, $V_0 = \$3600$ and $r = 7\%$ p.a.
2. Write the rule for compound interest and substitute in the values.	$V_n = V_0 R^n$, $R = 1 + \dfrac{r}{100}$ $$5100 = 3600 \times \left(1 + \frac{\frac{7}{4}}{100} \right)^n$$
3. Use CAS to solve for n.	$\text{solve} \left(5100 = 3600 \times \left(1 + \dfrac{\frac{7}{4}}{100} \right)^n, n \right)$ $n = 20.08$ quarters
4. Write the answer using more meaningful units.	As the interest is compounded annually, $n = 20.08$ quarters.
5. Write the answer as a sentence.	It will take 20.08 years for $3600 to amount to $5100.

Note: Where it is the most efficient method, CAS should be used to solve the following problems.

1. **WE8** Determine the interest rate per annum (correct to 2 decimal places) that would enable an investment of $4000 to grow to $5000 over 2 years if interest is compounded quarterly.

2. Determine the interest rate per annum (correct to 2 decimal places) that would enable an investment of $7500 to grow to $10 000 over 3 years if interest is compounded monthly.

3. **MC** Lillian wishes to have $24 000 in a bank account after 6 years so that she can buy a new car. The account pays interest at 15.5% p.a. compounded quarterly. The amount (correct to the nearest dollar) that Lillian should deposit in the account now, if she is to reach her target, is:

   A. $3720   **B.** $9637   **C.** $10 109
   D. $12 117   **E.** $22 320

4. Determine the interest rate per annum (correct to 2 decimal places) that would enable investments of:

   a. $2000 to grow to $3000 over 3 years if interest is compounded 6-monthly

   b. $12 000 to grow to $15 000 over 4 years (with interest compounded quarterly).

5. Determine the interest rate per annum (correct to 2 decimal places) that would enable investments of:

   a. $25 000 to grow to $40 000 over $2\frac{1}{2}$ years (compounded monthly)

   b. $43 000 to grow to $60 000 over $4\frac{1}{2}$ years (compounded fortnightly)

   c. $1400 to grow to $1950 over 2 years (compounded weekly).

6. **MC** Select the minimum interest rate per annum (compounded quarterly) needed for $2300 to grow to at least $3200 in 4 years.

   A. 6% p.a.   **B.** 7% p.a.   **C.** 9% p.a.   **D.** 8% p.a.   **E.** 10% p.a.

7. **WE9** Calculate how long it will take $3000 to amount to $4500 at 7% p.a. with interest compounded annually.

8. Calculate how long it will take $7300 to amount to $10 000 at 7.5% p.a. with interest compounded annually.

9. **WE10** Calculate the number of interest-bearing periods, n, required and hence the time it will take $4700 to amount to $6100 at a rate of 9% p.a., with interest compounded quarterly.

10. Calculate the number of interest-bearing periods, n, required and hence the time it will take $3800 to amount to $6300 at a rate of 15% p.a., with interest compounded quarterly.

11. Calculate how long it will take (with interest compounded annually) for:

    a. $2000 to amount to $3173.75 at 8% p.a.
    b. $9250 to amount to $16 565.34 at 6% p.a.

12. Calculate how long it will take (with interest compounded annually) for:

    a. $850 to amount to $1000 at 7% p.a.   **b.** $12 000 to amount to $20 500 at 13.25% p.a.

13. Calculate the number of interest-bearing periods, n, required, and hence the time in more meaningful terms when:

    a. $V_n = \$2100$, $V_0 = \$1200$, $r = 3\%$ per half-year   **b.** $V_n = \$13\,500$, $V_0 = \$8300$, $r = 2.5\%$ per quarter

    c. $V_n = \$16\,900$, $V_0 = \$9600$, $r = 1\%$ per month.

14. Calculate the number of interest-bearing periods, n, required, and hence the time in more meaningful terms when:

    a. $V_n = \$24\,000$, $V_0 = \$16\,750$, $r = 0.25\%$ per fortnight

    b. $7800 is to amount to $10 000 at a rate of 8% p.a. (compounded quarterly)

    c. $800 is to amount to $1900 at a rate of 11% p.a. (compounded quarterly).

15. Wanda has invested $1600 in an account at a rate of 10.4% p.a., interest compounded quarterly. Calculate how many years it will take to reach $2200.

16. **MC** Select the smallest number of interest periods, n, required for $6470 to grow to at least $9000 in an account with interest paid at 6.5% p.a. and compounded half-yearly.

    A. 10   **B.** 11   **C.** 12   **D.** 20   **E.** 22

17. Determine how many years it would take for:

    a. $1400 to accrue $300 interest at 8% p.a., with interest compounded monthly

    b. $8000 to accrue $4400 interest at 9.6% p.a., with interest compounded fortnightly.

18. Jennifer and Miguel each want to save $15 000 for a car. Jennifer has $11 000 to invest in an account with her bank, which pays 8% p.a., with interest compounded quarterly. Miguel's credit union has offered him 11% p.a., with interest compounded quarterly.

    a. Calculate how long it will take Jennifer to reach her target.

    b. Calculate how much Miguel will need to invest in order to reach his target at the same time as Jennifer. Assume their accounts were opened at the same time.

6.5 Exam questions

Question 1 (1 mark)

Source: VCE 2020, Further Mathematics Exam 1, Section A, Q27; © VCAA.

MC Gen invests $10 000 at an interest rate of 5.5% per annum, compounding annually.

After how many years will her investment first be more than double its original value?

    A. 12   **B.** 13   **C.** 14   **D.** 15   **E.** 16

Question 2 (3 marks)

Source: VCE 2020, Further Mathematics Exam 2, Section A, Q9; © VCAA.

Samuel opens a savings account.

Let B_n be the balance of this savings account, in dollars, n months after it was opened.

The month-to-month value of B_n can be determined using the recurrence relation shown below.

$$B_0 = 5000, \quad B_{n+1} = 1.003\,B_n$$

a. Write down the value of B_4, the balance of the savings account after four months. Round your answer to the nearest cent.

b. Calculate the monthly interest rate percentage for Samuel's savings account.

c. After one year, the balance of Samuel's savings account, to the nearest dollar, is $5183.

If Samuel had deposited an additional $50 at the end of each month immediately after the interest was added, how much extra money would be in the savings account after one year?

Round your answer to the nearest dollar.

▶ **Question 3 (1 mark)**

Source: VCE 2019, Further Mathematics Exam 1, Section A, Q18; © VCAA.

MC The value of a compound interest investment, in dollars, after n years, V_n, can be modelled by the recurrence relation shown below.

$$V_0 = 100\,000, \quad V_{n+1} = 1.01\,V_n$$

The interest rate, per annum, for this investment is

A. 0.01% **B.** 0.101% **C.** 1% **D.** 1.01% **E.** 101%

More exam questions are available online.

6.6 Nominal and effective annual interest rate

LEARNING INTENTION

At the end of this subtopic you should be able to:
- differentiate between nominal and effective annual interest rates
- calculate the effective annual interest rate.

6.6.1 Calculating the effective interest rate

The **nominal interest rate** is the compound interest rate advertised by a financial institution. It is usually expressed as a percentage per annum. The nominal interest rate disregards the compounding period.

When money is borrowed or invested with compounding periods greater than one year, the value of the loan or investment increases with the number of compounding periods. The **effective annual interest rate** is used to compare the annual nominal interest between loans or investments with these different compounding periods.

It takes the compounding period into account and is therefore a more realistic measure of interest rates.

> **Effective annual interest rate formula**
>
> $$r_{\text{eff}} = \left(1 + \frac{r}{100n}\right)^n - 1 = \left(1 + \frac{i}{n}\right)^n - 1$$
>
> where r_{eff} **is the effective annual interest rate (%)**
> r **is the nominal annual interest rate**
> i **is the nominal annual interest rate, as a decimal**
> n **is the number of compounding periods per year.**

Note: While the formula for effective annual interest rate is shown, it is not an expectation of the course to apply it. The effective rates are completed on CAS.

For a loan of $100 at 10% p.a. compounding quarterly over 2 years, the effective annual interest rate:

$$r_{eff} = \left(1 + \frac{i}{n}\right)^{n} - 1$$

$$= \left(1 + \frac{0.10}{4}\right)^{4} - 1$$

$$= 0.1038$$

$$= 10.38\%$$

This means that the effective annual interest rate is actually 10.38% and not 10%. The comparison between the two can be shown in the following table.

Period	Amount owing ($)	Annual effective rate calculation ($)
1	$100\left(1 + \dfrac{0.10}{4}\right) = 102.50$	
2	$102.50\left(1 + \dfrac{0.10}{4}\right) = 105.06$	
3	$105.06\left(1 + \dfrac{0.10}{4}\right) = 107.69$	
4 (year 1)	$107.69\left(1 + \dfrac{0.10}{4}\right) = 110.38$	$100\left(1 + \dfrac{10.38}{100}\right)^{1} = 110.38$
5	$110.38\left(1 + \dfrac{0.10}{4}\right) = 113.14$	
6	$113.14\left(1 + \dfrac{0.10}{4}\right) = 115.97$	
7	$115.97\left(1 + \dfrac{0.10}{4}\right) = 118.87$	
8 (year 2)	$118.87\left(1 + \dfrac{0.10}{4}\right) = 121.84$	$100\left(1 + \dfrac{10.38}{100}\right)^{2} = 121.84$

tlvd-3925

WORKED EXAMPLE 11 Calculating the effective annual interest rate

Jason decides to borrow money for a holiday. If he takes a personal loan over 4 years with equal quarterly repayments compounding at 12% p.a., calculate the effective annual rate of interest (correct to 2 decimal places).

THINK

WRITE

Method 1:

1. Write the values for i and n.

$n = 4$ (since quarterly)
$i = 0.12$ (12% as a decimal)

2. Write the formula for effective annual rate of interest.

$r_{eff} = \left(1 + \dfrac{i}{n}\right)^{n} - 1$

3. Substitute $n = 4$ and $i = 0.12$.

$$r_{\text{eff}} = \left(1 + \frac{0.12}{4}\right)^4 - 1$$
$$= 0.1255$$
$$= 12.55\%$$

4. Write the answer.

The effective annual interest rate is 12.55% p.a. for a loan of 12%, correct to 2 decimal places.

TI	THINK	DISPLAY/WRITE	CASIO	THINK	DISPLAY/WRITE

Method 2:

1. On a Calculator page, press MENU and then select:
   8: Finance
   5: Interest Conversion
   2: Effective Interest Rate
   Complete the entry line as:
   eff (12, 4)
   Then press ENTER.
   Note: The first entry in the brackets is the nominal interest rate, followed by the number of compounding periods per year.

1. On the Main screen, tap:
   Action
   Financial
   Interest Conversion
   convEff
   Complete the entry line as:
   convEff (4, 12)
   Then press EXE.
   Note: The first entry in the brackets is the number of compounding periods per year followed by the nominal interest rate.

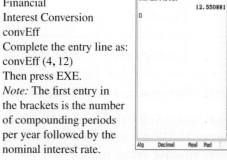

2. The answer appears on the screen.

The effective interest rate is 12.55% p.a., correct to 2 decimal places.

2. The answer appears on the screen.

The effective interest rate is 12.55% p.a., correct to 2 decimal places.

 Resources

Interactivity Effective annual interest rate (int-6288)

6.6 Exercise

Students, these questions are even better in jacPLUS

 Receive immediate feedback and access sample responses

 Access additional questions

 Track your results and progress

Find all this and MORE in jacPLUS

1. **WE11** Brad wants to take his family to Africa. He decides to take out a loan over 4 years with equal quarterly repayments compounding at 11% p.a. Calculate the effective annual rate of interest (correct to 1 decimal place).

2. Caitlin wants to landscape her garden. She decides to take out a loan over 3 years with equal quarterly repayments compounding at 14% p.a. Calculate the effective annual rate of interest (correct to 1 decimal place).

3. William wants to purchase a new video recorder. If William pays $125 monthly instalments over 3 years at an interest rate of 11.5% p.a. compound interest, calculate the effective annual interest rate he pays. Give your answer correct to 2 decimal places.

4. Use the information given in the table for each of the items to answer the question.

	Item	Cash price ($)	Deposit ($)	Monthly instalment ($)	Compound interest rate	Term of loan
a	Television	$875	$150		8% p.a.	2 years
b	New car	$23 990	$2000		10% p.a.	5 years
c	Clothing	$550	$100		7.5% p.a.	1 year
d	Refrigerator	$1020	$50		$6\frac{3}{4}$% p.a.	2 years

Calculate the effective annual interest rate for each item.

5. A camera valued at $1200 is purchased via a loan with a compound interest rate of 18.41% p.a. to be paid over an 18-month period. Calculate the effective annual interest rate.

6. The bank approves a personal loan of $5000. An interest rate of 12.5% p.a. is charged, with repayments to be made over a 9-month period in equal weekly instalments. Calculate the effective annual interest rate.

7. MC For a compound interest rate of 4.85% p.a. on an investment with monthly compounding periods for 2 years, the effective rate of interest is:
   A. 4.96%
   B. 4.85%
   C. 9.3%
   D. 5.12%
   E. 9.6%

8. MC A reducing balance loan was used to purchase a home theatre system valued at $2500. If the loan is paid off with quarterly repayments over 3 years at 9.6% p.a., then the effective annual interest rate is closest to:
   A. 10.03%
   B. 8.75%
   C. 5.2%
   D. 9.6%
   E. 9.95%

9. Sarah had two options to invest $10 000 for 2 years. The two options are:
   i. compounding interest of 9% p.a. compounding monthly
   ii. an effective annual interest rate of 9.3%.

   Use calculations to show which option she should take.

10. Calculate the effective annual interest rate that would require the same amount to be paid back as a $25 000 loan at 8.5% p.a. compounding monthly for 5 years.

6.6 Exam questions

▶ **Question 1 (1 mark)**

Source: VCE 2020, Further Mathematics Exam 1, Section A, Q28; © VCAA.

MC The nominal interest rate for a loan is 8% per annum.

When rounded to two decimal places, the effective interest rate for this loan is **not**
- **A.** 8.33% per annum when interest is charged daily.
- **B.** 8.32% per annum when interest is charged weekly.
- **C.** 8.31% per annum when interest is charged fortnightly.
- **D.** 8.30% per annum when interest is charged monthly.
- **E.** 8.24% per annum when interest is charged quarterly.

▶ **Question 2 (1 mark)**

MC A comparison of two loans was made with the following details:

Loan A: Nominal rate of 8% with interest compounding daily (standard year).

Loan B: Nominal rate of 8.1% with interest compounding biannually.

Which loan has the greater effective interest and by what margin?
- **A.** Loan B, Margin = 8.33%
- **B.** Loan B, Margin = 0.07%
- **C.** Loan A, Margin = 8.33%
- **D.** Loan A, Margin = 0.07%
- **E.** Loan B, Margin = 8.26%

▶ **Question 3 (1 mark)**

MC A compound interest rate of 5.85% p. a. is charged on a hire-purchase item, with monthly repayments over 6 years.

What is the effective interest rate closest to?
- **A.** 6.0%
- **B.** 8.9%
- **C.** 9.3%
- **D.** 10.6%
- **E.** 11.53%

More exam questions are available online.

6.7 Review

6.7.1 Summary

doc-38037

6.7 Exercise

Multiple choice

1. **MC** A bank pays simple interest of 6% p.a. on an investment of $5000. The recurrence relation, V_n, that can be used to model this investment after n years is:

   A. $V_{n+1} = V_n + 5000$
   B. $V_{n+1} = 300V_n$
   C. $V_{n+1} = V_n + 5000, \ V_0 = 5000$
   D. $V_{n+1} = V_n + 5000, \ V_0 = 300$
   E. $V_{n+1} = V_n + 300, \ V_0 = 5000$

2. **MC** Clayton has invested $360 in a bank for 3 years at 8% simple interest each year. At the end of the 3 years, the total amount he will receive is:

   A. $86.40   B. $236.80   C. $28.80   D. $388.80   E. $446.40

3. **MC** A loan of $5000 is taken over 5 years. The amount of interest accrued is $1125. The simple interest rate per year on this loan is:

   A. 3%   B. 4.5%   C. 3.75%   D. 5%   E. 3.5%

4. **MC** A loan of $10 000 is taken over 10 years. The amount of interest accrued is $2000. The simple interest rate per year on this loan is:

   A. 3%   B. 4.5%   C. 2%   D. 5%   E. 2.5%

5. **MC** The following recurrence model was used to model an investment for n years.

$$V_0 = 2000, \ V_{n+1} = 1.05V_n$$

   The annual rate of interest for this investment is:

   A. 1% p.a.   B. 5% p.a.   C. 10.5% p.a.   D. 50% p.a.   E. 105% p.a.

6. **MC** An investment of $100 000 is compounded monthly at a rate of 12% p.a. The recurrence relation for V_n, the value of the investment after n months, is:

   A. $V_{n+1} = 1.01 \, V_n, \ V_0 = 100 \, 000$
   B. $V_{n+1} = 1.12 \, V_n, \ V_0 = 100 \, 000$
   C. $V_{n+1} = V_n + 12, \ V_0 = 100 \, 000$
   D. $V_{n+1} = 12 \, V_n, \ V_0 = 100 \, 000$
   E. $V_n = 12 \, n, \ V_0 = 100 \, 000$

7. **MC** An investment of $4500 earns compound interest at a rate of 6.4% p.a. and is invested for 5 years. The balance in the account at the end of the investment period, if interest is compounded quarterly, is:

   A. $6181.40   B. $4871.71   C. $6136.50   D. $15 561.27   E. $8592.20

8. **MC** After $4\frac{1}{2}$ years, $1200 has grown to $1750 in an account where interest is compounded monthly. The annual interest rate is:

   A. 7.0%   B. 0.7%   C. 8.41%   D. 3.2%   E. 38%

9. **MC** A sum of $850 is invested at 8% p.a. compound interest, credited fortnightly. For the balance to grow to $1200, the investment should be left for a minimum of:

   A. 112 years.   **B.** 113 years.   **C.** 4 years 8 fortnights.
   D. 4 years 9 fortnights.   **E.** 5 years.

10. **MC** In an account that pays compound interest at 12% p.a., credited daily, the effective rate of interest is:

    A. 12%   **B.** 12.75%   **C.** 13.8%   **D.** 15%   **E.** 12.95%

Short answer

11. Cynthia invested $270 with a building society in a fixed deposit account that paid 8% p.a. simple interest for 4 years. Calculate the amount Cynthia received at the end of the 4 years.

12. If $725 is invested for 3 years and earns $206.65 interest, calculate the annual interest rate.

13. Jack put some money away for $4\frac{1}{2}$ years in a bank account that is paying $3\frac{3}{4}$% p.a. simple interest. He found on his bank statement he had earned $67.50. Calculate the amount Jack invested.

14. Alex invested $1800 in an insurance company bond that pays $12\frac{1}{2}$% p.a. simple interest provided he keeps the bond for 4 years. Calculate Alex's total return from the bond at the end of the 4 years.

15. Calculate the amount that must be invested at 9.25% p.a., with interest compounded 6-monthly, if it is to grow to $5000 over 4 years.

16. Determine how much interest $950 would earn if it was invested for 3 years at 12% p.a. with interest credited daily.

Extended response

17. An investment bond is offered to the public at 10% simple interest per year. Louis buys a bond worth $4000 that will mature in $2\frac{1}{2}$ years. Determine how much Louis will receive in total at the end of the $2\frac{1}{2}$ years.

18. If $5400 is to be invested for 5 years, determine which of the options below would be best.

    a. 12% p.a. simple interest
    b. Compound interest at 11.8% p.a., credited quarterly
    c. Compound interest at 11.7% p.a., credited monthly

19. Geoff wants to buy a windsurfer. Its retail price is $3995. He needs to save up until he has enough cash to pay for the windsurfer. Geoff's first option for financing the purchase is to place the balance of his savings account, $1983.50, into a term deposit offering 5.6% per annum for a 2-year term.

    a. Calculate the total value of his investment at the end of 2 years.
    b. Geoff uses the term deposit investment towards the purchase of the windsurfer. Determine the extra fortnightly savings needed over the next 2 years to make up the balance of $3995.

    Another option for Geoff is to place his $1983.50 into a building society, which offers 5.4% interest, compounded monthly.

    c. Calculate how long it will take Geoff to accumulate enough funds for the windsurfer.

    Geoff decided to invest with a building society. Eight months later he received a $1000 bonus from his employer.

    d. Calculate the amount of money Geoff has in his account at the time that he received his bonus.
    e. If Geoff immediately deposited the entire bonus into his building society account, determine how much longer he would need to wait to get enough funds for the windsurfer.

20. Bridie has invested her money in an account to save for a holiday. The amount of money in her account after n months, V_n, can be modelled using the recurrence relation $V_{n+1} = 1.005 V_n$, $V_0 = 6500$.

   a. Calculate the annual interest rate.
   b. Calculate the amount of money Bridie saved before she opens the account.
   c. Calculate the amount of money Bridie has in her account after two years.

6.7 Exam questions

Question 1 (2 marks)

Source: VCE 2021, Further Mathematics Exam 2, Section A Core, Q8; © VCAA.

For renovations to the coffee shop, Sienna took out a reducing balance loan of $570 000 with interest calculated fortnightly.

The balance of the loan, in dollars, after n fortnights, S_n, can be modelled by the recurrence relation

$$S_0 = 570\,000, \quad S_{n+1} = 1.001\,S_n - 1193$$

   a. Calculate the balance of this loan after the first fortnightly repayment is made. **(1 mark)**
   b. Show that the compound interest rate for this loan is 2.6% per annum. **(1 mark)**

Question 2 (1 mark)

Source: VCE 2017, Further Mathematics Exam 1, Section A, Core, Q19; © VCAA.

MC Shirley would like to purchase a new home. She will establish a loan for $225 000 with interest charged at the rate of 3.6% per annum, compounding monthly.

Each month, Shirley will pay only the interest charged for that month.

After three years, the amount that Shirley will owe is

   A. $73 362   **B.** $170 752   **C.** $225 000   **D.** $239 605   **E.** $245 865

Question 3 (5 marks)

Source: VCE 2016, Further Mathematics Exam 2, Core, Q5; © VCAA.

Ken has opened a savings account to save money to buy a new caravan.

The amount of money in the savings account after n years, V_n, can be modelled by the recurrence relation shown below.

$$V_0 = 15\,000, \quad V_{n+1} = 1.04 \times V_n$$

   a. How much money did Ken initially deposit into the savings account? **(1 mark)**
   b. Use recursion to write down calculations that show that the amount of money in Ken's savings account after two years, V_2, will be $16 224. **(1 mark)**
   c. What is the annual percentage compound interest rate for this savings account? **(1 mark)**
   d. The amount of money in the account after n years, V_n, can also be determined using a rule.

   i. Complete the rule below by writing the appropriate numbers in the boxes provided. **(1 mark)**

   $$V_n = \boxed{}^{\,n} \times \boxed{}$$

   ii. How much money will be in Ken's savings account after 10 years? **(1 mark)**

Question 4 (4 marks)

Source: VCE 2014, Further Mathematics Exam 2, Module 4, Q3; © VCAA.

The cricket club had invested $45 550 in an account for four years.

After four years of compounding interest, the value of the investment was $60 000.

    a. How much interest was earned during the four years of this investment? **(1 mark)**

    b. Interest on the account had been calculated and paid quarterly. **(1 mark)**

    What was the annual rate of interest for this investment?

    Write your answer, correct to one decimal place.

    c. The $60 000 was re-invested in another account for 12 months.

    The new account paid interest at the rate of 7.2% per annum, compounding monthly.

    At the end of each month, the cricket club added an additional $885 to the investment.

        i. The equation below can be used to determine the account balance at the end of the first month, immediately after the $885 was added.

        Complete the equation by filling in the boxes. **(1 mark)**

$$\text{Account balance} = 60\,000 \times \left(1+\boxed{}\right) + \boxed{}$$

        ii. What was the account balance at the end of 12 months?

        Write your answer, correct to the nearest dollar. **(1 mark)**

Question 5 (4 marks)

Source: VCE 2013, Further Mathematics Exam 2, Module 4, Q3; © VCAA.

Hugo paid $7500 for a second bike under a hire-purchase agreement.

A flat interest rate of 8% per annum was charged.

He will fully repay the principal and the interest in 24 equal monthly instalments.

    a. Determine the monthly instalment that Hugo will pay.

    Write your answer in dollars, correct to the nearest cent. **(1 mark)**

    b. Find the effective rate of interest per annum charged on this hire-purchase agreement.

    Write your answer as a percentage, correct to two decimal places. **(1 mark)**

    c. Explain why the effective interest rate per annum is higher than the flat interest rate per annum. **(1 mark)**

    d. The value of his second bike, purchased for $7500, will be depreciated each year using the reducing balance method of depreciation.

    One year after it was purchased, this bike was valued at $6375.

    Determine the value of the bike five years after it was purchased.

    Write your answer, correct to the nearest dollar. **(1 mark)**

More exam questions are available online.

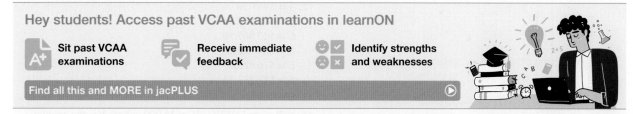

Hey students! Access past VCAA examinations in learnON

Sit past VCAA examinations   Receive immediate feedback   Identify strengths and weaknesses

Find all this and MORE in jacPLUS

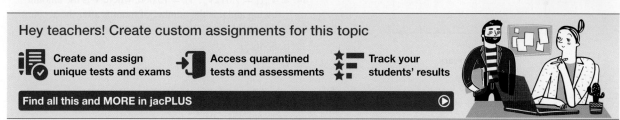

Hey teachers! Create custom assignments for this topic

Create and assign unique tests and exams   Access quarantined tests and assessments   Track your students' results

Find all this and MORE in jacPLUS

Answers

Topic 6 Modelling compound interest investments and loans using recursion

6.2 Simple interest

6.2 Exercise

1. a. $V_{n+1} = V_n + 86.70, \ V_0 = 1020$
   b. $1106.70, $1193.40, $1280.10, $1366.80, $1453.50

2. a. $B_{n+1} = B_n + 48.13, \ B_0 = 713$
   b. $761.13, $809.26, $857.39, $905.52, $953.65

3. a. $A_{n+1} = A_n + 187.50, \ A_0 = 1500$
   b. $1875
   c. $375

4. a. $A_{n+1} = A_n + 132, \ A_0 = 2400$
   b. $2730
   c. $330

5. a. $714, $748, $782, $816 b. $13 975, $14 950, $15 925

6. a. $13 912.50
   b.
   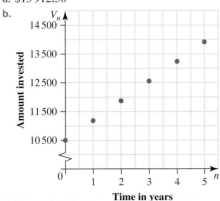
   Time in years

7. a. For an initial investment of $6200, the building society is offering a flat rate of interest set at $565 per annum.
   b. 9.11%

8. a. $2418.50
   b.
   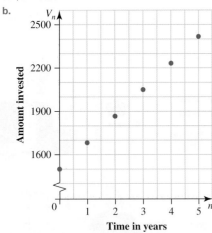
   Time in years

9. $3070

10. 3 years

11. B

12. a. $360   b. $2700   c. $131.25

6.2 Exam questions

Note: Mark allocations are available with the fully worked solutions online.

1. D

2. C

3. D

6.3 Compound interest as a geometric recurrence relation

6.3 Exercise

1. a. $V_{n+1} = 1.06V_n, \ V_0 = 7500$, where V is the amount invested and n is the number of years after the initial investment.
   b. $9468.58

2. a. $V_{n+1} = 1.075V_n, \ V_0 = 3250$, where V is the amount invested and n is the number of years after the initial investment.
   b. $4665.80

3. a. $V_{n+1} = 1.005V_n, \ V_0 = 600$, where V is the amount invested and n is the number of months after the initial investment.
   b. $618.23

4. a. $V_{n+1} = 1.01V_n, \ V_0 = 2500$, where V is the amount borrowed and n is the number of months after the initial loan.
   b. $50.25

5. B

6. C

7. D

8. A

9. D

10. E

11. $A = 8\%$
    $B = 6210
    $C = 6210
    $D = 6706.80
    $E = 7822.81
    $F = 7822.81

12. $A = 967.50
    $B = $13 867.50$
    $C = $13 867.50$
    $D = $16 025.63$
    $E = $16 025.63$
    $F = $17 227.55$

13. $5650.67

14. a. $V_{n+1} = 1.012V_n, \ V_0 = 15 000$, where V is the amount owing and n is the number of months after the account due date.
    b. $546.51

15. D

6.3 Exam questions

Note: Mark allocations are available with the fully worked solutions online.

1. C

2. a. $203
   b. $A_{n+1} = 1.015 \times A_n$
      $A_0 = 428$
   c. $26.26

3. $28\,775.56

6.4 Compound interest using a rule

6.4 Exercise

1. a. $V_5 = 2500\left(1 + \dfrac{7.5}{100}\right)^5$

   b. $V_5 = \$3589.07$; $I = \$1089.07$

2. $9657.36

3. a. 12
   b. 1.75%
   c. $5172.05
   d.

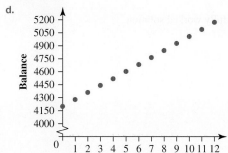

Even though it is difficult to see on the graph, the graph is exponential.

4. a. 24
   b. 0.4583%
   c. $8369.92
   d.

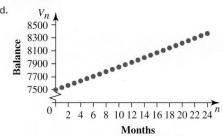

Even though it is difficult to see on the graph, the graph is exponential.

5. $3448.40

6. $4290.73

7. a. $583.20   b. $1630.47
   c. $4472.27   d. $3764.86

8. a. i. $2150   ii. $150
   b. i. $2311.25   ii. $311.25
   c. i. $3086.60   ii. $1086.60

9. a. 5   b. 20   c. 8
   d. 72   e. 9   f. 15

10. a. 1.5%   b. 2%   c. 1.5%   d. 1.75%

11. a. $1749.60
    b. $1757.49
    c. $1759.33
    d. $1760.05
    e. The balance increases as the compounding periods become more frequent.

12. a. $605.60   b. $1314.84   c. $795.77
    d. $1043.10   e. $1641.82

13. E

14. B

15. a. $3542.13   b. $2052.54
    c. $2969.18   d. $5000

16. a. $2069.61
    b. $1531.33
    c. $3564.10
    d. $5307.05
    e. $930.39, $468.67, $2035.90 and $4692.95

6.4 Exam questions

Note: Mark allocations are available with the fully worked solutions online.

1. E   2. D   3. B

6.5 Calculating rate or time for compound interest

6.5 Exercise

1. 11.31%

2. 9.63%

3. B

4. a. 13.98%   b. 5.62%

5. a. 18.95%   b. 7.41%   c. 16.59%

6. 8.34%, C

7. Approximately 6 years

8. Approximately 5 years

9. Approximately 12 quarters or 3 years

10. Approximately 14 quarters or 3.5 years

11. a. 6 years   b. 10 years

12. a. 3 years   b. 5 years

13. a. $19, 9\frac{1}{2}$ years   b. 20, 5 years   c. $57, 4\frac{3}{4}$ years

14. a. 145, 5 years 15 fortnights

    b. $13, 3\frac{1}{4}$ years

    c. 32 quarters, 8 years

15. $13, 3\frac{1}{4}$ years

16. B

17. a. $n = 30$ months, $2\frac{1}{2}$ years

 b. $n = 119$ fortnights; 4 years 15 fortnights

18. a. 4 years   b. $9718.11

6.5 Exam questions

Note: Mark allocations are available with the fully worked solutions online.

1. B
2. a. $5060.27

 b. 0.3%

 c. $610

3. C

6.6 Nominal and effective annual interest rate

6.6 Exercise

1. 11.5%
2. 14.8%
3. 12.13%
4. a. 8.30%

 b. 10.47%

 c. 7.76%

 d. 6.96%

5. 20.05%
6. 13.30%
7. A
8. E
9. The effective annual interest rate of 9.3% is the better option (the compound interest rate has an effective annual interest rate of 9.38%).
10. 8.84%

6.6 Exam questions

Note: Mark allocations are available with the fully worked solutions online.

1. C
2. D
3. A

6.7 Review

6.7 Exercise

Multiple choice

1. E   2. E   3. B   4. C   5. B   6. A
7. A   8. C   9. D   10. B

Short answer

11. $356.40   12. 9.5%   13. $400
14. $2700   15. $3482.46   16. $411.58

Extended response

17. $5000
18. C

19. a. $2205.65   b. $34.41   c. 13 years
 d. $2056.04   e. 5 years

20. a. 6% p.a.   b. $6500   c. $7326.54

6.7 Exam questions

Note: Mark allocations are available with the fully worked solutions online.

1. a. $569 377

 b. 2.6%

2. D
3. a. $15 000

 b. $16 224

 c. 4%

 d. i. $V_n = 1.04^n \times 15\,000$

 ii. $22 203.66

4. a. $14 450

 b. 6.9%

 c. i. $60\,000\,(1 + 0.006) + 885$

 ii. $75 443

5. a. $362.50

 b. 15.36%

 c. See the fully worked solution.

 d. $3328

7 Modelling reducing balance loans, annuities and perpetuities using recursion

LEARNING SEQUENCE

Fully worked solutions for this topic are available online.

7.1 Overview

7.1.1 Introduction

When we invest money with a financial institution, the institution pays us interest because it is using our money to lend to others. Conversely, when we borrow money from an institution, we are using the institution's money and so it charges us interest.

In reducing balance loans, interest is usually charged every month by the financial institution and repayments are made by the borrower on a regular basis. These repayments nearly always amount to more than the interest for the same period of time and so the amount still owing is reduced. Since the amount still owing is continually decreasing and interest is calculated on the current balance but debited monthly, the amount of interest charged also decreases throughout the life of the loan. This means that less of the amount borrowed is paid off in the early stages of the loan compared to the end.

KEY CONCEPTS

This topic covers the following key concepts from the VCE Mathematics Study Design:

- use of a first-order linear recurrence relation to model and analyse (numerically and graphically) the amortisation of a reducing balance loan, including the use of a recurrence relation to determine the value of the loan or investment after *n* payments for an initial sequence from first principles
- use of a table to investigate and analyse the amortisation of a reducing balance loan on a step-by-step basis, the payment made, the amount of interest paid, the reduction in the principal and the balance of the loan
- use of technology with financial modelling functionality to solve problems involving reducing balance loans, such as repaying a personal loan or a mortgage, including the impact of a change in interest rate on repayment amount, time to repay the loan, total interest paid and the total cost of the loan
- use of a first-order linear recurrence relation to model and analyse (numerically and graphically) the amortisation of an annuity, including the use of a recurrence relation to determine the value of the annuity after *n* payments for an initial sequence from first principles
- use of a table to investigate and analyse the amortisation of an annuity on a step-by-step basis, the payment made, the interest earned, the reduction in the principal and the balance of the annuity
- use of technology to solve problems involving annuities including determining the amount to be invested in an annuity to provide a regular income paid, for example, monthly or quarterly
- simple perpetuity as a special case of an annuity that lasts indefinitely
- use of a first-order linear recurrence relation to model and analyse (numerically and graphically) annuity investment, including the use of a recurrence relation to determine the value of the investment after payments have been made for an initial sequence from first principles
- use of a table to investigate and analyse the growth of an annuity investment on a step-by-step basis after each payment is made, the payment made, the interest earned and the balance of the investment
- use of technology with financial modelling functionality to solve problems involving annuity investments, including determining the future value of an investment after a number of compounding periods, the number of compounding periods for the investment to exceed a given value and the interest rate or payment amount needed for an investment to exceed a given value in a given time.

Source: VCE Mathematics Study Design (2023–2027) extracts © VCAA; reproduced by permission.

7.2 Modelling reducing balance loans with recurrence relations

LEARNING INTENTION

At the end of this subtopic you should be able to:
- apply first-order linear recurrence relations to model reducing balance loans
- create a table to investigate and analyse the amortisation of a reducing balance loan or an annuity, and interpret amortisation tables.

7.2.1 Recurrence relations of the form $V_0 = a$, $V_{n+1} = RV_n \pm d$

In Topic 5, we learned to write a sequence as a recurrence relation. These sequences were expressed as either linear or geometric. Many situations can be modelled using a combination of geometric and linear recurrence relations.

These can be expressed in the form of $V_0 = a$, $V_{n+1} = RV_n \pm d$, where V_0 is the initial term, V_n is the nth term of the sequence, r is the percentage increase or decrease, $R = 1 \pm \dfrac{r}{100}$ and d is the value added or subtracted.

WORKED EXAMPLE 1 Modelling with a recurrence relation

The population of a small town is known to have a consistent increase in population of 3.5% each year. The population of the town at the beginning of 2018 is 23 500 people. A large factory in the town announces in 2018 that it needs to relocate 250 people in its workforce out of town at the end of each year for the next 3 years.

a. Write a recurrence relation to describe this situation.

b. Use your calculator to determine the population of the town at the end of 2020.

c. State whether the relocations at the factory in the three years caused the population to decrease from the 23 500 people at the beginning of 2018.

THINK	WRITE
a. 1. State the initial values for V_0, r, R and d.	a. $V_0 = 23\,500$ $r = 3.5\%$ $\quad = 0.035$ $R = 1 + 0.035$ $\quad = 1.035$ $d = 250$
2. Write the general form of the recurrence relation. The percentage is increasing so $R = 1 + \dfrac{3.5}{100} = 1.035$. 250 is subtracted as this many people are being relocated.	$V_0 = a$, $V_{n+1} = RV_n \pm d$
3. Substitute the values for V_0, r, R and d.	$V_0 = 23\,500$, $V_{n+1} = 1.035V_n - 250$

▶

b. 1. On a calculator screen, type 23 500, press ENTER/EXE, then type ×1.035 − 250. Continue to press ENTER/EXE 3 times.	**b.**	
	23 500	23 500
	23 500 × 1.035 − 250	24 072.5
	24 072.5 × 1.035 − 250	24 665.04
	24 665.04 × 1.035 − 250	25 278.31

2. Write the answer.

At the end of 2020, the population of the town is 25 278 people, to the nearest whole number.

c. Write the conclusion.

c. The population has continued to increase from 23 500 at the beginning of 2018 to 25 278 at the end of 2020.

7.2.2 Modelling reducing balance loans with a recurrence relation

A recurrence relation can be used to model a reducing balance loan.

Formula for calculating a reducing balance loan with a recurrence relation

$$V_0 = a, \ V_{n+1} = RV_n - d$$

where

V_0 **is the initial amount borrowed**

V_n **is the balance of the loan after** n **payments**

$R = 1 + \dfrac{r}{100}$ (r **is the interest rate per compounding period**)

d **is the payment made per compounding period.**

---

tlvd-3927

WORKED EXAMPLE 2 Reducing balance loan with a recurrence relation

A loan of \$2500 is taken out at a rate of 7.5% p.a. (interest compounded monthly) and is to be paid back monthly with instalments of \$509.42.
a. Write a recurrence relation to describe this situation.
b. Use your calculator to determine the number of monthly repayments needed to pay off this loan.

THINK	WRITE
a. 1. State the initial values for V_0, r, R and d.	**a.** $V_0 = 2500$ $r = \dfrac{7.5}{12}$ $= 0.625$ $R = 1 + \dfrac{0.625}{100}$ $= 1.006\ 25$ $d = 509.42$
2. Write the general form of the recurrence relation.	$V_0 = a, \ V_{n+1} = RV_n - d$
3. Substitute in the values for V_0, r, R and d.	$V_0 = 2500, \ V_{n+1} = 1.006\ 25V_n - 509.42$

b. 1. On a calculator screen, type 2500, press ENTER/EXE and then type:
×1.006 25 − 509.42.
Continue to press ENTER/EXE until the output is less than 0.
Count the number of iterations of the recurrence relation until the balance is less than 0.

2. Write the answer.

b.

2500	2500
$2500 \times 1.006\,25 - 509.42$	2006.21
$2006.21 \times 1.006\,25 - 509.42$	1509.32
$1509.32 \times 1.006\,25 - 509.42$	1009.34
$1009.34 \times 1.006\,25 - 509.42$	506.23
$506.23 \times 1.006\,25 - 509.42$	−0.03

After 5 months, the account balance reads −0.03, so it would take 5 months to pay off the loan.
Note: As the above calculations have been rounded to 2 decimal places, this can impact on the final balance. If there was an overpayment of 3 cents, the final payment would be adjusted by 3 cents.

7.2.3 Amortisation of a reducing balance loan using a table

The process of paying off a loan by regular payments over a period of time is known as **amortisation**. The amortisation of a loan can be tracked on a step-by-step basis by following the payments made, the interest added and reduction in the principal. For each payment period, an amortisation table provides your loan balance, interest charged on your loan, and the amount of principal that you pay off.

Consider an amortisation table for Worked example 2.
- The *scheduled payment* of $509.42 remains fixed.
- The *interest rate* of 0.625% per month remains fixed.
- The *interest* is calculated on the previous month's balance.
- The *principal reduction* is the *scheduled payment – interest for that month*.
- The *balance of the loan* is the *previous month's balance – principal reduction*.

Payment number (*n*)	Payment ($)	Interest ($)	Principal reduction ($)	Balance of loan ($)
0	0.00	0.00	0.00	2500.00
1	509.42	$0.006\,25 \times 2500 = 15.63$	$509.42 - 15.63 = 493.80$	$2500 - 493.80 = 2006.20$
2	509.42	$0.006\,25 \times 2006.20 = 12.54$	$509.42 - 12.54 = 496.88$	$2006.20 - 496.88 = 1509.32$
3	509.42	$0.006\,25 \times 1509.32 = 9.43$	$509.42 - 9.43 = 499.99$	$1509.32 - 499.99 = 1009.33$
4	509.42	$0.006\,25 \times 1009.33 = 6.31$	$509.42 - 6.31 = 503.11$	$1009.33 - 503.11 = 506.22$
5	509.42	$0.006\,25 \times 506.22 = 3.17$	$509.42 - 3.17 = 506.25$	$506.22 - 506.25 = -0.03$

WORKED EXAMPLE 3 Creating an amortisation table

A family takes out a loan of $5500 at 8% p.a. to pay for solar heating for their pool. The loan is to be paid back in half-yearly instalments of $1515.20. An incomplete amortisation table for this loan is shown below.

a. Calculate the interest rate per payment period.
b. Determine how much of the first payment of $1515.20 is interest.
c. Calculate the balance of the loan after two payments have been made.
d. Calculate the principal reduction after 3 payments have been made.
e. Calculate the total interest paid on this loan.
f. State the amount of the final payment to ensure a balance of zero at the end of the loan period.

Payment number (n)	Payment ($)	Interest ($)	Principal reduction ($)	Balance of loan ($)
0	0.00	0.00	0.00	5500.00
1	1515.20		1295.20	4204.80
2	1515.20	168.19	1347.01	
3	1515.20	114.31		1456.90
4	1515.20	58.28	1456.92	−0.02

THINK

a. 1. Write the value of the annual interest rate.

   2. Calculate the half-yearly interest rate.

   3. Write the answer.

b. 1. Calculate 4% of $5500.
   2. Write the answer.

c. 1. The balance of the loan after two payments have been made is the previous balance minus the balance reduction.

   2. Write the answer.

d. 1. Principal reduction = payment − interest
   2. Write the answer.

e. 1. The total interest paid is the sum of all interest payments.

   2. Write the answer.

f. 1. The last payment results in an overpayment of 2 cents.

WRITE

a. Annual interest rate = 8%

Half-yearly interest rate = $\frac{8}{2}$ = 4%

The interest rate per payment period is 4%.

b. $0.04 \times 5500 = 220$
The interest component of the first payment is $220.

c. Previous balance − balance reduction
= 4204.80 − 1347.01
= 2857.79

The balance of the loan after two payments have been made is $2857.79.

d. $1515.20 − 114.31 = 1400.89$

The principal reduction after 3 payments is $1400.89.

e. $220 + 168.19 + 114.31 + 58.28 = 560.78$

The total interest paid is $560.78.

f. The last payment should be $1515.18.

1. **WE1** The enrolment numbers at a school have been growing steadily at a rate of 1.5% per annum for a number of years. The school decides to accept a further 50 international students every year. The total number of enrolments at the beginning of 2018 is 565 students.

   a. Write a recurrence relation to describe this situation, where V_n is the enrolment number after n years.
   b. Use your calculator to determine the total number of enrolments at the beginning of 2021.
   c. Determine the number of new students enrolling at the school in 2019.

2. At a childcare centre, the sand in the sandpit decreases on average by 35% every year. The centre has a standing order with the local garden supplier to top up the sand by $1.5\,\text{m}^3$ once a year.
   The following recurrence relation can be used to model this situation, where A_n is the amount of sand in the sandpit after n years.

   $$A_0 = 3.5, \quad A_{n+1} = 0.65A_n + 1.5$$

   a. Calculate the initial amount of sand in the sandpit.
   b. Determine how much sand will be in the sandpit after 4 years.
   c. Determine after how many years the sand in the sandpit will exceed $4\,\text{m}^3$.

3. **MC** The recurrence relation $P_0 = 10\,000$, $P_{n+1} = 1.03P_n - 250$ will generate a sequence that:
   A. is an example of linear decay.
   B. is an example of geometric growth.
   C. is an example of both geometric growth and linear growth.
   D. is an example of both geometric growth and linear decay.
   E. is an example of both geometric decay and linear decay.

4. **WE2** A loan of $15\,000 is taken out at a rate of 12.5% p.a. (interest compounded half-yearly) and is to be paid back half-yearly with $3074.44 instalments.

   a. Write a recurrence relation to describe this situation.
   b. Use your calculator to determine the number of half-yearly repayments needed to pay off this loan.

5. Nadjme invests $1500 with interest paid at a rate of 5.5% p.a. (interest compounded monthly) and each month she adds another $200 from her monthly pay to her investment.

   a. Write a recurrence relation to describe this situation.
   b. Use your calculator to determine the balance of her investment after 5 months.

6. **WE3** Sourav takes out a loan of $2400 at 9% p.a. to buy a new laptop. The loan is to be paid back monthly in instalments of $489.64. The incomplete amortisation table for this loan is shown below.

Payment number (n)	Payment ($)	Interest ($)	Principal reduction ($)	Balance of loan ($)
0	0.00	0.00	0.00	2400.00
1	489.64		471.64	1928.36
2	489.64	14.46	475.18	1453.18
3	489.64	10.90	478.74	
4	489.64	7.31		492.11
5	489.64	3.69	485.95	

a. Calculate the interest rate per payment period.
b. Determine how much of the first payment of $489.64 is interest.
c. Calculate the balance of the loan after three payments have been made.
d. Calculate the principal reduction of the 4th payment.
e. Calculate the total interest paid on this loan.
f. Calculate the balance left at the end of 5 months to pay out the loan.

7. A loan of $25 000 is taken out over 10 years at a rate of 5.5% p.a. (interest debited monthly) and is to be paid back monthly with $271.32 instalments. Complete the table below for the fourth payment.

Payment number (n)	Payment ($)	Interest ($)	Principal reduction ($)	Balance of loan ($)
0	0.00	0.00	0.00	25 000.00
1	271.32	114.58	156.74	24 843.26
2	271.32	113.86	157.46	24 685.80
3	271.32	113.14	158.18	24 527.62
4	271.32			

8. Paul wanted to buy a new road bike so he took out a $10 000 loan over 6 years at a rate of 8% p.a. (interest debited quarterly) to be paid back quarterly with instalments of $528.71. Complete the table below for the first three payments and state how much interest he has paid and how much he owes at the end of this time.

Payment number (n)	Payment ($)	Interest ($)	Principal reduction ($)	Balance of loan ($)
0	0.00	0.00	0.00	10 000.00
1	528.71			
2	528.71			
3	528.71			

9. Kali took out a loan for $23 000 to pay for her wedding over 7 years at a rate of 6.5% p.a. (interest debited monthly), which is to be paid back monthly with instalments of $341.54.

a. Write down a recurrence relation to model this loan.
b. Use your recurrence relation to determine how much Kali owes after her third payment was made.

10. In order to buy a new air-conditioning unit, Noah took out a personal loan of $2000, with an interest rate of 12% p.a. He will repay the loan with 8 equal monthly payments of $261.

a. Write down a recurrence relation to model this loan.
b. Use the recurrence relation to determine how much is still owed after 4 months (4 payments).
c. Determine whether the loan is paid out exactly after 8 months. If not, determine how much Noah will need to add to the last payment to fully repay the loan.

11. Link borrows $18 000, taken out over 2 years and to be repaid in monthly instalments. (*Note:* As the interest rate increases, the monthly repayment increases if the loan period is to remain the same.)
Calculate the amount still owing after 5 months if interest is debited monthly at a rate of:

a. 6% p.a. and the repayment is $532.90
b. 9% p.a. and the repayment is $608.04
c. 12% p.a. and the repayment is $688.66.

12. Nithya wants to buy a secondhand car, and her bank offers her a personal loan of $8000 at an interest rate of 13% p.a., with interest debited fortnightly and with fortnightly repayments of $124.11 over 3 years.
After 8 weeks, assuming there are exactly 52 weeks in a year, she still owes:

A. $7661.03
B. $7402.32
C. $7227.69
D. $7488.99
E. $6899.45

13. Ben took out a loan for $20 000 for a small start-up business at 16.5% p.a. The contract required that he repay the loan over 5 years with monthly instalments of $421.02. After 8 months Ben still owes:

A. $12 344.05   B. $14 7661.22   C. $9653.13   D. $19 001.67   E. $18 774.05

7.2 Exam questions

Question 1 (1 mark)
Source: VCE 2019, Further Mathematics Exam 1, Section A, Q20; © VCAA.

 Consider the following amortisation table for a reducing balance loan.

Payment number	Payment	Interest	Principal reduction	Balance
0	0.00	0.00	0.00	300 000.00
1	1050.00	900.00	150.00	299 850.00
2	1050.00	899.55	150.45	299 699.65
3	1050.00	899.10	150.90	299 548.65

The annual interest rate for this loan is 3.6%.

Interest is calculated immediately before each payment.

For this loan, the repayments are made

A. weekly.   B. fortnightly.   C. monthly.   D. quarterly.   E. yearly.

Question 2 (1 mark)
Source: VCE 2018, Further Mathematics Exam 1, Section A, Q22; © VCAA.

 Adam has a home loan with a present value of $175 260.56

The interest rate for Adam's loan is 3.72% per annum, compounding monthly.

His monthly repayment is $3200.

The loan is to be fully repaid after five years.

Adam knows that the loan cannot be exactly repaid with 60 repayments of $3200.

To solve this problem, Adam will make 59 repayments of $3200. He will then adjust the value of the final repayment so that the loan is fully repaid with the 60th repayment.

The value of the 60th repayment will be closest to
   A. $368.12   **B.** $2831.88   **C.** $3200.56   **D.** $3557.09   **E.** $3568.12

▶ **Question 3 (1 mark)**
Source: VCE 2017, Further Mathematics Exam 1, Section A, Core, Q17; © VCAA.

MC The value of a reducing balance loan, in dollars, after n months, V_n, can be modelled by the recurrence relation shown below.

$$V_0 = 26\,000, \qquad V_{n+1} = 1.003\,V_n - 400$$

What is the value of this loan after five months?
   A. $24\,380.31   **B.** $24\,706.19   **C.** $25\,031.10   **D.** $25\,355.03   **E.** $25\,678.00

More exam questions are available online.

7.3 Solving reducing balance loan problems using finance solver

LEARNING INTENTION

At the end of this subtopic you should be able to:
- apply technology (Finance Solver) to solve problems involving reducing balance loans
- calculate the number of repayments, principal and interest needed to pay off a loan.

7.3.1 The Finance Solver

In the previous subtopic, recurrence relations were used to help solve reducing balance loan problems. The iterative process is a very simple way of generating the terms in a recurrence relation. There are limitations to this method as repeating the same process for a loan taken out monthly for 5 years is equivalent to 60 iterations. The Finance Solver application on your CAS calculator can be used to solve problems where it is impractical to use the recurrence relation.

The Finance Solver application on the CAS calculator has a particular set of fields that are similar to those in the following table.

N	The total number of repayment periods
I%	The interest rate per annum
PV	The initial value of the loan – **this is positive as the money is coming to you**
PMT (Pmt)	The periodic payment – **this is negative if you are repaying a loan**
FV	The future value of the loan – **positive, negative or zero**
P/Y (PpY)	The number of payment periods in a year
C/Y (CpY)	The number of compounding periods in a year (nearly always the same as P/Y)

Remember:

When entering values into Finance Solver, pay particular attention to the sign of the values.

Cash flowing towards you is considered to be positive; cash flowing away from you is considered to be negative.

The values of N, I, P/Y and C/Y are always positive but particular attention must be paid to the signs of PV, PMT and FV.

If you have taken out a loan, that is a positive number. (Consider it as money coming to you.)

If you have to make a payment, that is a negative number. (Consider this as money leaving you.)

If the FV is:
- negative, there is still money to be paid
- zero, the loan has been paid out
- positive, the borrower has overpaid.

Fill in all the fields except the unknown field, place your cursor in that field and press ENTER/EXE.

tlvd-3929

WORKED EXAMPLE 4 Calculating the amount owing on a loan

A loan of $50 000 is taken out over 20 years at a rate of 6% p.a. (interest debited monthly) and is to be repaid with monthly instalments of $358.22. Determine the amount still owing after 10 years.

THINK

1. Using Finance Solver on your CAS calculator, enter the appropriate values into the fields.

N	120	The number of payments for 10 years at 12 interest calculations per year
I%	6	The annual interest rate
PV	50 000	The value of the loan
PMT (Pmt)	−358.22	The repayment per month
FV	?	This is the unknown value: **place the cursor in this cell**.
P/Y (PpY)	12	There are 12 payments per year.
C/Y (CpY)	12	There are 12 compounding periods per year.

Press ENTER/tap Solve.

2. Write the answer.

WRITE

N = 120 payments
I = 6% p.a.
Amount borrowed = $50 000
Repayment per month = $358.22

The balance of the loan after 10 years is $32 264.98.

TI	THINK	DISPLAY/WRITE	CASIO	THINK	DISPLAY/WRITE

TI | THINK

1. On a Calculator page, press MENU.
   8: Finance
   1: Finance Solver
   Complete the fields as shown, place the cursor in the field marked FV and then press ENTER.

DISPLAY/WRITE

CASIO | THINK

1. On the Financial screen, tap Compound Interest. Complete the fields as shown, place the cursor in the field marked FV and then tap Solve.

DISPLAY/WRITE

2. The answer appears on the screen.

The balance of the loan after 10 years is $ $32 264.98.

2. The answer appears on the screen.

The balance of the loan after 10 years is $ $32 264.98.

Note that, even though 10 years is the halfway point of the term of the loan, more than half of the original $50 000 is still owing.

7.3.2 Number of repayments

The situation often arises in reducing balance loans that a potential borrower knows how much needs to be borrowed as well as the amount that can be repaid each month. The borrower then wants to know how long the loan needs to be to accommodate these conditions. The number of repayments, N, can be calculated using the Finance Solver on the CAS calculator.

tlvd-3930

WORKED EXAMPLE 5 Calculating the number of payments needed to pay off a loan

Rob wants to borrow $2800 for a new sound system at 7.5% p.a., with interest adjusted monthly.

a. Calculate Rob's monthly repayment if the loan is fully repaid in $1\frac{1}{2}$ years.

b. Calculate the total interest charged.

THINK

a. 1. Using Finance Solver on your CAS calculator, enter the appropriate values into the fields.

N	18	The number of payments for $1\frac{1}{2}$ years with 12 interest calculations per year
I%	7.5	The annual interest rate
PV	2800	The value of the loan
PMT (Pmt)	?	This is the unknown value; **place the cursor in this cell.**
FV	0	Fully paid means a future value of zero.
P/Y (PpY)	12	There are 12 payments per year.
C/Y (CpY)	12	There are 12 compounding periods per year.

Press ENTER/tap Solve.

2. Write the answer as a sentence.

b. 1. Total interest = total repayments
             − amount borrowed

2. Write the answer as a sentence.

WRITE

a. N = 18 payments
  I = 7.5% p.a.
  Amount borrowed = $2800
  Future value = 0

The regular monthly payment is $164.95 over 18 months.

b. Total interest = 164.95 × 18 − 2800
             = 2969.10 − 2800
             = $169.10

The total interest on the $2800 loan over 18 months is $169.10.

TI	THINK	DISPLAY/WRITE	CASIO	THINK	DISPLAY/WRITE

a. 1. On a Calculator page, press MENU.
8: Finance
1: Finance Solver
Complete the fields as shown, place the cursor in the field marked Pmt and then press ENTER.

Finance Solver

N:	18.	▶
I(%):	7.5	▶
PV:	2800.	▶
Pmt:	−164.95467953346	▶
FV:	0.	▶
PpY:	12	⬍

Finance Solver info stored into
tvm.n, tvm.i, tvm.pv, tvm.pmt, …

a. 1. On the Financial screen, tap Compound Interest.
Complete the fields as shown, place the cursor in the field marked PMT and then tap Solve.

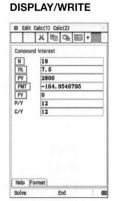

2. The answer appears on the screen. The monthly repayment is $164.95.

2. The answer appears on the screen. The monthly repayment is $164.95.

WORKED EXAMPLE 6 Calculating the instalment, principal and interest

Josh borrows $12 000 for some home office equipment. He agrees to repay the loan over 4 years with monthly instalments at 7.8% p.a. (adjusted monthly). Determine:

a. the instalment value

b. the principal repaid and interest paid during the 10th repayment.

THINK

a. 1. Using Finance Solver on your CAS, enter the appropriate values.

N	48	The number of payments for 10 years at 12 interest calculations per year
I%	7.8	The annual interest rate
PV	12 000	The value of the loan
PMT (Pmt)	?	This is the unknown value; **place the cursor in this cell**.
FV	0	The loan is paid out so the future value is zero.
P/Y (PpY)	12	There are 12 payments per year.
C/Y (CpY)	12	There are 12 compounding periods per year.

WRITE

a. N = 48 payments
I = 7.8% p.a.
Amount borrowed = $12 000
Future value = 0

2. Write the answer as a sentence.

The monthly repayment over a 4-year period is $291.83.

b. 1. Determine the amount owing after 9 months.

N	9
I%	7.8
PV	12 000
PMT (Pmt)	−291.83
FV	**? place the cursor in this cell.**
P/Y (PpY)	12
C/Y (CpY)	12

Press ENTER/tap Solve.

b. $N = 9$

$I = 7.8\%$ p.a.

Amount borrowed $= \$12\,000$

Payment per month $= \$291.83$

2. Write the amount owing after 9 months.

Amount owing after 9 months is $10 024.73.

3. Determine the amount owing after 10 months. Change N = 9 to N = 10 and press ENTER/tap Solve.

Amount owing after 10 months is $9798.06.

4. Principal repaid $= V_9 - V_{10}$ (Amount owing after 9 months minus the amount owing after 10 months.)

Principal repaid $= 10\,024.73 - 9798.06$
$\qquad\qquad = \$226.67$

5. Interest paid = repayment − principal repaid

Total interest $= \$291.83 - 226.67$
$\qquad\qquad\quad = \$65.16$

6. Write the answer as a sentence.

In the 10th repayment, $226.67 principal is repaid and $65.16 interest is paid.

Sometimes we may want to determine the time for only part of the loan term. The procedure that is followed is the same as in Worked example 5; however, V_n is zero only if we are calculating the time to repay the loan in full. Otherwise we should consider the amount still owing at that time.

WORKED EXAMPLE 7 Calculating the number of payments

tlvd-3931

Some time ago, Peter borrowed $14 000 to buy a car. Interest on this reducing balance loan has been charged at 9.2% p.a. (adjusted monthly) and he has been paying $446.50 each month to service the loan.

Currently he still owes $9753.92. Determine how long ago Peter borrowed the money.

THINK

1. Enter the following values into Finance Solver on your CAS:

N	? place the cursor in this cell.
I%	9.2
PV	14 000
PMT (Pmt)	−446.50
FV	−9.753.92
P/Y (PpY)	12 (monthly payment)
C/Y (CpY)	12 (monthly compounds)

Press ENTER/tap Solve.

2. Interpret the results.

3. Write the answer as a sentence.

WRITE

$N = 11.999\,995\,037\,662$

$N = 12$ months
Time $= 1$ year
Peter has had the loan for the past 12 months.

WORKED EXAMPLE 8 Calculating the number of repayments

A loan of \$11 000 is being repaid by monthly instalments of \$362.74 with interest being charged at 11.5% p.a. (debited monthly). Currently, the amount owing is \$7744.05. Determine how much longer it will take to:

a. reduce the amount outstanding to \$2105.11

b. repay the loan in full.

THINK

a. 1. Using the Finance Solver on your CAS, enter the following values:

N	? place the cursor in this cell.
I%	11.5
PV	7744.05
PMT (Pmt)	−362.74
FV	−2105.11
P/Y (PpY)	12
C/Y (CpY)	12

Press ENTER/tap Solve.

2. Interpret the results and write a statement.

b. 1. Repeat part **a**, entering the appropriate values into Finance Solver. Enter FV: $= 0$ to represent the fully repaid loan.

2. Write the answer as a sentence.

WRITE

a. $N = 17.999\,988\,603\,29$

$N = 18$ months
It will take another $1\frac{1}{2}$ years to reduce the amount owing to \$2105.11.

b. $N = 23.999\,534\,856\,457$
   $= 24$ months

It will take another 2 years to repay the loan in full.

7.3.3 Effects of changing the repayment

Since most loans are taken over a long period of time, it is probable that a borrower's financial situation will change during this time. For instance, a borrower may receive a pay rise and so their take-home pay is greater per week or fortnight. They may then choose to increase the value of the repayments made to service the loan.

It may also be that the borrower's financial situation deteriorates, in which case they may request from their financial institution that the repayment value be decreased and the loan period subsequently extended.

In this section we will look at how changing the repayment value affects the term of the loan and the total interest paid.

WORKED EXAMPLE 9 Calculating repayments, interest and the term of the loan

A reducing balance loan of $16 000 has a term of 5 years. It is to be repaid by monthly instalments at a rate of 8.4% p.a. (debited monthly).
a. Determine the repayment value.
b. Calculate the term of the loan if the repayment is increased to $393.62.
c. Calculate the total interest paid for repayments of $393.62.
d. Determine by how much the interest figure in part c differs from that paid for the original offer.

THINK

a. 1. Using Finance Solver on your CAS, enter the following values:

N	60
I%	8.4
PV	16 000
PMT (Pmt)	? place the cursor in this cell.
FV	0
PpY	12
CpY	12

Press ENTER/tap Solve.

2. Write the answer as a sentence.

b. 1. Using Finance Solver on your CAS, enter the following values:

N	? place the cursor in this cell.
I%	8.4
PV	16 000
PMT (Pmt)	−393.62
FV	0
P/Y (PpY)	12
C/Y (CpY)	12

2. Solve for N.

WRITE

a. Pmt = 327.49

$16 000 to be paid off in 5 years at 8.4% p.a. will need monthly repayments of $327.49.

b.

N = 47.999 695 088 867

3. Interpret the results.		$N = 48$ months
		$\text{Time} = \dfrac{48}{12}$ years
		$= 4$ years
4. Write the answer as a sentence.		The new term of the loan would be 4 years.

c. Interpret paid = total repayments − principal repaid

c. When term $= 4$ years, $d = 393.62$
$\text{Interest} = 48 \times 393.62 - 16\,000$
$= \$\,2893.76$

d. 1. i. Review the known quantities.
ii. Calculate the interest difference.

d. When term $= 5$ years, $d = 327.49$
$\text{Interest} = 60 \times 327.49 - 16\,000$
$= \$3649.40$
$\text{Interest difference} = 3649.40 - 2893.76$
$= \$755.64$

2. Write the answer as a sentence.

If the repayment is increased from \$327.49 to \$393.62 per month, then \$755.64 is saved in interest payments.

If a borrower does increase the value of each repayment and if all other variables remain the same, the term of the loan is reduced. Conversely, a decrease in the repayment value increases the term of the loan. Consequently there are two stages to the loan, each with a different repayment.

tlvd-3932

WORKED EXAMPLE 10 Changing the term of the loan and interest paid

Brad borrowed \$22 000 to start a business and agreed to repay the loan over 10 years with quarterly instalments of \$783.22 and interest debited at 7.4% p.a. However, after 6 years of the loan, Brad was struggling to meet his repayments and decided to decrease the repayment value to \$648.68. Determine:
a. the actual term of the loan
b. the total interest paid
c. the extra cost incurred by the decrease in the repayment value.

THINK

a. 1. To determine how much is left on the loan after 6 years, use Finance Solver on your CAS to enter the following values:

N	24
I%	7.4
PV	22 000
PMT (Pmt)	−783.22
FV	? place the cursor in this cell.
P/Y (PpY)	4
C/Y (CpY)	4

WRITE

a. $N = 24$
Interest rate $= 7.4\%$ p.a.
The amount borrowed $= \$22\,000$
The quarterly repayment $= \$783.22$
$FV = \$10\,761.83$

2. Write the answer as a sentence.

At the end of 6 years, the balance of the loan is $10 761.83.

3. Now determine the N value to reduce $10 761.83 (the remaining part of the loan) to zero.

Balance is $10 761.88 and the repayments are $648.68.

4. Using Finance Solver, enter the appropriate values.

N = 19.999 770 539 209

N	? place the cursor in this cell.
I%	7.4
PV	10 761.9
PMT (Pmt)	−648.68
FV	0
P/Y (PpY)	4
C/Y (CpY)	4

Solve for N.

5. Interpret the results.

N = 20 quarters
Time = 5 years

6. Determine the total term of the loan.

Total term = 6 + 5
= 11 years

b. For the two-repayment scenario:
Interest paid = total repayments −
principal repaid
In this case, in two stages.

b. For the two-repayment scenario:
Interest = 783.22 × 24 + 648.68 × 20 − 22 000
= $9770.88

c. 1. For the same repayment scenario:
d = 783.22 for 10 years.

c. For the same repayment scenario:
Interest = 783.22 × 40 − 22 000
= $9328.80

2. Determine the difference between the two scenarios.

Interest difference = 9770.88 − 9328.80
= $447.08

3. Write the answer as a sentence.

Brad has added $447.08 in interest by decreasing his repayment amount.

7.3 Exercise

Students, these questions are even better in jacPLUS

 Receive immediate feedback and access sample responses

 Access additional questions

 Track your results and progress

Find all this and MORE in jacPLUS

1. **WE4** A loan of $65 000 is taken out over 20 years at a rate of 12% p.a. (interest debited monthly) and is to be repaid with monthly instalments of $715.71. Determine the amount still owing after:

a. 5 years

b. 10 years

c. 15 years.

2. A loan of $52 000 is taken out over 15 years at a rate of 13% p.a. (interest debited fortnightly) and is to be repaid with fortnightly instalments of $303.37. Determine the amount still owing after:

   a. 4 years
   b. 8 years
   c. 12 years.

3. **WE5** Macca wants to borrow $3500 to buy a motorbike at 6.8% p.a., with interest adjusted monthly.

   a. Calculate Macca's monthly repayment if the loan is fully repaid in 2 years.
   b. Calculate the total interest charged.

4. Willow wants to borrow $2400 for a new computer at 6.3% p.a., with interest adjusted monthly.

   a. Calculate Willow's monthly repayment if the loan is fully repaid in 2 years.
   b. Calculate the total interest charged.

5. **WE6** Grace has borrowed $18 000 to buy a car. She agrees to repay the reducing balance loan over 5 years with monthly instalments at 8.1% p.a. (adjusted monthly). Determine:

   a. the instalment value
   b. the principal repaid and the interest paid during:

      i. the 10th repayment
      ii. the 50th repayment.

6. Sarah borrows $8000 for an overseas holiday. She agrees to repay the loan over 4 years with monthly instalments at 6.8% p.a. (adjusted monthly). Determine:

   a. the instalment value
   b. the principal repaid and interest paid during the:

      i. 10th repayment
      ii. 40th repayment.

7. Jim has a reducing balance loan of $3500 that he is using for a holiday and has agreed to repay it by monthly instalments of $206.35 at a rate of 7.6% p.a. (interest debited monthly). Determine:

   a. the number of repayments needed to repay in full and the time in years
   b. the total interest charged.

8. **WE7** Simon borrowed $8000. Interest on this reducing balance loan has been charged at 8.7% p.a. (adjusted monthly) and he has been paying $368.45 each month to service the loan. Currently he still owes $5489.56. Determine how long ago Simon borrowed the money.

9. Aditya borrowed $11 000. Interest on this reducing balance loan has been charged at 6.5% p.a. (adjusted monthly) and he has been paying $409.50 each month to service the loan. Currently he still owes $5565.48. Calculate how long ago Aditya borrowed the money.

10. A loan of $2400 is taken out with a reducing balance interest rate of 4.5% per annum, with interest debited monthly. The borrower wishes to pay instalments of $154.82 per month. Calculate, correct to the nearest month, the term of such a loan.

11. **WE8** A loan of $15 000 is being repaid by monthly instalments of $423.82, with interest being charged at 11.5% p.a. (debited monthly). Currently, the amount owing is $8357.65. Calculate how much longer it will take to:

    a. reduce the amount outstanding to $2450.15
    b. repay the loan in full.

12. A loan of \$9000 is being repaid by monthly instalments of \$273.56, with interest being charged at 8.8% p.a. (debited monthly). Currently, the amount owing is \$6900.86. Calculate how much longer it will take to:

   a. reduce the amount outstanding to \$1670.48
   b. repay the loan in full.

13. **WE9** A reducing balance loan of \$18 000 has a term of 5 years. It is to be repaid by monthly instalments at a rate of 7.8% p.a. (debited monthly).

   a. Determine the repayment value.
   b. Calculate the term of the loan if the repayment is increased to \$390.50.
   c. Calculate the total interest paid for repayments of \$390.50.
   d. Determine by how much the interest figure in part c differs from that paid for the original offer.

14. A reducing balance loan of \$25 000 has a term of 5 years. It is to be repaid by fortnightly instalments at a rate of 6.5% p.a. (debited fortnightly).

   a. Determine the repayment value.
   b. Calculate the term of the loan if the repayment is increased to \$245.
   c. Calculate the total interest paid for repayments of \$245.
   d. Determine by how much the interest figure in part c differs from that paid for the original offer.

15. **WE10** James borrowed \$21 000 for some home renovations and agreed to pay the loan over 7.5 years, with quarterly instalments of \$899.41 and interest debited at 6.8% p.a. However, after 6 years of the loan, James decided to decrease the repayment value to \$685.05.
   Calculate:

   a. the actual term of the loan
   b. the total interest paid
   c. the extra interest charged by decreasing the repayment value.

16. Gabriel borrowed \$17 000 for some new furniture and agreed to pay the loan over 8 years with quarterly instalments of \$670.29 and interest debited at 5.9% p.a. However, after 6 years of the loan, Gabriel decided to increase the repayment value to \$1724.02. Calculate:

   a. the actual term of the loan
   b. the total interest paid
   c. the interest savings achieved by increasing the repayment value.

17. Aimee has borrowed \$5500 for some new outdoor furniture. She is to repay the reducing balance loan by quarterly instalments of \$861.29, with interest debited quarterly at 9.4% p.a.
   Calculate:

   a. how long it will take Aimee to repay the loan in full
   b. the total interest charged.

18. Melpomeni's loan of \$22 000 was taken out some time ago. Interest has been charged at 7.8% p.a. (adjusted monthly) and monthly repayments of \$443.98 have serviced the loan. If the amount still owing is \$14 209.88:

   a. determine how long ago the loan was taken out
   b. calculate the term of the loan.

19. Some time ago, Elizabeth took out a loan of \$25 000. Interest has been charged at 10.5% p.a. (adjusted monthly) and monthly repayments of \$537.35 have serviced the loan.
   If the amount still owing is \$11 586.64:

   a. determine how long ago the loan was taken out
   b. Calculate the term of the loan.

20. **MC** A reducing balance loan of $80 000 is taken out at
7.9% p.a. (adjusted monthly) to finance the purchase of a
boat. It is to be repaid with monthly instalments of $639.84.
The loan will be paid in full in:

A. 10 years.
B. 15 years.
C. 20 years.
D. 22 years.
E. 25 years.

21. Gila's reducing balance loan of $9000 is to be repaid by monthly instalments of $230.43, with interest
charged at 10.5% p.a. (debited monthly).

a. Currently, the amount owing is $8069.78. Determine how much longer it will take to:

i. reduce the amount owing to $3822.20
ii. repay the loan in full.

b. Some time later the amount owing has fallen to $3226.06. Determine how much longer it will take to:

i. reduce the amount owing to $1341.23
ii. repay the loan in full.

22. Megan wanted to borrow $50 000 and was offered a reducing
balance loan over 20 years at 6.9% p.a. (adjusted monthly) with
monthly instalments.

a. Calculate the monthly repayment value.
b. Calculate the term of the loan if instead the repayment was:

i. increased to $577.97
ii. increased to $486.33
iii. decreased to $361.85
iv. decreased to $352.90.

c. In each case in part b above, calculate the total interest paid.
d. For each case above, calculate the interest difference from the
original offer.

23. Jack borrowed $20 000 and agreed to repay the loan over 10 years, with quarterly instalments of $750.48
and interest debited quarterly at 8.6% p.a. However, after 5 years he decided to increase the repayment value.
Calculate the following if the quarterly payment was increased to:

a. 901.48

i. The actual term of the loan
ii. The total interest paid
iii. The interest saving achieved by increasing the quarterly repayment

b. 1154.34

i. The actual term of the loan
ii. The total interest paid
iii. The interest saving achieved by increasing the quarterly repayment.

24. Robin borrowed $25 000 and agreed to repay this reducing balance loan over 10 years with quarterly instalments of $975.06, with interest charged at 9.5% p.a. After 4 years Robin increased her repayment value to $1167.17. The term of her loan will be closest to:

A. 6 years.   B. 7 years.   C. 8 years.
D. 9 years.   E. 10 years.

25. A loan of $25 000 is repaid in 2 stages over $8\frac{3}{4}$ years, with quarterly instalments. For the first 4 years the repayment was $975.06; it was increased to $1167.17 for the remaining time. The total amount of interest charged would be closest to:

A. $9000   B. $10 000   C. $11 000   D. $12 000   E. $13 000

26. Anne is repaying a $26 000 loan over 8 years with monthly instalments of $383.61 at 9.2% p.a., debited monthly on the outstanding balance. She has made 2 years' worth of repayments but would like to repay the loan in full in the next 5 years.
Calculate:

a. the amount that she still owes
b. the monthly repayment value needed to repay in full.

7.3 Exam questions

Question 1 (1 mark)
Source: VCE 2021, Further Mathematics Exam 1, Section A, Q22; © VCAA.
MC Joanna deposited $12 000 in an investment account earning interest at the rate of 2.8% per annum, compounding monthly.

She would like this account to reach a balance of $25 000 after five years.

To achieve this balance, she will make an extra payment into the account each month, immediately after the interest is calculated.

The minimum value of this payment is closest to
A. $113.85   B. $174.11   C. $580.16   D. $603.22   E. $615.47

Question 2 (1 mark)
Source: VCE 2015, Further Mathematics Exam 1, Section B, Module 4, Q8; © VCAA.
MC Cindy took out a reducing balance loan of $8400 to finance an overseas holiday.

Interest was charged at a rate of 9% per annum, compounding quarterly.

Her loan is to be fully repaid in six years, with equal quarterly payments.

After three years, Cindy will have reduced the balance of her loan by approximately
A. 9%   B. 35%   C. 43%   D. 50%   E. 57%

Question 3 (1 mark)
Source: VCE 2014, Further Mathematics Exam 1, Section B, Module 4, Q5; © VCAA.
MC A bank approves a $90 000 loan for a customer.

The loan is to be repaid fully over 20 years in equal monthly payments.

Interest is charged at a rate of 6.95% per annum on the reducing monthly balance.

To the nearest dollar, the monthly payment will be
A. $478   B. $692   C. $695   D. $1409   E. $1579

More exam questions are available online.

7.4 The effect of rate and repayment changes on reducing balance loans

LEARNING INTENTION

At the end of this subtopic you should be able to:
- investigate the frequency of repayments on reducing balance loans
- investigate the effect of changing the interest rate
- investigate the effect of interest-only loans.

7.4.1 Frequency of repayments

In this section we investigate the effect on the actual term of the loan and on the total amount of interest charged by making more frequent repayments. While the value of the repayment will change, the actual outlay will not.

For example, a $3000 quarterly (every 3 months) repayment will be compared to a $1000 monthly repayment. That is, the same amount is repaid during the same period of time in each case. So the only variable will be how often repayments are made.

In all cases in this section, interest will be charged just before a repayment is made, although this may not be the case in practice.

WORKED EXAMPLE 11 Investigating the effect of changing the frequency of repayments

Tessa wants to buy a dress shop. She borrows $15 000 at 8.5% p.a. (debited prior to each repayment) of the reducing balance. She can afford quarterly repayments of $928.45.
One-third of the quarterly repayment of $928.45 gives the equivalent monthly repayment of $309.48. The equivalent fortnightly repayment is $142.84.
Calculate:
a. the term of the loan and the amount still owing prior to the last payment if Tessa made payments monthly
b. the term of the loan and the amount still owing prior to the last payment if Tessa made payments fortnightly.

THINK	WRITE
a. 1. Identify the given values. Enter the appropriate values using Finance Solver. Remember that PpY = 12 and CpY = 12 for monthly repayments.	a. Initial loan = 15 000 Monthly repayment = 309.48 Interest rate = 8.5% N = ?

2. Solve for N.

$N = 59.582\,518\,723\,273$

3. The value obtained for N is 59.58, which means that a 60th repayment is required. That is, N = 60.

$N = 60$ months

Term of loan $= \dfrac{60}{12}$

$= 5$ years

4. To calculate the amount still owing prior to the last payment, determine V_n when N = 59. Enter the appropriate values using Finance Solver:

$FV = -179.273\,603\,537\,66$

N	59
I%	8.5
PV	15 000
PMT (Pmt)	−309.48
FV	**? place the cursor in this cell.**
PpY	12
CpY	12

Press ENTER/tap solve.

5. Write the amount still owing.

The amount still owing prior to the last payment is $179.27.

b. 1. Enter the appropriate values using Finance Solver on your CAS.
Remember that PpY = 26 and CpY = 26 for fortnightly repayments.

b. For fortnightly repayments:
Initial loan = 15 000
Fortnightly repayment = 142.84
Interest rate = 8.5%

2. Solve for N.

$N = ?$

3. The value obtained for N is 128.85, which means that a 129th repayment is required. That is, N = 129.

$N = 129$ fortnights
Term of loan = 4 years, 25 fortnights

4. To calculate the amount still owing prior to the last payment, determine F_V (or FV) when N = 128. Enter the appropriate values using Finance Solver:

$FV = -120.636\,212\,821\,95$

N	128
I%	8.5
PV	15 000
PMT	−142.84
FV	**? place the cursor in this cell.**
PpY	26
CpY	26

5. Write the amount still owing.

The amount still owing prior to the last payment is $120.64.

It can be seen from Worked example 11 that while the same outlay is maintained, there may be a slight decrease in the term of a loan when repayments are made more often. Let us now calculate the savings for such a loan. In this situation, we should consider the final (partial) payment separately, because the amount of interest that it attracts is less than a complete repayment, d.

The total interest paid is now calculated as usual:

Total interest paid

total interest = total repayments – principal repaid

WORKED EXAMPLE 12 Comparing the interest paid by making more frequent repayments

In Worked example 11, Tessa's $15 000 loan at 8.5% p.a. gave the following three scenarios:
1. **Quarterly repayments of $928.45 for 5 years**
2. **Monthly repayments of $309.48 for 59 months, with $179.27 still outstanding**
3. **Fortnightly repayments of $142.84 for 128 fortnights, with $120.64 still owing**

Compare the total interest paid by Tessa if she repaid her loan:
a. **quarterly**　　　　　　　　　b. **monthly**　　　　　　　　　c. **fortnightly.**

THINK	WRITE
a. For quarterly repayments: Total interest = total repayments 　　　　　　　　　　– principal repaid	a. For quarterly repayments: Total interest = $928.45 \times 20 - 15\,000$ 　　　　　　　 = \$3569
b. 1. For monthly repayments, calculate the interest rate and the amount owing after 59 months. (Refer to Worked example 11.)	b. For monthly repayments: $r = \dfrac{8.5}{12}\%$ 　 $= 0.7083\%$ $V_{59} = \$179.27$
2. Calculate the interest on the 59th repayment to determine the final repayment.	Interest on the 59th repayment = 0.7083% of $179.27 　　　　　　　　　　　　　　 $= 0.007\,083 \times 179.27$ 　　　　　　　　　　　　　　 $= \$1.27$ Final repayment $= 179.27 + 1.27$ 　　　　　　　　 $= \$180.54$
3. Calculate the total interest paid.	Total interest $= 309.48 \times 59 + 180.54 - 15\,000$ 　　　　　　 $= \$3439.86$
c. 1. For fortnightly repayments, calculate the interest rate and the amount owing after 128 fortnights. (Refer to Worked example 11.)	c. For fortnightly repayments: $r = \dfrac{8.5}{26}\%$ 　 $= 0.3269\%$ The 128th repayment $= 120.64$
2. Calculate the interest on the 128th repayment to determine the final repayment.	Interest on the 128th repayment = 0.3269% of $120.64 　　　　　　　　　　　　　　 $= 0.003\,269 \times 120.64$ 　　　　　　　　　　　　　　 $= \$0.39$ Final repayment $= 120.64 + 0.39$ 　　　　　　　　 $= \$121.03$

▶

3. Calculate the total interest paid.	Total interest = $142.84 \times 128 + 121.03 - 15\,000$ = \$3404.55
4. Calculate the interest saving with monthly repayments over quarterly repayments.	Monthly interest saving = $3569 - 3439.86$ = \$129.14
5. Calculate the interest saving if repayments are made fortnightly instead of quarterly.	Fortnightly interest saving = $3569 - 3404.55$ = \$164.45
6. Write a comparison statement.	Tessa saves \$164.45 if she repays fortnightly rather than quarterly and \$129.14 if she repays monthly rather than quarterly.

The slight time saving calculated in Worked example 12 when repayments were made more often have now been transformed to money saving. The saving increases as the frequency of repayment increases. This is because the amount outstanding is reduced more often and so the amount of interest added is slightly lower.

A saving of \$164 over 5 years, out of more than \$18 000 repaid, might not seem much, but the saving increases as the term of the loan increases and as the amount borrowed increases.

7.4.2 Changing the rate

Of all the variables associated with reducing balance loans, the one that is most likely to change during the term of a loan is the interest rate. These rates rarely stay the same for the life of a loan; for most loans, the rate will change several times.

The Reserve Bank of Australia is the main monetary authority of the federal government and, as such, is the overall guiding influence on monetary factors in the Australian economy. Consequently, it indirectly controls the lending interest rates of financial institutions here.

There is usually some variation in rates between institutions; for example, a lower rate may be designed to attract more customers. Within each institution there are rate variations for different types of reducing balance loans. Banks advertise their loan rates to attract customers.

In this section, we investigate the effect that changing the interest rate has on the term of the loan and on the total interest paid. It should be remembered that as the interest rate increases, so too will the term (if d remains constant) of the loan since more interest needs to be paid.

First, let us simply compare loan situations by varying only the rate.

WORKED EXAMPLE 13 Investigating the effect of changing the interest rate

A reducing balance loan of \$18 000 has been taken out over 5 years at 8% p.a. (adjusted monthly), with monthly repayments of \$364.98.
a. Calculate the total interest paid.
b. If, instead, the rate was 9% p.a. (adjusted monthly) and the repayments remained the same, calculate:
   i. the term of the loan   **ii. the total amount of interest paid.**

THINK	WRITE
a. For 8% p.a.: Total interest = total repayments − principal repaid	**a.** For 8% p.a.: Total interest = $364.98 \times 60 - 18\,000$ = \$3898.80
b. i. 1. Using Finance Solver, enter the appropriate values. Remember that PpY and CpY will both equal 12 for monthly repayments.	**b. i.** For 9% p.a. $V_0 = 18\,000, \quad d = 364.98, \quad I = 9, \quad V_n = 0,$ $N = ?$
2. Solve for N.	$N = 61.810\,665\,384\,123$
3. N = 61.81 means 61 full repayments plus a final smaller payment.	$N = 62$ months
ii. 1. Calculate the amount owing after 61 repayments to calculate the amount still owing. Enter the following values:	**ii.** Term = 5 years, 2 months

N	61
I%	9
PV	18 000
FV	? place the cursor here.
PMT	−364.98
PpY	12
CpY	12

THINK	WRITE
2. Solve for FV and interpret the result.	FV = −293.886 672 877 The amount still owing after 61 repayments is \$293.88.
3. Calculate the interest on the 62nd repayment to determine the final repayment. $r = \dfrac{9}{12} = 0.75\%$	Interest on final repayment = 0.75% of \$293.8 = \$2.20 Final repayment = 293.88 + 2.20 = \$296.08
4. Total interest = total repayments − principal repaid	Total interest paid = $364.98 \times 61 + 296.08 - 18\,000$ = \$4559.86

In Worked example 13, the rate was increased by only 1% p.a. on \$18 000 for only 5 years, yet the amount of interest paid has increased from \$3898.80 to \$4559.86, a difference of \$661.06. This difference takes on even more significant proportions over a longer period of time and with a larger principal.

Let us now consider varying the rate during the term of the loan.

tlvd-4004

Natsuko and Hymie took out a loan for home renovations. The loan of $42 000 was due to run for 10 years and attract interest at 7% p.a., debited quarterly on the outstanding balance. Repayments of $1468.83 were made each quarter.
After 4 years, the rate changed to 8% p.a. (debited quarterly). The repayment value didn't change.
a. Calculate the amount outstanding when the rate changed.
b. Calculate the actual term of the loan.
c. Compare the total interest paid to what it would have been had the rate remained at 7% p.a. for the 10 years.

THINK	WRITE
a. 1. Rate changes after 4 years; that is, N = 16. State the initial loan, repayments, interest rate and the number of payments.	**a.** $V_0 = 42\,000$, $d = 1468.83$, $I = 7\%$, $n = 16$, $v_n = ?$ Initial loan = 42 000 Repayment = 1468.83 I = 7%, N = 16%
2. Use your CAS to calculate the amount owing after 16 payments.	FV = −28 584.356 811 602 The amount outstanding when the rate changed is $28 584.36.
b. 1. Calculate N to repay $28 584.36 in full at the new rate. Use Finance Solver to enter the following values: I = 8, PV = 28 584.36, Pmt = −1468.83, FV = 0, PpY = 4 and CpY = 4.	**b.** The new interest rate of 8%: N = −24.896 005 939 422 $\quad$ = 25 quarters Time = $6\frac{1}{4}$ years
2. Calculate the total term of the loan — that is, time at 7% plus time at 8%.	Term = 4 years + $6\frac{1}{4}$ years $\quad$ = $10\frac{1}{4}$ years
c. 1. Calculate the amount owing after 24 repayments to calculate the amount still owing. (That is, on your calculator change the value of N to 24 and solve for FV.)	**c.** FV = −1291.669 456 564 649 The amount still owing after the 24th repayment is $1291.61.
2. Calculate the interest on the outstanding amount to determine the final repayment. $r = \dfrac{8}{4} = 2\%$	Interest on final payment = 2% of $1291.61 $\quad$ = $25.83 Final repayment = 1291.61 + 25.83 $\quad$ = $1317.44

3. Calculate the total interest for the rate change scenario. The number of repayments is 40 at \$1468.83 plus 1 at \$1317.44.	For the rate change scenario, total interest $= 1468.83 \times 40 + 1317.44 - 42\,000$ $= \$18\,070.64$
4. Calculate the total interest if the rate remained at 7%.	For the rate at 7% only, total interest $= 1468.83 \times 40 - 42\,000$ $= \$16\,753.20$
5. Determine the interest difference between the two scenarios.	Interest difference $= 18\,070.64 - 16\,753.20$ $= \$1317.44$
6. Write a comparison statement.	An extra \$1317.44 interest will be paid due to the interest rate change from 7% p.a. to 8% p.a.

In the situations studied so far, the repayment value remained the same, even though the rate varied. In practice, this is what happens if the rate decreases and therefore the term of the loan decreases. However, when the rate increases, financial institutions will generally increase the repayment value to maintain the original term of the loan. This was discussed in section 7.3.2. If this is not done, the term of the loan can increase quite dramatically. In fact, this may occur to such an extent that the repayments are insufficient to cover the interest added, so that the amount outstanding *increases*.

Consider a \$44\,000 loan over 15 years at 10% p.a. (monthly).

Monthly repayments $=$ \$472.83

After 5 years the amount owing is \$35\,779.02.

Suppose the interest rate rises dramatically to 16% p.a.

After a further 10 years under these conditions, the amount owing is \$37\,014.72.

That is, the amount owing has increased.

This situation is not beneficial to either the lender or the borrower.

7.4.3 Interest-only loans

Interest-only loans are loans where the borrower makes only the minimum repayment equal to the interest charged on the loan. As the initial principal and amount owing is the same for the period of this loan, we could use either the simple interest formula or CAS to solve problems of this type. When using Finance Solver, the present value (PV) and future value (FV) are entered as the same amount. Note that the future value is negative to indicate money owed to the bank.

This type of loan is used by two kinds of borrowers: investors in shares and/or property or families that are experiencing financial difficulties and seek short-term relief from high repayment schedules.

Jade wishes to borrow $40 000 to invest in shares. She uses an interest-only loan to minimise her repayments and hopes to realise a capital gain if she can sell the shares at a higher value. The term of the loan is 6.9% p.a. compounded monthly, with monthly repayments equal to the interest charged.

a. Calculate the monthly interest-only repayment.

b. If, in 3 years, she sells the shares for $50 000, calculate the profit she would make on this investment strategy.

THINK	WRITE
a. 1. Identify the initial loan, the interest rate and N, where N is equal to one payment period.	**a.** Initial loan = 40 000 Interest rate = 6.9% N = 1 month
2. Evaluate I using the simple interest formula.	$I = \dfrac{40\,000 \times 6.9 \times \frac{1}{12}}{100}$ $= \$230$
3. Write the answer.	The monthly repayment to pay the interest only for the loan is $230.
b. 1. Determine whether the capital gain on the shares exceeds the amount paid in interest.	**b.** Capital gain = selling price − purchase price $= \$50\,000 - \$40\,000$ $= \$10\,000$ Total interest charged = repayment × number of payments $= \$230 \times 36$ $= \$8280$ Profit = capital gain − loan cost $= \$10\,000 - \$8280$ $= \$1720$
2. Write the answer as a sentence.	Jade will make a profit of $1720.

1. **WE11,12** Bella wants to buy a shoe shop. She borrows $18 000 at 8.3% p.a. (debited prior to each repayment) of the reducing balance. She can afford quarterly repayments of $1108.80 and this will pay the loan in full in exactly 5 years.

   One third of the quarterly repayment gives the equivalent monthly repayment of $369.60.

   The equivalent fortnightly repayment is $170.58.

   a. If Bella made monthly repayments, calculate:

      i. the term of the loan
      ii. the amount still owing prior to the last payment.

   b. If Bella made fortnightly repayments, calculate:

      i. the term of the loan
      ii. the amount still owing prior to the last payment.

   Bella's $18 000 loan at 8.3% p.a. gave the following scenarios:

   A. Quarterly repayments of $1108.8 for 5 years
   B. Monthly repayments of $369.60 for 59 months, with $217.37 outstanding
   C. Fortnightly repayments of $170.58 for 128 fortnights, with $149.63 still owing

   c. Compare the total interest paid by Bella if she repaid her loan:

      i. quarterly   ii. monthly   iii. fortnightly.

2. Simba wants to invest money in shares. He borrows $25 000 at 7.2% p.a. (debited prior to each repayment) of the reducing balance. He can afford quarterly repayments of $1292.02, which will pay the loan in full in exactly 6 years.

   One third of the quarterly repayment gives the equivalent monthly repayment of $430.67.

   The equivalent fortnightly repayment is $198.77.

   Calculate:

   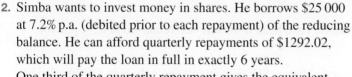

   a. If Simba made monthly repayments, calculate:

      i. the term of the loan
      ii. the amount still owing prior to the last payment.

   b. If Simba made fortnightly repayments, calculate:

      i. the term of the loan
      ii. the amount still owing prior to the last payment.

   Simba's $25 000 loan at 7.2% p.a. gave the following scenarios:

   A. Quarterly repayments of $1292.02 for 6 years
   B. Monthly repayments of $430.67 for 71months with $246.17 outstanding
   C. Fortnightly repayments of $198.77 for 154 fortnights with $164.36 still owing

   c. Compare the total interest paid by Simba if he repaid his loan:

      i. quarterly   ii. monthly   iii. fortnightly.

3. **WE13** A reducing balance loan of $20 000 has been taken out over 5 years at 10% p.a. (adjusted monthly), with monthly repayments of $424.94.

   a. Calculate the total interest paid.
   b. If, instead, the rate was 11% p.a. (adjusted monthly) and the repayments remained the same, calculate:

      i. the term of the loan
      ii. the total amount of interest paid.

4. A reducing balance loan of $16 500 has been taken out over 4 years at 8% p.a. (adjusted monthly), with monthly repayments of $402.81.

   a. Calculate the total interest paid.
   b. If, instead, the rate was 9% p.a. (adjusted monthly) and the repayments remained the same, calculate:

      i. the term of the loan
      ii. the total amount of interest paid.

5. **WE14** Jessica took out a loan for a new kitchen. The loan of $32 000 was due to run for 10 years and attract interest at 8% p.a., debited quarterly on the outstanding balance. Repayments of $1169.78 were made each quarter. After 4 years the rate changed to 9% p.a. (debited quarterly). The repayment value didn't change.

   a. Calculate the amount outstanding when the rate changed.
   b. Calculate the actual term of the loan.
   c. Compare the total interest paid to what it would have been if the rate had remained at 8% p.a. for 10 years.

6. Roger took out a loan to get his large pine trees removed after they were affected during a massive storm. The loan of $28 000 was due to run for 8 years and attract interest at 6.5% p.a., debited monthly on the outstanding balance. Repayments of $374.81 were made each month. After 3 years the rate changed to 7.5% p.a. (debited monthly).
   The repayment value didn't change.

   a. Calculate the amount outstanding when the rate changed.
   b. Calculate the actual term of the loan.
   c. Compare the total interest paid to what it would have been if the rate had remained at 6.5% p.a. for 8 years.

7. **WE15** Jodie wishes to borrow $30 000 to invest in shares. She uses an interest-only loan to minimise her repayments and hopes to realise a capital gain when she sells the shares at a higher value. The term of the loan is 5.8% p.a. compounded monthly, with monthly repayments equal to the interest charged.

   a. Calculate the monthly interest-only repayments.
   b. If, in 3 years, she sells the shares for $35 000, calculate the profit she would make on this investment strategy.

8. Max wishes to borrow $50 000 to invest in shares. He uses an interest-only loan to minimise his repayments and hopes to realise a capital gain when he sells the shares. The term of the loan is 7.5% p.a. compounded monthly, with monthly repayments equal to the interest charged.

   a. Calculate the monthly interest-only repayments.
   b. If, in 3 years, he sells the shares for $63 000, calculate the profit he would make on this investment.

9. Harper borrows $2000 to purchase new cricket gear. The loan is to be paid in full over 1 year with quarterly payments at an interest rate of 7.5% p.a.

a. Calculate the quarterly payment required.
b. Complete an amortisation table for the loan using a first-order recurrence relation for each payment.
c. Calculate the principal outstanding after the third payment.
d. Calculate the total amount paid on the loan.
e. Calculate the total interest paid on the loan.

10. Grace takes a loan out for $3750 to purchase new curtains for her house. The loan is to be paid in full over 1 year, with bi-monthly payments at a rate of 6% p.a.

a. Calculate the bi-monthly payment required.
b. Complete an amortisation table for the loan using a first-order recurrence relation.
c. If the payments were monthly instead of bi-monthly, calculate the payments required.
d. Complete an amortisation table for the loan with monthly payments.
e. Calculate the interest paid on the loan for both monthly and bi-monthly payments and comment on the answers.

11. Phul has a reducing balance loan of $40 000. The loan has interest charged at 8% p.a. (debited before each repayment) and can be repaid by quarterly instalments of $1462.23 over exactly 10 years. The equivalent monthly repayment is $487.41 and the equivalent fortnightly repayment is $224.96.

Determine the term of the loan and the amount still owing before the final repayment, if repayments are made:

a. monthly
b. fortnightly.

12. A loan of $25 000 attracts interest at 8.25% p.a. on the outstanding balance. The following four scenarios are available:

a. Half-yearly repayments of $3101.48 for 5 years

b. Quarterly repayments of $1550.74 for $4\frac{3}{4}$ years, with $1217.93 still owing

c. Monthly repayments of $516.91 for 58 months, with $512.33 still owing

d. Fortnightly repayments of $238.58 for 127 fortnights, with $140.27 still owing

Compare the total interest paid if the loan is repaid:

i. half-yearly
ii. quarterly
iii. monthly
iv. fortnightly.

The following information refers to questions 13 and 14.

Betty has borrowed $65 000 to finance her plant and flower nursery. Betty chooses to repay the loan, which attracts interest at 9.3% p.a. on the outstanding balance, by fortnightly repayments of $309.66 rather than the equivalent monthly repayment of $670.92.

13. **MC** The term of the loan will be:

A. 14 years 1 month.
B. 14 years 25 fortnights.
C. 15 years.
D. 15 years 1 fortnight.
E. 15 years 1 month.

14. **MC** The amount Betty will save is closest to:
A. $240   B. $260   C. $270   D. $300   E. $320

15. A reducing balance loan of $25 000 has been taken out over 5 years at 8% p.a. (adjusted monthly), with monthly repayments of $506.91.

a. Calculate the total interest paid.
b. If, instead, the rate was 9% p.a. (adjusted monthly) with the same repayments maintained, calculate:

   i. the term of the loan now   ii. the total interest paid.
c. If, instead, the rate was 10% p.a. (adjusted monthly) with the same repayments maintained, calculate:

   i. the term of the loan now   ii. the total interest paid.

The following information relates to questions 16 and 17.

Clint's $28 000 loan for his house extensions has interest debited every month at 12% p.a. of the outstanding balance. The loan was due to run for 10 years and he was to make repayments of $401.72 per month to service the loan.

After he had made 50 repayments, his credit union reduced the interest rate to 10.75% p.a. (adjusted monthly) for the remainder of the loan.

16. **MC** If Clint maintained the monthly repayment, the term of the loan would be:

A. $9\frac{3}{4}$ years.

B. $10\frac{1}{4}$ years.

C. $10\frac{3}{4}$ years.

D. 11 years.

E. 12 years.

17. **MC** The total interest paid by Clint would lie between:

A. $18 700 and $18 800.   B. $18 800 and $18 900.   C. $18 900 and $19 000.
D. $46 800 and $46 900.   E. $46 900 and $47 000.

18. The Risky brothers want to invest in $140 000 worth of shares. They take out an interest-only loan from the bank. The loan is at 10.8% p.a. compounded quarterly with quarterly repayments.

a. Calculate the quarterly repayment amount.
b. If in 1 year they sell the shares for $152 000, calculate the amount of profit or loss they will have made with this investment strategy.

19. **MC** The Bigs have had a new addition to the family and John, the father, takes 12 months of leave from work to stay at home. To financially cope, they ask their bank manager for an interest-only loan for this period on the outstanding amount on their home loan, which is currently $210 000. If the terms of the interest-only loan are 6.79% p.a. compounded fortnightly, the fortnightly repayments will be closest to:

A. $548.42
B. $1188.25
C. $14 637
D. $4879
E. $21 032.18

7.4 Exam questions

Question 1 (1 mark)

Source: VCE 2021, Further Mathematics Exam 1, Section A, Core, Q24; © VCAA.

MC Bob borrowed $400 000 to buy an apartment.

The interest rate for this loan was 3.14% per annum, compounding monthly.

A scheduled monthly repayment that allowed Bob to fully repay the loan in 20 years was determined. Bob decided, however, to make interest-only repayments for the first two years.

After these two years the interest rate changed. Bob was still able to pay off the loan in the 20 years by repaying the scheduled amount each month.

The interest rate, per annum, for the final 18 years of the loan was closest to

A. 1.85%
B. 2.21%
C. 2.79%
D. 3.14%
E. 4.07%

Question 2 (1 mark)

Source: VCE 2017, Further Mathematics Exam 1, Section A, Core, Q24; © VCAA.

MC Xavier borrowed $245 000 to pay for a house.

For the first 10 years of the loan, the interest rate was 4.35% per annum, compounding monthly.

Xavier made monthly repayments of $1800.

After 10 years, the interest rate changed.

If Xavier now makes monthly repayments of $2000, he could repay the loan in a further five years.

The new annual interest rate for Xavier's loan is closest to

A. 0.35%
B. 4.1%
C. 4.5%
D. 4.8%
E. 18.7%

▶ **Question 3 (1 mark)**

Source: VCE 2013, Further Mathematics, Exam 1, Section B, Module 4, Q9; © VCAA.

MC The following information relates to the repayment of a home loan of $300 000.
- The loan is to be repaid fully with monthly payments of $2500.
- Interest compounds monthly.
- After the first monthly payment has been made, the amount owing on the loan is $299 000.

Which one of the following statements is true?
- **A.** After two months, $297 995 is still owing on the loan.
- **B.** $1000 of interest has been paid in the first month.
- **C.** The loan will be fully repaid in less than 15 years.
- **D.** Halfway through the term of the loan, the amount still owing will be $150 000.
- **E.** Payments of $2750 rather than $2500 per month will reduce the time to repay the loan fully by more than three years.

More exam questions are available online.

7.5 Annuities and perpetuities

LEARNING INTENTION

At the end of this subtopic you should be able to:
- model annuities and perpetuities using a recurrence relation
- solve annuities problems using Finance Solver
- calculate the principal (V_0) and interest rate (r).

7.5.1 Using a recurrence relation to model annuities

An **annuity** is an investment account that at some future date will provide a regular stream of income for a set period of time. Annuities may be in the form of bursaries or scholarships, or retirement funds.

A recurrence relation for an annuity

$$V_0 = a, \ V_{n+1} = RV_n - d$$

where

V_0 **is the intial amount borrowed**

V_n **is the balance of the loan after** n **payments**

$R = 1 + \dfrac{r}{100}$ **(**r **is the interest rate per compounding period)**

d **is the payment made.**

As part of her postgraduate studies, Margaret has been awarded a 'writer in residence' scholarship of \$8500 for 4 months at a local college. The money is invested in an annuity that pays 4.5% p.a., compounded monthly. Margaret is paid \$2140 per month from the annuity.

a. Write a recurrence relation to describe this situation.
b. Use your calculator to determine how much money is left in the annuity at the end of 4 months.
c. Calculate the adjustment to be made to Margaret's last payment.
d. Complete the amortisation table below for this annuity.

Payment number (n)	Payment (\$)	Interest (\$)	Principal reduction (\$)	Balance of loan (\$)
0	0.00	0.00	0.00	8500.00
1	2140	A	2108.13	6391.88
2	2140	23.97	B	4275.84
3	2140	16.03	2123.97	C
4	D	8.07	2131.93	0.00

THINK

a. 1. State the initial values for V_0, r, R and d.

   2. Write the general form of the recurrence relation.

   3. Substitute in the values for V_0, r, R and d.

b. 1. On a calculator screen, type 8500 and press ENTER/EXE.
Then type $\times 1.003\,75 - 2140$.
Press ENTER/EXE 4 times.

   2. Write the answer.

c. If the balance is greater than zero, the last payment must be adjusted accordingly.

d. A is interest calculated on \$8500.
B is principal reduction.
C is balance of the loan.
To calculate D: as there was \$19.95 left after 4 payments, the final payment is adjusted by \$19.95 to leave the balance of the investment at \$0.

WRITE

a. $V_0 = 8500$

$$r = \frac{4.5}{12}$$

$$= 0.375$$

$$R = 1 + \frac{0.375}{100}$$

$$= 1.003\,75$$

$$d = 2000$$

$V_0 = a, \ V_{n+1} = RV_n - d$

$V_0 = 8500, \ V_{n+1} = 1.003\,75V_n - 2140$

b.

8500	8500
$8500 \times 1.003\,75 - 2140$	6391.88
$6391.88 \times 1.003\,75 - 2140$	4275.84
$4275.84 \times 1.003\,75 - 2140$	2151.88
$2151.88 \times 1.003\,75 - 2140$	19.95

After 4 months, the balance of the annuity is \$19.95.

c. Margaret's last payment should be increased by \$19.95.
Last payment = \$2159.95

d. $A = 8500 \times 0.003\,75 = 31.88$
$B = 2140 - 23.97 = 2116.03$
$C = 4275.84 - 2123.97 = 2151.87$
$D = 2140 + 19.95 = 2159.95$

7.5.2 Solving annuity problems using Finance Solver

The Finance Solver application on your CAS calculator can be used to solve annuity problems.

N	The total number of payments
I%	The interest rate per annum
PV	The present value of the investment or loan – **this is negative for an annuity**
PMT (Pmt)	The periodic payment – **this is positive for an annuity providing periodic payments to you**
FV	The future value of the loan or investment – **this is positive or zero for an annuity providing periodic payments to you**
P/Y (PpY)	The number of payment periods in a year
C/Y (CpY)	The number of compounding periods in a year (nearly always the same as P/Y)

tlvd-4005

WORKED EXAMPLE 17 Using Finance Solver for an annuity

Jacob invests \$120 000 in an annuity paying 7.5% per month. Using Finance Solver on your calculator, determine:
a. how long this annuity will last if the monthly payment is \$4000
b. how much Jacob should receive each month if he wishes the annuity to last for 20 years.

THINK

a. 1. Using Finance Solver on your CAS, enter the appropriate values:

N	?	The number of payments for 10 years at 12 interest calculations per year. **Place the cursor in this cell.**
I%	7.5	The annual interest rate
PV	−120 000	The value of the loan
PMT (Pmt)	4000	The value of the payment
FV	0	The investment is paid out so the future value is zero.
P/Y (PpY)	12	There are 12 payments per year.
C/Y (CpY)	12	There are 12 compounding periods per year.

2. Press ENTER/tap Solve.

3. Write the answer as a sentence.

b. 1. Calculate the payment each month if the annuity lasts for 20 years.

WRITE

a. I = 7.5 % p.a.
Amount invested = \$120 000
Future value = 0

N = 33.32
The annuity will last 33 months, to the nearest month.

b. N = 20 × 12
    = 240
I = 7.5% p.a.
Amount of annuity = \$12 000

N	240
I%	7.5
PV	−120 000
PMT (PMT)	? place the cursor in this cell.
FV	0
P/Y (PpY)	12
C/Y (CpY)	12

2. Using Finance Solver, Press ENTER/tap Solve.

3. Write the answer.

Payment = 966.71

The payment per month is $966.71.

7.5.3 Perpetuities

A **perpetuity** is an annuity where a permanently invested sum of money provides regular payments that continue forever.

Many *scholarships* or *grants* offered to students at universities are provided by funds known as perpetuities.

The funds last for an indefinite period of time as long as the amount paid out is no more than the interest earned on the initial lump sum deposited. The type of investment that is used to earn the interest is usually a bond, which offers a fixed interest amount, paid on a regular basis, over a long period of time. Wealthy people who wish to encourage and support a worthwhile cause usually set up these perpetuities.

The balance of the amount invested does not change and is the same for an indefinite period.

Perpetuity formula

$$d = \frac{V_0 r}{100}$$

where

d = **the amount of the regular payment per period ($)**

V_0 = **the principal ($)**

r = **the interest rate earned per period (%).**

When calculating perpetuities, remember:
1. The period of the regular payment must be the same as the period of the given interest rate.
2. Finance Solver can be used in calculations involving perpetuities. As the principal does not change, the present value (PV or negative cash flow) and the future value (FV or positive cash flow) are entered as the same amount, but with opposite signs.

MC Robert wishes to use part of his wealth to set up a scholarship fund that helps young students from his town further their education at university. Robert invests $200 000 in a bond that offers a long-term guaranteed interest rate of 4% p.a.

If the interest is calculated once a year, then the annual amount provided as scholarship will be:
A. $188 000
B. $288 000
C. $666.67
D. $8000
E. $4000

THINK	WRITE
1. Write the perpetuity formula.	$d = \dfrac{V_0 r}{100}$
2. List the values of V_0 and r.	$V_0 = \$200\,000$ and $r = 4\%$ p.a.
3. Substitute the values into the formula and calculate the amount provided.	$d = \dfrac{\$200\,000 \times 4}{100}$ $= 8000$
4. Write the answer.	The annual amount provided for the scholarship is $8000. Therefore, **D** is the correct answer.

7.5.4 Determining V_0 and r

As was the case with earlier sections in this topic, there are calculations where we need to determine the principal (V_0) or interest rate (r) needed to provide a certain regular payment (d).

For example, how much needs to be invested at 3% p.a. interest to provide a $10 000 annual grant, or what interest rate is needed so that $100 000 will provide a $4000 yearly scholarship indefinitely?

Other calculations involve calculating what extra amount could be granted annually as a scholarship if the interest is compounded monthly in each year rather than once a year and if the scholarship is paid in two equal six-monthly instalments.

Transposing the perpetuity formula

$$V_0 = \frac{100 \times d}{r} \quad \text{and}$$

$$r = \frac{100 \times d}{V_0}$$

If the frequency of the payments each year is not the same as the compounding period of the given interest rate, then Finance Solver is to be used with different values for PpY and CpY.

Remember the following:
1. The principal must be known in order to use Finance Solver.
2. Finance Solver gives the interest rate per annum.

WORKED EXAMPLE 19 Perpetuities and changing interest rates

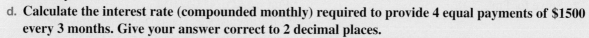

A Rotary Club has $100 000 to set up a perpetuity as a grant for the local junior sporting clubs. The club invests in bonds that return 5.2% p.a. compounded annually.

a. Determine the amount of the annual grant.
b. Calculate the interest rate (compounded annually) required if the perpetuity is to provide $6000 each year.

The Rotary Club wants to investigate other possible arrangements for the structure of the grant.

c. Determine how much extra the annual grant would amount to if the original interest rate was compounded monthly.

d. Calculate the interest rate (compounded monthly) required to provide 4 equal payments of $1500 every 3 months. Give your answer correct to 2 decimal places.

THINK	WRITE
a. 1. Write the perpetuity formula and list the values of V_0 and r.	a. $d = \dfrac{V_0 r}{100}$ $V_0 = \$100\,000$ $r = 5.2\%$ p.a.
2. Substitute the values into the formula and calculate the value of the annual grant.	$d = \dfrac{\$100\,000 \times 5.2}{100}$ $= 5200$
3. Write the answer as a sentence.	The amount of the annual grant is $5200.
b. 1. Write the perpetuity formula and list the values of V_0 and d.	b. $r = \dfrac{100 \times d}{V_0}$ $V_0 = \$100\,000$ $d = \$6000$
2. Substitute the values into the formula and calculate the interest rate.	$r = \dfrac{100 \times 6000}{100\,000}$ $= 6$
3. Write the answer.	For a $100 000 perpetuity to provide $6000 a year, the bond needs to offer an interest rate of 6% p.a.
c. 1. As the frequency of the payment is not the same as the compounding period, the perpetuity formula cannot be used. Use Finance Solver and enter the values as follows.	c.

n (N)	1
r (I%)	5.2
V_0 (PV)	$-100\,000$
d (Pmt)	**? place the cursor in this cell.**
V_n (FV)	$100\,000$
PpY	1 (one payment per year)
CpY	12 (there are 12 compound periods per year)

2. Solve for Pmt.

Pmt = 5325.741 057 054
If the interest was compounded monthly, the annual grant would amount to $5325.74.

3. Compare the sizes of the two grants and write the answer.

The extra amount is:
$5325.74 − $5200 = $125.74
If the interest is compounded monthly, the annual grant would increase by $125.74.

d. 1. As the frequency of the payment is not the same as the compounding period, the perpetuity formula cannot be used. Use Finance Solver and enter the values as follows.

d.

n (N)	1
r (I%)	**? place the cursor in this cell.**
V_0 (PV)	$-100\,000$
d (Pmt)	1500
V_n (FV)	$100\,000$
PpY	4 (four payments per year)
CpY	12 (there are 12 compound periods per year)

2. Solve for I to determine the required interest rate.

I = 5.970 247 527 183
r = 5.97

3. Write the answer.

An interest rate of 5.97% p.a. compounded annually is needed to provide four equal payments of $1500, correct to 2 decimal places.

A benefactor of a college has been approached to provide a Year 7 scholarship of $1000 per term. He is able to get a financial institution to offer a long-term interest rate of 8% per annum. Calculate the principal that needs to be invested.

THINK	WRITE
1. Write the perpetuity formula and list the values of d and r. Both d and r need to be expressed in the same period of time.	$V_0 = \dfrac{100 \times d}{r}$ $d = \$1000$ per term (4 terms per year) $R = 8\%$ p.a. $\quad = \dfrac{8}{4}$ $\quad = 2\%$ per term
2. Substitute the values into the formula and calculate the value of the annual grant.	$V_0 = \dfrac{100 \times 1000}{2}$ $\quad = 50\,000$
3. Write the answer.	The principal that needs to be invested to provide a scholarship of $1000 per term at an annual interest rate of 8% is $50\,000$.

Note that Finance Solver cannot be used in Worked example 20 as the principal is not known. (Both PV and FV are unknowns.)

7.5 Exercise

Students, these questions are even better in jacPLUS

 Receive immediate feedback and access sample responses

 Access additional questions

 Track your results and progress

Find all this and MORE in jacPLUS

1. **WE16** Abdo has been granted a small scholarship of $12 500 to attend a sporting institute for one year. The money is invested in an annuity that pays 6% p.a., compounded quarterly. Abdo is paid $3243 per quarter from the annuity.

   a. Write a recurrence relation to describe this situation.
   b. Use your calculator to determine how much money is left in the annuity at the end of the year.
   c. Calculate the adjustment to be made to Abdo's last payment.
   d. Complete the amortisation table below for this annuity.

Payment number (n)	Payment ($)	Interest ($)	Principal reduction ($)	Balance of loan ($)
0	0.00	0.00	0.00	12 500
1	3243	**A**	3055.50	9444.50
2	3243	141.67	**B**	6343.17
3	3243	95.15	3147.85	**C**
4	**D**	47.93	3195.07	0

2. **MC** Maya has decided to study full time for 5 months. She has managed to save $20 000 and deposits this in an annuity fund, which pays 5.5% p.a. compounded monthly. Her payment from this fund each month is $3500.
   At the end of 5 months, the money left in her account will be:

   A. $5843.76   **B.** $4213.23   **C.** $3156.98   **D.** $2801.38   **E.** $0

3. **WE17** Alice invests $160 000 in an annuity paying 9.5% p.a., compounding monthly. Using Finance Solver on your calculator, determine:

   a. how long this annuity will last if the monthly payment is $6000
   b. how much Alice should receive each month if she wishes the annuity to last for 30 years.

4. **MC** Genevieve wants to invest her inheritance from her grandmother of $100 000 in an annuity that offers an interest rate of 4.5% p.a. compounded quarterly.
   If she wishes to take a quarterly payment of $3000, this annuity will last for:

   A. 42 months.   **B.** 54 months.   **C.** 33 months.   **D.** 26 months.   **E.** 13 months.

5. **WE18** **MC** Chris wants to invest $150 000 in a bond that offers a long-term guaranteed interest rate of 5% p.a. If the interest is calculated once a year, the annual interest earned is:

   A. $142 500   **B.** $7500   **C.** $130 000   **D.** $6500   **E.** $8000

6. **MC** Freda chose to invest $95 000 in a bond that offers a long-term guaranteed interest rate of 3.8% p.a. If the interest is calculated once a year, then the annual interest earned is:

   A. $25 000   **B.** $4845   **C.** $91 390   **D.** $3610   **E.** $5430

7. **WE19** A charity has $75 000 to set up a perpetuity as a grant to help homeless people.
   The charity invests in bonds that return 4.8% p.a. compounded annually.

   a. Determine the amount of the annual grant.
   b. Calculate the interest rate (compounded annually) required if the perpetuity is to provide $4800 each year.

   The charity wants to investigate other possible arrangements for the structure of the grant.

   c. Determine how much extra the annual grant would amount to if the original interest rate was compounded monthly.
   d. Calculate the interest rate (compounded monthly) required to provide 4 equal payments of $1200 every 3 months. Give your answer correct to 2 decimal places.

8. A family wants to use $125 000 to set up a perpetuity as a scholarship fund at their old school.
   The charity invests in bonds that return 5.7% p.a. compounded annually.

   a. Determine the amount of the annual grant.
   b. Calculate the interest rate (compounded annually) required if the perpetuity is to provide $8000 each year.

   The charity wants to investigate other possible arrangements for the structure of the grant.

   c. Determine how much extra the annual grant would amount to if the original interest rate was compounded monthly.
   d. Calculate the interest rate (compounded monthly) required to provide 4 equal payments of $2000 every 3 months. Give your answer correct to 2 decimal places.

9. **WE20** A benefactor of a school had been approached to provide a Year 12 scholarship of $5000 per term. They are able to get a financial institution to offer a long-term interest rate of 6.25% per annum. Calculate the principal that needs to be invested.

10. A benefactor of a college had been approached to provide a Year 11 scholarship of $2000 per term. They are able to get a financial institution to offer a long-term interest rate of 7.5% per annum. Calculate the principal that needs to be invested.

11. **MC** The owner of a technology company uses part of their wealth to set up research grants that help young Australian scientists with their endeavours. They invest $350 000 in a bond that offers a long-term guarantee of 5% p.a.

If the interest is calculated once a year, then the annual amount provided as a research grant will be:

A. $70 000 **B.** $17 500 **C.** $1750 **D.** $7000 **E.** $3500

12. Use the perpetuity formula to calculate the annual payment as specified in each of the following situations.

a. $400 000 invested at 4% p.a., paid once a year

b. $300 000 invested at 1% per quarter, paid 4 times each year

13. Use the perpetuity formula to calculate the annual payment as specified in each of the following situations.

a. $100 000 invested at 12% p.a., calculated monthly, paid out monthly

b. $2 million invested at 6% p.a., compounded quarterly, paid out every 3 months

14. Check your answers to questions 12 and 13 using Finance Solver on your CAS.

15. An AFL club has $80 000 to set up a perpetuity as a grant for the local senior sporting clubs. The club invests in bonds that return 4% p.a. compounded annually.

a. Determine the amount of the annual grant.

b. Calculate the interest rate (compounded annually) required if the perpetuity is to provide $5000 each year.

The club wishes to investigate other possible arrangements for the structure of the grant.

c. Determine how much extra the annual grant would amount to if the original interest rate was compounded quarterly.

d. Calculate the interest rate (compounded monthly) required to provide 12 equal monthly payments of $400.

16. Prachi invested $15 000 in a fund that will earn on average 10.5% p.a. over a 1-year period, with interest calculated monthly. On top of the initial investment, Prachi also contributes $250 at the start of each month. Complete the table to determine the value of his investment at the end of the sixth month.

Time period (n)	Principal ($)	Interest earned ($)	Balance ($)
1	15 000		
2			
3			
4			
5			
6			

17. Use the perpetuity formula to:

a. calculate the interest rate (p.a.) required for each of the following perpetuities.

   i. $400 000 provides $5000 per annum, with interest compounded annually

   ii. Half a million dollars provides $1000 each month, with interest compounded monthly

   iii. $800 000 provides $30 000 every six months, with interest compounded biannually

   iv. $100 000 provides $200 per fortnight, with interest calculated fortnightly

b. Check your answers to parts i–iv in a using Finance Solver on your CAS.

18. Use Finance Solver on your CAS to calculate the interest rate required for each of the following perpetuities. Give your answers correct to 2 decimal places.

   a. $400 000 provides $5000 per annum with interest compounded monthly.
   b. Half a million dollars provides $1000 each month with interest compounded annually.
   c. $800 000 provides $30 000 every six months with interest compounded quarterly.
   d. $100 000 provides $200 per fortnight with interest compounded monthly.

19. A benefactor of a college has been approached to provide a Year 9 scholarship of $200 per month. They are able to get a financial institution to offer a long-term interest rate of 3.6% per annum, compounded monthly.

   a. Calculate the principal that needs to be invested.
   b. **MC** The total amount given by the perpetuity over a 50-year period is:

   A. $10 000   B. $2400   C. $57 000   D. $66 700   E. $120 000

20. Use the perpetuity formula to calculate the initial sum to be invested in a perpetuity specified as follows:

   a. A $1200 per annum grant from a fund offering 6% p.a. compounded annually
   b. A $10 000 per annum grant from a fund offering 4.5% p.a. compounded annually
   c. A $300 per month scholarship from a fund offering 0.5% per month
   d. A $120 per month grant from a fund offering 3% p.a. compounded monthly

7.5 Exam questions

Question 1 (3 marks)

Source: VCE 2020, Further Mathematics Exam 2, Section A, Q10; © VCAA.

Samuel now invests $500 000 in an annuity from which he receives a regular monthly payment. The balance of the annuity, in dollars, after n months, A_n, can be modelled by a recurrence relation of the form

$$A_0 = 500\,000, \qquad A_{n+1} = kA_n - 2000$$

   a. Calculate the balance of this annuity after two months if $k = 1.0024$ **(1 mark)**
   b. Calculate the annual compound interest rate percentage for this annuity if $k = 1.0024$ **(1 mark)**
   c. For what value of k would this investment act as a simple perpetuity? **(1 mark)**

Question 2 (1 mark)

Source: VCE 2020, Further Mathematics Exam 1, Section A, Q25; © VCAA.

MC The graph below represents the value of an annuity investment, A_n, in dollars, after n time periods.

A recurrence relation that could match this graphical representation is

   A. $A_0 = 200\,000$, $A_{n+1} = 1.015A_n - 2500$
   B. $A_0 = 200\,000$, $A_{n+1} = 1.025A_n - 5000$
   C. $A_0 = 200\,000$, $A_{n+1} = 1.03A_n - 5500$
   D. $A_0 = 200\,000$, $A_{n+1} = 1.04A_n - 6000$
   E. $A_0 = 200\,000$, $A_{n+1} = 1.05A_n - 8000$

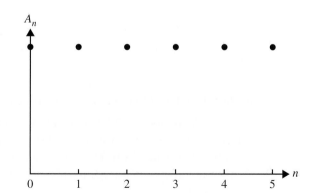

Source: VCE 2016, Further Mathematics Exam 1, Section A, Q24; © VCAA.

MC Mai invests in an annuity that earns interest at the rate of 5.2% per annum compounding monthly.

Monthly payments are received from the annuity.

The balance of the annuity will be $130 784.93 after five years.

The balance of the annuity will be $66 992.27 after 10 years.

The monthly payment that Mai receives from the annuity is closest to

- **A.** $1270
- **B.** $1400
- **C.** $1500
- **D.** $2480
- **E.** $3460

More exam questions are available online.

7.6 Annuity investments

LEARNING INTENTION

At the end of this subtopic you should be able to:
- model annuity investments using first-order linear recurrence relations.

7.6.1 A savings plan

A savings plan, like a Christmas Club account, is an investment where an initial sum as well as regular deposits are made. The interest earned is calculated regularly on the balance of the investment, which increases with each regular deposit (annuity). This is similar to reducing balance loans, with the main difference being that it is an investment rather than a loan, so the principal amount is *growing*.

An **annuity investment** is an investment that has regular deposits made over a period of time. The annuity investment can be modelled using a recurrence relation.

> ### Annuity investments
>
> $$V_0 = a, \ V_{n+1} = RV_n + d$$
>
> **where**
>
> $V_0 = $ **initial investment**
>
> $V_{n+1} = $ **amount after $n+1$ payments**
>
> $V_n = $ **amount at time n**
>
> $R = 1 + \dfrac{r}{100}$
>
> $r = $ **interest rate per period**
>
> $d = $ **deposit amount.**

Notice that the only difference to the recurrence relation for a reducing balance loan is that d is positive rather than negative.

WORKED EXAMPLE 21 Determining the balance of the investment

An initial amount of $1000 was invested in an account for 5 years at a rate of 5.04% p.a. (interest calculated monthly), and an additional deposit of $100 was made each month. Use a recurrence relation to determine the balance of the investment at the end of 5 months.

THINK	WRITE
1. State the initial values for V_0, r, and d.	$V_0 = 1000$ $r = \dfrac{5.04}{12}$ $= 0.42$ $R = 1 + \dfrac{0.42}{100}$ $= 1.0042$ $d = 100$
2. Using the iterative process on your calculator, type 1000 and then press ENTER/EXE. Type $\times 1.0042 + 100$ and press ENTER/EXE five times.	$\begin{array}{ll} 1000 & 1000 \\ 1000 \times 1.0042 + 100 & 1104.20 \\ 1104.20 \times 1.0042 + 100 & 1208.84 \\ 1208.84 \times 1.0042 + 100 & 1313.91 \\ 1313.91 \times 1.0042 + 100 & 1419.43 \\ 1419.43 \times 1.0042 + 100 & 1525.39 \end{array}$
3. Write the answer.	After 5 months, the balance of the investment will be $1525.39.

An amortisation table can be completed for the annuity investment example in Worked example 21.

Time period (n)	Payments ($)	Interest ($)	Principal increase ($)	Balance of investment ($)
0	0	0	0	1000
1	100	$1000 \times 0.0042 = 4.20$	104.20	1104.20
2	100	$1104.20 \times 0.0042 = 4.64$	104.64	1208.84
3	100	$1208.84 \times 0.0042 = 5.08$	105.08	1313.91
4	100	$1313.92 \times 0.0042 = 5.52$	105.52	1419.43
5	100	$1419.43 \times 0.0042 = 5.96$	105.96	1525.39

The Finance Solver application on your CAS calculator can be used to solve annuity investment problems.

N	The total number of payments
I%	The interest rate per annum
PV	The present value of the investment or loan – **this is negative for an annuity investment**
PMT (Pmt)	The periodic payment to the bank – **this is negative for an annuity investment**
FV	The future value of the investment – **this is positive for an annuity investment**
P/Y (PpY)	The number of payment periods in a year
C/Y (CpY)	The number of compounding periods in a year (nearly always the same as P/Y)

tlvd-4007

WORKED EXAMPLE 22 Investigating the effect of increasing deposits

Helen currently has $2000 in a savings account that is averaging an interest rate of 8% p.a. compounded annually. She wants to calculate the amount that she will receive in 5 years when she plans to go on an overseas trip.

a. If she deposits $6000 each year, calculate (correct to the nearest $1000) the amount available for her overseas trip.

b. If she places her $2000 and increases her deposits to $7000 each year into a different savings account that can offer 9% p.a. compounded annually, calculate (correct to the nearest $1000) the amount available for her overseas trip.

c. Calculate the extra amount saved by investing $7000 each year at 9% p.a. compared with $6000 each year at 8% p.a.

THINK

a. 1. Use Finance Solver and enter the values as follows.

N	5
I (%)	8
PV	−2000
Pmt	−6000
FV	? place the cursor in this cell.
PpY	1 (one payment per year)
CpY	1

2. Solve for FV.
3. Write the answer, rounding correctly.

b. 1. Using Finance Solver, enter the values as follows.

N	5
I%	9
PV	−2000
Pmt	−7000
FV	? place the cursor in this cell.
PpY	1 (one payment per year)
CpY	1

2. Solve for FV.

3. Write the answer, rounding correctly.

c. The extra amount saved is the difference between the amounts found in parts a and b.

WRITE

a. $N = 5$
$r = 8\%$
Money invested $= \$2000$
Payment per year $= \$6000$

FV = 38 138.26
In 5 years' time, the investment will be $38 000, correct to the nearest 1000.

b. $N = 5$
$r = 9\%$
Money invested $= \$2000$
Payment per year $= \$7000$

The future value of the investment is $44 970.22.

The final balance of the investment after 5 years, if Helen deposits $7000 each year into an account offering 9% p.a., would be $45 000, correct to the nearest $1000.

c. The extra amount is:
$45 000 − $38 000 = $7000

7.6.2 Planning for retirement

When do you want to retire, and how many years will you want to spend in retirement? How much money will you need? The average life expectancy in Australia is 82.90 years. So, most people will need to plan on 20 to 25 years in retirement. A common retirement income target is 60 to 65 per cent of your pre-retirement income, bearing in mind that you won't have to pay for expenses such as commuting, work clothes and, hopefully, mortgage repayments by then.

Furthermore, you may be entitled to a part pension and/or tax concessions. So, if you earn $60 000 a year now, you may think of a retirement income of about $36 000 per year. Remember, these figures are in today's dollars.

Planning for retirement is an issue that you'll need to revise regularly, maybe with a financial planner. The annuities formula and Finance Solver can be used to calculate how much money is needed under different financial situations.

WORKED EXAMPLE 23 Investing for retirement

Andrew is aged 45 and is planning to retire at 65 years of age. He estimates that he needs $480 000 to provide for his retirement. His current superannuation fund has a balance of $60 000 and is delivering 7% p.a. compounded monthly.
a. Determine the monthly contributions needed to meet the retirement lump sum target.
b. If in the final ten years before retirement, Andrew doubles his monthly contribution calculated in part a. Determine the new lump sum amount available for retirement.
c. Determine how much extra Andrew could expect if the interest rate from part b is increased to 9% p.a. (for the final 10 years) compounded monthly. Round the answer correct to the nearest $1000.

THINK

a. 1. Using Finance Solver, enter the values as follows.

N	240
I%	7
PV	−60 000
Pmt	? place the cursor in this cell.
FV	480 000
PpY	12
CpY	12

2. Solve for Pmt.
3. Write the answer.

WRITE

a. $N = 20 \times 12$
$\quad = 240$
$r = 7\%$
Money invested $= \$60\,000$
Future value $= \$480\,000$

Payment $= \$456.26$

The monthly contribution to achieve a retirement lump sum of $480 000 is $456.26.

b. 1. We need to calculate the balance after the first 10 years with $d = \$456.26$ and N = 120. Enter the values into Finance Solver and solve for FV.

N	120
I%	9
PV	−2000
Pmt	−7000
FV	**? place the cursor in this cell.**
PpY	1 (one payment per year)
CpY	1

2. Solve for FV.

3. We need to calculate the balance for the second 10 years, with double monthly contributions. Enter the values into Finance Solver and solve for FV.

N	120
I%	7
PV	−199 550.58
Pmt	−912.52
FV	**? place the cursor in this cell.**
PpY	12
CpY	12

4. Solve for FV.
5. Write the answer.

c. 1. Enter the values into Finance Solver and solve for FV.

N	120
I%	9
PV	−199 550.58
Pmt	−912.52
FV	**? place the cursor in this cell.**
PpY	12
CpY	12

2. Solve for FV.
3. Calculate the difference in answers.

4. Write the answer, rounding appropriately.

b. N = 10×12
$\quad = 120$
$r = 7\%$
Money invested = $60 000
Monthly contributions = $456.26

After 10 years, the balance of the investment is $199 550.58.

Balance after 10 years is $199 550.58.
N = 10×12
$\quad = 120$
$r = 7\%$
Monthly contributions = 456.26×2
$\qquad\qquad\qquad\quad = \$912.52$
Money invested = $199 550.58

Future value = $558 972.44
New lump sum = $558 972.44

c. The balance after 10 years is $199 550.58.
N = 10×12
$\quad = 120$
Money contributed = 456.26×2
$\qquad\qquad\qquad = \$912.52$
The interest rate is 9%.

The future value = $665 755.37
The difference is:
$665\,755.37 - 558\,972.44$
$= 106\,782.93$

If the interest rate is increased to 9% for the final 10 years, Andrew could expect an extra $107 000, correct to the nearest $1000.

7.6 Exercise

1. **WE21** Clifford invested $7000 in a managed fund that will earn on average 10% p.a. over a 3-year period, with interest calculated monthly. Clifford contributes $150 monthly to his investment.

   a. Use a recurrence relation to determine the value of his investment at the end of the fifth month.
   b. Complete the following table by calculating the values of A, B, C and D to show the value of the investment at the end of each of the first 5 months.

Time period (n)	Payments ($)	Interest ($)	Principal increase ($)	Balance of investment ($)
0	0	0	0	7000
1	150	A	208.10	7208.10
2	150	59.83	B	7417.93
3	150	61.57	211.57	7629.50
4	150	C	213.32	7842.82
5	150	65.10	215.01	D

2. Pauline invested $10 000 in a fund that will earn on average 12% p.a. over a 1-year period with interest calculated monthly. If Pauline contributes $200 per month to her initial investment:

   a. express this situation in the form of a recurrence relation
   b. determine the value of her investment at the end of the sixth month.

3. An initial deposit of $5000 was made on an investment taken out over 10 years at a rate of 7.5% p.a. (interest calculated monthly) and an additional deposit of $100 is made each month.
   Complete the table below for the first three deposits and hence calculate how much interest had been earned over this time.

Time period (n)	Payments ($)	Interest ($)	Principal increase ($)	Balance of investment ($)
0	0	0	0	5000
1	100			
2	100			
3	100			

4. An initial deposit of $10 000 was made on an investment taken out over 7 years at a rate of 8% p.a. (interest calculated monthly) and an additional deposit of $150 is made each month.
   Complete the last row of the table below and hence determine the value of the investment at the end of the fourth month.

Time period ($)	Payments ($)	Interest ($)	Principal increase ($)	Balance of investment ($)
0	0	0	0	10 000
1	150	66.67	216.67	10 216.67
2	150	68.04	218.04	10 434.71
3	150	69.58	219.58	10 654.27
4				

5. **WE22** Sharyn has $2500 in a savings account that is averaging an interest rate of 6.5% p.a. compounded annually. She wants to calculate the amount she will receive in 5 years when she plans to buy a car.
   a. If she deposits $6500 each year, determine (correct to the nearest $1000) the amount available for her car.
   b. If she places her $2500 and deposits $8000 each year into a different savings account that can offer 7% p.a. compounded annually, determine (correct to the nearest $1000) the amount available for her car.
   c. Calculate the extra amount she will save by investing $8000 each year at 7.5% p.a. compared with $6500 each year at 6.5% p.a.

6. Rhonda has $4000 in a savings account that is averaging an interest rate of 5.5% p.a. compounded annually. She wants to calculate the amount she will receive in 5 years when she plans to go on a holiday.
   a. If she deposits $6000 each year, determine (correct to the nearest $1000) the amount available for her holiday.
   b. If she places her $4000 and deposits $8000 each year into a different savings account that can offer 6.5% p.a. compounded annually, determine (correct to the nearest $1000) the amount available for her holiday.
   c. Calculate the extra amount saved by investing $8000 each year at 6.5% p.a. compared with $6000 each year at 5.5% p.a.

7. An initial deposit of $20 000 was made on an investment taken out over 3 years at a rate of 6% p.a. (interest calculated monthly), and an additional deposit of $200 is made each month.
   Complete the table below for the first five deposits and calculate the amount of interest that was earned:
   a. in the second month
   b. in the fifth month
   c. over the first five months.

Time period (n)	Payments ($)	Interest ($)	Principal increase ($)	Balance of investment ($)
0	0	0	0	20 000
1	200			
2	200			
3	200			
4	200			
5	200			

8. Barbara currently has $60 000 in an investment account that is averaging an interest rate of 6% p.a., compounded annually. She wants to calculate the amount that she will receive after 20 years.
   a. If she deposits $9000 each year, determine (correct to the nearest $1000) the amount available for her after 20 years.
   b. If she places her $60 000 and increases her deposits to $10 000 each year into a different savings account that can offer 8% p.a. compounded annually, determine (correct to the nearest $1000) the amount available for her after 20 years.
   c. Calculate the extra amount she will save by investing $10 000 each year at 8% p.a. compared with $9000 each year at 6% p.a.

9. **WE23** Peter is 48 and is planning to retire at 65 years of age. He estimates that he needs $550 000 to retire at 65. His current superannuation fund has a balance of $55 000 and is delivering 8% p.a. compounded monthly.
   a. Determine the monthly contributions needed to meet the retirement lump sum target.
   b. If in the final ten years before retirement Peter doubles his monthly contribution calculated in part a, determine the new lump sum amount available to him for retirement.
   c. How much extra could Peter expect if the interest rate from part b is increased to 10% p.a. (for the final 10 years) compounded monthly? Round the answer correct to the nearest $1000.

10. Patricia is 54 and is planning to retire at 65 years of age. She estimates that she needs $600 000 to retire at 65. Her current superannuation fund has a balance of $115 000 and is delivering 9.5% p.a. compounded monthly.

   a. Determine the monthly contributions needed to meet Patricia's retirement lump sum target.
   b. If in the final five years before retirement Patricia doubles her monthly contribution calculated from part a, determine the new lump sum amount available for retirement.
   c. Determine how much extra Patricia could expect if the interest rate from part b is increased to 11% p.a. (for the final 5 years) compounded monthly. Round the answer correct to the nearest $1000.

11. Justin is aged 38 and is planning to retire at 60 years of age. He estimates that he needs $680 000 to provide for his retirement. His current superannuation fund has a balance of $40 000 and is delivering 5% p.a. compounded monthly.

   a. Determine the monthly contributions needed to meet the retirement lump sum target.
   b. If in the final ten years before retirement Justin doubles his monthly contribution calculated from part a, determine the new lump sum amount target for his retirement.
   c. Determine how much extra Justin could expect if the interest rate from part b is increased to 8% p.a. compounded monthly (for the final 10 years). Round the answer correct to the nearest $1000.

12. Mr Rookie is aged 25 and is planning to retire at 55 years of age. He estimates that he needs $880 000 to provide for his retirement. His current superannuation fund has a balance of $600 and is delivering 7% p.a. compounded monthly.

   a. Determine the monthly contributions needed to meet the retirement lump sum target.
   b. If in the final ten years before retirement, Mr Rookie adds a further $300 to his monthly contribution calculated from part a, calculate the new lump sum for retirement.

7.6 Exam questions

Question 1 (1 mark)

Source: VCE 2017, Further Mathematics Exam 1, Section A, Core, Q23; © VCAA.

MC Four lines of an amortisation table for an annuity investment are shown below.

The interest rate for this investment remains constant, but the payment value may vary.

Payment number	Payment	Interest	Principal addition	Balance of investment
17	100.00	27.40	127.40	6977.50
18	100.00	27.91	127.91	7105.41
19	100.00	28.42	128.42	7233.83
20				7500.00

The balance of the investment after payment number 20 is $7500.

The value of payment number 20 is closest to

   A. 29   **B.** 100   **C.** 135   **D.** 237   **E.** 295

▶ **Question 2 (3 marks)**

Source: VCE 2017, Further Mathematics Exam 2, Section A, Core, Q7; © VCAA.

Alex sold his mechanics' business for $360 000 and invested this amount in a perpetuity.

The perpetuity earns interest at the rate of 5.2% per annum.

Interest is calculated and paid monthly.
  a. What monthly payment will Alex receive from this investment? **(1 mark)**
  b. Later, Alex converts the perpetuity to an annuity investment.
     This annuity investment earns interest at the rate of 3.8% per annum, compounding monthly.
     For the first four years Alex makes a further payment each month of $500 to his investment.
     This monthly payment is made immediately after the interest is added.
     After four years of these regular monthly payments, Alex increases the monthly payment.
     This new monthly payment gives Alex a balance of $500 000 in his annuity after a further two years.

     What is the value of Alex's new monthly payment?
     Round your answer to the nearest cent. **(2 marks)**

▶ **Question 3 (2 marks)**

Source: VCE 2015, Further Mathematics Exam 2, Module 4, Q3; © VCAA.

Jane and Michael decide to set up an annual music scholarship.

To fund the scholarship, they invest in a perpetuity that pays interest at a rate of 3.68% per annum.

The interest from this perpetuity is used to provide an annual $460 scholarship.
  a. Determine the minimum amount they must invest in the perpetuity to fund the scholarship. **(1 mark)**
  b. For how many years will they be able to provide the scholarship? **(1 mark)**

More exam questions are available online.

7.7 Review

7.7.1 Summary

doc-38038

7.7 Exercise

Multiple choice

1. **MC** A loan of $14 000 is taken out over 4 years at 9.75% p.a. (debited fortnightly) on the outstanding balance. The fortnightly repayment needed to repay the loan in full, to the nearest dollar, is:

   A. $135 B. $145 C. $163 D. $170 E. $319

2. **MC** Rachel repaid a reducing balance loan of $22 000 in 5 years by quarterly repayments and with interest charged quarterly at 8.2% p.a. on the outstanding balance. The total amount of interest she paid at the end of the first quarter is closest to:

   A. $1804 B. $451 C. $150 D. $9000 E. $10 000

3. **MC** The number of monthly repayments required to repay a $41 000 reducing balance loan in full, if the repayments are $588.39 and interest is debited monthly at 10.5% p.a., will be closest to:

   A. 500 B. 600 C. 90 D. 100 E. 110

4. **MC** A reducing balance loan of $56 000 is repaid by quarterly instalments of $1332.24 over 15 years at an interest rate of 5% p.a. (adjusted quarterly). If, instead, repayments of $1500 per quarter were made throughout the loan (other variables remaining unchanged), the term of the loan would be:

   A. 15 years B. 14 years C. $12\frac{3}{4}$ years D. $12\frac{1}{2}$ years E. 12 years

Questions 5 to 7 refer to the following information.

A reducing balance loan of $24 000 attracting interest at 6.5% p.a. can be repaid over 5 years by either quarterly repayments or fortnightly repayments.

5. **MC** The fortnightly repayment value is between:

   A. $120 and $140. B. $140 and $160. C. $160 and $180.
   D. $180 and $200. E. $200 and $220.

6. **MC** The quarterly repayment value is between:

   A. $1200 and $1220. B. $1250 and $1300. C. $1350 and $1360.
   D. $1400 and $1420. E. $1460 and $1470.

7. **MC** If, instead, a rival institution offered a rate of 5.5% p.a., the quarterly repayment value that would enable the loan to be repaid in full in the same time would be between:

   A. $1350 and $1375. B. $1375 and $1400. C. $1400 and $1425.
   D. $1425 and $1450. E. $1450 and $1500.

8. **MC** Elena invests $2000 in a new superannuation fund that is paying interest of 6.0% p.a. compounded monthly, with $360 monthly contributions from her employer. Select the recurrence relations used to model this situation.

A. $V_{n+1} = 1.5V_n + 360, V_0 = 2000$

B. $V_{n+1} = 1.005V_n - 360, V_0 = 2000$

C. $V_{n+1} = 1.5V_n - 360, V_0 = 2000$

D. $V_{n+1} = 1.005V_n + 360, V_0 = 2000$

E. $V_{n+1} = 1.05V_n + 360, V_0 = 2000$

9. **MC** Kerry Green borrows $305 000 to invest in an apartment. She wishes to use the bank's money to purchase the property. If the terms of an interest-only loan are 6.47% p.a. compounded monthly, the monthly repayment is closest to:

A. $1644.45

B. $19 733.50

C. $30 611.62

D. $1644.46

E. none of the above.

10. **MC** Claire is aged 48 and is planning to retire at 65. Her annual salary is $60 000 and her employer contributions are 10% of her gross monthly income. The superannuation fund has been returning an interest rate of 9.6% p.a. compounded monthly. Claire's current balance is $92 200, which she wants to grow to $800 000.

The extra amount that Claire will have to contribute each month to ensure this final payout is achieved is closest to:

A. $0

B. $1990

C. $650

D. $150

E. $240

Short answer

Questions 11–13 refer to the following information.

Helmut and Su-Li want to buy a boat, so they borrow $70 000 at 7% p.a. (adjusted monthly) on the balance outstanding and agree to repay the loan over 20 years with instalments of $542.71 per month.

11. Calculate the total amount of interest to be paid.

12. Determine the amount still owing after 10 years.

13. If the repayment value is increased to $600 per month after 10 years, calculate the term of the loan.

14. Frieda is repaying a $55 000 housing loan with interest calculated quarterly at 7% p.a. of the amount outstanding. Quarterly repayments of $1487.93 are being made to service the loan.

Twenty repayments have already been made. Frieda decides to change her repayments to $250 per fortnight. The bank responds by adjusting interest fortnightly.

Calculate the difference that these changes would make to the overall term of the loan.

15. Bert and Ernie have just finished repaying a 5-year, $8000 reducing balance loan, which they needed in order to re-carpet their home. During the first year, interest was debited monthly at 8.5% p.a.; during the next 2 years, the rate fell to 8% p.a. but rose to 8.3% p.a. for the remaining period of the loan.

Before signing their contract, Bert and Ernie were also given the option of fixing the interest rate at 8.25% p.a. (debited monthly) for the term of the loan.

By considering the total amount of interest that they paid, determine whether Bert and Ernie made the right decision in choosing the variable rather than the fixed interest rate for the loan.

16. The actors' guild provides a donation of $1.2 million to be set up as a perpetuity that earns 7.2% p.a. in interest, compounded annually.

    a. Calculate the amount of the annual grant.

    b. Calculate the interest rate compounded monthly that would provide the same annual grant.

Extended response

17. Use Finance Solver to calculate the unknown value in each of the following annuity investments.

    a. A superannuation fund is to grow from $40 000 to $400 000 in 10 years, with an expected interest rate of 7.7% p.a. compounded quarterly. Calculate the quarterly contributions needed to achieve this growth.

    b. A superannuation fund's performance is often measured by the interest rate returned on superannuation investments. Calculate the required interest rate, compounded monthly, if a current balance of $40 000 is to mature to $400 000, given there are 20 years to go with monthly contributions of $200 per month.

18. Alisha has saved $10 000 and has several options for how to spend the money when she turns 18. She chooses to investigate these options. Answer the following, recalling simple and compound interest formulas from your previous studies.

    a. Option 1 is for her to use the sum of $10 000 towards a new car valued at $27 000. The car dealership will accept $10 000 as a deposit. The remaining amount can be borrowed from the dealer at 6% p.a (simple interest) and is to be repaid in 4 years' time. Calculate the total cost of the car.

    b. Another option is to buy a secondhand car for $12 000. Alisha can borrow the extra $2000 at 7% p.a. compounded annually for 3 years with a single payment at the end of the term.

        i. Calculate the amount of interest Alisha will have to pay at the end of the 3 years for the $2000 loan.

        ii. Calculate the total cost of the car.

    c. The third option is to buy a reliable car for $5000 and use the other $5000 to buy a collectable painting, the value of which will increase with time.

        i. The value of the painting is expected to increase at a rate of 5% p.a. Calculate the value of the painting in 4 years' time.

        ii. The $5000 car is expected to depreciate at 10% p.a. using the straight line depreciation model. Determine its future value in 4 years. (Straight line depreciation means the item decreases in value at a fixed rate each year.)

    d. The other two cars, being upmarket models, depreciate at 15% p.a. on reducing value.

        i. Calculate the value of the investment in 4 years for the $27 000 car.

        ii. Calculate the value of the investment in 4 years for the $12 000 car.

        iii. State which of the three options loses the most money.

    e. Use the information in parts **a–d** to fill in the following table.

Description	$27 000 car	$12 000 car	$5000 car and painting
Cost of loan (interest charged only)			
Depreciation of goods			
Total cost			
Less benefits (for example, increase in value)			
Total cost			

19. Nathan borrowed $6000 towards purchasing a motorbike. The reducing balance loan is to be repaid in full over 2 years with quarterly repayments. The interest rate is 4.5% p.a., compounding quarterly.

    a. Calculate the quarterly repayments required.
    b. Use a first-order recurrence relation to complete an amortisation table for the loan.
    c. Calculate the total amount paid on the loan.
    d. Calculate the total interest paid on the loan.
    e. Calculate the effective annual interest rate correct to 2 decimal places.

7.7 Exam questions

Question 1 (1 mark)

Source: VCE 2021, Further Mathematics Exam 1, Section A Core, Q18; © VCAA.

MC Deepa invests $500 000 in an annuity that provides an annual payment of $44 970.55.

Interest is calculated annually.

The first five lines of the amortisation table are shown below.

Payment number	Payment ($)	Interest ($)	Principal reduction ($)	Balance ($)
0	0.00	0.00	0.00	500 000.00
1	44 970.55	20 000.00	24 970.55	475 029.45
2	44 970.55	19 001.18	25 969.37	449 060.08
3	44 970.55	17 962.40		422 051.93
4	44 970.55	16 882.08	28 088.47	393 963.46

The principal reduction associated with payment number 3 is
   A. $17 962.40   **B.** $25 969.37   **C.** $27 008.15   **D.** $28 088.47   **E.** $44 970.55

Question 2 (1 mark)

Source: VCE 2021, Further Mathematics Exam I, Section A Core, Q19; © VCAA.

MC Deepa invests $500 000 in an annuity that provides an annual payment of $44 970.55

Interest is calculated annually.

The first five lines of the amortisation table are shown below.

Payment number	Payment ($)	Interest ($)	Principal reduction ($)	Balance ($)
0	0.00	0.00	0.00	500 000.00
1	44 970.55	20 000.00	24 970.55	475 029.45
2	44 970.55	19 001.18	25 969.37	449 060.08
3	44 970.55	17 962.40		422 051.93
4	44 970.55	16 882.08	28 088.47	393 963.46

The number of years, in total, for which Deepa will receive the regular payment of $44 970.55 is closest to
   A. 12   **B.** 15   **C.** 16   **D.** 18   **E.** 20

Question 3 (3 marks)

Source: VCE 2020, Further Mathematics Exam 2, Section A, Q8; © VCAA.

Samuel has a reducing balance loan.

The first five lines of the amortisation table for Samuel's loan are shown below.

Payment number	Payment ($)	Interest ($)	Principal reduction ($)	Balance ($)
0	0.00	0.00	0.00	320 000.00
1	1600.00	960.00	640.00	319 360.00
2	1600.00	958.08	641.92	318 718.08
3	1600.00	956.15		318 074.23
4	1600.00			

Interest is calculated monthly and Samuel makes monthly payments of $1600.

Interest is charged on this loan at the rate of 3.6% per annum.

a. Using the values in the amortisation table

   i. calculate the principal reduction associated with payment number 3 **(1 mark)**

   ii. calculate the balance of the loan after payment number 4 is made.
      Round your answer to the nearest cent. **(1 mark)**

b. Let S_n be the balance of Samuel's loan after n months.
   Write down a recurrence relation, in terms of S_0, S_{n+1} and S_n, that could be used to model the
   month-to-month balance of the loan. **(1 mark)**

Question 4 (1 mark)

Source: VCE 2018, Further Mathematics Exam 1, Section A, Q23; © VCAA.

MC Five lines of an amortisation table for a reducing balance loan with monthly repayments are shown below.

Repayment number	Repayment	Interest	Principal Reduction	Balance of Loan
25	$2200.00	$972.24	$1227.76	$230 256.78
26	$2200.00	$967.08	$1232.92	$229 023.86
27	$2200.00	$961.90	$1238.10	$227 785.76
28	$2200.00	$1002.26	$1197.74	$226 588.02
29	$2200.00	$996.99	$1203.01	$225 385.01

The interest rate for this loan changed immediately before repayment number 28.

This change in interest rate is best described as

   A. an increase of 0.24% per annum.

   B. a decrease of 0.024% per annum.

   C. an increase of 0.024% per annum.

   D. a decrease of 0.0024% per annum.

   E. an increase of 0.00024% per annum.

Source: VCE 2016, Further Mathematics Exam 2, Core, Q7; © VCAA.

Ken has borrowed $70 000 to buy a new caravan.

He will be charged interest at the rate of 6.9% per annum, compounding monthly.

    a. For the first year (12 months), Ken will make monthly repayments of $800.

        i. Find the amount that Ken will owe on his loan after he has made 12 repayments. **(1 mark)**

        ii. What is the total interest that Ken will have paid after 12 repayments? **(1 mark)**

    b. After three years, Ken will make a lump sum payment of $L in order to reduce the balance of his loan.
    This lump sum payment will ensure that Ken's loan is fully repaid in a further three years.
    Ken's repayment amount remains at $800 per month and the interest rate remains at 6.9% per annum,
    compounding monthly.
    What is the value of Ken's lump sum payment, $L?
    Round your answer to the nearest dollar. **(2 marks)**

More exam questions are available online.

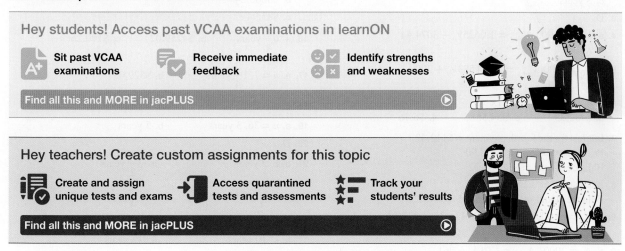

Answers

Topic 7 Modelling reducing balance loans, annuities and perpetuities using recursion

7.2 Modelling reducing balance loans with recurrence relations

7.2 Exercise

1. a. $V_0 = 565$, $V_{n+1} = 1.015V_n + 50$
   b. 743 students
   c. 59 new students
2. a. $A_0 = 3.5\,\text{m}^3$
   b. $4.15\,\text{m}^3$
   c. 3 years
3. D
4. a. $V_0 = 15\,000$, $V_{n+1} = 1.0625V_n - 3074.44$
   b. 6
5. a. $V_0 = 1500$, $V_{n+1} = 1.004\,583V_n + 200$
   b. $2543.90
6. a. 0.75%　　b. $18　　c. $974.44
   d. $482.33　　e. $54.36　　f. $6.16
7. Interest for the fourth payment = $112.42
   Principal reduction = $158.90
   Balance of loan = $24\,368.72
8. Interest = $580.15
   Balance at end of 3 payments: $8994.02
9. a. $V_0 = 23\,000$, $V_{n+1} = 1.005\,42 \times V_n - 341.54$
   b. $22\,345.83
10. a. $A_0 = 2000$, $A_{n+1} = 1.01A_n - 261$
    b. $1021.44
    c. $3.15
11. a. $15\,763.24　　b. $15\,599.06　　c. $15\,405.32
12. A
13. E

7.2 Exam questions

Note: Mark allocations are available with the fully worked solutions online.
1. B
2. E
3. A

7.3 Solving reducing balance loan problems using finance solver

7.3 Exercise

1. a. $59\,633.49　　b. $49\,884.16　　c. $32\,172.59
2. a. $46\,102.98　　b. $36\,196.88　　c. $19\,556.12
3. a. $156.39　　　　b. $253.36
4. a. $106.69　　　　b. $160.56
5. a. $365.84
   b. i. $259.59, $106.25　　ii. $339.75, $26.09

6. a. $190.83
   b. i. $153.09, $37.74　　ii. $181.36, $9.47
7. a. $n = 18$, $1\frac{1}{2}$ years　　b. $214.30
8. $n \approx 8$ months
9. $n \approx 15$ months
10. 16 months
11. a. $n \approx 16$ months　　b. $n \approx 22$ months
12. a. $n \approx 22$ months　　b. $n \approx 28$ months
13. a. $363.25　　　　b. $n \approx 55$ months
    c. $3477.50　　　d. $317.50
14. a. $225.49　　　　b. $n \approx 118$ fortnights
    c. $3910　　　　d. $403.70
15. a. $n \approx 32$ quarters $(8\ \text{years})$
    b. $6066.24
    c. $83.94
16. a. $6\frac{3}{4}$ years　　b. $4259.02　　c. $190.26
17. a. $n = 7$, $1\frac{3}{4}$ years　　b. $529.03
18. a. $n = 24$, 2 years　　b. 5 years
19. a. $n = 36$, 3 years　　b. 5 years
20. D
21. a. i. 2 years　　ii. $3\frac{1}{2}$ years
    b. i. $\frac{3}{4}$ year　　ii. $1\frac{1}{4}$ years
22. a. $384.65
    b. i. 10 years　　ii. 13 years
       iii. 23 years　　iv. $24\frac{1}{2}$ years

	i	ii	iii	iv
c.	$19\,356.40	$25\,867.48	$49\,870.60	$53\,752.60
d.	$22\,959.60 less	$16\,448.52 less	$7554.60 more	$11\,436.60 more

23.

	i	ii	iii
a.	9 years	$9433.28	$585.92
b.	8 years	$8861.68	$1157.52

24. D
25. E
26. a. $21\,164.60　　b. $441.40

7.3 Exam questions

Note: Mark allocations are available with the fully worked solutions online.
1. B
2. C
3. C

7.4 The effect of rate and repayment changes on reducing balance loans

7.4 Exercise

1. a. i. Term of loan $= 5$ years
   ii. $217.37
   b. i. Term of loan $= 4$ years 25 fortnights
   ii. $149.63
   c. i. $4176 ii. $4025.27 iii. $3984.35

2. a. i. Term of loan $= 6$ years
   ii. $246.17
   b. i. Term of loan $= 5$ years 25 fortnights
   ii. $164.36
   c. i. $6008.48 ii. $5825.22 iii. $5775.40

3. a. $5496.40
   b. i. Term $= 5$ years and 2 months
   ii. $6294.91

4. a. $2834.88
   b. i. Term $= 4$ years and 2 months
   ii. $3286.88

5. a. $22 125.28 b. $10 \frac{1}{4}$ years c. $1070.37

6. a. $19 156.45
   b. Term $= 8$ years $+ 2$ months
   c. $660.18 more

7. a. $145
   b. Jodie made a loss of $220.

8. a. $312.50 b. Profit $= \$1750$

9. a. $523.66
   b. See the table at the foot of the page.*
   c. $514.00
   d. $2094.62
   e. $94.62

10. a. $647.06
    b. See the table at the foot of the page.**
    c. $322.75
    d. See the table at the foot of the page.***

*9. b.

Payment number (n)	Payment ($)	Interest ($)	Principal reduction ($)	Balance of loan ($)
0	0	0	0	2000
1	523.66	37.50	486.16	1513.84
2	523.66	28.38	495.28	1018.56
3	523.66	19.10	504.56	514.00
4	523.66	9.64	514.02	−0.02

**10. b.

Payment number (n)	Payment ($)	Interest ($)	Principal reduction ($)	Balance of loan ($)
0	0	0	0	3750
1	647.06	37.50	609.56	3140.44
2	647.06	31.40	615.66	2524.78
3	647.06	25.25	621.81	1902.97
4	647.06	19.03	628.03	1274.94
5	647.06	12.75	634.31	640.63
6	647.06	6.41	640.65	−0.02

***10. d.

Payment number (n)	Payment ($)	Interest ($)	Principal reduction ($)	Balance of loan ($)
0	0	0	0	3750
1	322.75	18.75	304	3446.00
2	322.75	17.23	305.52	3140.48
3	322.75	15.70	307.05	2833.43
4	322.75	14.17	308.58	2524.85
5	322.75	12.62	310.13	2214.71
6	322.75	11.07	311.68	1903.05
7	322.75	9.52	313.23	1589.81
8	322.75	7.95	314.80	1275.01
9	322.75	6.38	316.37	958.64
10	322.75	4.79	317.96	640.68
11	322.75	3.20	319.55	321.13
12	322.75	1.61	321.14	−0.01

e. Monthly: \$122.99; bi-monthly: \$132.34
The interest paid on the monthly payments is almost \$10 less than the interest paid on the bi-monthly payments.

11. a. 10 years, $A_{119} = \$102.61$
  b. 9 years 24 fortnights, $A_{257} = \$185.59$

12. See the table at the foot of the page.*

13. C

14. B

15. a. \$5414.60
  b. i. 5 years 2 months
     ii. \$6333.26
  c. i. 5 years 4 months
     ii. \$7331.64

16. A

17. C

18. a. \$3780
  b. \$3120

19. A

7.4 Exam questions

Note: Mark allocations are available with the fully worked solutions online.

1. B

2. B

3. A

7.5 Annuities and perpetuities

7.5 Exercise

1. a. $V_0 = 12\,500$, $V_{n+1} = 1.015V_n - 3243$
  b. 25 cents
  c. 25 cents would need to be added to Abdo's last payment.
  d. See the table at the foot of the page.**

2. D

3. a. 30 months   **b.** \$1345.37

4. A

5. B

6. D

7. a. \$3600   **b.** 6.4% p.a.
  c. \$3680.27   **d.** 6.37%

8. a. \$7125   **b.** 6.4% p.a.
  c. \$7314.12   **d.** 6.37%

9. \$320\,000

10. \$106\,666.67

11. B

12. a. \$16\,000 per year
  b. \$3000 per quarter

13. a. \$1000 per month
  b. \$30\,000 per quarter

14. (Q12)

  a. N: = 1   **b.** N: = 1
  I%: = 4   I%: = 4
  PV: = −400\,000   PV: = −300\,000
  Pmt: = 16\,000   Pmt: = 3000
  FV: = 400\,000   FV: = 300\,000
  PpY: = 1   PpY: = 4
  CpY: = 1   CpY: = 4
  PmtAt: = END   PmtAt: = END

(Q13)

  c. N: = 1   **d.** N: = 1
  I%: = 12   I%: = 6
  PV: = −100\,000   PV: = −2\,000\,000
  Pmt: = 1000   Pmt: = 30\,000
  FV: = 100\,000   FV: = 2\,000\,000
  PpY: = 12   PpY: = 4
  CpY: = 12   CpY: = 4
  PmtAt: = END   PmtAt: = END

15. a. \$3200 per year   **b.** 6.25%
  c. \$48.32   **d.** 6%

16. \$17\,088.13

17. a. i. 1.25%   **ii.** 2.40%   **iii.** 7.50%   **iv.** 5.20%

---

*12.

	Frequency	Total interest (\$)	Saving (\$)
a	Half-yearly	6014.80	–
b	Quarterly	5707.11	\$307.69 saving on half-yearly
c	Monthly	5496.63	\$518.17 saving on half-yearly
d	Fornightly	5440.38	\$574.42 saving on half-yearly

**1. d.

Payment number (n)	Payment (\$)	Interest (\$)	Principal reduction (\$)	Balance of loan (\$)
0	0.00	0.00	0.00	12\,500
1	3243	**A** = 187.50	3055.50	9444.50
2	3243	141.67	**B** = 3101.33	6343.17
3	3243	95.15	3147.85	**C** = 3195.32
4	**D** = 3243.25	47.93	3195.07	**0**

b. **i.** N: $= 1$
I%: $= 1.25$
PV: $= -400\,000$
Pmt: $= 5000$
FV: $= 400\,000$
PpY: $= 1$
CpY: $= 1$
PmtAt: $=$ END

ii. N: $= 1$
I%: $= 2.4$
PV: $= -500\,000$
Pmt: $= 1000$
FV: $= 500\,000$
PpY: $= 12$
CpY: $= 12$
PmtAt: $=$ END

iii. N: $= 1$
I%: $= 7.5$
PV: $= -800\,000$
Pmt: $= 30\,000$
FV: $= 800\,000$
PpY: $= 2$
CpY: $= 2$
PmtAt: $=$ END

iv. N: $= 1$
I%: $= 5.2$
PV: $= -100\,000$
Pmt: $= 200$
FV: $= 100\,000$
PpY: $= 26$
CpY: $= 26$
PmtAt: $=$ END

18. a. N: $= 1$
I%: $= 1.242\,895\,218$
PV: $= -400\,000$
Pmt: $= 5000$
FV: $= 400\,000$
PpY: $= 1$
CpY: $= 12$
PmtAt: $=$ END
$r = 1.24\%$ p.a.

b. N: $= 1$
I%: $= 2.426\,576\,795$
PV: $= -500\,000$
mt: $= 1000$
FV: $= 500\,000$
PpY: $= 12$
CpY: $= 1$
PmtAt $:=$ END
$r := 2.43\%$ p.a.

c. N: $= 1$
I%: $= 7.430\,975\,749$
PV: $= -800\,000$
Pmt: $= 30\,000$
FV: $= 800\,000$
PpY: $= 2$
CpY: $= 4$
PmtAt: $=$ END
$r: = 7.43\%$ p.a.

d. N: $= 1$
I%: $= 5.206\,067\,34$
PV: $= -100\,000$
Pmt: $= 200$
FV: $= 100\,000$
PpY: $= 26$
CpY: $= 12$
PmtAt: $=$ END
$r: = 5.21\%$ p.a.

19. a. $\$66\,666.67$ **b.** E

20. a. $\$20\,000$ **b.** $\$222\,222.22$
c. $\$60\,000$ **d.** $\$48\,000$

7.5 Exam questions

Note: Mark allocations are available with the fully worked solutions online.

1. a. $\$498\,398.08$
b. 2.88%
c. $k = 1.004\%$
2. B
3. C

7.6 Annuity investments

7.6 Exercise

1. a. $\$8057.92$
b. A. $= \$58.10$ B. $= \$209.83$
    C. $= \$63.32$ D. $= \$8057.83$
2. a. $V_0 = 10\,000$, $V_{n+1} = 1.01V_n + 200$
b. $\$11\,845.60$
3. $\$96.22$
4. $\$10\,875.23$
5. a. $\$40\,000$ **b.** $\$50\,000$ **c.** $\$10\,000$

6. a. $\$39\,000$ **b.** $\$51\,000$ **c.** $\$12\,000$
7. a. $\$101.50$ **b.** $\$106.05$ **c.** $\$515.08$
8. a. $\$523\,000$ **b.** $\$737\,000$ **c.** $\$214\,000$
9. a. $\$779.70$ **b.** $\$692\,647.66$ **c.** $\$124\,000$
10. a. $d = \$1185.74$ **b.** $\$690\,610.68$ **c.** $\$47\,000$
11. a. $\$1168.46$ **b.** $\$861\,442.14$ **c.** $\$238\,000$
12. a. $\$717.34$ **b.** $\$931\,925.99$

7.6 Exam questions

Note: Mark allocations are available with the fully worked solutions online.
1. D
2. a. $\$1560$
b. After 4 years, Alex's investment grows to $\$444\,872.94$. To grow to $\$500\,000$ in a further two years, Alex's new monthly payment will be $\$805.65$.
3. a. $\$12\,500$
b. An infinite number of years

7.7 Review

7.7 Exercise

Multiple choice

1. C
2. B
3. E
4. C
5. E
6. D
7. B
8. D
9. D
10. D

Short answer

11. $\$60\,250.40$
12. $\$46\,741.44$
13. 18 years and 8 months or 19 years (to the nearest whole year)
14. If changes not made, term $= 15$ years. With changes, term $= 13$ years 20 fortnights.
So difference in term $= 1$ year 6 fortnights.
15. Fixed: interest $= \$1790.20$
Variable: interest $= \$1785.37$, so correct decision is variable, $\$4.83$ better off.
16. a. $\$86\,400$ **b.** 6.97%

Extended response

17. a. $\$5287.59$ **b.** 9.39% p.a.
18. a. $\$31\,080$
b. i. $\$450.09$ **ii.** $\$12\,450.09$
c. i. $\$6077.53$ **ii.** $\$3000$
d. i. $\$14\,094.17$
    ii. $\$6264.08$
    iii. The $\$27\,000$ car option

e. See the table at the foot of the page.*

19. a. $788.46

   b. See the table at the foot of the page.**

   c. $6307.68

   d. $307.68

   e. 4.58% (correct to 2 decimal places)

7.7 Exam questions

Note: Mark allocations are available with the fully worked solutions online.

1. C

2. B

3. a. i. $643.85

   ii. $317 428.45

   b. $S_0 = 320\,000, S_{n+1} = 1.003 \times S_n - 1600$

4. A

5. a. i. $65 076.22

   ii. $4676.22

   b. $28 204

*18. e.

Description	$27 000 car	$12 000 car	$5000 car and painting
Cost of loan (interest charged only)	4080.00	450.09	—
Depreciation of goods	12 905.83	5735.92	2000.00
Total cost	31 080.00	12 450.09	10 000.00
Less benefits (for example, increase in values)	—	—	1077.53
Total cost	31 080.00	12 450.09	8922.47

**19. b.

Payment number (n)	Payment ($)	Interest ($)	Principal reduction ($)	Balance of loan ($)
0	0	0	0	6000
1	788.46	67.5	720.96	5279.04
2	788.46	59.39	729.07	4549.97
3	788.46	51.19	737.27	3812.70
4	788.46	42.89	745.57	3067.13
5	788.46	34.51	753.95	2313.17
6	788.46	26.02	762.44	1550.74
7	788.46	17.45	771.01	779.72
8	788.46	8.77	779.69	0.035

8 Matrices

Fully worked solutions for this topic are available online.

8.1 Overview

8.1.1 Introduction

Matrices were being used by ancient civilisations as far back as 300 BC. There is evidence in the form of clay tablets and texts that both the Chinese and the Babylonians used matrix methods to study and solve simultaneous linear equations. For centuries, mathematicians used matrices to solve problems, but it was not until the nineteenth century that the English mathematician Arthur Cayley published a formal definition of a matrix in his *Memoir on the Theory of Matrices* (1858), thus establishing matrices as a specific branch of mathematics. Matrices are used extensively in engineering, physics, geology, computer science, graphic design, economics, computer games and many other areas of applied mathematics. Computer programmers need an understanding of

matrices to create the extraordinary 3D visual effects in movie blockbusters and realistic computers games that are now expected in this genre of movies and games. Matrices are used to encrypt code and hence play a vital role in the security of websites and databases.

KEY CONCEPTS

This topic covers the following key concepts from the VCE Mathematics Study Design:
- matrix arithmetic: the order of a matrix, types of matrices (row, column, square, diagonal, symmetric, triangular, zero, binary and identity), the transpose of a matrix, and elementary matrix operations (sum, difference, multiplication of a scalar, product and power)
- inverse of a matrix, its determinant, and the condition for a matrix to have an inverse
- use of matrices to represent numerical information presented in tabular form, and the use of a rule for the a_{ij} element of a matrix to construct the matrix
- binary and permutation matrices, and their properties and applications
- communication and dominance matrices and their use in analysing communication systems and ranking players in round-robin tournaments.
- use of the matrix recurrence relation: S_0 = initial state matrix, $S_{n+1} = TS_n$ or $S_{n+1} = LS_n$ where T is a transition matrix, L is a Leslie matrix, and S_n is a column state matrix, to generate a sequence of state matrices (assuming the next state only relies on the current state)
- informal identification of the equilibrium state matrix in the case of regular transition matrices (no noticeable change from one state matrix to the next state matrix)
- use of transition diagrams, their associated transition matrices and state matrices to model the transitions between states in discrete dynamical situations and their application to model-and-analyse practical situations such as the modelling and analysis of an insect population comprising eggs, juveniles and adults
- use of the matrix recurrence relation S_0 = initial state matrix, $S_{n+1} = TS_n + B$ to extend modelling to populations that include culling and restocking.

Source: VCE Mathematics Study Design (2023–2027) extracts © VCAA; reproduced by permission.

8.2 Matrix representation

LEARNING INTENTION

At the end of this subtopic you should be able to:
- identify the order of a matrix
- classify types of matrices (row, column, square, diagonal, symmetric, triangular, zero, binary)
- generate matrices to store data.

8.2.1 Types of matrices

A **matrix** (plural *matrices*) is a rectangular array of numbers arranged in rows and columns. The numbers in a matrix are called the **elements** of the matrix.

Order and elements of a matrix

The matrix shown here has 3 *rows* and 2 *columns*. We say that it is a 3×2 **rectangular matrix** and its **order** is 3 by 2.

$$A = \begin{bmatrix} 2 & 3 \\ -1 & 1 \\ 0 & 4 \end{bmatrix}$$

Capital letters will be used to represent matrices.

Order of a matrix

In general, a matrix with m rows and n columns is known as an $m \times n$ matrix.

The elements in a matrix are referred to by the row and then by the column position.

The element in the second row and the first column of matrix A above is -1. This is represented as $a_{21} = -1$ or the 2, 1 element.

The elements of a matrix

In general, the elements of matrix A are referred to as a_{ij} where i refers to the row position and j refers to the column position.

$$\text{That is, } A = \begin{bmatrix} a_{11} & a_{12} & a_{13} & a_{14} & \cdots \\ a_{21} & a_{22} & a_{23} & a_{24} & \cdots \\ a_{31} & a_{32} & a_{33} & a_{34} & \cdots \\ \cdots & \cdots & \cdots & \cdots & a_{mn} \end{bmatrix} \text{ for the matrix of order } m \times n.$$

The 2, 1 element is a_{21} and the 3, 2 element is a_{32}.

Matrix types

A matrix with one row is called a **row matrix** or *row vector*.

$$B = \begin{bmatrix} 2 & 1 & 3 \end{bmatrix} \text{ is a } 1 \times 3 \text{ row matrix.}$$

A matrix with one column is called a **column matrix** or *column vector*.

$$C = \begin{bmatrix} 2 \\ 1 \\ 3 \end{bmatrix}$$ is a 3×1 column matrix.

A matrix with an equal number of rows and columns is called a **square matrix**.

$$D = \begin{bmatrix} 2 & 5 & 4 \\ 1 & 2 & -1 \\ 3 & -2 & 0 \end{bmatrix}$$ is a 3×3 square matrix.

$$E = \begin{bmatrix} 4 & 2 \\ -5 & 7 \end{bmatrix}$$ is a 2×2 square matrix.

A square matrix that has only non-zero elements on the main diagonal is called a **diagonal matrix**.

$$F = \begin{bmatrix} 7 & 0 & 0 \\ 0 & 5 & 0 \\ 0 & 0 & 6 \end{bmatrix}$$ is a 3×3 diagonal matrix.

A **zero matrix** consists only of elements that are 0.

$$O = \begin{bmatrix} 0 & 0 \\ 0 & 0 \end{bmatrix}$$ is a 2×2 zero matrix.

Two matrices are *equal* if they are of the same order and all corresponding elements are equal.

$$\overset{A}{\begin{bmatrix} 1 & 0 & 2 & 4 \\ 2 & -1 & 3 & 5 \end{bmatrix}} = \overset{B}{\begin{bmatrix} 1 & 0 & 2 & 4 \\ 2 & -1 & 3 & 5 \end{bmatrix}}$$

These two 2×4 matrices are equal, as the corresponding elements are equal.

That is, $A = B$ as $a_{11} = b_{11} = 1$, $a_{12} = b_{12} = 0$, ...

The **transpose** of a matrix can be found by switching its rows and columns.

We denote the transpose of matrix A as A^T.

If $A = \begin{bmatrix} 1 & 0 & 2 & 4 \\ 2 & -1 & 3 & 5 \end{bmatrix}$ then $A^T = \begin{bmatrix} 1 & 2 \\ 0 & -1 \\ 2 & 3 \\ 4 & 5 \end{bmatrix}$.

A **symmetrical matrix** needs to be a square matrix.

Matrix A is symmetric if $A = A^T$, where A^T is the **transpose matrix**.

The elements with respect to the main diagonal are symmetric in a symmetric matrix, so $a_{ij} = a_{ji}$ for all indices i and j.

The following matrix is a 3×3 symmetric matrix, where the black numbers represent the main diagonal.

$$\begin{bmatrix} 1 & 3 & -7 \\ 3 & 5 & 9 \\ -7 & 9 & 2 \end{bmatrix}$$

A **triangular matrix** is a type of square matrix.

A triangular matrix can be labelled a **lower triangular matrix** if all elements above the main diagonal are zero. However, if all elements below the main diagonal are zero, then it is called an **upper triangular matrix**.

The following are both 3×3 triangular matrices.

$$\begin{bmatrix} 3 & 0 & 0 \\ 8 & 5 & 0 \\ 7 & -2 & 4 \end{bmatrix} \qquad \begin{bmatrix} 1 & -3 & -1 \\ 0 & 5 & 8 \\ 0 & 0 & 2 \end{bmatrix}$$

Lower triangular matrix   Upper triangular matrix

A matrix that consists only of elements that are either 0 or 1 is known as a **binary matrix**. These are also referred to as (0, 1), Boolean or logical matrices.

WORKED EXAMPLE 1 Classifying matrices

For each of the following matrices, give the order and the appropriate name (if it can be categorised). Where possible, write the 2, 1 element and the position of the number 3 in x_{ij} form.

$$A = \begin{bmatrix} 3 & 0 & 0 \\ 0 & 5 & 0 \\ 0 & 0 & 4 \end{bmatrix}, B = \begin{bmatrix} 7 & 3 & 1 \end{bmatrix}, C = \begin{bmatrix} 0 & 1 \\ 0 & 1 \end{bmatrix}, D = \begin{bmatrix} 9 & 7 & -5 \\ 7 & -1 & 4 \\ -5 & 4 & 3 \end{bmatrix}, E = \begin{bmatrix} -2 & 12 & 3 \\ 0 & 4 & -2 \\ 0 & 0 & 6 \end{bmatrix}$$

THINK	WRITE
1. A has 3 rows and 3 columns, with the numbered elements in the main diagonal. A zero is in the 2, 1 position and a 3 is in position 1, 1.	A is a 3×3 diagonal matrix. The element in the 2, 1 position is a 0 and the number 3 is represented by a_{11}.
2. B has 1 row and 3 columns. There is no element in the 2, 1 position and 3 is in position 1, 2.	B is a 1×3 row matrix. There is no element in the 2, 1 position and the number 3 is represented by b_{12}.
3. C has 2 rows and 2 columns. All elements are 0 or 1. A zero is in the 2, 1 position and there is no number 3 in the matrix.	C is a 2×2 binary matrix. The element in the 2, 1 position is a 0 and there is no number 3 in the matrix.
4. D has 3 rows and 3 columns. The elements are symmetric about the main diagonal. A 7 is in the 2, 1 position and a 3 is in position 3, 3.	D is a 3×3 symmetrical matrix. The element in the 2, 1 position is a 7 and the number 3 is represented by d_{33}.
5. E has 3 rows and 3 columns; it is a square matrix. The elements have zeros below the main diagonal. A zero is in the 2, 1 position and a 3 is in position 1, 3.	E is a 3×3 upper triangular matrix. The element in the 2, 1 position is a 0 and the number 3 is represented by e_{13}.

8.2.2 Using matrices to store data

Matrices are useful for recording and storing bivariate data (data that depend on two categories). They can also keep track of the coefficients of systems of numbers in a simple two-dimensional format. For storing data, whether the data are organised in a row or column need not be specified; they are shown in the same manner as the original data format.

Distance between 3 local townships (kilometres)			
	Town A	**Town B**	**Town C**
Town A	0	23	17
Town B	23	0	43
Town C	17	43	0

$$\begin{bmatrix} 0 & 23 & 17 \\ 23 & 0 & 43 \\ 17 & 43 & 0 \end{bmatrix}$$

Labelling convention is not important as the matrix is simply being used to store data.

However, when performing mathematical processes using the data stored in matrices, the defining of the rows and columns must follow formal conventions. A common example where the columns and rows must follow a well-established convention is two-way frequency tables.

Two-way frequency tables follow the convention of organising explanatory variable headings in the columns and response variable headings in the rows.

		Primary	Secondary
Attitude	Fewer	5	2
	Same	29	9
	More	33	36
	Total	67	47

$$\begin{bmatrix} 5 & 2 \\ 29 & 9 \\ 33 & 36 \\ 67 & 47 \end{bmatrix}$$

Matrices are usually presented without headings or labels on the rows and columns.

WORKED EXAMPLE 2 Generating matrices

tlvd-3933

a. **Generate a matrix to show the number of major country roads between five nearby townships in the network shown.**

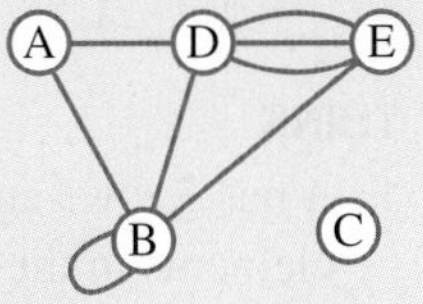

b. **Generate a matrix to represent the following contingency table for party preferences of pensioners and working-class people.**

		Social class	
		Pensioners	Working class
Party preference	Labor	18	12
	Liberal	16	11
	Total	34	23

c. **Generate a 2×2 matrix to represent the information provided in the following scenario. Omit the totals from your matrix.**
In a survey, 139 pensioners and 102 working-class people were asked whether they approved or disapproved of a proposed freeway. 37 pensioners and 79 working-class people approved of the freeway.

THINK	WRITE
a. 1. Set up a blank 5×5 matrix. This means there are 25 entries inside the matrix. Label the rows and columns for accuracy.	**a.** $\begin{array}{c} \\ A \\ B \\ C \\ D \\ E \end{array}\begin{array}{c} \begin{array}{ccccc} A & B & C & D & E \end{array} \\ \begin{bmatrix} & & & & \\ & & & & \\ & & & & \\ & & & & \\ & & & & \end{bmatrix} \end{array}$
2. Consider vertex A. It is connected to vertices B and D once each, so put a 1 in the corresponding columns of row 1 (shown in red) and in the corresponding rows of column 1 (shown in blue).	$\begin{array}{c} \\ A \\ B \\ C \\ D \\ E \end{array}\begin{array}{c} \begin{array}{ccccc} A & B & C & D & E \end{array} \\ \begin{bmatrix} & 1 & & 1 & \\ 1 & & & & \\ & & & & \\ 1 & & & & \\ & & & & \end{bmatrix} \end{array}$

3. Consider vertex B. It is connected to itself once and vertices A, D and E once each. Put a 1 in the corresponding columns of row 2 (shown in red) and in the corresponding rows of column 2 (shown in blue). *Note:* Some elements, such as m_{12} and m_{21}, will have been previously filled.

$$\begin{array}{c c} & \begin{array}{c c c c c} A & B & C & D & E \end{array} \\ \begin{array}{c} A \\ B \\ C \\ D \\ E \end{array} & \left[\begin{array}{c c c c c} & 1 & & 1 & \\ 1 & 1 & & 1 & 1 \\ & & & & \\ 1 & 1 & & & \\ & 1 & & & \end{array}\right] \end{array}$$

4. Vertex C is not connected to any other vertex, so fill in column 3 and row 3 with zeros (shown in red).

$$\begin{array}{c c} & \begin{array}{c c c c c} A & B & C & D & E \end{array} \\ \begin{array}{c} A \\ B \\ C \\ D \\ E \end{array} & \left[\begin{array}{c c c c c} & 1 & 0 & 1 & \\ 1 & 1 & 0 & 1 & 1 \\ 0 & 0 & 0 & 0 & 0 \\ 1 & 1 & 0 & & \\ & 1 & 0 & & \end{array}\right] \end{array}$$

5. Vertex D is connected to vertices A and B once. This was recorded in previous steps. Vertex D is also connected to vertex E three times, so put a 3 in elements m_{45} and m_{54} (shown in red).

$$\begin{array}{c c} & \begin{array}{c c c c c} A & B & C & D & E \end{array} \\ \begin{array}{c} A \\ B \\ C \\ D \\ E \end{array} & \left[\begin{array}{c c c c c} & 1 & 0 & 1 & \\ 1 & 1 & 0 & 1 & 1 \\ 0 & 0 & 0 & 0 & 0 \\ 1 & 1 & 0 & & 3 \\ & 1 & 0 & 3 & \end{array}\right] \end{array}$$

6. Vertex E has all its connections already entered, as they were considered in previous steps, so complete the matrix by placing a 0 in all unoccupied places, indicating no connections between the vertices.

$$\begin{array}{c c} & \begin{array}{c c c c c} A & B & C & D & E \end{array} \\ \begin{array}{c} A \\ B \\ C \\ D \\ E \end{array} & \left[\begin{array}{c c c c c} 0 & 1 & 0 & 1 & 0 \\ 1 & 1 & 0 & 1 & 1 \\ 0 & 0 & 0 & 0 & 0 \\ 1 & 1 & 0 & 0 & 3 \\ 0 & 1 & 0 & 3 & 0 \end{array}\right] \end{array}$$

7. Remove the labels to complete the matrix. Check your result by comparing the entries in the matrix with the original network representation. This is best done on a vertex-by-vertex basis.

$$\left[\begin{array}{c c c c c} 0 & 1 & 0 & 1 & 0 \\ 1 & 1 & 0 & 1 & 1 \\ 0 & 0 & 0 & 0 & 0 \\ 1 & 1 & 0 & 0 & 3 \\ 0 & 1 & 0 & 3 & 0 \end{array}\right]$$

b. Set up a 3×2 matrix and record the data as shown, with labels removed.

b. $$\left[\begin{array}{c c} 18 & 12 \\ 16 & 11 \\ 34 & 23 \end{array}\right]$$

c. 1. Set up a 2×2 matrix to record the data. For accuracy, you may label the rows and columns, but remember to remove them in your final answer. Record the independent variable (pensioners/working class) in the columns, and the dependent variable (approve/disapprove) in the rows.

c.
$$\begin{array}{c c} & \begin{array}{c c} W & P \end{array} \\ \begin{array}{c} \text{approve} \\ \text{disapprove} \end{array} & \left[\begin{array}{c c} & \\ & \end{array}\right] \end{array}$$

2. We know that, of the 139 pensioners, 37 approved of the freeway and the remainder ($139 - 37 = 102$) disapproved of the freeway. Enter this information into m_{12} and m_{22}.

$$\begin{array}{c c} & \begin{array}{c c} W & P \end{array} \\ \begin{array}{c} \text{approve} \\ \text{disapprove} \end{array} & \left[\begin{array}{c c} & 37 \\ & 102 \end{array}\right] \end{array}$$

3. We also know that, of the 102 working-class people, 79 people approved of the freeway and the remainder $(102 - 79 = 23)$ disapproved of the freeway. Put this information into m_{11} and m_{21}. Remove any labels on the rows or columns to complete the matrix.

$$\begin{bmatrix} 79 & 37 \\ 23 & 102 \end{bmatrix}$$

Matrices are not just useful for storing data. They can be added, subtracted, multiplied (but not divided) and generally manipulated to extract greater information from the data.

For example, we can use matrices to quickly calculate percentages in the two-way frequency table used in Worked example 2. This will be examined more closely later in the topic.

8.2 Exercise

Students, these questions are even better in jacPLUS

 Receive immediate feedback and access sample responses

 Access additional questions

 Track your results and progress

Find all this and MORE in jacPLUS

1. **WE1** For each of the following matrices, give the order and the appropriate name (if it can be categorised). Where possible, write the 2, 1 element and the position of the number 3 in x_{ij} form.

$$A = \begin{bmatrix} -4 & 0 & 0 \\ 0 & 6 & 0 \\ 0 & 0 & 3 \end{bmatrix}, \; B = \begin{bmatrix} 2 & -3 \\ 3 & 0 \end{bmatrix}, \; C = \begin{bmatrix} 1 & 1 \\ 0 & 1 \end{bmatrix}, \; D = \begin{bmatrix} 3 & 1 & -5 \\ 1 & -12 & 6 \\ -5 & 6 & 13 \end{bmatrix}, \; E = \begin{bmatrix} -2 & 0 & 0 \\ 5 & 4 & 0 \\ -10 & 3 & 6 \end{bmatrix}$$

2. For each of the following matrices, give the order and the appropriate name (if it can be categorised). Where possible, write the 1, 2 element and the position of the number 3 in x_{ij} form.

$$A = \begin{bmatrix} 3 & -5 & 2 \\ 0 & 5 & -12 \\ 0 & 0 & 15 \end{bmatrix}, \; B = \begin{bmatrix} -1 & -3 & 1 \end{bmatrix}, \; C = \begin{bmatrix} 7 & 10 \\ 3 & 1 \end{bmatrix}, \; D = \begin{bmatrix} 12 & 7 & -17 \\ 7 & -9 & 3 \\ -17 & 3 & 12 \end{bmatrix}$$

3. For each of the following:
   i. state the order and type of the matrix
   ii. where possible, write the 2, 1 element
   iii. state the position of the number 3 in x_{ij} form.

   a. $A = \begin{bmatrix} 1 & -1 \\ 2 & 3 \\ 2 & 0 \\ 0 & -3 \end{bmatrix}$

   b. $B = \begin{bmatrix} 7 & -1 & 0 \\ 0 & 2 & 1 \\ 0 & 0 & 3 \end{bmatrix}$

   c. $C = \begin{bmatrix} 2 \\ 1 & 4 & 2 \\ 3 \end{bmatrix}$

   d. $D = \begin{bmatrix} 1 \\ 3 \\ -3 \end{bmatrix}$

   e. $E = \begin{bmatrix} 3 & 1 & 2 \end{bmatrix}$

   f. $F = \begin{bmatrix} -1 & -4 & -12 \\ -4 & 15 & 3 \\ -12 & 3 & 4 \end{bmatrix}$

4. For each of the following matrices, give the order and the appropriate name (if it can be categorised). Where possible, write the 2, 2 element and the position of the number 7 in x_{ij} form.

$$A = \begin{bmatrix} -13 & 0 & 0 \\ -11 & 4 & 0 \\ 10 & 5 & 7 \end{bmatrix}, \ B = \begin{bmatrix} 0 & 7 & 1 \end{bmatrix}, \ C = \begin{bmatrix} 1 & 0 \\ 0 & 1 \end{bmatrix}, \ D = \begin{bmatrix} 6 & 0 & 5 \\ 0 & 0 & -9 \\ 5 & -9 & 7 \end{bmatrix}$$

5. i. State the a_{21} element in each of the matrices given.

a. $\begin{bmatrix} 4 & -1 & 30 \\ 1 & -5 & 1 \\ 5 & 6 & 3 \end{bmatrix}$

b. $\begin{bmatrix} 1 \\ 2 \\ -3 \end{bmatrix}$

c. $\begin{bmatrix} 0 & 2 & 4 \end{bmatrix}$

d. $\begin{bmatrix} 2 & -4 & -3 \end{bmatrix}$

e. $\begin{bmatrix} 2 & -1 \\ 3 & -3 \\ -5 & 0 \\ 6 & 3 \end{bmatrix}$

f. $\begin{bmatrix} 0.2 & 0.5 & 3.1 & 2.9 \\ 3.5 & 2.1 & 0.1 & 0.8 \end{bmatrix}$

ii. Consider the matrices given in part i. State in which of the matrices a_{32} exists.

6. **MC** The value of a_{21} in the matrix is:

$$\begin{bmatrix} 2.4 & 3.6 \\ -0.5 & 1.6 \\ 1.6 & 0 \\ -2.5 & 2.4 \end{bmatrix}$$

A. 3.6   B. 1.6   C. −0.5   D. 0   E. 2.4

7. **MC** The number 3 in the following matrix can be represented using the notation:

$$\begin{bmatrix} 2 & 2 & 1 \\ 1 & 0 & 0 \\ 1 & 3 & 2 \\ 2 & 5 & 4 \end{bmatrix}$$

A. a_{23} B. a_{32} C. a_{22} D. a_{12} E. none of these.

8. Using the values 0 and 1 only, state the 4×4 diagonal matrix.

9. **WE2** Represent the following network as a matrix.

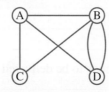

10. Represent the following contingency table for sport preference as a matrix.

		Grades	
		Grade 11	**Grade 12**
Sport preference	**AFL**	26	22
	Cricket	15	9
	Total	41	31

11. The figure shown represents the number of routes between four towns.

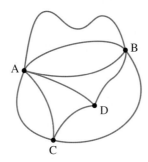 

Represent the possible routes as a 4×4 matrix in the following form.

Number of
routes to

Number of
routes from
$\begin{bmatrix} & & \\ & & \\ & & \end{bmatrix}$

12. Represent the following contingency table as a 3×3 matrix.

		Types of renters		
		Singles	Couples	Total
Rent preference	Live independently	12	23	35
	Share with friends	9	16	25
	Total	21	39	60

13. Represent the final score at an AFL match as a matrix.

	Goals	Behinds	Points
Geelong	15	10	100
Carlton	12	15	87

14. The following information represents the goods to be delivered from a warehouse to individual suppliers.

To supplier A	5 bicycles	12 helmets	6 tyres
To supplier B	7 bicycles	2 helmets	15 tyres
To supplier C	15 bicycles	7 helmets	0 tyres

Present the information as a matrix, with the suppliers placed in the columns.

15. Represent the following coordinates of Cartesian points as a 4×2 matrix.
$$(3, 2), (-4, -1), (4, -1), (4, 1)$$

Question 1 (1 mark)

Source: VCE 2021, Further Mathematics Exam 1, Section B, Module 1, Q1; © VCAA.

MC If matrix $M = \begin{bmatrix} 3 & 2 \\ 8 & 9 \\ 13 & 7 \end{bmatrix}$, then its transpose, M^T, is

A. $\begin{bmatrix} 2 & 3 \\ 9 & 8 \\ 7 & 13 \end{bmatrix}$
B. $\begin{bmatrix} 2 & 9 & 7 \\ 3 & 8 & 13 \end{bmatrix}$
C. $\begin{bmatrix} 7 & 9 & 2 \\ 13 & 8 & 3 \end{bmatrix}$
D. $\begin{bmatrix} 3 & 8 & 13 \\ 2 & 9 & 7 \end{bmatrix}$
E. $\begin{bmatrix} 13 & 8 & 3 \\ 7 & 9 & 2 \end{bmatrix}$

Question 2 (1 mark)

Source: VCE 2017, Further Mathematics Exam 1, Section B, Module 1, Q1; © VCAA.

MC Kai has a part-time job.

Each week, he earns money and saves some of this money.

The matrix below shows the amounts earned (E) and saved (S), in dollars, in each of three weeks.

$$\begin{array}{c} \\ Week\,1 \\ Week\,2 \\ Week\,3 \end{array} \begin{array}{cc} E & S \\ \begin{bmatrix} 300 & 100 \\ 270 & 90 \\ 240 & 80 \end{bmatrix} \end{array}$$

How much did Kai save in week 2?

A. $80   **B.** $90   **C.** $100   **D.** $170   **E.** $270

Question 3 (1 mark)

Source: VCE 2016, Further Mathematics Exam 1, Section B, Module 1, Q5; © VCAA.

MC Let $M = \begin{bmatrix} 1 & 2 & 3 & 4 \\ 3 & 4 & 5 & 6 \end{bmatrix}$

The element in row i and column j of M is m_{ij}.

The elements of M are determined by the rule

A. $m_{ij} = i + j - 1$
B. $m_{ij} = 2i - j + 1$
C. $m_{ij} = 2i + j - 2$
D. $m_{ij} = i + 2j - 2$
E. $m_{ij} = i + j + 1$

More exam questions are available online.

8.3 Addition, subtraction and scalar operations with matrices

LEARNING INTENTION

At the end of this subtopic you should be able to:
- perform elementary matrix operations (sum, difference, multiplication of a scalar).

8.3.1 Addition and subtraction

Matrices can be added or subtracted by applying the usual rules of arithmetic on corresponding elements of the matrices.

Addition and subtraction of matrices

1. Addition and subtraction of matrices can be performed only if the matrices are of the same order.
2. Addition and subtraction of matrices is performed by adding or subtracting corresponding elements.

tlvd-3934

WORKED EXAMPLE 3 Adding and subtracting matrices

Given the matrices:

$$A = \begin{bmatrix} 2 & 4 \\ 3 & -2 \end{bmatrix}, B = \begin{bmatrix} 1 & 6 \\ 8 & 2 \end{bmatrix}, C = \begin{bmatrix} 2 & 3 & -1 \\ 4 & -2 & 5 \end{bmatrix}, D = \begin{bmatrix} 5 & 2 & 3 \\ 4 & -2 & -2 \end{bmatrix}, \text{determine, if possible:}$$

a. $A + B$ b. $B + A$ c. $B - C$ d. $B - A$ e. $D - C$.

THINK

a. 1. A and B are both of the same order, 2×2, so they can be added.

2. Add the numbers in the corresponding positions of each matrix.

b. 1. B and A are both of the same order, 2×2, so they can be added.

2. Add the numbers in the corresponding positions of each matrix.

c. The subtraction cannot be performed since the order of B is 2×2 and the order of C is 2×3.

d. Subtract the numbers in the corresponding positions of each matrix, as B and A are of the same order.

e. 1. D and C are both of the same order, 2×3, so they can be subtracted.

2. Subtract the numbers in the corresponding positions of each matrix.

WRITE

a. $A + B = \begin{bmatrix} 2 & 4 \\ 3 & -2 \end{bmatrix} + \begin{bmatrix} 1 & 6 \\ 8 & 2 \end{bmatrix}$

$= \begin{bmatrix} 2+1 & 4+6 \\ 3+8 & -2+2 \end{bmatrix}$

$= \begin{bmatrix} 3 & 10 \\ 11 & 0 \end{bmatrix}$

b. $B + A = \begin{bmatrix} 1 & 6 \\ 8 & 2 \end{bmatrix} + \begin{bmatrix} 2 & 4 \\ 3 & -2 \end{bmatrix}$

$= \begin{bmatrix} 1+2 & 6+4 \\ 8+3 & 2+-2 \end{bmatrix}$

$= \begin{bmatrix} 3 & 10 \\ 11 & 0 \end{bmatrix}$

c. $B - C$ cannot be calculated since B and C are of different orders.

d. $B - A = \begin{bmatrix} 1 & 6 \\ 8 & 2 \end{bmatrix} - \begin{bmatrix} 2 & 4 \\ 3 & -2 \end{bmatrix}$

$= \begin{bmatrix} 1-2 & 6-4 \\ 8-3 & 2--2 \end{bmatrix}$

$= \begin{bmatrix} -1 & 2 \\ 5 & 4 \end{bmatrix}$

e. $D - C = \begin{bmatrix} 5 & 2 & 3 \\ 4 & -2 & -2 \end{bmatrix} - \begin{bmatrix} 2 & 3 & -1 \\ 4 & -2 & 5 \end{bmatrix}$

$= \begin{bmatrix} 5-2 & 2-3 & 3--1 \\ 4-4 & -2--2 & -2-5 \end{bmatrix}$

$= \begin{bmatrix} 3 & -1 & 4 \\ 0 & 0 & -7 \end{bmatrix}$

Notice in parts **a** and **b** of Worked example 3 that the resulting matrices are both the same. In general, $A + B = B + A$ when A and B are of the same order; that is, matrix addition is **commutative**. Try this for other matrices. Using the matrices from above, try $A - B$.

What do you notice about $B - A$ and $A - B$?

8.3.2 Scalar multiplication

As we have seen in arithmetic, repeated addition can be more efficiently calculated using multiplication. For example, $2 + 2 + 2 = 3 \times 2$.

A similar approach applies to matrices. Consider the matrix $A = \begin{bmatrix} 1 & 5 \\ 4 & -2 \end{bmatrix}$.

To add $A + A + A$, the simplest approach is to calculate $3A$. This is done by multiplying each element of A by 3.

$$A + A + A = \begin{bmatrix} 1 & 5 \\ 4 & -2 \end{bmatrix} + \begin{bmatrix} 1 & 5 \\ 4 & -2 \end{bmatrix} + \begin{bmatrix} 1 & 5 \\ 4 & -2 \end{bmatrix}$$

$$= \begin{bmatrix} 1+1+1 & 5+5+5 \\ 4+4+4 & -2+-2+-2 \end{bmatrix}$$

$$= \begin{bmatrix} 3 & 15 \\ 12 & -6 \end{bmatrix}$$

$$3A = 3\begin{bmatrix} 1 & 5 \\ 4 & -2 \end{bmatrix}$$

$$= \begin{bmatrix} 3\times 1 & 3\times 5 \\ 3\times 4 & 3\times -2 \end{bmatrix}$$

$$= \begin{bmatrix} 3 & 15 \\ 12 & -6 \end{bmatrix}$$

In the term $3A$, the number 3 is called the **scalar**, and the term $3A$ is an example of **scalar multiplication** of matrices.

Scalar multiplication applies to matrices of any order. The scalar quantity can be any number, positive or negative, fraction or decimal, real or imaginary. This can be generalised as follows.

$$c \times \begin{bmatrix} a_{11} & a_{12} & a_{13} & a_{14} & \dots \\ a_{21} & a_{22} & a_{23} & a_{24} & \dots \\ a_{31} & a_{32} & a_{33} & a_{34} & \dots \\ \dots & \dots & \dots & \dots & a_{mn} \end{bmatrix} = \begin{bmatrix} c \times a_{11} & c \times a_{12} & c \times a_{13} & c \times a_{14} & \dots \\ c \times a_{21} & c \times a_{22} & c \times a_{23} & c \times a_{24} & \dots \\ c \times a_{31} & c \times a_{32} & c \times a_{33} & c \times a_{34} & \dots \\ \dots & \dots & \dots & \dots & c \times a_{mn} \end{bmatrix}$$

To multiply a matrix by a scalar, each element in the matrix is multiplied by the scalar. The order of the matrix remains unchanged.

WORKED EXAMPLE 4 Multiplying a matrix by a scalar

Given the matrices:

$$A = \begin{bmatrix} 2 & 6 \\ -1 & 3 \end{bmatrix} \text{ and } B = \begin{bmatrix} 3 & 6 \\ 1 & 3 \end{bmatrix},$$

calculate:

a. $2A$

b. $0.4B$

c. $3A + 4A$

d. $A + \dfrac{1}{3}B$

e. $3(A + B)$.

THINK	WRITE

a. Multiply each element of A by 2.

a. $2A = 2\begin{bmatrix} 2 & 6 \\ -1 & 3 \end{bmatrix}$

$= \begin{bmatrix} 2\times2 & 2\times6 \\ 2\times-1 & 2\times3 \end{bmatrix}$

$= \begin{bmatrix} 4 & 12 \\ -2 & 6 \end{bmatrix}$

b. Multiply each element of B by 0.4.

b. $0.4B = 0.4\begin{bmatrix} 3 & 6 \\ 1 & 3 \end{bmatrix}$

$= \begin{bmatrix} 0.4\times3 & 0.4\times6 \\ 0.4\times1 & 0.4\times3 \end{bmatrix}$

$= \begin{bmatrix} 1.2 & 2.4 \\ 0.4 & 1.2 \end{bmatrix}$

c. 1. $3A + 4A$ simplifies to $7A$.

c. $3A + 4A = 7A$

2. Multiply each element of A by 7.

$7A = 7\begin{bmatrix} 2 & 6 \\ -1 & 3 \end{bmatrix}$

$= \begin{bmatrix} 14 & 42 \\ -7 & 21 \end{bmatrix}$

d. 1. Calculate $\dfrac{1}{3}B$ by multiplying each element of B by $\dfrac{1}{3}$.

d. $\dfrac{1}{3}B = \dfrac{1}{3}\begin{bmatrix} 3 & 6 \\ 1 & 3 \end{bmatrix}$

$= \begin{bmatrix} \dfrac{3}{3} & \dfrac{6}{3} \\ \dfrac{1}{3} & \dfrac{3}{3} \end{bmatrix}$

$= \begin{bmatrix} 1 & 2 \\ \dfrac{1}{3} & 1 \end{bmatrix}$

2. Complete the addition by adding A to $\dfrac{1}{3}B$.

$A + \dfrac{1}{3}B = \begin{bmatrix} 2 & 6 \\ -1 & 3 \end{bmatrix} + \begin{bmatrix} 1 & 2 \\ \dfrac{1}{3} & 1 \end{bmatrix}$

$= \begin{bmatrix} 3 & 8 \\ -\dfrac{2}{3} & 4 \end{bmatrix}$

e. 1. Add A and B.

e. $A + B = \begin{bmatrix} 2 & 6 \\ -1 & 3 \end{bmatrix} + \begin{bmatrix} 3 & 6 \\ 1 & 3 \end{bmatrix}$

$= \begin{bmatrix} 5 & 12 \\ 0 & 6 \end{bmatrix}$

2. Multiply this matrix by 3.

$3(A + B) = 3\begin{bmatrix} 5 & 12 \\ 0 & 6 \end{bmatrix}$

$= \begin{bmatrix} 15 & 36 \\ 0 & 18 \end{bmatrix}$

TI	THINK	DISPLAY/WRITE	CASIO	THINK	DISPLAY/WRITE

TI | THINK

1. To define Matrices A and B, on a Calculator page open the Maths expression palette and select the matrix template. Fill in the number of rows and columns as 2×2. Press ENTER.

DISPLAY/WRITE

CASIO | THINK

1. To define Matrices A and B, on a Main screen tap:
Keyboard
Math 2
Tap the 2×2 template.

DISPLAY/WRITE

2. Use the tab key to enter the values into the matrix as shown; press tab to bring the cursor outside the matrix.
Press:
Ctrl
Var
a
Then press ENTER.

Repeat for Matrix B.

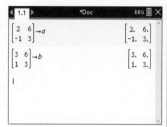

2. Enter the values into the matrix; tap outside the matrix to bring the cursor outside.
Tap:
Store (arrow symbol) a
(Remember to use Var a)
Then tap EXE.

Repeat for Matrix B.

3. Complete the entry lines as:
$2 \times a$, then press ENTER.
$0.4 \times b$, then press ENTER.
$3 \times a + 4 \times a$, then press ENTER.
$a + \frac{1}{3} \times b$, then press ENTER.
$3 \times (a + b)$, then press ENTER.

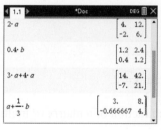

3. Complete the entry lines as: $2 \times a$, then tap EXE.
$0.4 \times b$, then tap EXE.
$3 \times a + 4 \times a$, then tap EXE.
$a + \frac{1}{3} \times b$, then tap EXE.
$3 \times (a + b)$, then tap EXE.

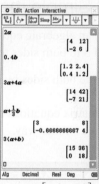

4. The answers are shown on the screen.

a. $2A = \begin{bmatrix} 4 & 12 \\ -2 & 6 \end{bmatrix}$

b. $0.4B = \begin{bmatrix} 1.2 & 2.4 \\ 0.4 & 1.2 \end{bmatrix}$

c. $3A + 4B = \begin{bmatrix} 14 & 42 \\ -7 & 21 \end{bmatrix}$

d. $A + \frac{1}{3}B = \begin{bmatrix} 3 & 8 \\ -0.667 & 4 \end{bmatrix}$

e. $3(A + B) = \begin{bmatrix} 15 & 36 \\ 0 & 18 \end{bmatrix}$

4. The answers are shown on the screen.

a. $2A = \begin{bmatrix} 4 & 12 \\ -2 & 6 \end{bmatrix}$

b. $0.4B = \begin{bmatrix} 1.2 & 2.4 \\ 0.4 & 1.2 \end{bmatrix}$

c. $3A + 4B = \begin{bmatrix} 14 & 42 \\ -7 & 21 \end{bmatrix}$

d. $A + \frac{1}{3}B = \begin{bmatrix} 3 & 8 \\ -0.667 & 4 \end{bmatrix}$

e. $3(A + B) = \begin{bmatrix} 15 & 36 \\ 0 & 18 \end{bmatrix}$

Properties of addition of matrices

The following list is a summary of properties of addition of matrices. These properties hold true when A, B and C are $m \times n$ matrices, k and c are constants and O is a zero matrix (a matrix with all elements equal to zero).

Property	Example
Commutative (does not matter which order the matrices are operated on)	$A + B = B + A$
Associative (does not matter where the brackets are placed)	$(A + B) + C = A + (B + C)$ $(kc)A = k(cA)$
Identity	$A + O = A = O + A$
Inverse	$A + -A = O = -A + A$
Distributive	$kA + kB = k(A + B)$ $kA + cA = (k + c)A$

Note: For subtraction, matrices *do not* obey the **associative** or commutative laws.

For example, $A - B \neq B - A$ and $(A - B) - C \neq A - (B - C)$.

8.3.3 Simple matrix equations

To solve an algebraic equation such as $4x - 3 = 5$:
1. add 3 to both sides to obtain $4x - 3 + 3 = 5 + 3$ or $4x = 8$
2. divide both sides by 4 $\left(\text{or multiply by } \dfrac{1}{4}\right)$ to obtain $x = 2$.

Simple matrix equations that require the addition or subtraction of a matrix or multiplication of a scalar can be solved in a similar way.

WORKED EXAMPLE 5 Solving matrix equations

Solve the following matrix equations.

a. $5E = \begin{bmatrix} 20 & 15 & -5 \\ 5 & 0 & 12 \end{bmatrix}$

b. $D + \begin{bmatrix} 2 & -4 \\ 3 & 2 \end{bmatrix} = \begin{bmatrix} 7 & 2 \\ 1 & 5 \end{bmatrix}$

c. If $A = \begin{bmatrix} 2 & 6 \\ -1 & 3 \end{bmatrix}$ and $B = \begin{bmatrix} 3 & 6 \\ 1 & 3 \end{bmatrix}$, determine C if $2C + A = 3B$.

THINK

a. **1.** To get E by itself, multiply both sides by $\dfrac{1}{5}$.

2. Simplify the matrix E.

WRITE

a. $5E = \begin{bmatrix} 20 & 15 & -5 \\ 5 & 0 & 12 \end{bmatrix}$

$E = \dfrac{1}{5} \begin{bmatrix} 20 & 15 & -5 \\ 5 & 0 & 12 \end{bmatrix}$

$E = \begin{bmatrix} 4 & 3 & -1 \\ 1 & 0 & 2.4 \end{bmatrix}$

b. 1. To get D by itself, subtract $\begin{bmatrix} 2 & -4 \\ 3 & 2 \end{bmatrix}$ from both sides.

b. $D + \begin{bmatrix} 2 & -4 \\ 3 & 2 \end{bmatrix} = \begin{bmatrix} 7 & 2 \\ 1 & 5 \end{bmatrix}$

$$D = \begin{bmatrix} 7 & 2 \\ 1 & 5 \end{bmatrix} - \begin{bmatrix} 2 & -4 \\ 3 & 2 \end{bmatrix}$$

2. Simplify the matrix D.

$$D = \begin{bmatrix} 5 & 6 \\ -2 & 3 \end{bmatrix}$$

c. 1. First solve algebraically to get C by itself.

c. $2C + A = 3B$

$2C = 3B - A$

$C = \dfrac{1}{2}(3B - A)$

2. Calculate the value of $3B - A$.

$3B - A = \begin{bmatrix} 9 & 18 \\ 3 & 9 \end{bmatrix} - \begin{bmatrix} 2 & 6 \\ -1 & 3 \end{bmatrix}$

$= \begin{bmatrix} 7 & 12 \\ 4 & 6 \end{bmatrix}$

3. Multiply this by $\dfrac{1}{2}$ to solve for C.

$C = \dfrac{1}{2}\begin{bmatrix} 7 & 12 \\ 4 & 6 \end{bmatrix}$

$= \begin{bmatrix} 3.5 & 6 \\ 2 & 3 \end{bmatrix}$

The matrix operations involving addition, subtraction and scalar multiplication can be applied to practical situations such as stock inventory, price discounting and marking up of store prices.

WORKED EXAMPLE 6 Applying elementary matrix operations

A retail chain of three stores has an inventory of three models each of televisions and Bluetooth headphones, represented as the following matrices. The first column represents the televisions and the second column represents the Bluetooth headphones. Each row represents a different model.

$$\text{Store A} = \begin{bmatrix} 12 & 21 \\ 5 & 12 \\ 3 & 7 \end{bmatrix}, \text{ store B} = \begin{bmatrix} 23 & 32 \\ 8 & 15 \\ 1 & 11 \end{bmatrix}, \text{ store C} = \begin{bmatrix} 5 & 17 \\ 2 & 12 \\ 0 & 14 \end{bmatrix}$$

a. If the third row represents the most expensive models, state which store has the highest number of models of expensive televisions.

b. Give the matrix that would represent the total stock of televisions and headphones for all three stores.

c. The wholesale price (in dollars) of each model of television and headphones is presented in the following matrix.

$$\begin{bmatrix} 100 & 30 \\ 250 & 80 \\ 400 & 200 \end{bmatrix}$$

If the wholesale prices are marked up by 50%, show the recommended retail prices in a 3×2 matrix.

d. Store C wishes to have a sale. If it discounts all retail prices by 10%, represent the discounted prices as a 3×2 matrix.

THINK	WRITE
a. The first column represents the televisions and the third row represents the most expensive model (a_{31}). Store A has 3 expensive televisions, store B has 1 expensive television and store C has 0 expensive televisions.	**a.** Store A has the greatest number of expensive televisions.

b. The total stock of televisions and headphones for all three stores is given by the sum of the three matrices. Add the matrices and simplify.

b. Total inventory $=$ store A $+$ store B $+$ store C

$$= \begin{bmatrix} 12 & 21 \\ 5 & 12 \\ 3 & 7 \end{bmatrix} + \begin{bmatrix} 23 & 32 \\ 8 & 15 \\ 1 & 11 \end{bmatrix} + \begin{bmatrix} 5 & 17 \\ 2 & 12 \\ 0 & 14 \end{bmatrix}$$

$$= \begin{bmatrix} 12+23+5 & 21+32+17 \\ 5+8+2 & 12+15+12 \\ 3+1+0 & 7+11+14 \end{bmatrix}$$

$$= \begin{bmatrix} 40 & 70 \\ 15 & 39 \\ 4 & 32 \end{bmatrix}$$

c. 1. A markup of 50% represents 150% of the wholesale prices.

c. $50\% + 100\% = 150\%$

2. Multiply each element of the wholesale price matrix by 1.5 (150%).

$$150\% \text{ of } \begin{bmatrix} 100 & 30 \\ 250 & 80 \\ 400 & 200 \end{bmatrix} = 1.5 \times \begin{bmatrix} 100 & 30 \\ 250 & 80 \\ 400 & 200 \end{bmatrix}$$

$$= \begin{bmatrix} 150 & 45 \\ 375 & 120 \\ 600 & 300 \end{bmatrix}$$

d. 1. A discount of 10% represents 90% of the retail price.

d. $100\% - 10\% = 90\%$

2. Multiply each element of the retail price matrix by 0.9 (90%).

$$90\% \text{ of } \begin{bmatrix} 150 & 45 \\ 375 & 120 \\ 600 & 300 \end{bmatrix} = 0.9 \times \begin{bmatrix} 150 & 45 \\ 375 & 120 \\ 600 & 300 \end{bmatrix}$$

$$= \begin{bmatrix} 135 & 40.5 \\ 337.5 & 108 \\ 540 & 270 \end{bmatrix}$$

 Resources

 Interactivity Adding and subtracting matrices (int-6463)

1. **WE3** Given the matrices

$$A = \begin{bmatrix} 4 & 7 \\ 3 & -4 \end{bmatrix}, \ B = \begin{bmatrix} 0 & -3 \\ 5 & 2 \end{bmatrix}, \ C = \begin{bmatrix} 5 & 7 & -2 \\ 1 & 6 & -5 \end{bmatrix}, \ D = \begin{bmatrix} -7 & 7 & -2 \\ -5 & 6 & -8 \end{bmatrix}$$

   determine, if possible:

   a. $A + B$
   b. $B + A$
   c. $B - C$
   d. $B - A$
   e. $D - C$.

2. Given the matrices

$$A = \begin{bmatrix} 12 & 9 \\ -8 & -10 \end{bmatrix}, \ B = \begin{bmatrix} 15 & 20 & -21 \\ 18 & 10 & -13 \end{bmatrix}, \ C = \begin{bmatrix} 8 & 17 & -4 \\ 1 & 12 & -9 \end{bmatrix}, \ D = \begin{bmatrix} -9 & 17 & -12 \\ -12 & 5 & -11 \end{bmatrix}$$

   determine, if possible:

   a. $A + B$
   b. $B + C$
   c. $B - C$
   d. $B - D$
   e. $D - C$.

3. Given the matrices

$$A = \begin{bmatrix} 1 & 5 \\ -3 & 2 \end{bmatrix}, \ B = \begin{bmatrix} 3 & 0 \\ 5 & -2 \end{bmatrix}, \ C = \begin{bmatrix} 1 & 5 & 0 \\ 3 & -3 & 3 \end{bmatrix}, \ D = \begin{bmatrix} 4 & 2 & 2 \\ 3 & -2 & -1 \end{bmatrix}$$

   calculate, if possible:

   a. $A + B$
   b. $A + A$
   c. $B - C$
   d. $B - A$
   e. $D - C$.

4. **MC** Given the following matrices, the solution to $F - E$ is:

$$E = \begin{bmatrix} 1.2 & -0.5 \\ 3.6 & 5.0 \\ -3.5 & 2.2 \end{bmatrix} \text{ and } F = \begin{bmatrix} 0.2 & -0.5 \\ 2.4 & 2.5 \\ 0 & 1.1 \end{bmatrix}$$

   A. $\begin{bmatrix} 1.0 & 0 \\ 1.2 & 2.5 \\ -3.5 & 1.1 \end{bmatrix}$
   B. $\begin{bmatrix} 1.4 & -1.0 \\ 6.0 & 7.5 \\ -3.5 & 3.3 \end{bmatrix}$
   C. $\begin{bmatrix} 1.0 & -1.0 \\ 1.2 & 2.5 \\ 3.5 & 1.1 \end{bmatrix}$

   D. $\begin{bmatrix} -1.0 & 0 \\ -1.2 & -2.5 \\ 3.5 & -1.1 \end{bmatrix}$
   E. $\begin{bmatrix} 1.0 & -1.0 \\ 1.2 & 2.5 \\ -3.5 & 1.1 \end{bmatrix}$

5. The cost price (C) and sale price (S) of four items at an electrical appliance store are given as follows.

$$C = \begin{bmatrix} 23.50 & 45.00 & 87.50 & 140.00 \end{bmatrix} \text{ and } S = \begin{bmatrix} 49.50 & 135.00 & 169.00 & 299.95 \end{bmatrix}$$

   Calculate and show, as a matrix, the profit made on the sale of these four items.

6. **WE4** Consider the matrix $A = \begin{bmatrix} -6 & 8 \\ 6 & -4 \end{bmatrix}$.

   Calculate:

   a. $3A$
   b. $-2A$
   c. $\dfrac{1}{2}A$
   d. $0.4A$.

7. Given $A = \begin{bmatrix} 2 & -6 \\ 5 & 0 \end{bmatrix}$ and $B = \begin{bmatrix} -8 & 0 \\ 7 & -3 \end{bmatrix}$, calculate:

    a. $A + 2B$ b. $2A - B$.

8. Given the matrices

$$A = \begin{bmatrix} 4 & 5 \\ 1 & 2 \end{bmatrix} \text{ and } B = \begin{bmatrix} 4 & 8 \\ 2 & 6 \end{bmatrix},$$

    calculate:

    a. $3A$ b. $0.1B$ c. $2A + 3A$

    d. $A + \dfrac{1}{4}B$ e. $2(A + B)$ f. $\dfrac{1}{2}(3A - B)$.

9. Given the matrices

$$C = \begin{bmatrix} 1 & 3 & 0 & 2 & 2 \\ 2 & 2 & 4 & 1 & 3 \\ 1 & 0 & -2 & 1 & -1 \\ 2 & 1 & 3 & 1 & 0 \\ 1 & 1 & 5 & 1 & 1 \end{bmatrix} \text{ and } D = \begin{bmatrix} 4 & 6 & 0 & 1 & -2 \\ 2 & 3 & -4 & 1 & 3 \\ 3 & 0 & 2 & 1 & 1 \\ 2 & 2 & -3 & 1 & 10 \\ 5 & 1 & 8 & 1 & 0 \end{bmatrix},$$

    use CAS to evaluate the following:

    a. $4C$ b. $-0.1D$ c. $3C - 3D$

    d. $2(D + 3C)$ e. $\dfrac{1}{2}(C - 3D)$.

10. **WE5** Solve the following matrix equations.

    a. $4E = \begin{bmatrix} 24 & 40 & -36 \\ 48 & -8 & -80 \end{bmatrix}$

    b. $D + \begin{bmatrix} 2 & 6 \\ -3 & -7 \end{bmatrix} = \begin{bmatrix} 8 & 6 \\ 2 & -5 \end{bmatrix}$

    c. If $A = \begin{bmatrix} 3 & 7 \\ -2 & 2 \end{bmatrix}$ and $B = \begin{bmatrix} 4 & 5 \\ 2 & 6 \end{bmatrix}$, determine the matrix C if $2C + A = 4B$.

11. If $\begin{bmatrix} a & 2 \\ 6 & -6 \end{bmatrix} + \begin{bmatrix} -3 & b \\ c & 2 \end{bmatrix} = \begin{bmatrix} 6 & -3 \\ -2 & d \end{bmatrix}$, determine the values of a, b, c and d.

12. Solve the following matrix equations.

    a. $6E = \begin{bmatrix} 24 & 42 & -6 \\ 6 & 0 & 12 \end{bmatrix}$

    b. $D + \begin{bmatrix} 1 & 5 \\ 4 & 2 \end{bmatrix} = \begin{bmatrix} 6 & 2 \\ 8 & 5 \end{bmatrix}$

    c. If $A = \begin{bmatrix} 4 & 5 \\ 1 & 2 \end{bmatrix}$ and $B = \begin{bmatrix} 4 & 8 \\ 2 & 6 \end{bmatrix}$, calculate C if $2C + A = 4B$.

13. The order of five matrices is given as follows:

$$A = 3 \times 2, \; B = 2 \times 3, \; C = 3 \times 2, \; D = 3 \times 3 \text{ and } E = 3 \times 2$$

    State which of the following cannot be calculated.

    a. $-0.3C$ b. $D + A$ c. $2C - 3A$

    d. $2(E + 3C)$ e. $\dfrac{1}{2}(A - 3B)$ f. $A + C - E$

14. **WE6** A chain of three car yards has an inventory of front-wheel-drive and 4WD cars. They have new and used vehicles of both types. The inventory is represented in matrix form, where the new vehicles are placed in the first column and the front-wheel-drive vehicles are placed in the first row.

$$\text{Store A} = \begin{bmatrix} 6 & 4 \\ 4 & 7 \end{bmatrix}, \text{ store B} = \begin{bmatrix} 5 & 6 \\ 2 & 5 \end{bmatrix}, \text{ store C} = \begin{bmatrix} 3 & 3 \\ 3 & 0 \end{bmatrix}$$

a. Determine which store has the highest number of new 4WDs. State the label of the element.
b. Give the matrix that would represent the total stock of each type of vehicle for all three stores combined.

15. A student has kept records of their test results in matrix form. In semester 1, for English tests they got 72%, 76% and 81% and for Foundation Maths tests they got 84%, 68% and 82%. In semester 2, for English tests they got 78%, 76% and 89% and for Foundation Maths tests they got 74%, 77% and 85%.

a. Write the semester results in two separate 3×2 matrices.
b. State the matrix equation that would give the average for each test of the two subjects.
c. Calculate the average result from the two semesters and present it as a 3×2 matrix.

16. A car dealership has three new sales centres where they stock three models each of 4WDs and sedans, represented as the following matrices. The first column represents the 4WDs and the second column represents the sedans. Each row represents a different model.

$$\text{Centre A} = \begin{bmatrix} 7 & 16 \\ 5 & 8 \\ 2 & 4 \end{bmatrix}, \text{ centre B} = \begin{bmatrix} 13 & 12 \\ 8 & 15 \\ 3 & 1 \end{bmatrix}, \text{ centre C} = \begin{bmatrix} 14 & 7 \\ 12 & 12 \\ 9 & 4 \end{bmatrix}$$

a. If the third row represents the most expensive models, determine which centre has the highest number of models of expensive sedans.
b. Give the matrix that would represent the total stock of 4WDs and sedans for all three centres.
The wholesale price (in dollars) of each model of 4WDs and sedans is presented in the following matrix.

$$\begin{bmatrix} 20\,000 & 13\,000 \\ 25\,000 & 18\,000 \\ 40\,000 & 28\,000 \end{bmatrix}$$

c. If the wholesale prices are marked up by 100% to get the selling price, calculate the recommended retail prices.
d. Centre B wishes to have a clearance sale. If it discounts all retail prices by 10%, represent the discounted prices as a matrix.

17. In a township, the percentage of households that own no pets is 25%, those with one pet is 40%, those with two pets is 20% and those with more than two pets is 15%.

a. Set up a 1×4 matrix to represent the percentage ownership of pets.
b. Write an equation that will enable you to calculate the number of households for each category, given that there are 800 households in the town.
c. Evaluate the number of households for each category as a 1×4 matrix.

18. **a.** Place the following football team's total season score in a suitable matrix format.

Away games total score:
154 goals   214 behinds   1138 points
Home games total score:
207 goals   180 behinds   1422 points

b. Present the average game score for away games and home games as a suitable matrix. There are 11 away games and 11 home games.

19. The stock on hand (S) of items in a warehouse is summarised as the following matrix:

$$S = \begin{bmatrix} 23 & 150 & 112 & 5 & 1090 \\ 35 & 210 & 145 & 4 & 800 \\ 45 & 230 & 130 & 8 & 1200 \\ 32 & 700 & 230 & 4 & 400 \\ 77 & 420 & 160 & 7 & 850 \end{bmatrix}$$

a. For the company's plans to reduce costs, complete the following matrix equation if the new stock levels are to be set at 40% of the current stocks.

New stock level (N) = _____ of current stock levels (matrix _____)
= _____ × _____

b. Evaluate the new stock levels and display as a matrix. (Round answers to the nearest whole number.)

20. Use CAS to store three 3×2 matrices as A, B and C (they can contain any numbers). Check to see whether the following are true.

a. $A + B = B + A$
b. $A - B = B - A$
c. $(A - B) - C = A - (B - C)$
d. $3A + 3B = 3(A + B)$

8.3 Exam questions

Question 1 (1 mark)

Source: VCE 2019, Further Mathematics Exam 1, Section B, Module 1, Q1; © VCAA.

MC Consider the following four matrix expressions.

$$\begin{bmatrix} 8 \\ 12 \end{bmatrix} + \begin{bmatrix} 4 \\ 2 \end{bmatrix} \qquad \begin{bmatrix} 8 \\ 12 \end{bmatrix} + \begin{bmatrix} 4 & 0 \\ 0 & 2 \end{bmatrix}$$

$$\begin{bmatrix} 8 & 0 \\ 12 & 0 \end{bmatrix} + \begin{bmatrix} 4 \\ 2 \end{bmatrix} \qquad \begin{bmatrix} 8 & 0 \\ 12 & 0 \end{bmatrix} + \begin{bmatrix} 4 & 0 \\ 0 & 2 \end{bmatrix}$$

How many of these four matrix expressions are defined?

A. 0
B. 1
C. 2
D. 3
E. 4

Question 2 (1 mark)

Source: VCE 2019, Further Mathematics Exam 1, Section B, Module 1, Q3; © VCAA.

MC Consider the matrix P, where $p = \begin{bmatrix} 3 & 2 & 1 \\ 5 & 4 & 3 \end{bmatrix}$

The element in row i and column j of matrix P is p_{ij}.

The elements in matrix P are determined by the rule

A. $p_{ij} = 4 - j$ **B.** $p_{ij} = 2i + 1$ **C.** $p_{ij} = i + j + 1$ **D.** $p_{ij} = i + 2j$ **E.** $p_{ij} = 2i - j + 2$

Question 3 (1 mark)

Source: VCE 2017, Further Mathematics Exam 1, Section B, Module 1, Q6; © VCAA.

MC The table below shows information about two matrices, A and B.

Matrix	Order	Rule
A	3×3	$a_{ij} = 2i + j$
B	3×3	$b_{ij} = i - j$

The element in row i and column j of matrix A is a_{ij}.

The element in row i and column j of matrix B is b_{ij}.

The sum $A + B$ is

A. $\begin{bmatrix} 5 & 7 & 9 \\ 8 & 10 & 12 \\ 11 & 13 & 15 \end{bmatrix}$ **B.** $\begin{bmatrix} 5 & 8 & 11 \\ 7 & 10 & 13 \\ 9 & 12 & 15 \end{bmatrix}$ **C.** $\begin{bmatrix} 3 & 6 & 9 \\ 3 & 6 & 9 \\ 3 & 6 & 9 \end{bmatrix}$

D. $\begin{bmatrix} 3 & 3 & 3 \\ 6 & 6 & 6 \\ 9 & 9 & 9 \end{bmatrix}$ **E.** $\begin{bmatrix} 3 & 6 & 3 \\ 6 & 3 & 9 \\ 3 & 9 & 3 \end{bmatrix}$

More exam questions are available online.

8.4 Multiplying matrices

LEARNING INTENTION

At the end of this subtopic you should be able to:
- apply multiplication rules to matrices
- identify types of matrices (permutation and identity)
- identify properties of multiplication of matrices.

8.4.1 The multiplication rule

In the previous section, we discussed the multiplication of a matrix by a scalar, whereby each element of the matrix is multiplied by the scalar, and the order of the matrix remains unchanged. Now we wish to investigate the multiplication of a matrix by another matrix.

Two matrices can be multiplied if the number of columns in the first matrix is equal to the number of rows in the second matrix. Consider the following matrix multiplication

$$\begin{bmatrix} 1 & 2 & 3 \\ 2 & 4 & 5 \end{bmatrix} \times \begin{bmatrix} 1 & 2 & 3 & 4 \\ 2 & 0 & 1 & 6 \\ 3 & -4 & 0 & 8 \end{bmatrix}$$

The number of columns in the first matrix is 3 and the number of rows in the second matrix is 3. It is possible for these two matrices to be multiplied. A way to remember this is to write the order of each matrix next to the other, as shown.

$$2 \times 3 \times 3 \times 4$$

The order of the first matrix is 2×3 and the order of the second matrix is 3×4.

If the two inner numbers are the same, it is possible to multiply the matrices. If they are not the same, then it will not be possible to multiply the matrices.

The resulting matrix will be of the order 2×4 — the outer two numbers in the multiplication shown above.

The order of the product of two matrices

If matrix A is of order $m \times n$ and matrix B is of order $n \times p$, then $A \times B$ exists and its order is $m \times p$.

The following matrix multiplication is not possible because the number of columns in the first matrix is not equal to the number of rows in the second matrix.

$$\begin{bmatrix} 1 & 2 & 3 \\ 2 & 0 & 5 \\ 3 & -4 & 2 \end{bmatrix} \times \begin{bmatrix} 1 & 2 & 3 \\ 2 & 4 & 6 \end{bmatrix}$$

$$3 \times 3 \times 2 \times 3$$

The two inner numbers are not the same.

The following example shows how two 2×2 matrices are multiplied. (The resulting matrix will be of order 2×2.)

$$\begin{bmatrix} 1 & 2 \\ 3 & 4 \end{bmatrix} \times \begin{bmatrix} 5 & 7 \\ 6 & 8 \end{bmatrix}$$

Each row of the first matrix is multiplied by each column of the second matrix:

$$\begin{bmatrix} r_1 c_1 & r_1 c_2 \\ r_2 c_1 & r_2 c_2 \end{bmatrix} \text{ gives } \begin{bmatrix} 1 \times 5 + 2 \times 6 & 1 \times 7 + 2 \times 8 \\ 3 \times 5 + 4 \times 6 & 3 \times 7 + 4 \times 8 \end{bmatrix}$$

$$= \begin{bmatrix} 5 + 12 & 7 + 16 \\ 15 + 24 & 21 + 32 \end{bmatrix}$$

$$= \begin{bmatrix} 17 & 23 \\ 39 & 53 \end{bmatrix}$$

As long as the number of columns in the first matrix is equal to the number of rows in the second matrix, the method highlighted above can be used to multiply the matrices.

The procedure is repeated for two 3×3 matrices as shown.

$$\text{If } A = \begin{bmatrix} a_{11} & a_{12} & a_{13} \\ a_{21} & a_{22} & a_{23} \\ a_{31} & a_{32} & a_{33} \end{bmatrix} \text{ and } B = \begin{bmatrix} b_{11} & b_{12} & b_{13} \\ b_{21} & b_{22} & b_{23} \\ b_{31} & b_{32} & b_{33} \end{bmatrix} \text{ then } A \times B = \begin{bmatrix} a_{11} & a_{12} & a_{13} \\ a_{21} & a_{22} & a_{23} \\ a_{31} & a_{32} & a_{33} \end{bmatrix} \times \begin{bmatrix} b_{11} & b_{12} & b_{13} \\ b_{21} & b_{22} & b_{23} \\ b_{31} & b_{32} & b_{33} \end{bmatrix}$$

$$= \begin{bmatrix} a_{11} \times b_{11} + a_{12} \times b_{21} + a_{13} \times b_{31} & a_{11} \times b_{12} + a_{12} \times b_{22} + a_{13} \times b_{32} & a_{11} \times b_{13} + a_{12} \times b_{23} + a_{13} \times b_{33} \\ a_{21} \times b_{11} + a_{22} \times b_{21} + a_{23} \times b_{31} & a_{21} \times b_{12} + a_{22} \times b_{22} + a_{23} \times b_{32} & a_{21} \times b_{13} + a_{22} \times b_{23} + a_{23} \times b_{33} \\ a_{31} \times b_{11} + a_{32} \times b_{21} + a_{33} \times b_{31} & a_{31} \times b_{12} + a_{32} \times b_{22} + a_{33} \times b_{32} & a_{31} \times b_{13} + a_{32} \times b_{23} + a_{33} \times b_{33} \end{bmatrix}$$

The rows of the first matrix are multiplied by the columns of the second matrix.

When preparing to multiply two matrices, the order of the resultant matrix should be established first.

tlvd-3935

WORKED EXAMPLE 7 Multiplying matrices

Consider the following matrices.

$$A = \begin{bmatrix} 1 & 4 \\ 3 & 0 \end{bmatrix}, \ B = \begin{bmatrix} 2 & 1 \\ 1 & -2 \\ 3 & 4 \end{bmatrix}, \ C = \begin{bmatrix} 3 & 2 \\ 4 & 1 \end{bmatrix}$$

a. **Write the order of each matrix.**
b. **Determine which of the following products exist:**
   i. AB ii. AC iii. BA iv. CA
c. **Write the order of the products that exist.**
d. **Calculate the products that exist.**

THINK	WRITE
a. 1. Matrix A has 2 rows and 2 columns.	a. Matrix A is a 2×2 matrix.
2. Matrix B has 3 rows and 2 columns.	Matrix B is a 3×2 matrix.
3. Matrix C has 2 rows and 2 columns.	Matrix C is a 2×2 matrix.
b. i. The number of columns in matrix $A \neq$ the number of rows in matrix B, so AB does not exist.	b. i. AB does not exist.
ii. The number of columns in matrix $A =$ the number of rows in matrix C, so AC does exist.	ii. AC exists.
iii. The number of columns in matrix $B =$ the number of rows in matrix A, so BA exists.	iii. BA exists.
iv. The number of columns in matrix $C =$ the number of rows in matrix A, so CA exists.	iv. CA exists.
c. Identify the order of the products that exist by looking at the two outer numbers.	c. The order of AC is 2×2. The order of BA is 3×2. The order of CA is 2×2.
d. 1. Multiply the rows of matrix A by the columns of matrix C.	d. $AC = \begin{bmatrix} 1 & 4 \\ 3 & 0 \end{bmatrix} \times \begin{bmatrix} 3 & 2 \\ 4 & 1 \end{bmatrix}$
	$= \begin{bmatrix} 1 \times 3 + 4 \times 4 & 1 \times 2 + 4 \times 1 \\ 3 \times 3 + 0 \times 4 & 3 \times 2 + 0 \times 1 \end{bmatrix}$
2. Simplify AC.	$= \begin{bmatrix} 19 & 6 \\ 9 & 6 \end{bmatrix}$

3. Multiply the rows of matrix B by the columns of matrix A.

$$BA = \begin{bmatrix} 2 & 1 \\ 1 & -2 \\ 3 & 4 \end{bmatrix} \times \begin{bmatrix} 1 & 4 \\ 3 & 0 \end{bmatrix}$$

4. Simplify BA.

$$= \begin{bmatrix} 2 \times 1 + 1 \times 3 & 2 \times 4 + 1 \times 0 \\ 1 \times 1 + -2 \times 3 & 1 \times 4 + -2 \times 0 \\ 3 \times 1 + 4 \times 3 & 3 \times 4 + 4 \times 0 \end{bmatrix}$$

$$= \begin{bmatrix} 5 & 8 \\ -5 & 4 \\ 15 & 12 \end{bmatrix}$$

5. Multiply the rows of matrix C by the columns of matrix A.

$$CA = \begin{bmatrix} 3 & 2 \\ 4 & 1 \end{bmatrix} \times \begin{bmatrix} 1 & 4 \\ 3 & 0 \end{bmatrix}$$

$$= \begin{bmatrix} 3 \times 1 + 2 \times 3 & 3 \times 4 + 2 \times 0 \\ 4 \times 1 + 1 \times 3 & 4 \times 4 + 1 \times 0 \end{bmatrix}$$

6. Simplify CA.

$$= \begin{bmatrix} 9 & 12 \\ 7 & 16 \end{bmatrix}$$

TI | THINK	DISPLAY/WRITE

b.1. On a Calculator page, define matrices A, B and C.

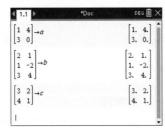

2. Complete the entry lines as:
$a \times b$, then press ENTER.
$a \times c$, then press ENTER.
$b \times a$, then press ENTER.
$c \times a$, then press ENTER.

3. The answers are shown on the screen.

i. AB – does not exist.

ii. $AC = \begin{bmatrix} 19 & 6 \\ 9 & 6 \end{bmatrix}$

i. $BA = \begin{bmatrix} 5 & 8 \\ -5 & 4 \\ 15 & 12 \end{bmatrix}$

ii. $CA = \begin{bmatrix} 9 & 12 \\ 7 & 16 \end{bmatrix}$

CASIO | THINK	DISPLAY/WRITE

1. On a Main screen, define matrices A, B and C.
Note: To add an extra row to a 2×2 matrix, tap on the 1×2 icon. To add an extra column to a 2×2 matrix, tap on the 2×1 icon.

2. Complete the entry lines as:
$a \times b$, then tap EXE.
$a \times c$, then tap EXE.
$b \times a$, then tap EXE.
$c \times a$, then tap EXE.

3. The answers are shown on the screen.

i. AB – The calculator states 'Invalid dimension' for this calculation.

ii. $AC = \begin{bmatrix} 19 & 6 \\ 9 & 6 \end{bmatrix}$

i. $BA = \begin{bmatrix} 5 & 8 \\ -5 & 4 \\ 15 & 12 \end{bmatrix}$

ii. $CA = \begin{bmatrix} 9 & 12 \\ 7 & 16 \end{bmatrix}$

In Worked example 7, you would have noticed that the results of the multiplication of AC and CA did not produce the same answer. It can be said that matrix multiplication is not commutative.

In general, for two matrices A and B, $AB \neq BA$.

However, there is one exception where a matrix multiplication is commutative. The following situation demonstrates this.

$$\text{Let } A = \begin{bmatrix} 1 & 4 \\ 3 & 5 \end{bmatrix} \text{ and } B = \begin{bmatrix} 1 & 0 \\ 0 & 1 \end{bmatrix}$$

$$AB = \begin{bmatrix} 1 & 4 \\ 3 & 5 \end{bmatrix} \times \begin{bmatrix} 1 & 0 \\ 0 & 1 \end{bmatrix} \qquad BA = \begin{bmatrix} 1 & 0 \\ 0 & 1 \end{bmatrix} \times \begin{bmatrix} 1 & 4 \\ 3 & 5 \end{bmatrix}$$

$$= \begin{bmatrix} 1 \times 1 + 4 \times 0 & 1 \times 0 + 4 \times 1 \\ 3 \times 1 + 5 \times 0 & 3 \times 0 + 5 \times 1 \end{bmatrix} \qquad = \begin{bmatrix} 1 \times 1 + 0 \times 3 & 1 \times 4 + 0 \times 5 \\ 0 \times 1 + 1 \times 3 & 0 \times 4 + 1 \times 5 \end{bmatrix}$$

$$= \begin{bmatrix} 1 & 4 \\ 3 & 5 \end{bmatrix} \qquad = \begin{bmatrix} 1 & 4 \\ 3 & 5 \end{bmatrix}$$

The identity matrix

Look closely at matrix B. A square matrix with the number 1 for all the elements on the main diagonal and 0 for all the other elements is called an **identity matrix**. The identity matrix can only be defined for square matrices (matrices of order 1×1, 2×2, 3×3, 4×4 and so on). The identity matrix is commonly referred to as I. Note that the identity matrix is a specific type of diagonal matrix.

From this, it can be stated that $AI = IA = A$. The identity matrix I acts in a similar way to the number 1 when numbers are being multiplied (for example, $2 \times 1 = 1 \times 2 = 2$).

The identity matrix

$AI = IA = A$ **where A is a square matrix and I is the identity matrix of the same order as A.**

8.4.2 The permutation matrix

We mentioned at the start of the topic that a binary matrix is a matrix that consists of elements that are all either 0 or 1. Just as the identity matrix is a special type of a binary matrix, a **permutation matrix** is another special type. A permutation matrix is an $n \times n$ matrix that is a row or column permutation of the identity matrix. Each row and column of a permutation matrix contains the digit 1 exactly once. Permutation matrices reorder the rows or columns of another matrix via multiplication.

If you have a matrix Q and a permutation matrix P, then:
- QP is a column permutation of Q
- PQ is a row permutation of Q.

$$\text{Given } Q = \begin{bmatrix} 1 & 2 & 3 \\ 4 & 5 & 6 \\ 7 & 8 & 9 \end{bmatrix} \text{ and } P = \begin{bmatrix} 0 & 0 & 1 \\ 1 & 0 & 0 \\ 0 & 1 & 0 \end{bmatrix}, \text{ then:}$$

- $QP = \begin{bmatrix} 2 & 3 & 1 \\ 5 & 6 & 4 \\ 8 & 9 & 7 \end{bmatrix}$, which is a **column permutation** of Q.

This is because the permutation matrix $P = \begin{bmatrix} 0 & 0 & 1 \\ 1 & 0 & 0 \\ 0 & 1 & 0 \end{bmatrix}$ has 1s in the positions a_{13}, a_{21} and a_{32}, so column $1 \to 3$, column $2 \to 1$ and column $3 \to 2$.

- $PQ = \begin{bmatrix} 7 & 8 & 9 \\ 1 & 2 & 3 \\ 4 & 5 & 6 \end{bmatrix}$, which is a **row permutation** of Q.

This is because the permutation matrix $P = \begin{bmatrix} 0 & 0 & 1 \\ 1 & 0 & 0 \\ 0 & 1 & 0 \end{bmatrix}$ has 1s in the position a_{13}, a_{21} and a_{32}, so row $3 \to 1$, row $1 \to 2$ and row $2 \to 3$.

WORKED EXAMPLE 8 Applying the permutation matrix

Given matrix $R = \begin{bmatrix} 7 & 3 & -1 \\ 6 & 5 & 2 \\ -3 & 9 & 8 \end{bmatrix}$ and the permutation matrix, $P = \begin{bmatrix} 0 & 1 & 0 \\ 1 & 0 & 0 \\ 0 & 0 & 1 \end{bmatrix}$, answer the following.

a. State what type permutation is RP.
b. State which column does not change when performing a column permutation of R.
c. When performing a row permutation of R, state the row to which row 2 goes.

THINK	WRITE
a. Look at the order of the matrix multiplication.	a. Since the permutation matrix is the second matrix in the multiplication of RP, it is a column permutation of R.
b. 1. Look at the positions of the 1s in the permutation matrix to determine the change in columns.	b. The positions of the 1s are a_{12}, a_{21} and a_{33}, so column $1 \to 2$, $2 \to 1$ and $3 \to 3$.
2. Write the answer.	The column that doesn't change is column 3.
c. 1. Look at the positions of the 1s in the permutation matrix to determine the change in rows.	c. The position of the 1s are a_{12}, a_{21} and a_{33}, so row $2 \to 1$, $1 \to 2$ and $3 \to 3$.
2. Write the answer.	Row 2 goes to row 1.

8.4.3 Properties of multiplication of matrices

The following list is a summary of properties of multiplication of matrices. These properties hold true when A, B and C are $m \times n$ matrices, I is an identity matrix and O is a zero matrix.

Property	Example
Associative	$(AB)C = A(BC)$
Identity	$AI = A = IA$
Distributive	$(A + B)C = AC + BC$ $C(A + B) = CA + CB$
Zero matrix	$AO = O = OA$

WORKED EXAMPLE 9 Applying the rules of multiplication of matrices

Soundsmart has three types of televisions priced at \$350, \$650 and \$890 and three types of Bluetooth headphones priced at \$69, \$120 and \$250. The store owner wishes to mark up the prices of the televisions by 12% and mark down the prices of the headphones by 10%.
a. Show the prices of the televisions and the headphones as a suitable matrix.
b. Show the matrix obtained by marking up the prices of the televisions by 12% and marking down the prices of the headphones by 10%.
c. Use matrix multiplication to calculate the new prices (correct to the nearest dollar).

THINK	WRITE
a. Place the televisions and the headphones in columns and the prices in the rows. This results in a 3×2 matrix.	a. $\begin{bmatrix} 350 & 69 \\ 650 & 120 \\ 890 & 250 \end{bmatrix}$
b. 1. A 12% markup is equivalent to 112% or 1.12. A markdown of 10% is equivalent to 90% or 0.9.	b. $12\% + 100\% = 112\%,\ 100\% - 10\% = 90\%$
2. We need to multiply the first column of the price matrix from **a** by 1.12 and the second column by 0.9. This is a diagonal matrix.	$\begin{bmatrix} 1.12 & 0 \\ 0 & 0.90 \end{bmatrix}$
c. 1. Multiply the two matrices to calculate the new prices. The order of the first matrix is 3×2 and the order of the second matrix is 2×2. The resulting matrix will be of the order 3×2.	c. New prices $= \begin{bmatrix} 350 & 69 \\ 650 & 120 \\ 890 & 250 \end{bmatrix} \times \begin{bmatrix} 1.12 & 0 \\ 0 & 0.90 \end{bmatrix}$ $= \begin{bmatrix} 350 \times 1.12 + 69 \times 0 & 350 \times 0 + 69 \times 0.9 \\ 650 \times 1.12 + 120 \times 0 & 650 \times 0 + 120 \times 0.9 \\ 890 \times 1.12 + 250 \times 0 & 890 \times 0 + 250 \times 0.9 \end{bmatrix}$ $= \begin{bmatrix} 392 & 62.1 \\ 728 & 108 \\ 996.8 & 225 \end{bmatrix}$
2. Round the answers to the nearest dollar.	The marked-up prices for the three types of televisions will be \$392, \$728 and \$997. The marked-down prices for the three types of Bluetooth headphones will be \$62, \$108 and \$225.

WORKED EXAMPLE 10 Applying the multiplication rule to matrices

The number of desktop and notebook computers sold by four stores is given in the table.

	Desktop	Notebook
Store A	10	4
Store B	4	5
Store C	5	10
Store D	3	2

If the desktop computers were priced at \$1500 each and the notebook computers at \$2300 each, use matrix operations to calculate:

a. **the total sales figures of each computer at each store**
b. **the total sales figures for each store**
c. **the store that had the highest sales figures for**

    i. **desktop computers**   ii. **total sales.**

THINK	WRITE

a. 1. Set up a 4×2 matrix to represent the sales figures and a 2×2 matrix to determine the total sales of each computer at each store.

a. Sales matrix $= \begin{bmatrix} 10 & 4 \\ 4 & 5 \\ 5 & 10 \\ 3 & 2 \end{bmatrix}$

Price matrix $= \begin{bmatrix} 1500 & 0 \\ 0 & 2300 \end{bmatrix}$

2. Multiply the two matrices. The resultant matrix displays the total sales of each computer at each store.

Total sales figures at each store

$= \begin{bmatrix} 10 & 4 \\ 4 & 5 \\ 5 & 10 \\ 3 & 2 \end{bmatrix} \times \begin{bmatrix} 1500 & 0 \\ 0 & 2300 \end{bmatrix}$

$= \begin{bmatrix} 15\,000 & 9200 \\ 6000 & 11\,500 \\ 7500 & 23\,000 \\ 4500 & 4600 \end{bmatrix}$

b. 1. Use the 4×2 sales matrix from part **a** and a 2×1 column matrix to calculate the total sales figure for each store.

b. Matrix to determine the total sales $= \begin{bmatrix} 1500 \\ 2300 \end{bmatrix}$

2. Multiply the matrices. The resultant matrix displays the total sales at each store.

Note: CAS can be used to perform the multiplication.

Total sales $= \begin{bmatrix} 10 & 4 \\ 4 & 5 \\ 5 & 10 \\ 3 & 2 \end{bmatrix} \times \begin{bmatrix} 1500 \\ 2300 \end{bmatrix}$

$= \begin{bmatrix} 24\,200 \\ 17\,500 \\ 30\,500 \\ 9100 \end{bmatrix}$

The total sales figure for store A was \$24 200, store B was \$17 500, store C was \$30 500 and store D was \$9100.

c. i. Examine the matrix from part **a**. The highest sales figure for desktop computers is the highest number in the first column.

c. i. Store A has the highest sales figure for desktop computers: \$15 000.

ii. Examine the matrix from part **b**. The store with the highest total sales figure is the row with the highest number.

ii. Store C has the highest total sales figure: $30 500.

8.4 Exercise

1. **WE7** Given $A = \begin{bmatrix} 8 & 2 \\ -6 & 6 \end{bmatrix}$ and $B = \begin{bmatrix} 1 & 2 \\ 3 & -1 \end{bmatrix}$, calculate:

   a. AB
   b. BA
   c. B^2.

2. For the matrices $A = \begin{bmatrix} 1 & -2 & 3 \\ 2 & 0 & 4 \end{bmatrix}$, $B = \begin{bmatrix} 1 & 1 \\ 2 & 2 \\ 3 & 3 \end{bmatrix}$ and $C = \begin{bmatrix} 1 & 1 \\ 1 & 0 \end{bmatrix}$:

   a. write the order of the three matrices
   b. determine which of these products exist.

   i. AB
   ii. AC
   iii. BA
   iv. CA
   v. BC

3. Consider the following matrices.

$$A = \begin{bmatrix} 2 & 4 \\ -3 & 3 \end{bmatrix}, \; B = \begin{bmatrix} 2 & -2 & 0 \\ 3 & 1 & 4 \end{bmatrix} \text{ and } C = \begin{bmatrix} 0 & -2 \\ 3 & 5 \end{bmatrix}$$

   a. Write the order of the three matrices.
   b. Determine which of the following products exist.

   i. AB
   ii. BA
   iii. AC
   iv. CA
   v. BC
   vi. CB

   c. Write the order of the products that exist.
   d. Calculate the products that exist.

4. a. For $M = \begin{bmatrix} 4 & 5 \\ 3 & 2 \end{bmatrix}$ and $N = \begin{bmatrix} 1 & 3 \\ 2 & -1 \end{bmatrix}$, show all your working and calculate the products:

   i. MN
   ii. NM.

   b. For $M = \begin{bmatrix} a & b \\ c & d \end{bmatrix}$ and $N = \begin{bmatrix} 1 & 0 \\ 0 & 1 \end{bmatrix}$, show all your working and calculate the products:

   i. MN
   ii. NM.

   c. Explain why $MN = NM$ (from part **b** above), but $MN \neq NM$ (from part **a**).

5. **WE8** Given the matrix $R = \begin{bmatrix} 4 & 3 & -2 \\ 1 & 5 & -7 \\ -4 & 9 & 6 \end{bmatrix}$ and the permutation matrix $P = \begin{bmatrix} 0 & 1 & 0 \\ 0 & 0 & 1 \\ 1 & 0 & 0 \end{bmatrix}$, answer the following.

  a. State what type of permutation is RP.
  b. State which column does not change when performing a column permutation of R.
  c. When performing a row permutation of R, state the row to which row 1 goes.

6. Given the matrix $R = \begin{bmatrix} 2 & 3 & -7 \\ 4 & 6 & -2 \\ -1 & 9 & 8 \end{bmatrix}$ and the permutation matrix, $P = \begin{bmatrix} 0 & 1 & 0 \\ 0 & 0 & 1 \\ 1 & 0 & 0 \end{bmatrix}$, answer the following.

  a. State what type of permutation is PR.
  b. With a row permutation of R, state where row 1 moves to.
  c. If a row permutation of R resulted in $R' = \begin{bmatrix} -1 & 9 & 8 \\ 2 & 3 & -7 \\ 4 & 6 & -2 \end{bmatrix}$, then determine the permutation matrix that achieved this.

7. **MC** If $A = B \times C$, then the element a_{31} is the result of:

  A. multiplying the third row by the first column.
  B. multiplying the first row by the third column.
  C. multiplying the third column by the first row.
  D. multiplying the first column by the third row.
  E. multiplying the third row by the first row.

8. **MC** Consider the following matrices.

$$A = \begin{bmatrix} 2 & 3 & 1 \\ -1 & 4 & 5 \end{bmatrix}, \ B = \begin{bmatrix} 4 & 1 \\ 3 & 0 \\ 2 & -1 \end{bmatrix}, \ C = \begin{bmatrix} 3 & 1 \\ 2 & 5 \end{bmatrix}, \ D = \begin{bmatrix} 1 & 4 & 1 \\ 0 & 5 & 2 \\ 3 & 6 & -3 \end{bmatrix}, \ E = \begin{bmatrix} 1 & 2 & 3 & 4 \\ 0 & 1 & -1 & 4 \end{bmatrix}$$

  a. Select which one of the following products does not exist.

    A. AB B. BC C. CA D. DE E. CE

  b. Select the order of the matrix BE.

    A. 3×4 B. 2×2 C. 4×3 D. 3×2 E. 2×4

  c. Select which of the following products gives a matrix of order 2×2.

    A. BA B. BC C. CA D. AB E. BD

  d. Select the matrix CE.

    A. $\begin{bmatrix} 1 & 5 & 0 & 16 \\ 5 & 12 & 13 & 28 \end{bmatrix}$ B. $\begin{bmatrix} 2 & 9 & 1 & 28 \\ 3 & 7 & 8 & 16 \end{bmatrix}$ C. $\begin{bmatrix} 3 & 2 \\ 7 & 9 \\ 8 & 1 \\ 16 & 28 \end{bmatrix}$

    D. $\begin{bmatrix} 2 & 3 \\ 9 & 7 \\ 1 & 8 \\ 28 & 16 \end{bmatrix}$ E. $\begin{bmatrix} 3 & 7 & 8 & 16 \\ 2 & 9 & 1 & 28 \end{bmatrix}$

e. Select the matrix D^2.

A. $\begin{bmatrix} 4 & 30 & 12 \\ 6 & 37 & 16 \\ 12 & 60 & 24 \end{bmatrix}$
B. $\begin{bmatrix} 4 & 30 & 6 \\ 6 & 37 & 4 \\ -6 & 24 & 24 \end{bmatrix}$
C. $\begin{bmatrix} 4 & 30 & 6 \\ 6 & 37 & 4 \\ 0 & 24 & 24 \end{bmatrix}$

D. $\begin{bmatrix} 4 & 30 & 6 \\ 6 & 37 & 4 \\ 6 & 24 & 24 \end{bmatrix}$
E. $\begin{bmatrix} 4 & 2 & 30 & 30 & 6 & 12 \\ 6 & 6 & 37 & 24 & 4 & 8 \\ -6 & 12 & 24 & 36 & 24 & 12 \end{bmatrix}$

9. Simplify the expressions below for the following matrices.

$$A = \begin{bmatrix} 2 & 4 \\ -3 & 3 \end{bmatrix}, \ B = \begin{bmatrix} 4 & 5 \\ 3 & 2 \end{bmatrix}, \ C = \begin{bmatrix} 0 & -2 \\ 3 & 5 \end{bmatrix}, \text{ and } D = \begin{bmatrix} 1 & 3 \\ 2 & -1 \end{bmatrix}$$

a. $2A + AB$　　**b.** $A(B + C)$　　**c.** $AB + CD$　　**d.** $BA + DC$　　**e.** $2DB - D$

10. **MC** A matrix of order 3×3 is the product of three matrices S, E and N, in that order ($S \times E \times N$).

If matrix $E = \begin{bmatrix} 2 & 1 & 4 \\ 0 & 3 & 2 \\ 1 & 1 & 0 \end{bmatrix}$ then the order of matrix N is:

A. 3×2　　　　**B.** 2×3　　　　**C.** 3×3　　　　**D.** 2×2

E. Not enough information. Also need the order of matrix S.

11. **WE9** The Whitegoods Store has three types of washing machines, priced at $550, $750 and $990, and three types of dryers, priced at $160, $220 and $350. The owner of the store wishes to mark up the prices of the washing machines by 8% and mark down the prices of the dryers by 8%.

   a. Show the original prices of washing machines and dryers as a suitable matrix.

   b. Show the diagonal matrix that would mark up washing machines by 8% and mark down the dryers by 8%.

   c. Calculate the new prices (correct to the nearest dollar) using the matrices from **a** and **b**.

12. A supermarket has three types of lettuce priced at $1.50, $2.00 and $2.75 each, and three types of potatoes priced at $2.50, $3.00 and $3.25 per kilogram.
   The manager wants to mark up the prices of lettuces by 15% and mark down the prices of potatoes by 12%.

   a. Show the original prices of the lettuces and potatoes as a suitable matrix.

   b. Show the matrix obtained by marking down the prices of potatoes by 12% and marking up the price of lettuces by 15%.

   c. Use matrix multiplication to calculate the new prices (correct to the nearest cent).

13. A supermarket has three types of apples priced at $2.50, $3.50 and $4.00 per kilogram and three types of avocados priced at $0.90, $1.90 and $2.50. The manager wishes to mark up the prices of the avocados by 15% and mark down the prices of the apples by 15%.

   a. Show the original prices of the apples and avocados as a suitable matrix.

   b. Show the matrix obtained by marking down the prices of the apples by 15% and marking up the prices of the avocados by 15%.

   c. Use matrix multiplication to calculate the new prices (correct to the nearest cent).

14. A sports store has four types of tennis racquets priced at $25.00, $35.00, $95.00 and $140.00 and four types of footballs priced at $9.90, $19.90, $75.00 and $128.00. The manager wishes to mark down the prices of all items by 20% in preparation for a sale.

  a. Show the original prices of the tennis racquets and footballs as a suitable matrix.
  b. Show the matrix obtained by marking down the prices by 20%.
  c. Use matrix multiplication to calculate the sale prices (correct to the nearest 10 cents).

15. **WE10** The number of pies and cans of soft drinks sold to four year-level groups at the canteen on a particular day is given in the table below.

	Pies	**Soft drinks**
Year 12	10	26
Year 10/11	25	45
Year 8/9	22	30
Year 7	5	22

If the pies were priced at $2.50 each and the soft drinks at $1.00 each, use matrix operations to calculate:

  a. the total sales figures of each food item for each year-level group
  b. the total sales figures for each year-level group
  c. the year-level group that had the highest sales figures for

  i. pies   **ii.** total sales.

16. Given the matrix $D = \begin{bmatrix} 1 & 3 & -3 \\ 5 & 7 & 8 \\ -6 & 4 & 2 \end{bmatrix}$ and the permutation matrix, $P = \begin{bmatrix} 0 & 0 & 1 \\ 1 & 0 & 0 \\ 0 & 1 & 0 \end{bmatrix}$, answer the following questions.

  a. State what type of permutation is PD.
  b. With a row permutation of D, state where row 1 moves to.
  c. With a column permutation of D, state where column 3 moves to.
  d. **MC** Select which of the following is not a permutation matrix.

  A. $\begin{bmatrix} 0 & 0 & 1 \\ 1 & 0 & 0 \\ 0 & 1 & 0 \end{bmatrix}$ **B.** $\begin{bmatrix} 0 & 0 & 1 \\ 0 & 1 & 0 \\ 1 & 0 & 0 \end{bmatrix}$ **C.** $\begin{bmatrix} 1 & 0 & 0 \\ 0 & 0 & 1 \\ 0 & 1 & 0 \end{bmatrix}$

  D. $\begin{bmatrix} 0 & 0 & 1 \\ 1 & 0 & 0 \\ 0 & 0 & 1 \end{bmatrix}$ **E.** $\begin{bmatrix} 0 & 1 & 0 \\ 1 & 0 & 0 \\ 0 & 0 & 1 \end{bmatrix}$

  e. If a row permutation of D resulted in $D' = \begin{bmatrix} -6 & 4 & 2 \\ 1 & 3 & -3 \\ 5 & 7 & 8 \end{bmatrix}$, determine the permutation matrix that achieved this.

17. The number of iPads and iPad minis sold by four stores is given in the table shown.

	iPad	**iPad mini**
Store A	14	5
Store B	9	7
Store C	10	8
Store D	7	6

If the iPads were priced at $550 each and the iPad minis at $320 each, use matrix operations to calculate:

a. the total sales figures of iPads at each store
b. the total sales figures for each store
c. the store that had the highest sales figures for:

   i. iPads   ii. total sales.

18. Four golf shops, A, B, C and D, are supplied with boxes of different brands of golf balls (Longer, Higher, Further and Straighter), as shown in the table.

	Longer	Higher	Further	Straighter
Shop A	12	10	10	12
Shop B	15	25	15	25
Shop C	10	10	10	10
Shop D	8	20	5	18

The cost of the golf balls, per box, is:

Longer: $15   Higher: $20   Further: $30   Straighter: $32

a. Present the stock of golf balls at the four shops as a 4×4 matrix.
b. Write the costs as a matrix.
c. Use matrix multiplication to determine the total value of each golf ball brand at each shop.
d. Calculate the total value of golf balls at each shop.

19. The sales figures (in dollars) for three months, at three stores, of three brands of mobile phones are shown in the following tables.

February	Mobile A ($)	Mobile B ($)	Mobile C ($)
Store A	600	500	0
Store B	480	750	840
Store C	240	1000	0

March	Mobile A ($)	Mobile B ($)	Mobile C ($)
Store A	1200	1000	420
Store B	840	1500	1260
Store C	1200	1750	0

April	Mobile A ($)	Mobile B ($)	Mobile C ($)
Store A	1440	750	1680
Store B	600	1500	2100
Store C	1560	250	420

a. Write a matrix to represent the sales figures for each month.
b. Use a suitable matrix operation to show the total sales figures ($) for the three months by store and mobile phone brand.
c. If brand A mobile phones cost $120, brand B mobile phones cost $250 and brand C mobile phones cost $420, use an appropriate matrix operation to calculate how many mobile phones of each brand were sold in total by each store, for the three months.
d. Calculate the total sales for each store for the three months.

20. The Fibonacci numbers are a part of an interesting sequence of numbers that have been the focus of a great deal of study over the years. The first 12 numbers of the sequence are presented in the table.

Term	1	2	3	4	5	6	7	8	9	10	11	12
Number	1	1	2	3	5	8	13	21	34	55	89	144

The matrix $\begin{bmatrix} 1 & 1 \\ 1 & 0 \end{bmatrix}$ can be used to determine the terms of the Fibonacci sequence.

a. Let $A = \begin{bmatrix} 1 & 1 \\ 1 & 0 \end{bmatrix}$ and complete the following calculations using CAS.

   i. $A \times A$ (that is, A^2)   **ii.** A^3 **iii.** A^4 **iv.** A^5

b. Carefully study each of the answers found in part **a.** Explain how the elements in each answer and the power in the question relate to the Fibonacci numbers.

c. Using your answer from part **b** (that is, without performing any calculations), determine the elements in the matrix A^8.

d. Use CAS and matrix methods to calculate the 30th Fibonacci number.

8.4 Exam questions

▶ **Question 1 (1 mark)**

Source: VCE 2020, Further Mathematics Exam 1, Section B, Module 1, Q1; © VCAA.

MC The matrix $\begin{bmatrix} 1 & 0 & 0 \\ 0 & 1 & 1 \\ 0 & 0 & 1 \end{bmatrix}$ is an example of

  A. a binary matrix.   **B.** an identity matrix.   **C.** a triangular matrix.
  D. a symmetric matrix.   **E.** a permutation matrix.

▶ **Question 2 (1 mark)**

Source: VCE 2020, Further Mathematics Exam 1, Section B, Module 1, Q2; © VCAA.

MC Matrix $A = \begin{bmatrix} 1 & 2 \\ 0 & 3 \\ 1 & 0 \\ 4 & 5 \end{bmatrix}$ and matrix $B = \begin{bmatrix} 2 & 0 & 3 & 1 \\ 4 & 5 & 2 & 0 \end{bmatrix}$.

Matrix $Q = A \times B$.

The element in row i and column j of matrix Q is q_{ij}.

Element q_{41} is determined by the calculation

  A. $0 \times 0 + 3 \times 5$ **B.** $1 \times 1 + 2 \times 0$ **C.** $1 \times 2 + 2 \times 4$
  D. $4 \times 1 + 5 \times 0$ **E.** $4 \times 2 + 5 \times 4$

▶ **Question 3 (1 mark)**

Source: VCE 2019, Further Mathematics Exam 1, Section B, Module 1, Q2; © VCAA.

MC There are two rides called The Big Dipper and The Terror Train at a carnival.

The cost, in dollars, for a child to ride on each ride is shown in the table below.

Ride	Cost ($)
The Big Dipper	7
The Terror Train	8

Six children ride once only on The Big Dipper and once only on The Terror Train.

The total cost of the rides, in dollars, for these six children can be determined by which one of the following calculations?

A. $[6] \times [7 \quad 8]$

B. $[6] \times \begin{bmatrix} 7 \\ 8 \end{bmatrix}$

C. $[6 \quad 6] \times [7 \quad 8]$

D. $[6 \quad 6] \times \begin{bmatrix} 7 \\ 8 \end{bmatrix}$

E. $\begin{bmatrix} 6 \\ 6 \end{bmatrix} \times [7 \quad 8]$

More exam questions are available online.

8.5 The inverse of a matrix and its determinant

LEARNING INTENTION

At the end of this subtopic you should be able to:
- determine the inverse of a matrix
- calculate the determinant of a matrix
- apply the condition for a matrix to have an inverse
- apply matrix equations to solve for an unknown matrix.

8.5.1 Determining the multiplicative inverse of a square 2 × 2 matrix

Recall from arithmetic that any number multiplied by its reciprocal (multiplicative inverse) results in 1. For example, $8 \times \dfrac{1}{8} = 1$.

Now, consider the following matrix multiplication.

$$\begin{bmatrix} 2 & 3 \\ 3 & 5 \end{bmatrix} \times \begin{bmatrix} 5 & -3 \\ -3 & 2 \end{bmatrix} = \begin{bmatrix} 2 \times 5 + 3 \times -3 & 2 \times -3 + 3 \times 2 \\ 3 \times 5 + 5 \times -3 & 3 \times -3 + 5 \times 2 \end{bmatrix}$$

$$= \begin{bmatrix} 1 & 0 \\ 0 & 1 \end{bmatrix}$$

Notice that the answer is the identity matrix (I).

This means that one matrix is the **multiplicative inverse** of the other. In matrices, we use the symbol A^{-1} to denote the multiplicative inverse of A. A must be a square matrix.

The multiplicative inverse

If $AA^{-1} = A^{-1}A = I$, then A^{-1} is called the **multiplicative inverse of A.**

In matrices, we are often given a square 2×2 matrix and are asked to determine its multiplicative inverse. The following notes display how this is achieved.

Let A represent a matrix of the form $\begin{bmatrix} a & b \\ c & d \end{bmatrix}$.

1. Swap the elements on the main diagonal and multiply the elements on the other diagonal by -1.

   This results in the matrix $\begin{bmatrix} d & -b \\ -c & a \end{bmatrix}$.

2. Multiply this matrix by $\dfrac{1}{ad-bc}$.

$$A^{-1} = \frac{1}{ad-bc}\begin{bmatrix} d & -b \\ -c & a \end{bmatrix}$$

$$= \begin{bmatrix} \dfrac{d}{ad-bc} & \dfrac{-b}{ad-bc} \\ \dfrac{-c}{ad-bc} & \dfrac{a}{ad-bc} \end{bmatrix}$$

Note: The value of $ad-bc$ is known as the **determinant** of matrix A. It is commonly written as det A or $|A|$.

tlvd-3936

WORKED EXAMPLE 11 Calculating the determinant of a matrix

Calculate the determinant of each of the following matrices.

$$A = \begin{bmatrix} 3 & 4 \\ 2 & -3 \end{bmatrix}, B = \begin{bmatrix} 6 & 3 \\ 2 & 2 \end{bmatrix} \text{ and } C = \begin{bmatrix} 4 & 8 \\ 3 & 6 \end{bmatrix}$$

THINK	WRITE
1. For a matrix of the form $\begin{bmatrix} a & b \\ c & d \end{bmatrix}$, the determinant is given by $ad-bc$.	$\det A = ad - bc$
2. Calculate the determinant for each of the three matrices.	$\det A = 3 \times -3 - 4 \times 2$ $= -17$ $\det B = 6 \times 2 - 3 \times 2$ $= 6$ $\det C = 4 \times 6 - 8 \times 3$ $= 0$

TI | THINK	DISPLAY/WRITE	CASIO | THINK	DISPLAY/WRITE
1. On a Calculator page, define matrices A, B and C.	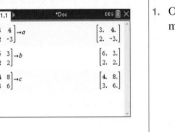	1. On a Main screen, define matrices A, B and C.	
2. Press MENU and select: 7: Matrix & Vector 3: Determinant Complete the entry line as: det(a), then press ENTER. Repeat for matrices B and C.	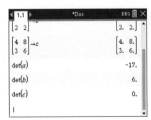	2. Tap: Action Matrix Calculation det Complete the entry line as: det(a), then tap EXE. Repeat for matrices B and C.	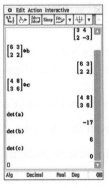

3.	The answers are shown on the screen.	a. $\det A = -17$		3.	The answers are shown on the screen.	a. $\det A = -17$
		b. $\det B = 6$				b. $\det B = 6$
		c. $\det C = 0$				c. $\det C = 0$

8.5.2 Singular and regular matrices

Notice that the determinant of matrix C in Worked example 11 was 0. Matrices with a determinant of 0 cannot have an inverse, because $\dfrac{1}{0}$ is undefined. A matrix with a determinant of 0 is called a **singular matrix**.

If the determinant of a matrix is not 0, it is called a regular matrix.

WORKED EXAMPLE 12 Determining the inverse of a matrix

If $A = \begin{bmatrix} 3 & 5 \\ 2 & 4 \end{bmatrix}$, determine its inverse, A^{-1}.

THINK

1. Calculate the determinant of A. (If the determinant is equal to 0, the inverse will not exist.)

2. Use the rule to determine the inverse.

   That is, $A^{-1} = \dfrac{1}{ad-bc}\begin{bmatrix} d & -b \\ -c & a \end{bmatrix}$. Swap the elements on the main diagonal and multiply the elements on the other diagonal by -1. Multiply this matrix by $\dfrac{1}{ad-bc}$.

WRITE

$\det A = ad - bc$
$= 3 \times 4 - 5 \times 2$
$= 2$

$A^{-1} = \dfrac{1}{2}\begin{bmatrix} 4 & -5 \\ -2 & 3 \end{bmatrix}$

Note: Fractional scalars can be left outside the matrix unless they give whole numbers when multiplied by each element.

TI	THINK	DISPLAY/WRITE	CASIO	THINK	DISPLAY/WRITE
1. On a Calculator page, define the Matrix A. Complete the entry line as: a^{-1}, then press ENTER.	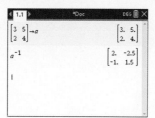	1. On a Main screen, define the Matrix A.	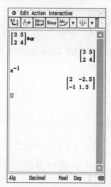
2. The answer is shown on the screen.	$A^{-1} = \begin{bmatrix} 2 & -2.5 \\ -1 & 1.5 \end{bmatrix}$	2. The answer is shown on the screen.	$A^{-1} = \begin{bmatrix} 2 & -2.5 \\ -1 & 1.5 \end{bmatrix}$

For matrices of a higher order than 2×2, for example, 3×3, 4×4 and so on, determining the inverse (and determinant) is more difficult and CAS is required.

8.5.3 Matrix equations

Recall from algebra that to solve an equation in the form $4x = 9$, we need to divide both sides by 4 (or multiply both sides by $\frac{1}{4}$) to obtain the solution $x = \frac{9}{4}$.

A matrix equation of the type $AX = B$ is solved in a similar manner. Both sides of the equation are multiplied by A^{-1}. Since the order of multiplying matrices is important, we must be careful of the position of the inverse (remember that the products of AX and XA are different).

Solving for X in the following situations:

1. For $AX = B$:

Pre-multiply by A^{-1}:
$$A^{-1}AX = A^{-1}B$$
$$IX = A^{-1}B \qquad \text{since } A^{-1}A = I$$
$$X = A^{-1}B \qquad \text{since } IX = X$$

2. For $XA = B$:

Post-multiply by A^{-1}:
$$XAA^{-1} = BA^{-1}$$
$$XI = BA^{-1} \qquad \text{since } AA^{-1} = I$$
$$X = BA^{-1} \qquad \text{since } XI = X$$

Matrix equations

1. **If $AX = B$, then $X = A^{-1}B$.**
2. **If $XA = B$, then $X = BA^{-1}$.**

▶ tlvd-3937

WORKED EXAMPLE 13 Applying matrix equations to solve for the unknown matrix

For the matrices
$$A = \begin{bmatrix} 2 & 5 \\ -1 & 3 \end{bmatrix} \text{ and } B = \begin{bmatrix} 16 & -3 \\ 3 & 7 \end{bmatrix}$$
solve for the unknown matrix X if:

a. $AX = B$ 　　　　　　　　　**b.** $AX = \begin{bmatrix} 13 \\ -1 \end{bmatrix}$.

THINK	WRITE
a. 1. We are required to pre-multiply by A^{-1} to get matrix X by itself.	**a.** $AX = B$ $A^{-1}AX = A^{-1}B$ $X = A^{-1}B$
2. Determine A^{-1}. First calculate the determinant. Then swap the elements on the leading diagonal of A and multiply the elements on the other diagonal by -1.	$\det A = ad - bc$ $= 2 \times 3 - 5 \times -1$ $= 11$ $A^{-1} = \dfrac{1}{ad - bc} \begin{bmatrix} d & -b \\ -c & a \end{bmatrix}$ $= \dfrac{1}{11} \begin{bmatrix} 3 & -5 \\ 1 & 2 \end{bmatrix}$

3. Write the equation to be solved and substitute in the matrices.

$$X = A^{-1}B$$

$$= \frac{1}{11}\begin{bmatrix} 3 & -5 \\ 1 & 2 \end{bmatrix}\begin{bmatrix} 16 & -3 \\ 3 & 7 \end{bmatrix}$$

4. Calculate the product of A^{-1} and B. Multiply each element by the fractional scalar as they will all result in whole numbers.

$$= \frac{1}{11}\begin{bmatrix} 33 & -44 \\ 22 & 11 \end{bmatrix}$$

$$= \begin{bmatrix} 3 & -4 \\ 2 & 1 \end{bmatrix}$$

b. 1. We are required to pre-multiply by A^{-1} to get matrix X by itself.

b. $$AX = \begin{bmatrix} 13 \\ -1 \end{bmatrix}$$

$$A^{-1}AX = \begin{bmatrix} 13 \\ -1 \end{bmatrix}$$

$$X = A^{-1}\begin{bmatrix} 13 \\ -1 \end{bmatrix}$$

2. Use A^{-1} from part a. Calculate the product of A^{-1} and $\begin{bmatrix} 13 \\ -1 \end{bmatrix}$. Multiply each element by the fractional scalar.

$$= \frac{1}{11}\begin{bmatrix} 3 & -5 \\ 1 & 2 \end{bmatrix}\begin{bmatrix} 13 \\ -1 \end{bmatrix}$$

$$= \frac{1}{11}\begin{bmatrix} 44 \\ 11 \end{bmatrix}$$

$$= \begin{bmatrix} 4 \\ 1 \end{bmatrix}$$

on Resources

 Interactivity Inverse matrices (int-6465)

8.5 Exercise

Students, these questions are even better in jacPLUS

Receive immediate feedback and access sample responses

Access additional questions

Track your results and progress

Find all this and MORE in jacPLUS

1. **WE11** Calculate the determinants of the following matrices.

$$A = \begin{bmatrix} 2 & 2 \\ 4 & 7 \end{bmatrix} \text{ and } B = \begin{bmatrix} 6 & 4 \\ 3 & -5 \end{bmatrix}$$

2. Calculate the determinants of the following matrices.

$$A = \begin{bmatrix} -6 & -8 \\ 15 & 12 \end{bmatrix} \text{ and } B = \begin{bmatrix} 6 & 8 \\ 3 & 4 \end{bmatrix}$$

3. Calculate the determinants of the following matrices.

a. $A = \begin{bmatrix} 9 & 8 \\ 7 & 5 \end{bmatrix}$ b. $B = \begin{bmatrix} \frac{1}{2} & -\frac{1}{6} \\ 1 & \frac{1}{3} \end{bmatrix}$ c. $C = \begin{bmatrix} 5 & 9 \\ 3 & 8 \end{bmatrix}$ d. $D = \begin{bmatrix} -3 & 6 \\ -2 & 4 \end{bmatrix}$

4. State which matrices from question 3 will not have an inverse. Explain your answer.

5. **WE12** If $A = \begin{bmatrix} 5 & 6 \\ 6 & 8 \end{bmatrix}$, determine its inverse, A^{-1}.

6. For the matrix $C = \begin{bmatrix} -5 & -1 \\ 10 & 4 \end{bmatrix}$, calculate:

   a. det C
   b. C^{-1}, the inverse of C.

7. Calculate the inverse matrix for each matrix (where possible) in question **3**.

8. If $C = \begin{bmatrix} 9 & 5 \\ 5 & 3 \end{bmatrix}$, determine its inverse, C^{-1}.

9. Given the two matrices below, show that A and B are inverses of each other.
$$A = \begin{bmatrix} 11 & 12 \\ 10 & 11 \end{bmatrix} \text{ and } B = \begin{bmatrix} 11 & -12 \\ -10 & 11 \end{bmatrix}$$

10. **MC** If $T = \begin{bmatrix} -2 & 3 \\ -2 & 5 \end{bmatrix}$, then T^{-1} is equal to:

   A. $-\dfrac{1}{4} \begin{bmatrix} 5 & -3 \\ 2 & -2 \end{bmatrix}$

   B. $\dfrac{1}{4} \begin{bmatrix} 5 & -3 \\ 2 & -2 \end{bmatrix}$

   C. $-\dfrac{1}{4} \begin{bmatrix} 5 & 3 \\ -2 & -2 \end{bmatrix}$

   D. $-\dfrac{1}{4} \begin{bmatrix} 2 & -2 \\ 3 & -5 \end{bmatrix}$

   E. $\dfrac{1}{4} \begin{bmatrix} 5 & 3 \\ 2 & 2 \end{bmatrix}$

11. **MC** If $P = \begin{bmatrix} 12 & 4 \\ -12 & -6 \end{bmatrix}$, then P^{-1} could be:

   A. $\begin{bmatrix} \dfrac{1}{2} & \dfrac{1}{6} \\ -\dfrac{1}{2} & -\dfrac{1}{4} \end{bmatrix}$

   B. $\begin{bmatrix} -\dfrac{1}{4} & -\dfrac{1}{6} \\ \dfrac{1}{2} & \dfrac{1}{2} \end{bmatrix}$

   C. $\begin{bmatrix} \dfrac{1}{20} & \dfrac{1}{30} \\ -\dfrac{1}{10} & -\dfrac{1}{10} \end{bmatrix}$

   D. $\begin{bmatrix} 0.3 & 0.2 \\ -0.5 & -0.5 \end{bmatrix}$

   E. $\begin{bmatrix} \dfrac{1}{4} & \dfrac{1}{6} \\ -\dfrac{1}{2} & -\dfrac{1}{2} \end{bmatrix}$

12. For the matrix $\begin{bmatrix} 0.1 & 0.2 \\ 0.25 & 0.45 \end{bmatrix}$:

   a. calculate the determinant

   b. state the inverse in the form $\begin{bmatrix} \dfrac{d}{ad-bc} & -\dfrac{b}{ad-bc} \\ -\dfrac{c}{ad-bc} & \dfrac{a}{ad-bc} \end{bmatrix}$.

13. Use CAS to calculate:
    i. the determinant
    ii. the inverse matrix for each of the following matrices.

a. $A = \begin{bmatrix} 5 & 3 \\ 6 & -2 \end{bmatrix}$

b. $B = \begin{bmatrix} -3 & 4 \\ 3 & -6 \end{bmatrix}$

c. $C = \begin{bmatrix} 5 & 3 \\ 7 & 4 \end{bmatrix}$

d. $D = \begin{bmatrix} 0.4 & 1.0 \\ 0.2 & 0.25 \end{bmatrix}$

e. $E = \begin{bmatrix} 5 & 2 \\ -10 & -4 \end{bmatrix}$

f. $F = \begin{bmatrix} 0.3 & 0.48 \\ 0.5 & 0.8 \end{bmatrix}$

g. $G = \begin{bmatrix} 4 & 6 & 8 \\ 6 & 4 & 6 \\ 8 & 6 & 4 \end{bmatrix}$

h. $H = \begin{bmatrix} 12 & 8 & 0 & 4 \\ 8 & 4 & 8 & 8 \\ 4 & 12 & 8 & 12 \\ 8 & 4 & 4 & 8 \end{bmatrix}$

i. $I = \begin{bmatrix} 0 & 1 & 3 & 4 \\ 2 & 0 & 3 & 4 \\ 2 & 1 & 0 & 4 \\ 2 & 1 & 3 & 0 \end{bmatrix}$

j. $J = \begin{bmatrix} 1 & 4 \\ 5 & 3 \\ 6 & 3 \end{bmatrix}$

14. **WE13** For the matrices $A = \begin{bmatrix} 7 & 5 \\ 4 & 3 \end{bmatrix}$ and $B = \begin{bmatrix} 6 & -3 \\ 3 & 7 \end{bmatrix}$, solve for the unknown matrix X, given:

a. $AX = B$

b. $AX = \begin{bmatrix} 13 \\ -1 \end{bmatrix}$.

15. For the matrices
    $A = \begin{bmatrix} 9 & 5 \\ 7 & 4 \end{bmatrix}$ and $B = \begin{bmatrix} -4 & 8 \\ -5 & 7 \end{bmatrix}$, solve for the unknown matrix X, given:

a. $AX = B$

b. $AX = \begin{bmatrix} 12 \\ -2 \end{bmatrix}$.

16. For the matrices
    $A = \begin{bmatrix} 5 & 5 \\ 0 & 6 \end{bmatrix}$, $B = \begin{bmatrix} 6 & -3 \\ 3 & 3 \end{bmatrix}$ and $C = \begin{bmatrix} 5 & 0 \\ 3 & 6 \end{bmatrix}$, solve for the unknown matrix X if:

a. $AX = B$

b. $XA = B$

c. $XC = A$

d. $CX = \begin{bmatrix} 100 \\ 120 \end{bmatrix}$.

17. If $F = \begin{bmatrix} 4 & 3 \\ 3 & -1 \end{bmatrix}$ and $G = \begin{bmatrix} 0.2 & -1.2 \\ 0.1 & 2.8 \end{bmatrix}$, solve the following matrix equations.

a. $FX = \begin{bmatrix} 5 \\ 7 \end{bmatrix}$

b. $GX = \begin{bmatrix} -1.0 \\ 2.9 \end{bmatrix}$

18. Solve the following matrix equation for X.
    $\begin{bmatrix} 2 & 3 \\ 3 & 6 \end{bmatrix} X = \begin{bmatrix} 1 & 4 \\ -3 & 2 \end{bmatrix}$

8.5 Exam questions

Question 1 (1 mark)
Source: VCE 2021, Further Mathematics Exam 1, Section B, Module 1, Q4; © VCAA.

MC Ramon and Norma are names that contain the same letters but in a different order.

The permutation matrix that can change $\begin{bmatrix} R \\ A \\ M \\ O \\ N \end{bmatrix}$ into $\begin{bmatrix} N \\ O \\ R \\ M \\ A \end{bmatrix}$ is

A. $\begin{bmatrix} 0 & 0 & 0 & 0 & 1 \\ 1 & 0 & 0 & 0 & 0 \\ 0 & 0 & 1 & 0 & 0 \\ 0 & 1 & 0 & 0 & 0 \\ 0 & 0 & 0 & 1 & 0 \end{bmatrix}$

B. $\begin{bmatrix} 0 & 0 & 0 & 0 & 1 \\ 0 & 0 & 0 & 1 & 0 \\ 0 & 0 & 1 & 0 & 0 \\ 1 & 0 & 0 & 0 & 0 \\ 0 & 1 & 0 & 0 & 0 \end{bmatrix}$

C. $\begin{bmatrix} 1 & 0 & 0 & 0 & 0 \\ 0 & 0 & 0 & 1 & 0 \\ 0 & 0 & 0 & 0 & 1 \\ 0 & 0 & 1 & 0 & 0 \\ 0 & 1 & 0 & 0 & 0 \end{bmatrix}$

D. $\begin{bmatrix} 0 & 0 & 0 & 0 & 1 \\ 0 & 0 & 0 & 1 & 0 \\ 1 & 0 & 0 & 0 & 0 \\ 0 & 1 & 0 & 0 & 0 \\ 0 & 0 & 1 & 0 & 0 \end{bmatrix}$

E. $\begin{bmatrix} 0 & 0 & 0 & 0 & 1 \\ 0 & 0 & 0 & 1 & 0 \\ 1 & 0 & 0 & 0 & 0 \\ 0 & 0 & 1 & 0 & 0 \\ 0 & 1 & 0 & 0 & 0 \end{bmatrix}$

Question 2 (1 mark)
Source: VCE 2021, Further Mathematics Exam 1, Section B, Module 1, Q3; © VCAA.

$$ax + 4y = 10$$
$$18x + by = 6$$

MC The set of simultaneous linear equations above does **not** have a unique solution when
A. $a = 2$, $b = 36$
B. $a = 3$, $b = 22$
C. $a = 4$, $b = 20$
D. $a = 5$, $b = 12$
E. $a = 6$, $b = 14$

Question 3 (1 mark)
Source: VCE 2015, Further Mathematics Exam 1, Section B, Module 6, Q7; © VCAA.

MC Matrix P has inverse matrix P^{-1}.

Matrix P is multiplied by the scalar w ($w \neq 0$) to form matrix Q.

Matrix Q^{-1} is equal to

A. $\dfrac{1}{w}P^{-1}$ **B.** $\dfrac{1}{w^2}P^{-1}$ **C.** wP^{-1} **D.** $w^2 P^{-1}$ **E.** P^{-1}

More exam questions are available online.

8.6 Dominance and communication matrices

8.6.1 Reachability

A **directed graph** (or digraph) is a graph or network where every edge has a direction. Directed graphs can be used to represent many situations, such as traffic flow, competitions between teams or the order of activities in a production line.

As the name suggests, **reachability** is the concept of how it is possible to go from one vertex in a directed network to another. The different pathways that link the vertices are analysed.

One-, two- and three-stage pathways

Consider the directed network shown, representing possible pathways (routes) from town A to town C.

By inspection, it can be seen that there are two pathways that go directly from A to C, without passing through B. That is, there are two **one-stage pathways** from A to C. A one-stage pathway is one that includes one edge only.

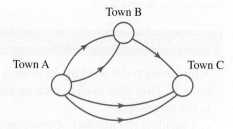

There are also two pathways that go from A to C via B. These are called **two-stage pathways**. A two-stage pathway is one that contains two edges only.

Notice that there are no routes entering town A but there are four leaving it. We say that the **indegree** of A is zero, while its **outdegree** is four. The indegree is the number of edges moving into a vertex and the outdegree is the number of edges moving away from a vertex. The indegree of B is two and its outdegree is one. The indegree of C is three and its outdegree is zero. A is the **source** and C is the **sink** of the network.

8.6.2 Matrix representation

The one-stage and two-stage pathways for a directed network can be represented in matrix form. The matrix here displays all of the possible one-stage pathways for the previous network. It is commonly known as the **adjacency matrix** and is denoted by A.

$$
\begin{array}{c}
 & \text{To} \\
 & \begin{array}{ccc} A & B & C \end{array} \\
\text{From} \begin{array}{c} A \\ B \\ C \end{array} & \left[\begin{array}{ccc} 0 & 2 & 2 \\ 0 & 0 & 1 \\ 0 & 0 & 0 \end{array} \right]
\end{array}
$$

$$
A = \left[\begin{array}{ccc} 0 & 2 & 2 \\ 0 & 0 & 1 \\ 0 & 0 & 0 \end{array} \right]
$$

Adjacency matrix

The matrix shows that there are 2 one-stage pathways from A to B and 2 one-stage pathways from A to C. There is also a one-stage pathway from B to C. Notice that the sum of each row is equal to the outdegree of each vertex and the sum of each column is equal to the indegree of each vertex. This can be a useful tip to ensure you have completed the adjacency matrix correctly.

The matrix here displays all of the possible two-stage pathways of the network. This matrix is denoted by A^2. There are 2 two-stage pathways from A to C (via B).

$$\begin{array}{c} & & \text{To} \\ & & \begin{array}{ccc} \text{A} & \text{B} & \text{C} \end{array} \\ \text{From} & \begin{array}{c} \text{A} \\ \text{B} \\ \text{C} \end{array} & \begin{bmatrix} 0 & 0 & 2 \\ 0 & 0 & 0 \\ 0 & 0 & 0 \end{bmatrix} \end{array}$$

$$A^2 = \begin{bmatrix} 0 & 0 & 2 \\ 0 & 0 & 0 \\ 0 & 0 & 0 \end{bmatrix}$$

Note that it is also possible to represent other stage pathways in matrix form; for example, three-stage is denoted by A^3, four-stage is denoted by A^4 and so on. This text will only concentrate on representing up to two-stage pathways in matrix form.

tlvd-4008

WORKED EXAMPLE 14 Calculating one-, two- and three-stage pathways

A businessperson working in town A wishes to meet with a colleague in town D. They also need to pick up some documents from town B to take to the meeting.

a. Calculate the number of two-stage paths that they could take and name them.

b. Calculate the number of three-stage paths that they could take and name them.

c. Represent the one-stage and two-stage pathways of the directed network in matrix form.

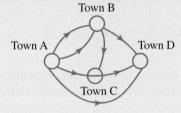

THINK	WRITE
a. 1. Determine the number of pathways from A to B (first stage).	**a.** There are two paths from A to B.
2. Determine the number of pathways from B to D (second of the two stages).	There is only one path from B to D.
3. Write the pathways.	There are 2 two-stage paths, A–B–D and A–B–D.
b. 1. Determine what two-stage pathways are possible from B to D.	**b.** The only two-stage path from B to D is B–C–D.
2. Write the pathways.	The only three-stage paths are A–B–C–D and A–B–C–D. There are two possible routes.
c. 1. Draw a matrix to show all the possible one-stage pathways throughout the network.	**c.** $$\begin{array}{c} & & \text{To} \\ & & \begin{array}{cccc} \text{A} & \text{B} & \text{C} & \text{D} \end{array} \\ \text{From} & \begin{array}{c} \text{A} \\ \text{B} \\ \text{C} \\ \text{D} \end{array} & \begin{bmatrix} 0 & 2 & 1 & 1 \\ 0 & 0 & 1 & 1 \\ 0 & 0 & 0 & 1 \\ 0 & 0 & 0 & 0 \end{bmatrix} \end{array}$$

2. The matrix can be represented without the labels along the side and the top.

$$A = \begin{bmatrix} 0 & 2 & 1 & 1 \\ 0 & 0 & 1 & 1 \\ 0 & 0 & 0 & 1 \\ 0 & 0 & 0 & 0 \end{bmatrix}$$

3. Repeat steps 1 and 2 and display all the two-stage pathways throughout the network.

$$A^2 = \begin{bmatrix} 0 & 0 & 2 & 3 \\ 0 & 0 & 0 & 1 \\ 0 & 0 & 0 & 0 \\ 0 & 0 & 0 & 0 \end{bmatrix}$$

8.6.3 Dominance

If an edge in a directed network moves from A to B, then it can be said that A is dominant, or has a greater influence, over B. If an edge moves from B to C, then B is dominant over C. However, we often wish to determine the dominant vertex in a network; that is, the vertex that holds the most influence over all the other vertices. This may be clearly seen by inspection, by examining the pathways between the vertices. It may be the vertex that has the most edges moving away from it. Generally speaking, if there are more ways to go from A to B than there are to go from B to A, then A is the dominant vertex.

In Worked example 14, town A is dominant over all the other vertices (towns) as it has edges moving to each of the other vertices. Similarly, B has edges moving to C and D, so B is dominant over C and D, and C is dominant over D. Using this inspection technique, we can list the vertices in order of dominance from A, then B, then C and finally D.

However, the dominant vertex in a directed network may not be easily determined by inspection. There may be an edge moving from A to B and another one from B to A. Which is the more dominant then? A more formal approach to determine a dominant vertex can be taken using matrix representation. Using the matrices from Worked example 15, this approach is outlined below.

Take the matrices that represent the one-stage pathways (the adjacency matrix, A) and two-stage pathways (A^2) and add them together. (When adding matrices, simply add the numbers in the corresponding positions.)

$$\begin{matrix} A & & + & & A^2 \\ \begin{bmatrix} 0 & 2 & 1 & 1 \\ 0 & 0 & 1 & 1 \\ 0 & 0 & 0 & 1 \\ 0 & 0 & 0 & 0 \end{bmatrix} & + & \begin{bmatrix} 0 & 0 & 2 & 3 \\ 0 & 0 & 0 & 1 \\ 0 & 0 & 0 & 0 \\ 0 & 0 & 0 & 0 \end{bmatrix} & = & \begin{bmatrix} 0 & 2 & 3 & 4 \\ 0 & 0 & 1 & 2 \\ 0 & 0 & 0 & 1 \\ 0 & 0 & 0 & 0 \end{bmatrix} \end{matrix}$$

The resulting matrix, which we will call the **dominance matrix**, consists of all the possible one- and two-stage pathways in the network. By taking the sum of each row in this matrix, we can determine the **dominant vertex**. The dominant vertex belongs to the row that has the highest sum.

The first row corresponds with vertex A and has a sum of 9. Row 2 (vertex B) has a sum of 3, row 3 (vertex C) has a sum of 1 and row 4 (vertex D) has a sum of 0. The highest sum is 9, so the dominant vertex is A. The order of dominance is the same as for the inspection technique described earlier.

This formal approach just described is not the only technique used to determine dominance in a network. Other approaches are possible, but this section will concentrate only on the inspection technique and summing the rows of the matrix that results from $A + A^2$.

The concept of dominance can be applied to various situations such as transportation problems, competition problems and situations involving relative positions.

WORKED EXAMPLE 15 Determining the dominance matrix

The results of a round robin (each competitor plays each other once) tennis competition are represented by the directed graph shown.

a. By inspection, determine the dominant vertex (dominant competitor) — that is, the winner. Rank the competitors in finishing order.

b. Confirm your answer to part a by determining the matrix, $A + A^2$, and summing the rows of this matrix.

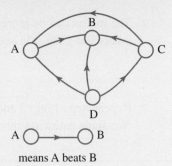

means A beats B

THINK

a. Vertex D has arrows moving to all other vertices, so D is the dominant vertex. Vertex C is dominant over B and A, and vertex A is dominant over B. List the competitors (vertices) in finishing order. (It is clear that B is the loser as all the arrows lead into it.)

b. 1. Determine the adjacency matrix, A (representing all of the one-stage pathways in the directed graph).

2. Determine A^2, the matrix representing all the two-stage pathways in the matrix.

3. Determine the resultant matrix when A and A^2 are added (add the numbers in the corresponding positions).

4. Determine the sum of each row of the resultant matrix.

5. Write the answer as a sentence. The row with the highest sum is the dominant vertex and the row with the lowest sum is the least dominant vertex.

WRITE

a. The finishing order from first to fourth is D, C, A and B.

b. $A = \begin{bmatrix} 0 & 1 & 0 & 0 \\ 0 & 0 & 0 & 0 \\ 1 & 1 & 0 & 0 \\ 1 & 1 & 1 & 0 \end{bmatrix}$

$A^2 = \begin{bmatrix} 0 & 0 & 0 & 0 \\ 0 & 0 & 0 & 0 \\ 0 & 1 & 0 & 0 \\ 1 & 2 & 0 & 0 \end{bmatrix}$

The resultant matrix is: $\begin{bmatrix} 0 & 1 & 0 & 0 \\ 0 & 0 & 0 & 0 \\ 1 & 2 & 0 & 0 \\ 2 & 3 & 1 & 0 \end{bmatrix}$

The sum of the first row (vertex A) is 1, the second row (vertex B) is 0, the third row (vertex C) is 3 and the final row (vertex D) is 6.

D is the winner, followed by C, then A and finally B.

8.6.4 Communication

In Worked example 15 we looked at results of a round robin tennis competition. This is where one competitor either wins or loses to another competitor. Thus, the network has arrows going only one way. In a **communication network**, the arrows can travel both ways. A communication network contains a set of people who can have a one-way or two-way communication link. A two-way communication link could be via phone. We can set up **communication matrices** from communication networks.

WORKED EXAMPLE 16 Determining the communication matrix from the communication networks

Determine the communication matrix from the following communication network.

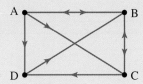

THINK

1. Set up the communication matrix with 'send call' being the vertical and 'receive call' being the horizontal.

2. Place 1 where A can communicate (with B, C and D).

3. Place 1 where B can communicate (with A and C).

4. Place 1 where C can communicate (with B and D).

5. Place 1 where D can communicate (with B).

WRITE

1.
$$\begin{array}{c} \\ A \\ B \\ C \\ D \end{array} \begin{array}{cccc} A & B & C & D \\ \left[\begin{array}{cccc} & & & \\ & & & \\ & & & \\ & & & \end{array} \right] \end{array}$$

2.
$$\begin{array}{c} \\ A \\ B \\ C \\ D \end{array} \begin{array}{cccc} A & B & C & D \\ \left[\begin{array}{cccc} 0 & 1 & 1 & 1 \\ & & & \\ & & & \\ & & & \end{array} \right] \end{array}$$

3.
$$\begin{array}{c} \\ A \\ B \\ C \\ D \end{array} \begin{array}{cccc} A & B & C & D \\ \left[\begin{array}{cccc} 0 & 1 & 1 & 1 \\ 1 & 0 & 1 & 0 \\ & & & \\ & & & \end{array} \right] \end{array}$$

4.
$$\begin{array}{c} \\ A \\ B \\ C \\ D \end{array} \begin{array}{cccc} A & B & C & D \\ \left[\begin{array}{cccc} 0 & 1 & 1 & 1 \\ 1 & 0 & 1 & 0 \\ 0 & 1 & 0 & 1 \\ & & & \end{array} \right] \end{array}$$

5.
$$\begin{array}{c} \\ A \\ B \\ C \\ D \end{array} \begin{array}{cccc} A & B & C & D \\ \left[\begin{array}{cccc} 0 & 1 & 1 & 1 \\ 1 & 0 & 1 & 0 \\ 0 & 1 & 0 & 1 \\ 0 & 1 & 0 & 0 \end{array} \right] \end{array}$$

6. Write the answer.

The communication matrix for the communication network shown is:

$$
\begin{array}{c}
\quad\ \ A\ \ B\ \ C\ \ D \\
\begin{array}{c} A \\ B \\ C \\ D \end{array}
\begin{bmatrix}
0 & 1 & 1 & 1 \\
1 & 0 & 1 & 0 \\
0 & 1 & 0 & 1 \\
0 & 1 & 0 & 0
\end{bmatrix}
\end{array}
$$

 Resources

 Interactivity The adjacency matrix (int-6466)

8.6 Exercise

1. **WE14** For the directed network shown, determine the number and name of the:

  a. **i.** one-stage paths

     ii. two-stage paths

     iii. three-stage paths from A to D.

  b. Represent the one-stage and two-stage pathways of the directed network in matrix form.

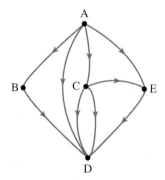

2. For the directed network shown, determine the number and name of the:

  a. **i.** one-stage paths

     ii. two-stage paths

     iii. three-stage paths from A to D.

  b. Represent the one-stage and two-stage pathways of the directed network in matrix form.

3. For each directed network shown, determine the number and name of the:

a.
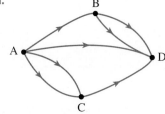

b.
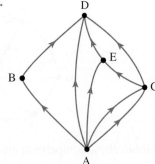

    i. one-stage paths
    ii. two-stage paths
    iii. three-stage paths from A to D.

4. Represent the one-stage and two-stage pathways of the directed networks in question 3 in matrix form.

5. The directed graph represents part of a river system where the water flows from the lake, L_1, to another lake, L_2. If fish eggs flow from L_1 to L_2, determine via how many different routes it is possible for the eggs to go. Name all the routes.

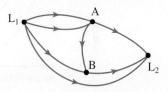

6. The bus routes between certain landmarks are shown in the diagram. Name all the different routes by which it is possible to reach:

a. D from A
b. A from D
c. B from D.

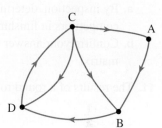

7. The directed network shown represents the pathways available to students as they move around their school. Name the different pathways by which the students can get from the:

a. office to the gym
b. common room to the science block
c. science block to the common room given that they wish to make a:

    i. one-stage trip
    ii. two-stage trip
    iii. three-stage trip.

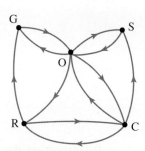

8. **WE15** Determine the dominant vertex for the following directed graph.

G = gym
S = science block
O = office
C = cafeteria
R = common room

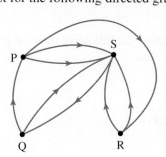

9. Determine the dominant vertex for the following directed graph.

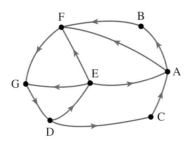

10. The results of a round robin chess competition are represented by the directed graph shown.

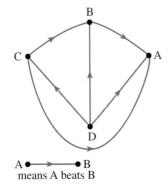

A •——→• B
means A beats B

a. By inspection, determine the dominant vertex (dominant competitor) — that is, the winner. Rank the competitors in finishing order.
b. Confirm your answer to part a by determining the matrix, $A + A^2$, and summing the rows of this resultant matrix.

11. The results of a round robin basketball competition are represented by the directed graph shown.

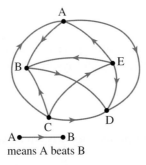

A •——→• B
means A beats B

a. By inspection, determine the dominant vertex (dominant team) — that is, the winner. Rank the teams in finishing order.
b. Confirm your answer to part a by determining the matrix, $A + A^2$, and summing the rows of this resultant matrix.

12. Determine the dominant vertex for each of the following directed graphs.

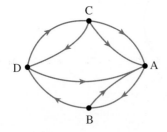

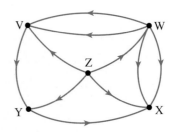

13. **MC** In the directed graph shown, the dominant vertex (by inspection) is:

A. Q
B. R
C. S
D. T
E. U

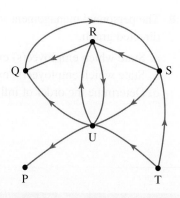

14. **WE16** Determine the communication matrix from the following communication network.

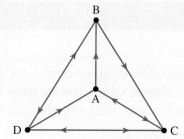

15. Given the communication matrix shown, answer the following questions.

$$
\begin{array}{c c c c c}
 & A & B & C & D \\
A & 0 & 0 & 1 & 0 \\
B & 1 & 0 & 0 & 1 \\
C & 1 & 1 & 0 & 1 \\
D & 1 & 1 & 0 & 0
\end{array}
$$

a. State who C can talk to.
b. State who B can receive calls from.
c. Explain why the main diagonal is all zeroes.
d. State who D cannot call.

16. Determine the communication matrix from the following communication network.

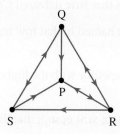

17. Given the following communication matrix, answer the following questions.

$$
\begin{array}{c c c c c}
 & A & B & C & D \\
A & 0 & 0 & 1 & 1 \\
B & 0 & 0 & 0 & 1 \\
C & 1 & 0 & 0 & 1 \\
D & 0 & 1 & 1 & 0
\end{array}
$$

a. State who C can talk to.
b. State who B can receive calls from.
c. State who can call only one person.
d. State who D cannot call.
e. State who can receive calls from everyone.

18. The personnel management roles of six employees are shown in the directed graph.

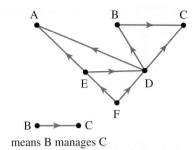

B •→• C
means B manages C

   a. State which employee(s) exerts the most influence in this group.
   b. State which employee(s) exerts the least influence in the group.
   c. Determine the order of influence of all six employees.

8.6 Exam questions

Question 1 (1 mark)

Source: VCE 2020, Further Mathematics Exam 1, Section B, Module 1, Q9; © VCAA.

MC Five competitors, Andy (*A*), Brie (*B*), Cleo (*C*), Della (*D*) and Eddie (*E*), participate in a darts tournament. Each competitor plays each of the other competitors once only, and each match results in a winner and a loser.

The matrix below shows the results of this darts tournament.

There are still two matches that need to be played.

$$
winner \quad
\begin{array}{c}
A \\ B \\ C \\ D \\ E
\end{array}
\overset{\begin{array}{ccccc} & & loser & & \\ A & B & C & D & E \end{array}}{
\begin{bmatrix}
0 & \ldots & 0 & 1 & 0 \\
\ldots & 0 & 1 & 0 & 1 \\
1 & 0 & 0 & \ldots & 1 \\
0 & 1 & \ldots & 0 & 0 \\
1 & 0 & 0 & 1 & 0
\end{bmatrix}}
$$

A '1' in the matrix shows that the competitor named in that row defeated the competitor named in that column.

For example, the '1' in row 2, column 3 shows that Brie defeated Cleo.

A '…' in the matrix shows that the competitor named in that row has not yet played the competitor named in that column.

The winner of this darts tournament is the competitor with the highest sum of their one-step and two-step dominances.

Which player, by winning their remaining match, will ensure that they are ranked first by the sum of their one-step and two-step dominances?
   A. Andy
   B. Brie
   C. Cleo
   D. Della
   E. Eddie

Question 2 (1 mark)

Source: VCE 2020, Further Mathematics Exam 1, Section B, Module 1, Q5; © VCAA.

MC The diagram below shows the direct communication links that exist between Sam (*S*), Tai (*T*), Umi (*U*) and Vera (*V*). For example, the arrow from Umi to Vera indicates that Umi can communicate directly with Vera.

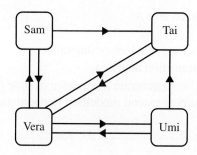

A communication matrix can be used to convey the same information.

In this matrix:

- a '1' indicates that a direct communication link exists between a sender and a receiver
- a '0' indicates that a direct communication link does not exist between a sender and a receiver.

The communication matrix could be

A.

$$\begin{array}{c} \\ sender \end{array} \begin{array}{c} \\ S \\ T \\ U \\ V \end{array} \overset{\displaystyle receiver}{\overset{\displaystyle S\ \ T\ \ U\ \ V}{\begin{bmatrix} 0 & 1 & 0 & 1 \\ 0 & 0 & 0 & 1 \\ 0 & 1 & 0 & 1 \\ 1 & 0 & 1 & 0 \end{bmatrix}}}$$

B.

$$\begin{array}{c} \\ sender \end{array} \begin{array}{c} \\ S \\ T \\ U \\ V \end{array} \overset{\displaystyle receiver}{\overset{\displaystyle S\ \ T\ \ U\ \ V}{\begin{bmatrix} 0 & 1 & 0 & 1 \\ 1 & 0 & 0 & 1 \\ 0 & 1 & 0 & 1 \\ 1 & 1 & 1 & 0 \end{bmatrix}}}$$

C.

$$\begin{array}{c} \\ sender \end{array} \begin{array}{c} \\ S \\ T \\ U \\ V \end{array} \overset{\displaystyle receiver}{\overset{\displaystyle S\ \ T\ \ U\ \ V}{\begin{bmatrix} 0 & 1 & 0 & 1 \\ 0 & 0 & 0 & 1 \\ 0 & 1 & 0 & 0 \\ 1 & 1 & 1 & 0 \end{bmatrix}}}$$

D.

$$\begin{array}{c} \\ sender \end{array} \begin{array}{c} \\ S \\ T \\ U \\ V \end{array} \overset{\displaystyle receiver}{\overset{\displaystyle S\ \ T\ \ U\ \ V}{\begin{bmatrix} 0 & 1 & 0 & 1 \\ 0 & 0 & 0 & 1 \\ 0 & 1 & 0 & 1 \\ 1 & 1 & 1 & 0 \end{bmatrix}}}$$

E.

$$\begin{array}{c} \\ sender \end{array} \begin{array}{c} \\ S \\ T \\ U \\ V \end{array} \overset{\displaystyle receiver}{\overset{\displaystyle S\ \ T\ \ U\ \ V}{\begin{bmatrix} 0 & 1 & 0 & 2 \\ 0 & 0 & 0 & 2 \\ 0 & 1 & 0 & 2 \\ 2 & 2 & 2 & 0 \end{bmatrix}}}$$

Question 3 (1 mark)

Source: VCE 2019, Further Mathematics Exam 1, Section B, Module 1, Q7; © VCAA.

MC The communication matrix below shows the direct paths by which messages can be sent between two people in a group of six people, *U* to *Z*.

$$\begin{array}{c} \\ sender \end{array} \begin{array}{c} \\ U \\ V \\ W \\ X \\ Y \\ Z \end{array} \overset{\displaystyle receiver}{\overset{\displaystyle U\ \ V\ \ W\ \ X\ \ Y\ \ Z}{\begin{bmatrix} 0 & 1 & 1 & 0 & 1 & 1 \\ 1 & 0 & 1 & 0 & 1 & 0 \\ 1 & 1 & 0 & 1 & 0 & 1 \\ 0 & 1 & 0 & 0 & 1 & 1 \\ 0 & 0 & 1 & 1 & 0 & 1 \\ 1 & 1 & 0 & 1 & 1 & 0 \end{bmatrix}}}$$

A '1' in the matrix shows that the person named in that row can send a message directly to the person named in that column.

For example, the '1' in row 4, column 2 shows that *X* can send a message directly to *V*.

In how many ways can *Y* get a message to *W* by sending it directly to **one** other person?

A. 0   **B.** 1   **C.** 2   **D.** 3   **E.** 4

More exam questions are available online.

8.7 Transition matrices and Leslie matrices

8.7.1 Powers of matrices

Andrei Markov was a Russian mathematician whose name is given to a technique that calculates probability associated with the state of various transitions (which can be represented in matrix form). The Markov system or chain answers questions such as, 'What is the probability that it will rain today given that it rained yesterday?' or 'What can be said about the long-term prospect of rainy days?'

Throughout this section, it will be necessary to evaluate a matrix raised to the power of a particular number, for example M^3. Only square matrices can be raised to a power, as the order of a non-square matrix does not allow for repeated matrix multiplication.

For example, a 2×3 matrix cannot be squared, because using the multiplication rule, we see the inner two numbers are not the same ($2 \times 3 \times 2 \times 3$).

tlvd-4009

WORKED EXAMPLE 17 Evaluating powers of matrices

Consider the following matrices.

$$A = \begin{bmatrix} 0.7 & 0.4 \\ 0.3 & 0.6 \end{bmatrix}, \quad B = \begin{bmatrix} 0.81 & 0.6 \\ 0.19 & 0.4 \end{bmatrix} \text{ and } C = \begin{bmatrix} 0.6 & 0.1 & 0.3 \\ 0.3 & 0.6 & 0.1 \\ 0.1 & 0.3 & 0.6 \end{bmatrix}$$

a. Use CAS to evaluate:
  i. A^3 **ii.** B^2 **iii.** C^0.

b. Use CAS to evaluate C^{40}, expressing the matrix in whole numbers multiplied by a fractional scalar.

THINK	WRITE
a. i. Enter matrix A in CAS and raise it to the power of 3.	**a. i.** $A^3 = \begin{bmatrix} 0.7 & 0.4 \\ 0.3 & 0.6 \end{bmatrix}^3$ $= \begin{bmatrix} 0.583 & 0.556 \\ 0.417 & 0.444 \end{bmatrix}$
ii. Enter matrix B in CAS and square it.	**ii.** $B^2 = \begin{bmatrix} 0.81 & 0.6 \\ 0.19 & 0.4 \end{bmatrix}^2$ $= \begin{bmatrix} 0.7701 & 0.726 \\ 0.2299 & 0.274 \end{bmatrix}$

iii. Enter matrix C in CAS and raise it to the power of 0.

iii. $C^0 = \begin{bmatrix} 5.6 & 0.1 & 0.3 \\ 0.3 & 0.6 & 0.1 \\ 0.1 & 0.3 & 0.6 \end{bmatrix}^0$

$= \begin{bmatrix} 1 & 0 & 0 \\ 0 & 1 & 0 \\ 0 & 0 & 1 \end{bmatrix}$

b. 1. Raise matrix C to the power of 40.

b. $C^{40} = \begin{bmatrix} 0.6 & 0.1 & 0.3 \\ 0.3 & 0.6 & 0.1 \\ 0.1 & 0.3 & 0.6 \end{bmatrix}^{40}$

$= \begin{bmatrix} 0.333\ 333 & 0.333\ 333 & 0.333\ 333 \\ 0.333\ 333 & 0.333\ 333 & 0.333\ 333 \\ 0.333\ 333 & 0.333\ 333 & 0.333\ 333 \end{bmatrix}$

2. Identify the fractional scalar common to each element and place it outside the matrix to give your answer using whole number elements.

$C^{40} = \dfrac{1}{3} \begin{bmatrix} 1 & 1 & 1 \\ 1 & 1 & 1 \\ 1 & 1 & 1 \end{bmatrix}$

TI	THINK	DISPLAY/WRITE	CASIO	THINK	DISPLAY/WRITE

a. 1. On a Calculator page, define matrices A, B and C.

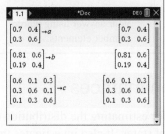

a. 1. On a Main screen, define matrices A, B and C.

2. Complete the entry lines as:
  i. a^3, then press ENTER
  ii. b^2, then press ENTER
  iii. c^0, then press ENTER.

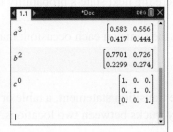

2. Complete the entry lines as:
  i. a^3, then tap EXE
  ii. b^2, then tap EXE
  iii. c^0, then tap EXE.

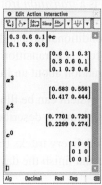

3. The answers are shown on the screen.

i. $a^3 = \begin{bmatrix} 0.583 & 0.556 \\ 0.417 & 0.444 \end{bmatrix}$

ii. $b^2 = \begin{bmatrix} 0.7701 & 0.726 \\ 0.2299 & 0.274 \end{bmatrix}$

iii. $c^0 = \begin{bmatrix} 1 & 0 & 0 \\ 0 & 1 & 0 \\ 0 & 0 & 1 \end{bmatrix}$

b. 1. Complete the entry line as: c^{40}, then press ENTER.

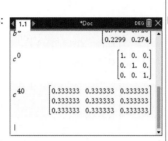

2. The answer is on the screen. Identify the fractional scalar common to each element and place it outside the matrix to give your answer using whole number elements.

$C^{40} = \dfrac{1}{3} \begin{bmatrix} 1 & 1 & 1 \\ 1 & 1 & 1 \\ 1 & 1 & 1 \end{bmatrix}$

3. The answers are shown on the screen.

i. $a^3 = \begin{bmatrix} 0.583 & 0.556 \\ 0.417 & 0.444 \end{bmatrix}$

ii. $b^2 = \begin{bmatrix} 0.7701 & 0.726 \\ 0.2299 & 0.274 \end{bmatrix}$

iii. $c^0 = \begin{bmatrix} 1 & 0 & 0 \\ 0 & 1 & 0 \\ 0 & 0 & 1 \end{bmatrix}$

b. 1. Complete the entry line as: c^{40}, then tap EXE.

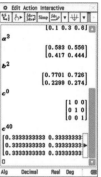

2. The answer is on the screen. Identify the fractional scalar common to each element and place it outside the matrix to give your answer using whole number elements.

$C^{40} = \dfrac{1}{3} \begin{bmatrix} 1 & 1 & 1 \\ 1 & 1 & 1 \\ 1 & 1 & 1 \end{bmatrix}$

8.7.2 Markov systems and transition matrices

A **Markov system** (or Markov chain) is a system used for estimating the distribution of states (condition or location) of an event, given information about the current states. It also investigates the manner of changes from one state to the next, according to fixed probabilities. Matrices can be used to model such situations where:

- there are defined sets of conditions or *states*
- there is a *transition* from one state to the next, where the next state's probability is *conditional* on the result of the preceding outcome
- the conditional probabilities for each outcome are the same on each occasion; that is, the same matrix is used for each transition
- information about an **initial state** is given.

A Markov system can be illustrated by means of a state transition statement, a table or a diagram. The following transition statements describe the movement of delivery trucks between two locations.

A group of delivery trucks transfer goods between two warehouses A and B. They start the day at either warehouse and finish the day parked at one of them.

70% of the trucks that start at A will park at A that night and 30% will park at B.

60% of the trucks that start at B will park at B that night and 40% will park at A.

This is how the statements are represented as a table and as a diagram. (In both, the values have been expressed as probabilities.)

	Transition from	
	Warehouse A	Warehouse B
Transition to — Warehouse A	0.7	0.4
Warehouse B	0.3	0.6

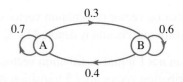

The statements, the table and the diagram all represent the same information. They can all be summarised as a **transition matrix** as shown. Throughout this section, the transition matrix will be denoted as T.

$$T = \begin{bmatrix} 0.7 & 0.4 \\ 0.3 & 0.6 \end{bmatrix}$$

Note that each of the columns of a transition matrix must add up to 1.

WORKED EXAMPLE 18 Representing a transition diagram as a transition matrix

Represent each of the following as a transition matrix.

a.

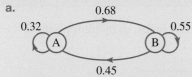

b. **There are a number of train carriages operating between two depots, North depot and South depot. At the end of each week, 40% of the carriages that started at North depot end up at South depot and 25% of the carriages that started at South depot end up at North depot.**

THINK

a. 1. Identify that there are two states, A and B. Enter the values in the correct manner — that is, *from* the column *to* the row. 0.32 from A to A and 0.68 from A to B. 0.55 from B to B and 0.45 from B to A.

   2. Check that each column adds up to 1 and remove any labels.

b. 1. Identify that there are two states, North depot and South depot. Enter the given percentage probabilities in decimal form.

   2. The missing values can be calculated knowing the columns must add up to 1. Remove any labels.

WRITE

a.
$$\begin{array}{c} \\ \text{To} \end{array} \begin{array}{cc} & \text{From} \\ & \begin{array}{cc} A & B \end{array} \\ \begin{array}{c} A \\ B \end{array} & \begin{bmatrix} 0.32 & 0.45 \\ 0.68 & 0.55 \end{bmatrix} \end{array}$$

$$T = \begin{bmatrix} 0.32 & 0.45 \\ 0.68 & 0.55 \end{bmatrix}$$

b.
$$\begin{array}{c} \\ \text{To} \end{array} \begin{array}{cc} & \text{From} \\ & \begin{array}{cc} \text{North} & \text{South} \end{array} \\ \begin{array}{c} \text{North} \\ \text{South} \end{array} & \begin{bmatrix} & 0.25 \\ 0.40 & \end{bmatrix} \end{array}$$

$$T = \begin{bmatrix} 0.60 & 0.25 \\ 0.40 & 0.75 \end{bmatrix}$$

8.7.3 Distribution vector and powers of the transition matrix

A distribution vector is a column vector with an entry for each state of the system. It is often referred to as the **initial state matrix** and is denoted by S_0.

If S_0 is an $n \times 1$ initial distribution vector state matrix involving n components and T is the transition matrix, then the distribution vector after 1 transition is the matrix product $T \times S_0$.

Distribution after 1 transition: $S_1 = T \times S_0$

The distribution one stage later is given by:

Distribution after 2 transition: $S_2 = T \times S_1$
$$= T \times (T \times S_0)$$
$$= (T \times T) \times S_0$$
$$= T^2 \times S_0$$

This pattern can continue to create a matrix recurrence relation.

Matrix recurrence relation
The matrix recurrence relation $S_{n+1} = TS_n$, where T is a transition matrix and S_n is a column state matrix, will generate a sequence of state matrices.

The distribution after n transitions can be obtained by pre-multiplying S_0 by T, n times, or by multiplying T^n by S_0.

Distribution after n transitions

$$S_n = T \times S_{n-1}$$
$$= T^n \times S_0$$

The sequence of states S_0, S_1, S_2, ..., S_n is called a Markov chain.

8.7.4 Applications to marketing

One common use for the above approach is in marketing, where organisations can predict their share of the market at any given moment. Marketing records show that when consumers are able to purchase certain goods — for example, groceries — from competing stores A and B, we can associate conditional probabilities with the likelihood that they will purchase from a given store or its competitor, depending on the store from which they had made their previous purchases over a set period, such as a month.

The following worked example highlights the application of transition matrices to marketing.

WORKED EXAMPLE 19 Applying a matrix recurrence relation

A survey shows that 75% of the time, customers will continue to purchase their groceries from store A if they purchased their groceries from store A in the previous month, while 25% of the time consumers will start purchasing their groceries from store B if they purchased their groceries from store A in the previous month. Similarly, the records show that 80% of the time, consumers will continue to purchase their groceries from store B if they purchased their groceries from store B in the previous month.

a. Use a matrix recurrence relation to determine how many customers are still purchasing their groceries from A and B at the end of two months, if 300 customers started at A and 300 started at B.
b. Calculate the percentage (in whole numbers) of customers who are purchasing their groceries at A and B at the end of 6 months, if 50% of the customers started at A.

THINK	WRITE
a. 1. Set up the transition matrix using the correct methods. Complete any missing probabilities, knowing the columns must add up to 1.	a. $T = \begin{array}{c} \\ A \\ B \end{array} \begin{array}{cc} A & B \\ \begin{bmatrix} 0.75 & 0.2 \\ 0.25 & 0.8 \end{bmatrix} \end{array}$ $T = \begin{bmatrix} 0.75 & 0.20 \\ 0.25 & 0.80 \end{bmatrix}$
2. Set up the initial state matrix. The elements for this matrix are the initial number of customers for each store.	$S_0 = \begin{bmatrix} 300 \\ 300 \end{bmatrix}$
3. The number of transitions is 2. Calculate the number of customers still purchasing at A and B by using the recurrence relation $S_{n+1} = TS_n$ and evaluating S_1 and S_2.	$S_1 = TS_0$ $= \begin{bmatrix} 0.75 & 0.20 \\ 0.25 & 0.80 \end{bmatrix} \begin{bmatrix} 300 \\ 300 \end{bmatrix}$ $= \begin{bmatrix} 285 \\ 315 \end{bmatrix}$ $S_2 = TS_1$ $= \begin{bmatrix} 0.75 & 0.20 \\ 0.25 & 0.80 \end{bmatrix} \begin{bmatrix} 285 \\ 315 \end{bmatrix}$ $= \begin{bmatrix} 276.75 \\ 323.25 \end{bmatrix}$
4. Interpret the answer represented by S_2.	After 2 months, 277 customers will be purchasing their groceries from store A and 323 will be purchasing their groceries from store B.

b. 1. Identify that after 6 months, $n = 6$. The initial state matrix contains the percentage of customers for each store at the start.

b. $S_0 = \begin{bmatrix} 50 \\ 50 \end{bmatrix}$

2. Use the formula to set up the matrix calculation.

$S_n = T^n \times S_0$

$S_6 = T^6 \times S_0$

$= \begin{bmatrix} 0.75 & 0.20 \\ 0.25 & 0.80 \end{bmatrix}^6 \times \begin{bmatrix} 50 \\ 50 \end{bmatrix}$

3. Use CAS to evaluate S_6.

$S_6 = \begin{bmatrix} 44.5982 \\ 55.4018 \end{bmatrix}$

4. Interpret the answer represented by S_6. (Round to the nearest per cent.)

After 6 months, 45% of the customers will be purchasing their groceries from store A and 55% of the customers will be purchasing their groceries from store B.

8.7.5 New state matrix with culling and restocking

The new state matrix $S_n = T^n S_0$ can be extended to include culling and restocking. This can be done by adding (restocking) or subtracting (culling) a matrix to our original new state matrix.

A matrix recurrence relation with culling and restocking

$$S_{n+1} = TS_n + B, \text{ where } B \text{ is a matrix (usually a column matrix)}$$

WORKED EXAMPLE 20 Applying the matrix recurrence relation with restocking

Betta Health Centres run concurrent Lift and Cycle fitness classes at all 3 of their gyms in FitTown.
A study shows that 80% of the clients who attended a Lift class one week will attend the Lift class the next week, while the other 20% will move to the Cycle class the next week. Similarly, 70% of the clients who attended a Cycle class one week will attend the Cycle class the next week, while the other 30% will move to the Lift class the next week.

The numbers are also affected by people joining and leaving the gym, with 2 additional people joining the Lift classes each week and 3 additional people joining the Cycle classes each week.
In the first week, 55 people attended the Lift classes and 62 people attended the Cycle classes.

Set up the matrix recurrence relation that can be used to determine how many people attended the Lift and Cycle classes in week 2.

THINK	WRITE
1. Set up a transition matrix to represent the changing numbers between the classes. Let L represent the Lift class and C represent the Cycle class.	$\begin{array}{cc} & \text{L} \quad \text{C} \end{array}$ $\begin{array}{c}\text{L}\\\text{C}\end{array}\begin{bmatrix} 0.8 & 0.3 \\ 0.2 & 0.7 \end{bmatrix}$
2. Determine the value of matrix B (represented by the additional people joining the classes).	$B = \begin{bmatrix} 2 \\ 3 \end{bmatrix}\begin{array}{l}\text{L}\\\text{C}\end{array}$
3. Set up the initial state matrix.	$S_1 = \begin{bmatrix} 55 \\ 62 \end{bmatrix}\begin{array}{l}\text{L}\\\text{C}\end{array}$
4. Enter the information into the matrix recurrence relation $S_2 = TS_1 + B$.	$S_2 = \begin{bmatrix} 0.8 & 0.3 \\ 0.2 & 0.7 \end{bmatrix}\begin{bmatrix} 55 \\ 62 \end{bmatrix} + \begin{bmatrix} 2 \\ 3 \end{bmatrix}$

WORKED EXAMPLE 21 Using matrix equations

A school was running extra Maths and English classes each week, with students given the choice of which extra classes they would attend. A matrix equation used to determine the number of students expected to attend extra classes is given by

$$S_{n+1} = \begin{bmatrix} 0.85 & 0.3 \\ 0.15 & 0.7 \end{bmatrix} S_n - \begin{bmatrix} 7 \\ 9 \end{bmatrix},$$

where S_n is the column matrix that lists the number of students attending in week n. The attendance matrix for the first week is given by

$$S_1 = \begin{bmatrix} 104 \\ 92 \end{bmatrix}\begin{array}{l}\text{Maths}\\\text{English}\end{array}$$

a. Calculate the number of students who are expected to attend extra English lessons in week 3.
b. Of the students who attended extra classes in week 3, determine how many are not expected to return for extra classes in week 4.

THINK	WRITE
a. 1. First calculate week 2 by using S_1.	**a.** $S_2 = \begin{bmatrix} 0.85 & 0.3 \\ 0.15 & 0.7 \end{bmatrix}\begin{bmatrix} 104 \\ 92 \end{bmatrix} - \begin{bmatrix} 7 \\ 9 \end{bmatrix}$ $= \begin{bmatrix} 109 \\ 71 \end{bmatrix}$
2. Calculate week 3 by using S_2 and then round off.	$S_3 = \begin{bmatrix} 0.85 & 0.3 \\ 0.15 & 0.7 \end{bmatrix}\begin{bmatrix} 109 \\ 71 \end{bmatrix} - \begin{bmatrix} 7 \\ 9 \end{bmatrix}$ $= \begin{bmatrix} 106.95 \\ 57.05 \end{bmatrix}$ $= \begin{bmatrix} 106 \\ 57 \end{bmatrix}$

3. Write the answer.

In week 3, it is expected that 57 students will attend extra English classes.

b. 1. Calculate week 4 using S_3, not using rounded numbers, but round off at the end.

b. $S_4 = \begin{bmatrix} 0.85 & 0.3 \\ 0.15 & 0.7 \end{bmatrix} \begin{bmatrix} 106.95 \\ 57.05 \end{bmatrix} - \begin{bmatrix} 7 \\ 9 \end{bmatrix}$

$= \begin{bmatrix} 101.0225 \\ 46.9775 \end{bmatrix}$

$= \begin{bmatrix} 101 \\ 46 \end{bmatrix}$

2. Calculate the total number of students doing extra classes in week 3 and subtract the total number of students doing extra classes in week 4.

Total in week 3 = 106 + 57 = 163 students.
Total in week 4 = 101 + 46 = 147 students.
Difference = 163 − 147 = 16 students.

3. Write the answer.

It is estimated that there will be 16 students who attended extra classes in week 3 but not in week 4.

8.7.6 Equilibrium state matrix

As higher and higher powers of T are taken, it can be seen that the values of the elements in the transition matrix show no noticeable difference, and approach a fixed matrix T^∞. We refer to T^∞ as the **equilibrium state** or **steady state** matrix. To test for steady state, a suitable value of n to test is 50. Then $n = 51$ should be tested. If the elements in the matrix haven't changed, then a steady state has been reached.

When there is no noticeable change from one state matrix to the next, the system is said to have reached its steady state.

If a Markov system is regular, then its long-term transition matrix is given by the square matrix whose columns are the same and equal to the steady state probability vector. This occurs as long as the transition matrix squared, T^2, has no zeros.

If T^2 contains any zeros, then it is not possible to reach a steady state.

$$T \qquad \begin{bmatrix} 0.75 & 0.2 \\ 0.25 & 0.8 \end{bmatrix}$$

$$T^2 \qquad \begin{bmatrix} 0.6125 & 0.31 \\ 0.3875 & 0.69 \end{bmatrix}$$

$$T^3 \qquad \begin{bmatrix} 0.536\,875 & 0.3705 \\ 0.463\,125 & 0.6295 \end{bmatrix}$$

$$T^4 \qquad \begin{bmatrix} 0.495\,281\,25 & 0.403\,775 \\ 0.504\,718\,75 & 0.596\,225 \end{bmatrix}$$

$$T^8 \qquad \begin{bmatrix} 0.449\,096\ldots & 0.440\,722\ldots \\ 0.550\,903\ldots & 0.559\,277\ldots \end{bmatrix}$$

$$T^{16} \qquad \begin{bmatrix} 0.444\,48\ldots & 0.444\,41\ldots \\ 0.555\,51\ldots & 0.555\,58\ldots \end{bmatrix}$$

$$T^{50} \qquad \begin{bmatrix} \dfrac{4}{9} & \dfrac{4}{9} \\ \dfrac{5}{9} & \dfrac{5}{9} \end{bmatrix}$$

$$T^{51} \qquad \begin{bmatrix} \dfrac{4}{9} & \dfrac{4}{9} \\ \dfrac{5}{9} & \dfrac{5}{9} \end{bmatrix}$$

These probabilities can be easily expressed as fractions, if n is very large.

8.7.7 Applications to weather

Predicting the long-term weather forecast is important to insurance companies who insure event organisers against losses if the event is 'rained on'. To do this they need to predict the long-term probability of rain. Suppose that for a Melbourne spring, long-run data suggests that there is a 65% chance that if today is dry, then the next day will also be dry.

Conversely, if today is wet, there is an 82% chance that the next day will also be wet. What is the long-term probability for it being a wet day if the initial day was dry? This style of problem is highlighted in the following worked example.

WORKED EXAMPLE 22 Using matrices to predict weather

An insurance company needs to measure its risk if it is to underwrite a policy for a planned major outdoor event. The company used the following information about the region.

Long-run data gathered about the region's weather suggests that there is a 75% chance that if today is dry, then the next day will also be dry.

Conversely, if today is wet, there is a 72% chance that the next day will also be wet. This information is given in the table.

	Today is dry	Today is wet
Next day is dry	0.75	0.28
Next day is wet	0.25	0.72

a. **Calculate the probability it will rain in three days' time if initially the day is dry.**
b. **Calculate the long-term probability of rain if initially the day is wet.**
c. **If the company insures only if they have the odds in their favour, determine if they will insure this event.**

THINK	WRITE
a. 1. Set up the transition matrix.	a. $T = \begin{array}{c} \text{dry} \\ \text{wet} \end{array} \begin{bmatrix} \overset{\text{dry}}{0.75} & \overset{\text{wet}}{0.28} \\ 0.25 & 0.72 \end{bmatrix}$ $= \begin{bmatrix} 0.75 & 0.28 \\ 0.25 & 0.72 \end{bmatrix}$
2. Set up the initial state matrix. For the initial day being dry, set dry as 1 and wet as 0.	$S_0 = \begin{bmatrix} 1 \\ 0 \end{bmatrix}$
3. Identify that in three days' time, $n = 3$. Substitute the matrices and use CAS to evaluate S_3.	$S_3 = T^3 \times S_0$ $= \begin{bmatrix} 0.75 & 0.28 \\ 0.25 & 0.72 \end{bmatrix}^3 \times \begin{bmatrix} 1 \\ 0 \end{bmatrix}$ $= \begin{bmatrix} 0.577\,275 \\ 0.422\,725 \end{bmatrix}$
4. Interpret the answer represented by S_3.	The probability of a dry day three days after a dry day is 57.7%.
b. 1. Identify that for a long-term steady state, n needs to be large, say $n = 50$. (Test that the same result is obtained when $n = 51$.)	b. $S_{50} = T^{50} \times S_0$

2. Set up the initial state matrix. For the initial day being wet, set dry as 0 and wet as 1.

$$S_0 = \begin{bmatrix} 0 \\ 1 \end{bmatrix}$$

3. Substitute the matrices and use CAS to evaluate S_{50} and S_{51}.

$$S_{50} = T^{50} \times S_0$$

$$= \begin{bmatrix} 0.75 & 0.28 \\ 0.25 & 0.72 \end{bmatrix}^{50} \times \begin{bmatrix} 0 \\ 1 \end{bmatrix}$$

$$= \begin{bmatrix} 0.528\,302 \\ 0.471\,698 \end{bmatrix}$$

$$S_{51} = T^{51} \times S_0$$

$$= \begin{bmatrix} 0.75 & 0.28 \\ 0.25 & 0.72 \end{bmatrix}^{51} \times \begin{bmatrix} 0 \\ 1 \end{bmatrix}$$

$$= \begin{bmatrix} 0.528\,302 \\ 0.471\,698 \end{bmatrix}$$

4. As there is no change from S_{50} to S_{51}, a steady state has been reached. Interpret the answer.

In the long run, there is a 47.2% chance it will be a wet day (or 25 in 53 chance) if the initial day is wet.

c. For it to be in the insurer's favour, there must be more than a 50% chance of it being dry.

c. In the long run, there is a 52.8% chance it will be a dry day. The odds are slightly in favour of the insurance company. Therefore, they will insure this event.

8.7.8 Applications to ecology

The **Leslie model** is used in ecology to model changes in population. It was invented by and named after Patrick Leslie. In this model, a population of a certain species is divided into age groups. The model uses birth and survival rates within those age groups to describe the growth of the female population of a species over time.

Consider the following information gathered about a population of female rabbits in a closed environment (one with no migration).

The birth rate is the average percentage of offspring born to that age group.

The survival rate is the average percentage of rabbits that survive to the next age group.

The assumption in this model is that no rabbit lives beyond 3 years.

Age group of rabbits (years)	0–<1	1–<2	2–<3
Number of rabbits (s)	80	40	60
Birth rate (b)	0.32	1.3	0.6
Survival rate (r)	0.45	0.77	0

A matrix recurrence relation can be used to generate a sequence of matrices to show the population of a species in n years.

A matrix recurrence relation with a Leslie matrix

$S_{n+1} = LS_n$, where L is a Leslie matrix and S_0 is the initial state matrix.

$S_1 = LS_0$

$S_2 = LS_1$ but $S_1 = LS_0$

So:

$S_2 = L \times LS_0$

$S_n = L^n S_0$

The Leslie matrix, L, is a square matrix that has the same number of rows and columns as the population has age groups. S_0 is a column matrix containing the initial population for each age group of the species.

For the information about the rabbit population provided in the table above, the recurrence relation would be as follows.

$$S_{n+1} = LS_n$$

$$L = \begin{bmatrix} 0.32 & 1.3 & 0.6 \\ 0.45 & 0 & 0 \\ 0 & 0.77 & 0 \end{bmatrix}, \quad S_0 = \begin{bmatrix} 80 \\ 40 \\ 60 \end{bmatrix}$$

The first row of the Leslie matrix, L, contains the birth rates for each group; the $(2, 1)$th element in the matrix is the survival rate for the age group 0–<1, and the $(3, 2)$th element in the matrix is the survival rate for the age group 1–<2. This is the last row, as no rats are expected to live beyond 3 years. S_0 contains the initial population of rabbits.

The population of rabbits in n years is:

$$S_n = \begin{bmatrix} 0.32 & 1.3 & 0.6 \\ 0.45 & 0 & 0 \\ 0 & 0.77 & 0 \end{bmatrix}^n \begin{bmatrix} 80 \\ 40 \\ 60 \end{bmatrix}$$

To determine the population of rabbits after 5 years,

$$S_5 = \begin{bmatrix} 0.32 & 1.3 & 0.6 \\ 0.45 & 0 & 0 \\ 0 & 0.77 & 0 \end{bmatrix}^5 \begin{bmatrix} 80 \\ 40 \\ 60 \end{bmatrix} = \begin{bmatrix} 113.6 \\ 36 \\ 30.8 \end{bmatrix}$$

$113.6 + 36 + 30.8 = 180.4$, so there are approximately 180 female rabbits after 5 years.

tlvd-4011

WORKED EXAMPLE 23 Applying a matrix recurrence relation with a Leslie matrix

An agriculturalist is studying the breeding and life cycle of hens in a breeding program. The hens have a three-stage life cycle: eggs, chickens and adults. Only adult females reproduce, producing on average 550 eggs in a lifetime. The agriculturalist's research has determined that on average, 72% of eggs survive to be chickens and 55% of chickens survive to adulthood. If this program has no adults, 20 eggs and 15 chickens at the start of the study, use a Leslie matrix model to determine:

a. the initial state matrix
b. the Leslie matrix
c. the number of adults after 4 cycles.

THINK	WRITE
a. 1. There are no adults, 20 eggs and 15 chickens.	**a.**
2. Write the initial state matrix.	$S_0 = \begin{bmatrix} 20 \\ 15 \\ 0 \end{bmatrix}$
b. 1. The reproduction rate is 0.72 for eggs and 0.55 for chickens.	**b.**
2. Write the Leslie matrix.	$L = \begin{bmatrix} 0 & 0 & 550 \\ 0.72 & 0 & 0 \\ 0 & 0.55 & 0 \end{bmatrix}$
	$S_n = L^n S_0$
c. 1. Write the rule.	$S_n = \begin{bmatrix} 0 & 0 & 550 \\ 0.72 & 0 & 0 \\ 0 & 0.55 & 0 \end{bmatrix}^n \begin{bmatrix} 20 \\ 15 \\ 0 \end{bmatrix}$
	$S_4 = L^4 S_0$
2. Substitute $n = 4$.	$S_4 = \begin{bmatrix} 0 & 0 & 550 \\ 0.72 & 0 & 0 \\ 0 & 0.55 & 0 \end{bmatrix}^4 \begin{bmatrix} 20 \\ 15 \\ 0 \end{bmatrix}$
3. Use CAS to calculate.	$S_4 = \begin{bmatrix} 0 & 0 & 550 \\ 0.72 & 0 & 0 \\ 0 & 0.55 & 0 \end{bmatrix}^4 \begin{bmatrix} 20 \\ 15 \\ 0 \end{bmatrix} = \begin{bmatrix} 0 \\ 3136.32 \\ 1796.85 \end{bmatrix}$
4. Write the answer.	After 4 cycles there would be approximately 1797 adults.

Other examples of the application of transition matrices are in population studies, stock inventory and sport.

8.7 Exercise

Students, these questions are even better in jacPLUS

 Receive immediate feedback and access sample responses

 Access additional questions

 Track your results and progress

Find all this and MORE in jacPLUS

The following matrices are required to answer questions 1 and 2. Define them in CAS to assist with your calculations.

$$A = \begin{bmatrix} 0.5 & 0.5 \\ 0.5 & 0.5 \end{bmatrix}, \ B = \begin{bmatrix} 0.75 & 0.25 \\ 0.25 & 0.75 \end{bmatrix}, \ C = \begin{bmatrix} 0.8 & 0.2 \\ 0.2 & 0.8 \end{bmatrix} \text{ and } D = \begin{bmatrix} 0.7 & 0.1 & 0.2 \\ 0.2 & 0.7 & 0.1 \\ 0.1 & 0.2 & 0.7 \end{bmatrix}$$

1. **WE17** Use CAS to answer the following questions.

   a. Evaluate the following correct to 2 decimal places.

   i. A^3 **ii.** B^2 **iii.** C^4 **iv.** C^8

   b. Evaluate C^{46}, expressing the elements of the matrix as fractions.

2. **a.** Use CAS to perform the following matrix operations.

    i. A^0 ii. B^0 iii. C^0 iv. D^0

  b. Describe the type of matrix produced in question **2a**.

3. **WE18** Represent the following as a transition matrix.

  a.

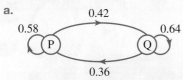

  b. There are a number of delivery trucks operating between warehouse A and warehouse B. At the end of each week, 31% of the trucks that started at warehouse A end up at warehouse B and 27% of trucks that started at warehouse B end up at warehouse A.

4. Represent each of the following as a transition matrix.

  a.

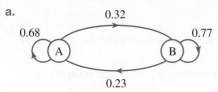

  b. There are a number of train carriages operating between North depot and South depot. At the end of each week, 20% of the carriages that started at North depot end up at South depot and 15% of the carriages that started at South depot end up at North depot.

5. **WE19** At a railway station, 250 goods wagons will be needed to carry goods from point A to point B. At the end of each week, 10% of the wagons that started the week at point A ended at point B, and 8% of the wagons that started at point B ended at point A.

  a. Write a transition matrix to represent this situation.

  b. Use a matrix recurrence relation to determine how many wagons are located at point A and point B at the end of two weeks, if 125 wagons started at point A and 125 wagons started at point B.

  c. Determine the percentage (in whole numbers) of wagons that will be at point A and point B at the end of 6 weeks, if 40% of the wagons started at A.

6. Write the transition matrices to represent the following tables of events and their transition probabilities. Complete any missing information.

  a.

		From	
		Dry	**Wet**
To	**Dry**	8596	70%
	Wet	1596	30%

  b.

		From		
		Gulius	**Grasby**	**Warlow**
	Gulius	0.65		0.03
To	**Grasby**	0.15	0.68	
	Warlow	0.12	0.97	

7. The following information relates to a survey conducted in October 2021 on supermarket shopping.

- 12% of store A customers will shop at store B the following month.
- 36% of store A customers will shop at store C the following month.
- 40% of store B customers will shop at store B the following month.
- 44% of store B customers will shop at store C the following month.
- 14% of store C customers will shop at store A the following month.
- 7% of store C customers will shop at store B the following month.

a. Represent this information as a transition matrix.

b. If the survey showed that 1200 customers shopped at store A, 800 customers shopped at store B and 1000 customers shopped at store C, determine the number of customers expected to be shopping at each store in March 2022.

c. Calculate the long-term share of the customers shopping at each store as a percentage (correct to 1 decimal place).

d. Express the answer for part c as a fraction of the customers shopping at each store.

8. If a train is late on one day, there is a 15% probability that the same train will be late the next day. If the train is on time one day, there is a 40% chance that it will be late the next day.

If, in a given week, a train arrives on time on Monday, calculate:

a. the probability that the train will be on time on the following Friday

b. the probability that the train will be late on the following Monday (assume that there are no trains operating on weekends).

9. Consider the following transition matrix.

$$T = \begin{bmatrix} 0.6 & 0.1 & 0 & 0.05 \\ 0.2 & 0.8 & 0 & 0 \\ 0.1 & 0.1 & 1 & 0 \\ 0.1 & 0 & 0 & 0.95 \end{bmatrix}$$

a. State the matrix T^2.

b. State why this matrix (T) has no steady state.

c. **MC** The transition matrix that matches the following transition diagram is:

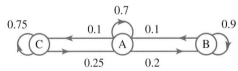

A. $\begin{bmatrix} 0.7 & 0.1 & 0.25 \\ 0.2 & 0.9 & 0 \\ 0.1 & 0 & 0.75 \end{bmatrix}$
 B. $\begin{bmatrix} 0.7 & 0.1 & 0.25 \\ 0.2 & 0.9 & 0.45 \\ 0.1 & 0.2 & 0.75 \end{bmatrix}$
 C. $\begin{bmatrix} 0.7 & 0.1 & 0.25 \\ 0.1 & 0.9 & 0 \\ 0.2 & 0 & 0.75 \end{bmatrix}$

D. $\begin{bmatrix} 0.7 & 0 & 0.25 \\ 0.2 & 0.9 & 0 \\ 0.1 & 0.1 & 0.75 \end{bmatrix}$
 E. $\begin{bmatrix} 0.7 & 0.1 & 0.75 \\ 0.2 & 0.9 & 0 \\ 0.1 & 0 & 0.25 \end{bmatrix}$

10. The following transition table describes the proportion of customers who purchase fuel from four nearby petrol stations each week.

	Petro	Texcal	Oilmart	CP
Petro	0.6	0.1	0	0.05
Texcal	0.2	0.8	0.05	0
Oilmart	0.1	0.1	0.9	0
CP	0.1	0	0.05	0.95

a. Represent the information as a transition matrix.

b. If the initial customer base is represented by the matrix $\begin{bmatrix} 4500 \\ 4000 \\ 2000 \\ 1500 \end{bmatrix}$, determine which petrol station has the largest customer base (include how many):

   i. after 2 weeks
   ii. after 8 weeks
   iii. in the long term. (*Hint:* Investigate large values of n.)

11. A farmer has two types of cows that they milk each day, the Holstein Friesian and the Jersey. They also breed them to increase the herd. A matrix recurrence relation used to determine the expected number of cows to be milked is given by

$$S_{n+1} = \begin{bmatrix} 0.60 & 0.35 \\ 0.40 & 0.65 \end{bmatrix} S_n + \begin{bmatrix} 1 \\ 2 \end{bmatrix},$$

where S_n is the column matrix that lists the number of cows on day n.
The number of cows milked on the first day is given by the matrix

$$S_1 = \begin{bmatrix} 106 \\ 72 \end{bmatrix} \begin{matrix} \text{Holstein Friesian} \\ \text{Jersey} \end{matrix}$$

a. Calculate the number of Jersey cows that are expected to be milked on day 4.
b. Determine how many more cows the farmer has to milk on day 4 compared to the first day.

12. **WE20** Top Climbers run concurrent Agility and Climb classes at all 3 of their centres in FitTown.
A study shows that 65% of the clients who attended an Agility class one week will attend the Agility class the next week, while the other 35% will move to the Climb class the next week.

Similarly, 77% of the clients who attended a Climb class one week will attend the Climb class the next week, while the other 23% will move to the Agility class the next week.

The numbers are also affected by people joining and leaving the gym, with 5 additional people joining the Agility classes each week and 2 people leaving the Climb classes each week.

In the first week 48 people attended the Agility classes and 41 people attended the Climb classes.

Set up the matrix recurrence relation that would be used to determine how many people attended the Agility and Climb classes in week 2.

13. A matrix recurrence relation representing the changing number of people who attend two different supermarkets on a weekly basis is given by

$$S_{n+1} = \begin{bmatrix} 0.61 & 0.18 \\ 0.39 & 0.82 \end{bmatrix} S_n + \begin{bmatrix} 16 \\ 7 \end{bmatrix}.$$

The first row in the state matrices represents supermarket A and the second row in the state matrices represents supermarket B.

- **a.** Calculate the percentage of people who attend supermarket A in week 1 and go back to supermarket A in week 2.
- **b.** Calculate the percentage of people who attend supermarket B in week 1 and go back to supermarket B in week 2.
- **c.** Calculate how many additional people go to each supermarket each week.

14. The Fine Drink Co. manufacture two drinks, Sweet Cola and Tangy Lemon. The manufacturing of the two products is determined by the matrix recurrence relation:

$$S_{n+1} = \begin{bmatrix} 0.84 & 0.22 \\ 0.16 & 0.78 \end{bmatrix} S_n + \begin{bmatrix} 110 \\ 71 \end{bmatrix},$$

where S_n is the column matrix that lists the number of drinks produced in week n. The number of drinks produced in the first week is given by the matrix

$$S_1 = \begin{bmatrix} 950 \\ 1050 \end{bmatrix} \begin{matrix} \text{Sweet Cola} \\ \text{Tangy Lemon} \end{matrix}$$

- **a.** Calculate the number of each type of drink produced in weeks 2, 3 and 4. Round all answers correct to a whole number.
- **b.** Determine how many more drinks in total were produced in week 4 than week 1.

15. **WE21** A school was running extra music lessons for the piano and guitar each week, with students given the choice of extra lessons to attend. A matrix recurrence relation used to determine the number of students expected to attend extra lessons is given by

$$S_{n+1} = \begin{bmatrix} 0.75 & 0.4 \\ 0.25 & 0.6 \end{bmatrix} S_n - \begin{bmatrix} 5 \\ 3 \end{bmatrix},$$

where S_n is the column matrix that lists the number of students attending in week n.

The attendance matrix for the first week is given by

$$S_1 = \begin{bmatrix} 34 \\ 26 \end{bmatrix} \begin{matrix} \text{Piano} \\ \text{Guitar} \end{matrix}$$

- **a.** Calculate the number of students who are expected to attend extra guitar lessons in week 2.
- **b.** Of the students who attended extra lessons in week 2, determine how many are not expected to return for extra lessons in week 3.

16. A farmer has their sheep moving between two paddocks, paddock A and paddock B, each week. A matrix recurrence relation used to determine the number of sheep expected to be in each of the paddocks (including the loss of sheep) is given by

$$S_{n+1} = \begin{bmatrix} 0.8 & 0.3 \\ 0.2 & 0.7 \end{bmatrix} S_n - \begin{bmatrix} 3 \\ 5 \end{bmatrix},$$

where S_n is the column matrix that lists the number of sheep in week n.
The number of sheep in each paddock in the first week is given by the matrix

$$S_1 = \begin{bmatrix} 136 \\ 108 \end{bmatrix} \begin{matrix} \text{Paddock A} \\ \text{Paddock B} \end{matrix}$$

a. Calculate the number of sheep who are expected to be in paddock B in week 4.
b. Calculate how many sheep have been lost between week 4 and week 5.

17. The population of the two main cities on an island is determined by the matrix recurrence relation

$$S_{n+1} = \begin{bmatrix} 0.95 & 0.08 \\ 0.05 & 0.92 \end{bmatrix} S_n + \begin{bmatrix} 1200 \\ 955 \end{bmatrix}$$

where S_n is the column matrix that lists the population produced in year n.
The population in 2021 is given by the matrix

$$S_1 = \begin{bmatrix} 55\,000 \\ 72\,000 \end{bmatrix} \begin{matrix} \text{City A} \\ \text{City B} \end{matrix}$$

Calculate the population of each city in 2022, 2023 and 2024. Round all answers correct to a whole number.

18. **WE22** An insurance company needs to measure its risk if it is to underwrite a policy for a planned major outdoor event. The company used the following information about the region where the event was to take place.

Long-run data gathered about the region's weather suggests that there is a 95% chance that if today is dry, then the next day will also be dry.
Conversely, if today is wet, there is a 45% chance that the next day will also be wet. This information is given in the table below.

	Today is dry	Today is wet
Next day is dry	0.95	0.55
Next day is wet	0.05	0.45

a. Calculate the probability it will rain in three days' time if initially the day is dry.
b. Calculate the long-term probability of rain if initially the day is wet.
c. If the company insures only if the odds are in its favour, explain whether the company will insure this event.

19. **MC** For $T = \begin{bmatrix} 0.5 & 0.2 & 0.3 \\ 0.3 & 0.8 & 0.3 \\ 0.2 & 0 & 0.4 \end{bmatrix}$ and $S_0 = \begin{bmatrix} 100 \\ 50 \\ 80 \end{bmatrix}$, the steady state distribution vector is:

A. $\begin{bmatrix} 0.3 \\ 0.6 \\ 0.1 \end{bmatrix}$ B. $\begin{bmatrix} 69 \\ 138 \\ 23 \end{bmatrix}$ C. $\begin{bmatrix} 30 \\ 60 \\ 100 \end{bmatrix}$ D. $\begin{bmatrix} 70 \\ 135 \\ 25 \end{bmatrix}$ E. $\begin{bmatrix} 100 \\ 50 \\ 80 \end{bmatrix}$

20. **WE26** A researcher has been collecting data on a colony of hamsters. The female hamsters live for a maximum of 3 years. In the first year of its life, an adult hamster has a fertility rate of 60%; in its second year, it has a fertility rate of 260%; and in its third year, it has a fertility rate of 50%.

The research shows that, on average, 50% of hamsters survive their first year and 60% of hamsters survive their second year. If this colony has 55 hamsters less than one year old, 32 hamsters less than two years old and 25 hamsters less than three years old, use a Leslie matrix model to determine:

a. the initial state matrix
b. the Leslie matrix
c. the total number of hamsters after 4 years.

21. There are two supermarkets to choose from in Matrixtown, and as the town grows, more shoppers shop at both supermarkets each week. A matrix recurrence relation used to determine the number of shoppers expected to shop in each of the supermarkets is given by

$$S_{n+1} = \begin{bmatrix} 0.80 & 0.25 \\ 0.20 & 0.75 \end{bmatrix} S_n + \begin{bmatrix} 6 \\ 4 \end{bmatrix},$$

where S_n is the column matrix that lists the number of shoppers in week n.
The number of shoppers who shopped at each supermarket in the first week is given by the matrix

$$S_1 = \begin{bmatrix} 256 \\ 194 \end{bmatrix} \begin{matrix} \text{Supermarket A} \\ \text{Supermarket B} \end{matrix}$$

a. Calculate the number of shoppers who are expected to shop in supermarket A in week 3.
b. Determine the difference in the number of shoppers at supermarket A compared to supermarket B after 4 weeks.

22. Many Victorians have been retiring and moving to Queensland, yet only some come from Queensland to retire in Victoria. The government of Victoria wants to investigate the expected long-term transition using the following information.

- 6000 Victorians moved to Queensland in 2021.
- 400 Queenslanders moved to Victoria in 2021.
- 73% of Victorians who moved to Queensland stayed.
- 27% of Victorians who moved to Queensland eventually returned to Victoria.
- 10% of Queenslanders who moved to Victoria stayed.
- 90% of Queenslanders who moved to Victoria returned to Queensland.

The government wants to budget for its proposed aged care program and needs to determine the long-term impact on numbers the 2021 transition will have. Calculate the net gain or loss of retirees to the state of Victoria.

8.7 Exam questions

▶ **Question 1 (1 mark)**

Source: VCE 2021, Further Mathematics Exam 1, Section B, Module 1, Q2; © VCAA.

MC Every Friday, the same number of workers from a large office building regularly purchase their lunch from one of two locations: the deli, *D*, or the cafe, *C*.

It has been found that:
- of the workers who purchase lunch from the deli on one Friday, 65% will return to purchase from the deli on the next Friday
- of the workers who purchase lunch from the cafe on one Friday, 55% will return to purchase from the cafe on the next Friday.

A transition matrix that can be used to describe this situation is

A. *this Firday*

$$\begin{array}{cc} D & C \end{array}$$
$$\begin{bmatrix} 0.55 & 0.35 \\ 0.45 & 0.65 \end{bmatrix} \begin{array}{c} D \\ C \end{array} \text{ next Friday}$$

B. *this Friday*

$$\begin{array}{cc} D & C \end{array}$$
$$\begin{bmatrix} 0.65 & 0.45 \\ 0.45 & 0.55 \end{bmatrix} \begin{array}{c} D \\ C \end{array} \text{ next Friday}$$

C. *this Friday*

$$\begin{array}{cc} D & C \end{array}$$
$$\begin{bmatrix} 0.65 & 0.55 \\ 0.45 & 0.55 \end{bmatrix} \begin{array}{c} D \\ C \end{array} \text{ next Friday}$$

D. *this Friday*

$$\begin{array}{cc} D & C \end{array}$$
$$\begin{bmatrix} 0.65 & 0.45 \\ 0.35 & 0.55 \end{bmatrix} \begin{array}{c} D \\ C \end{array} \text{ next Friday}$$

E. *this Friday*

$$\begin{array}{cc} D & C \end{array}$$
$$\begin{bmatrix} 0.65 & 0.55 \\ 0.35 & 0.45 \end{bmatrix} \begin{array}{c} D \\ C \end{array} \text{ next Friday}$$

▶ **Question 2 (1 mark)**

Source: VCE 2020, Further Mathematics Exam Section B, Module 1, Q7; © VCAA.

MC A small shopping centre has two coffee shops: Fatima's (*F*) and Giorgio's (*G*).

The percentage of coffee-buyers at each shop changes from day to day, as shown in the transition matrix T.

$$T = \begin{array}{c} \\ \\ \end{array} \begin{array}{cc} \text{today} & \\ F & G \end{array}$$
$$T = \begin{bmatrix} 0.85 & 0.35 \\ 0.15 & 0.65 \end{bmatrix} \begin{array}{c} F \\ G \end{array} \text{ tomorrow}$$

On a particular Monday, 40% of coffee-buyers bought their coffees at Fatima's.

The matrix recursion relation $S_{n+1} = TS_n$ is used to model this situation.

The percentage of coffee-buyers who are expected to buy their coffee at Giorgio's on Friday of the same week is closest to

- **A.** 31%
- **B.** 32%
- **C.** 34%
- **D.** 45%
- **E.** 68%

Source: VCE 2020, Further Mathematics Exam 1, Section B, Module 1, Q4; © VCAA.

MC In a particular supermarket, the three top-selling magazines are *Angel* (*A*), *Bella* (*B*) and *Crystal* (*C*). The transition diagram below shows the way shoppers at this supermarket change their magazine choice from week to week.

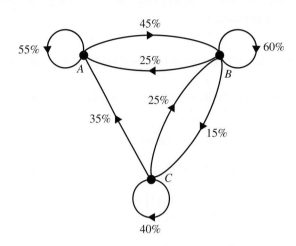

A transition matrix that provides the same information as the transition diagram is

A.   *this week*

$$
\begin{array}{ccc}
A & B & C
\end{array}
$$
$$
\begin{bmatrix}
0.55 & 0.70 & .035 \\
0.70 & 0.60 & 0.40 \\
0.35 & 0.40 & 0.40
\end{bmatrix}
\begin{array}{l}
A \\
B \ \ next\,week \\
C
\end{array}
$$

B.   *this week*

$$
\begin{array}{ccc}
A & B & C
\end{array}
$$
$$
\begin{bmatrix}
0.55 & 0.60 & .025 \\
0.45 & 0.15 & 0.35 \\
0 & 0.25 & 0.40
\end{bmatrix}
\begin{array}{l}
A \\
B \ \ next\,week \\
C
\end{array}
$$

C.   *this week*

$$
\begin{array}{ccc}
A & B & C
\end{array}
$$
$$
\begin{bmatrix}
0.55 & 0.25 & .035 \\
0.45 & 0.60 & 0.25 \\
0 & 0.15 & 0.40
\end{bmatrix}
\begin{array}{l}
A \\
B \ \ next\,week \\
C
\end{array}
$$

D.   *this week*

$$
\begin{array}{ccc}
A & B & C
\end{array}
$$
$$
\begin{bmatrix}
0.55 & 0.25 & 0.35 \\
0.45 & 0.60 & 0.25 \\
0.35 & 0.15 & 0.40
\end{bmatrix}
\begin{array}{l}
A \\
B \ \ next\,week \\
C
\end{array}
$$

E.   *this week*

$$
\begin{array}{ccc}
A & B & C
\end{array}
$$
$$
\begin{bmatrix}
0.55 & 0.25 & 0 \\
0.45 & 0.60 & 0.25 \\
0 & 0.15 & 0.75
\end{bmatrix}
\begin{array}{l}
A \\
B \ \ next\,week \\
C
\end{array}
$$

More exam questions are available online.

8.8 Review

8.8.1 Summary

Hey students! Now that it's time to revise this topic, go online to:

Access the topic summary

Review your results

Watch teacher-led videos

Practise VCAA exam questions

Find all this and MORE in jacPLUS

8.8 Exercise

Multiple choice

1. **MC** The a_{12} element in the following matrix is:

$$A = \begin{bmatrix} 2.4 & -0.5 \\ 3.6 & 1.6 \\ 1.6 & 0 \\ -2.5 & 2.4 \end{bmatrix}$$

   A. 3.6   B. 1.6   C. −0.5   D. 0   E. 2.4

2. **MC** The number 0 in the following matrix can be represented using the notation:

$$A = \begin{bmatrix} 2 & 2 & 1 \\ 1 & 0 & 2 \\ 1 & 3 & 2 \\ 2 & 5 & 4 \end{bmatrix}$$

   A. a_{23} B. a_{32} C. a_{22} D. a_{12} E. none of these.

3. **MC** Consider the following matrices.

$$E = \begin{bmatrix} 1.2 & -0.5 \\ 3.6 & 5.0 \\ -3.5 & 2.2 \end{bmatrix} \text{ and } F = \begin{bmatrix} 0.2 & -0.5 \\ 2.4 & 2.5 \\ 0 & 1.1 \end{bmatrix}$$

   The solution to $E - F$ is:

   A. $\begin{bmatrix} 1.0 & 0 \\ 1.2 & 2.5 \\ -3.5 & 1.1 \end{bmatrix}$ B. $\begin{bmatrix} 1.4 & -1.0 \\ 6.0 & 7.5 \\ -3.5 & 3.3 \end{bmatrix}$ C. $\begin{bmatrix} 1.0 & -1.0 \\ 1.2 & 2.5 \\ -3.5 & 1.1 \end{bmatrix}$

   D. $\begin{bmatrix} -1.0 & 0 \\ -1.2 & -2.5 \\ -3.5 & -11 \end{bmatrix}$ E. $\begin{bmatrix} 1.0 & -1.0 \\ 1.2 & 2.5 \\ -3.5 & 1.1 \end{bmatrix}$

4. **MC** If $A = B \times C$, then the element a_{13} is the result of:

   A. multiplying the third row by the first column.
   B. multiplying the third column by the first row.
   C. adding the first row to the third column.
   D. multiplying the first row by the third column.
   E. multiplying the first column by the third row.

Use the following matrices to answer questions 5 to 8.

$$A = \begin{bmatrix} 1 & 4 & 1 \\ 0 & 5 & 2 \\ 3 & 6 & -3 \end{bmatrix}, \ B = \begin{bmatrix} 3 & 1 \\ 2 & 5 \end{bmatrix}, \ C = \begin{bmatrix} 2 & 3 & 1 \\ -1 & 4 & 5 \end{bmatrix}, \ D = \begin{bmatrix} 1 & 2 & 3 & 4 \\ 0 & 1 & -1 & 4 \end{bmatrix}, \ E = \begin{bmatrix} 4 & 1 \\ 3 & 0 \\ 2 & -1 \end{bmatrix}$$

5. **MC** Select the product that does not exist.

   A. *AB* B. *BC* C. *CA* D. *ED* E. *CE*

6. **MC** The order of the matrix *BD* is:

   A. 2×2 B. 4×2 C. 4×4 D. 3×4 E. 2×4

7. **MC** Select the product that gives a matrix of order 2×2.

   A. *BA* B. *EC* C. *CA* D. *AC* E. *CE*

8. **MC** The matrix *ED* is:

   A. $\begin{bmatrix} 4 & 9 & 11 & 20 \\ 3 & 6 & 9 & 12 \\ 2 & 3 & 6 & 4 \end{bmatrix}$
   B. $\begin{bmatrix} 4 & 9 & 11 & 20 \\ 3 & 6 & 9 & 12 \\ 2 & -3 & -7 & 4 \end{bmatrix}$
   C. $\begin{bmatrix} 4 & 9 & 11 & 20 \\ 3 & 6 & 9 & 12 \\ 2 & 3 & 7 & 12 \end{bmatrix}$

   D. $\begin{bmatrix} 4 & 9 & 11 & 20 \\ 3 & 6 & 9 & 12 \\ 2 & 3 & 7 & 4 \end{bmatrix}$
   E. $\begin{bmatrix} 3 & 7 & 8 & 16 \\ 2 & 9 & 1 & 28 \end{bmatrix}$

9. **MC** The determinant of the matrix $\begin{bmatrix} 9 & 8 \\ 5 & 6 \end{bmatrix}$ is:

   A. $\dfrac{1}{14}$ B. $\dfrac{-1}{14}$ C. 14 D. 104 E. -8

10. **MC** If $P = \begin{bmatrix} 1 & 1 \\ 1 & 0 \end{bmatrix}$, its inverse P^{-1} is equal to:

    A. $\begin{bmatrix} 0 & -1 \\ -1 & 1 \end{bmatrix}$
    B. $\begin{bmatrix} 0 & 1 \\ 1 & -1 \end{bmatrix}$
    C. $\begin{bmatrix} -1 & -1 \\ -1 & 0 \end{bmatrix}$

    D. $\begin{bmatrix} -1 & 1 \\ 1 & 0 \end{bmatrix}$
    E. cannot be found.

11. **MC** The missing elements in the following transition matrix are:

$$T = \begin{bmatrix} 0.3 & 0.2 & \\ 0.3 & 0.7 & 0.3 \\ & 0.1 & 0.6 \end{bmatrix}$$

   A. $t_{31} = 0.3$ and $t_{13} = 0.5$ B. $t_{13} = 0.3$ and $t_{31} = 0.5$
   C. $t_{13} = 0.5$ and $t_{31} = 0.4$ D. $t_{31} = 0.4$ and $t_{13} = 0.1$
   E. $t_{13} = 0.4$ and $t_{13} = 0.1$

12. **MC** A transition analysis of the movement of a population of people at the Tamworth Country Festival is given by $S_4 = \begin{bmatrix} 0.5 & 0.2 \\ 0.5 & 0.8 \end{bmatrix}^4 \begin{bmatrix} 10\,000 \\ 12\,000 \end{bmatrix}$. The most correct answer for S_4 is:

   A. $\begin{bmatrix} 6300 \\ 15\,700 \end{bmatrix}$
   B. $\begin{bmatrix} 6288 \\ 15\,711 \end{bmatrix}$
   C. $\begin{bmatrix} 6320 \\ 15\,690 \end{bmatrix}$
   D. $\begin{bmatrix} 6400 \\ 15\,800 \end{bmatrix}$
   E. $\begin{bmatrix} 28.7 \\ 71.3 \end{bmatrix}$

13. **MC** Using the following table of events and their transition probabilities, determine the correct transition matrix.

	Percentage of customers and their choice of shopping the next time		
	Shop A	Shop B	Shop C
Shop A	85%		20%
Shop B	10%	60%	
Shop C	5%	20%	75%

A. $\begin{bmatrix} 0.85 & 0.2 & 0.2 \\ 0.1 & 0.6 & 0.3 \\ 0.05 & 0.2 & 0.75 \end{bmatrix}$

B. $\begin{bmatrix} 0.85 & 0.2 & 0.2 \\ 0.1 & 0.6 & 0.05 \\ 0.5 & 0.2 & 0.75 \end{bmatrix}$

C. $\begin{bmatrix} 0.85 & 0.2 & 0.2 \\ 0.1 & 0.6 & 0.05 \\ 0.05 & 0.2 & 0.75 \end{bmatrix}$

D. $\begin{bmatrix} 0.85 & -0.05 & 0.2 \\ 0.1 & 0.6 & 0.3 \\ 0.05 & 0.2 & 0.75 \end{bmatrix}$

E. $\begin{bmatrix} 85 & 20 & 20 \\ 10 & 60 & 5 \\ 5 & 20 & 75 \end{bmatrix}$

Use the following information for questions 14 and 15.

Four players, Alex (A), Lena (L), Nathan (N) and Rachel (R) took part in a round robin badminton competition. The results of the competition are summarised by adjacency matrix A and A^2 below.

$$A = \begin{array}{c} \\ A \\ L \\ N \\ R \end{array} \begin{array}{cccc} A & L & N & R \\ \begin{bmatrix} 0 & 0 & 0 & 0 \\ 1 & 0 & 0 & 1 \\ 1 & 1 & 0 & 1 \\ 1 & 0 & 0 & 0 \end{bmatrix} \end{array} \qquad A^2 = \begin{array}{c} \\ A \\ L \\ N \\ R \end{array} \begin{array}{cccc} A & L & N & R \\ \begin{bmatrix} 0 & 0 & 0 & 0 \\ 1 & 0 & 0 & 0 \\ 2 & 0 & 0 & 1 \\ 0 & 0 & 0 & 0 \end{bmatrix} \end{array}$$

14. **MC** Select the dominance matrix from the following matrices.

A. $\begin{array}{c} \\ A \\ L \\ N \\ R \end{array} \begin{array}{cccc} A & L & N & R \\ \begin{bmatrix} 0 & 0 & 0 & 0 \\ 2 & 0 & 0 & 1 \\ 2 & 1 & 0 & 1 \\ 1 & 0 & 0 & 0 \end{bmatrix} \end{array}$

B. $\begin{array}{c} \\ A \\ L \\ N \\ R \end{array} \begin{array}{cccc} A & L & N & R \\ \begin{bmatrix} 1 & 0 & 0 & 0 \\ 2 & 0 & 0 & 1 \\ 3 & 1 & 0 & 1 \\ 1 & 0 & 0 & 0 \end{bmatrix} \end{array}$

C. $\begin{array}{c} \\ A \\ L \\ N \\ R \end{array} \begin{array}{cccc} A & L & N & R \\ \begin{bmatrix} 1 & 0 & 0 & 0 \\ 2 & 0 & 0 & 1 \\ 3 & 1 & 0 & 1 \\ 1 & 0 & 0 & 0 \end{bmatrix} \end{array}$

D. $\begin{array}{c} \\ A \\ L \\ N \\ R \end{array} \begin{array}{cccc} A & L & N & R \\ \begin{bmatrix} 0 & 0 & 0 & 0 \\ 2 & 0 & 0 & 1 \\ 3 & 1 & 0 & 2 \\ 1 & 0 & 0 & 0 \end{bmatrix} \end{array}$

E. $\begin{array}{c} \\ A \\ L \\ N \\ R \end{array} \begin{array}{cccc} A & L & N & R \\ \begin{bmatrix} 0 & 0 & 0 & 0 \\ 1 & 0 & 0 & 2 \\ 2 & 0 & 1 & 3 \\ 0 & 0 & 0 & 1 \end{bmatrix} \end{array}$

15. **MC** The order (starting from the winner) in which competitors finished is:

A. Alex, Lena, Rachel, Nathan.
B. Lena, Nathan, Rachel, Alex.
C. Nathan, Rachel, Lena, Alex.
D. Nathan, Lena, Rachel, Alex.
E. Alex, Rachel, Lena, Nathan.

16. **MC** A railway goods distribution company has 300 wagons that transport coal from the coalfields at station A to the electricity generation plant at station B. At the end of each week, it is known that 20% of the wagons that started the week at station A end the week at station B and 30% of the wagons that started the week at station B end the week at station A.

The transition matrix that describes the distribution of wagons is:

A. $\begin{bmatrix} 0.8 & 0.2 \\ 0.3 & 0.7 \end{bmatrix}$

B. $\begin{bmatrix} 0.2 & 0.7 \\ 0.8 & 0.3 \end{bmatrix}$

C. $\begin{bmatrix} 0.8 & 0.3 \\ 0.2 & 0.7 \end{bmatrix}$

D. $\begin{bmatrix} 0.7 & 0.2 \\ 0.8 & 0.3 \end{bmatrix}$

E. $\begin{bmatrix} 120 \\ 180 \end{bmatrix}$

17. **MC** The communication matrix for the following communication network is:

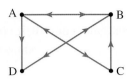

A.

	A	B	C	D
A	0	0	1	0
B	0	0	1	1
C	0	1	0	1
D	1	1	1	0

B.

	A	B	C	D
A	0	1	1	1
B	1	0	0	1
C	1	1	0	0
D	0	1	0	0

C.

	A	B	C	D
A	0	0	1	1
B	0	0	1	1
C	0	1	0	1
D	0	1	1	0

D.

	A	B	C	D
A	0	1	1	0
B	1	0	0	1
C	1	1	0	0
D	0	1	0	0

E.

	A	B	C	D
A	0	1	1	1
B	1	0	0	1
C	1	0	0	0
D	0	1	0	0

Short answer

18. Consider the following matrices.

$$A = \begin{bmatrix} 10 & 7 \\ 5 & -5 \\ 0 & 2 \end{bmatrix} \text{ and } B = \begin{bmatrix} 3 & -1 & 0 \\ 0 & 3 & 1 \\ 5 & 0 & 3 \end{bmatrix}$$

   a. State the order and type of matrix.
   b. Where possible, write the $3, 2$ element.
   c. State the position of the number 5 in x_{ij} form.

19. Given the matrices

$$A = \begin{bmatrix} 1 & 5 \\ 3 & 2 \end{bmatrix}, B = \begin{bmatrix} 3 & 0 \\ 5 & 2 \end{bmatrix} \text{ and } C = \begin{bmatrix} 1 & -5 & 0 \\ 3 & -3 & 3 \end{bmatrix}$$

   calculate, if possible:

   a. $A + B$ b. $B - C$ c. $2C$.

20. The matrix shown represents the prices (in dollars) of some mobile phone options. The first column displays the costs of two types of prepaid mobile phones and the second column represents two types of 12-month-plan mobile phones.

$$\begin{bmatrix} 249 & 29 \\ 680 & 49 \end{bmatrix}$$

The company wants to increase the price of the prepaid mobile phones by 12% and decrease the cost of the 12-month-plan mobile phones by 5%.

   a. Show the matrix used to represent the price changes as a 2×2 matrix.
   b. Use matrix multiplication to calculate the new prices.

21. In the matrix multiplication $A \times B = C$, the order of B is 2×3 and the order of C is 4×3. State the order of matrix A.

22. A survey shows that 85% of the time, customers will continue to purchase their groceries from store A if they purchased their groceries from store A in the previous month, while 15% of the time they will change to purchasing their groceries from store B if they purchased their groceries from store A in the previous month. Similarly, the records show that 60% of the time, customers will continue to purchase their groceries from store B if they purchased their groceries from store B in the previous month.

   a. Determine how many customers are still purchasing their groceries at stores A and B at the end of three months, if 200 customers started purchasing at store A and 200 started at store B.
   b. Calculate the percentage (in whole numbers) of customers who are purchasing their groceries at stores A and B at the end of 6 months, if 40% of the customers started at store A.

23. For each of the following matrices, give the order and the appropriate name (if it can be categorised). Where possible, write the 2, 2 element and the position of the number 7 in x_{ij} form.

$$A = \begin{bmatrix} 3 & 0 & 0 \\ 1 & 4 & 0 \\ 7 & 5 & 9 \end{bmatrix}, \quad B = \begin{bmatrix} 7 \\ -2 \\ 5 \end{bmatrix}, \quad C = \begin{bmatrix} 1 & 0 \\ 0 & 1 \end{bmatrix}, \quad D = \begin{bmatrix} -3 & 2 & -1 \\ 2 & 0 & -4 \\ -1 & -4 & 7 \end{bmatrix}$$

24. Given the matrix $R = \begin{bmatrix} 6 & -2 & -8 \\ 3 & 5 & -4 \\ -1 & 10 & 12 \end{bmatrix}$ and the permutation matrix, $P = \begin{bmatrix} 0 & 0 & 1 \\ 1 & 0 & 0 \\ 0 & 1 & 0 \end{bmatrix}$, answer the following.

   a. State what type of permutation PR is.
   b. With a row permutation of R, state where row 1 moves to.
   c. If a row permutation of R resulted in $R' = \begin{bmatrix} 3 & 5 & -4 \\ 6 & -2 & -8 \\ -1 & 10 & 12 \end{bmatrix}$, determine the permutation matrix that achieved this.

25. Given the following communication matrix, answer the following:

$$\begin{array}{c} \\ A \\ B \\ C \\ D \end{array} \begin{array}{c} \begin{array}{cccc} A & B & C & D \end{array} \\ \begin{bmatrix} 0 & 0 & 1 & 0 \\ 0 & 0 & 1 & 1 \\ 0 & 1 & 0 & 1 \\ 1 & 1 & 1 & 0 \end{bmatrix} \end{array}$$

   a. State who C can talk to.
   b. State who B can receive calls from.
   c. State who can call the most people.
   d. State who C cannot call.

Extended response

26. The owner of a chain of three local outdoor adventure stores is doing the bookkeeping for the months of January, February and March. They are particularly interested in the items they have available for hire. The numbers of overnight hires of sleeping mats, sleeping bags and tents for the three stores are given in the table.

	January			February			March		
	Sleeping mats	Sleeping bags	Tents	Sleeping mats	Sleeping bags	Tents	Sleeping mats	Sleeping bags	Tents
Store A	450	1200	1500	320	600	900	540	1300	1400
Store B	310	1000	1200	158	580	920	212	1080	1569
Store C	250	750	600	130	345	700	278	850	900

   a. Represent the number of hires from each store as a 3×3 matrix for each month.
   b. Write a matrix equation to calculate the total number of each hire option for the three stores during this three-month period.

The owner has a profit margin for each type of hire as follows: sleeping mats at \$0.50 each, sleeping bags at \$1.50 each and tents at \$4.50 each.

  c. Show the hire profit margins as a 3×1 matrix and use it to calculate the total profit for each of the stores at the end of this three-month period.

  d. Calculate the total profit made by the owner of the chain, during this three-month period.

The owner has just bought 120 new tents. They deliver an equal quantity to each outlet. However, over the years, they have determined that after a period of a week, 80% of hires from store A are returned to store A while 15% are returned to store B and the remainder to store C. For hires from store B, 70% are returned to store B, while 25% are returned to store C and the remainder to store A. Similarly for store C, 60% are returned to store C, while 30% are returned to store B and 10% are returned to store A.

  e. Represent this information as a transition matrix.

  f. Calculate the number of the new tents at each store after 2 weeks.

  g. Use appropriate calculations to suggest how the 120 tents could have been distributed among the stores initially, to avoid the slow transition of copies to a long-term steady state.

27. An ecologist is studying a colony of female locusts in a laboratory. Locusts have a three-stage life cycle: eggs, nymphs (young adults) and adults. Only adult females reproduce, producing 200 eggs before dying. On average, 2% of eggs survive to be nymphs and 5% of nymphs survive to adulthood. If this colony has 100 adults and no eggs or nymphs at the start of the study, use a Leslie matrix model to:

  a. determine the initial state matrix

  b. determine the Leslie matrix

  c. calculate the number of locusts after 1, 2, 3, 4, 5, 6, 7 and 8 years

  d. comment about the changing population of locusts over the 8 years.

28. There are two petrol stations to choose from in the town of Naralleville, and as the town grows, more people purchase fuel at the petrol stations each week. A matrix equation used to determine the number of customers expected to purchase fuel in each of the petrol stations is given by

$$P_{n+1} = \begin{bmatrix} 0.60 & 0.30 \\ 0.40 & 0.70 \end{bmatrix} S_n + \begin{bmatrix} 3 \\ 5 \end{bmatrix},$$

where P_n is the column matrix that lists the number of customers in week n. The number of customers who purchased petrol at each petrol station in the first week is given by the matrix

$$P_1 = \begin{bmatrix} 184 \\ 168 \end{bmatrix} \begin{matrix} \text{Station A} \\ \text{Station B} \end{matrix}$$

  a. Calculate the number of customers who are expected to purchase fuel at station A in week 3.

  b. Comment on the difference in the number of customers at station A after 4 weeks compared to 2 weeks.

8.8 Exam questions

▶ Question 1 (3 marks)

Source: VCE 2021, Further Mathematics Exam 2, Section B, Module 1, Q2; © VCAA.

The main computer system in Elena's office has broken down.

The five staff members, Ales (A), Brie (B), Chai (C), Dex (D) and Elena (E), are having problems sending information to each other.

Matrix M below shows the available communication links between the staff members.

$$
M = sender \begin{array}{c} \\ A \\ B \\ C \\ D \\ E \end{array}
\begin{array}{c} receiver \\ \begin{array}{ccccc} A & B & C & D & E \end{array} \\
\left[\begin{array}{ccccc}
0 & 1 & 0 & 0 & 1 \\
0 & 0 & 1 & 1 & 0 \\
1 & 0 & 0 & 1 & 0 \\
0 & 1 & 0 & 0 & 0 \\
0 & 0 & 0 & 1 & 0
\end{array} \right]
\end{array}
$$

In this matrix
- the '1' in row A, column B indicates that Alex can send information to Brie
- the '0' in row D, column C indicates that Dex cannot send information to Chai.
a. Which two staff members can send information directly to each other?
b. Elena needs to send documents to Chai.
   What is the sequence of communication links that will successfully get the information from Elena to Chai?
c. Matrix M^2 below is the square of matrix M and shows the number of two-step communication links between each pair of staff members.

$$
M^2 = sender \begin{array}{c} \\ A \\ B \\ C \\ D \\ E \end{array}
\begin{array}{c} receiver \\ \begin{array}{ccccc} A & B & C & D & E \end{array} \\
\left[\begin{array}{ccccc}
0 & 0 & 1 & 2 & 0 \\
0 & 1 & 0 & 1 & 0 \\
0 & 1 & 0 & 0 & 0 \\
0 & 0 & 1 & 1 & 0 \\
0 & 1 & 0 & 0 & 0
\end{array} \right]
\end{array}
$$

Only one pair of individuals has two different two-step communication links.

List each two-step communication link for this pair.

Question 2 (6 marks)
Source: VCE 2020, Further Mathematics Exam 2, Section B, Module 1, Q1 Q4; © VCAA.

The three major shopping centres in a large city, Eastmall (E), Grandmall (G) and Westmall (W), are owned by the same company.

The total number of shoppers at each of the centres at 1.00 pm on a typical day is shown in matrix V.

$$
\begin{array}{c} \begin{array}{ccc} E & \quad G & \quad W \end{array} \\ V = \begin{bmatrix} 2300 & 2700 & 2200 \end{bmatrix} \end{array}
$$

a. Write down the order of matrix V.
   Each of these centres has three major shopping areas: food (F), clothing (C) and merchandise (M).
   The proportion of shoppers in each of these three areas at 1.00 pm on a typical day is the same at all three centres and is given in matrix P below.

$$
P = \begin{bmatrix} 0.48 \\ 0.27 \\ 0.25 \end{bmatrix} \begin{array}{c} F \\ C \\ M \end{array}
$$

b. Grandmall's management would like to see 700 shoppers in its merchandise area at 1.00 pm.
   If this were to happen, how many shoppers, in total, would be at Grandmall at this time?

c. The matrix $Q = P \times V$ is shown below. Two of the elements of this matrix are missing.

$$Q = \begin{array}{c} \\ \end{array} \begin{array}{ccc} E & G & W \\ \end{array}$$

$$Q = \begin{bmatrix} 1104 & - & 1056 \\ 621 & - & 594 \\ 575 & 675 & 550 \end{bmatrix} \begin{array}{c} F \\ C \\ M \end{array}$$

  i. Complete **matrix Q above** by filling in the missing elements.
  (Answer on *matrix Q above.*)
  ii. The element in row i and column j of matrix Q is q_{ij}.
  What does the element q_{23} represents?

d. The average daily amount spent, in dollars, by each shopper in each of the three areas at Grandmall in 2019 is shown in matrix A_{2019} below.

$$A_{2019} = \begin{bmatrix} 21.30 \\ 34.00 \\ 14.70 \end{bmatrix} \begin{array}{c} F \\ C \\ M \end{array}$$

On one particular day, 135 shoppers spent the average daily amount on food, 143 shoppers spent the average daily amount on clothing 131 shoppers spent the average daily amount on merchandise.
Write a matrix calculation, using matrix A_{2019}, showing that the total amount spent by all these shoppers is $9663.20

e. In 2020, the average daily amount spent by each shopper was expected to change by the percentage shown in the table below.

Area	food	clothing	merchandise
Expected change	increase by 5%	decrease by 15%	decrease by 1%

The average daily amount, in dollars, expected to be spent in each area in 2020 can be determined by forming the matrix product $A_{2020} = K \times A_{2019}$
Write down matrix K.
$K =$

Question 3 (1 mark)

Source: VCE 2020, Further Mathematics Exam 1, Section B, Module 1, Q8; © VCAA.

MC The table below shows information about three matrices: A, B and C.

Matrix	Order
A	2×4
B	2×3
C	3×4

The transpose of matrix A, for example, is written as A^T.

What is the order of the product $C^T \times \left(A^T \times B\right)^T$?
  A. 2×3
  B. 3×4
  C. 4×2
  D. 4×3
  E. 4×4

Source: VCE 2020, Further Mathematics Exam 2, Section B, Module 1, Q2; © VCAA.

The preferred number of cafes (x) and sandwich bars (y) in Grandmall's food court can be determined by solving the following equations written in matrix form.

$$\begin{bmatrix} 5 & -9 \\ 4 & -7 \end{bmatrix} \begin{bmatrix} x \\ y \end{bmatrix} = \begin{bmatrix} 7 \\ 6 \end{bmatrix}$$

   a. The value of the determinant of the 2×2 matrix is 1.
      Use this information to explain why this matrix has an inverse.
   b. Write the three missing values of the inverse matrix that can be used to solve these equations.

$$\begin{bmatrix} - & 9 \\ - & - \end{bmatrix}$$

   c. Determine the preferred number of sandwich bars for Grandmall's food court.

Question 5 (5 marks)

Source: VCE 2019, Further Mathematics Exam 2, Section B, Module 1, Q1; © VCAA.

The car park at a theme park has three areas, *A, B* and *C.*

The number of empty (E) and full (F) parking spaces in each of the three areas at 1 pm on Friday are shown in matrix Q below.

$$Q = \begin{matrix} & E & F & \\ & \begin{bmatrix} 70 & 50 \\ 30 & 20 \\ 40 & 40 \end{bmatrix} & \begin{matrix} A \\ B \\ C \end{matrix} & area \end{matrix}$$

   a. what is the order of matrix Q?
   b. Write down a calculation to show that 110 parking spaces are full at 1 pm.
   c. Drivers must pay a parking fee for each hour of parking.
      Matrix P, below, shows the hourly fee, in dollars, for a car parked in each of the three areas.

$$\begin{matrix} & & area & \\ & A & B & C \\ P = & [1.30 & 3.50 & 1.80] \end{matrix}$$

The total parking fee, in dollars, collected from these 110 parked cars if they were parked for one hour is calculated as follows.

$$P \times L = [207.00]$$

Where matrix L is a 3×1 matrix.
Write down matrix L.
$L =$

d. The number of whole hours that each of the 110 cars had been parked was recorded at 1 pm. Matrix R, below, shows the number of cars parked for one, two, three of four hours in each of the areas A, B and C.

$$R = \begin{matrix} & \begin{matrix} & area & \\ A & B & C \end{matrix} & \\ \begin{bmatrix} 3 & 1 & 1 \\ 6 & 10 & 3 \\ 22 & 7 & 10 \\ 19 & 2 & 26 \end{bmatrix} & \begin{matrix} 1 \\ 2 \\ 3 \\ 4 \end{matrix} \, hours \end{matrix}$$

Matrix R^T is the transpose of matrix R.
Complete the matrix R^T below.

$$R^T = \begin{bmatrix} & & & \\ & & & \\ & & & \end{bmatrix}$$

e. The number of whole hours that each of the 110 cars had been parked was recorded at 1 pm. Matrix R, below, shows the number of cars parked for one, two, three of four hours in each of the areas A, B and C.

$$R = \begin{matrix} & \begin{matrix} & area & \\ A & B & C \end{matrix} & \\ \begin{bmatrix} 3 & 1 & 1 \\ 6 & 10 & 3 \\ 22 & 7 & 10 \\ 19 & 2 & 26 \end{bmatrix} & \begin{matrix} 1 \\ 2 \\ 3 \\ 4 \end{matrix} \, hours \end{matrix}$$

Explain what the element in row 3, column 2 of matrix R^T represents.

More exam questions are available online.

Hey students! Access past VCAA examinations in learnON

Sit past VCAA examinations

Receive immediate feedback

Identify strengths and weaknesses

Find all this and MORE in jacPLUS

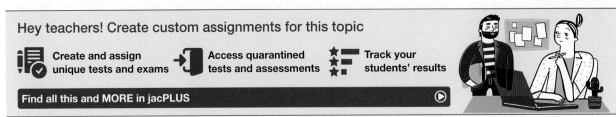

Hey teachers! Create custom assignments for this topic

Create and assign unique tests and exams

Access quarantined tests and assessments

Track your students' results

Find all this and MORE in jacPLUS

Answers

Topic 8 Matrices

8.2 Matrix representation

8.2 Exercise

1. A is a 3×3 diagonal matrix; 2, 1 is 0; 3 is represented by a_{33}.
   B is a 2×2 square matrix; 2, 1 is 3; 3 is represented by b_{21}.
   C is a 2×2 binary matrix; 2, 1 is 0; 3 is not in the matrix.
   D is a 3×3 symmetrical matrix; 2, 1 is 1; 3 is represented by d_{11}.
   E is a 3×3 lower triangular matrix; 2, 1 position is 5; 3 is represented by e_{32}.

2. A is a 3×3 upper triangular matrix; 1, 2 is -5; 3 is represented by a_{11}.
   B is a 1×3 row matrix; 1, 2 is -3; 3 is not in the matrix.
   C is a 2×2 square matrix; 1, 2 is 10; 3 is represented by c_{21}.
   D is a 3×3 symmetrical matrix; 1, 2 is 7; 3 is represented by d_{32} and d_{23}.

3. a. i. 4×2 rectangular matrix
      ii. 2
      iii. a_{22}
   b. i. 3×3 square matrix
      ii. 0
      iii. b_{33}
   c. Not a matrix
   d. i. 3×1 column matrix
      ii. 3
      iii. d_{21}
   e. i. 1×3 row matrix
      ii. Does not exist.
      iii. e_{11}
   f. i. 3×3 symmetrical matrix
      ii. -4
      iii. f_{32} and f_{23}

4. A is a (3×3) lower triangular matrix. The element in the 2, 2 position is 4 and the number 7 is represented by a_{33}.
   B is a (1×3) row matrix. The element in the 2, 2 position is 7 and the number 7 is represented by b_{12}.
   C is a (2×2) identity matrix. The element in the 2, 2 position is 1 and the number 7 is not in the matrix.
   D is a (3×3) symmetrical matrix. The element in the 2, 2 position is 0 and the number 7 is represented by d_{33}.

5. i. a. 1   b. 2
      c. Does not exist.   d. Does not exist.
      e. 3   f. 3.5
   ii. a and e

6. C

7. B

8. $\begin{bmatrix} 1 & 0 & 0 & 0 \\ 0 & 1 & 0 & 0 \\ 0 & 0 & 1 & 0 \\ 0 & 0 & 0 & 1 \end{bmatrix}$

9. $\begin{bmatrix} 0 & 1 & 1 & 1 \\ 1 & 0 & 1 & 2 \\ 1 & 1 & 0 & 0 \\ 1 & 2 & 0 & 0 \end{bmatrix}$
   Vertex A is connected once to vertices B, C and D.
   Vertex B is connected once to vertices A and C and connected twice to vertex D.
   Vertex C is connected once to vertices A and B.
   Vertex D is connected once to vertex A and twice to vertex B.

10. $\begin{bmatrix} 26 & 22 \\ 15 & 9 \\ 41 & 31 \end{bmatrix}$

11. $\begin{bmatrix} 0 & 3 & 2 & 1 \\ 3 & 0 & 1 & 1 \\ 2 & 1 & 0 & 1 \\ 1 & 1 & 1 & 0 \end{bmatrix}$

12. $\begin{bmatrix} 12 & 23 & 35 \\ 9 & 16 & 25 \\ 21 & 39 & 60 \end{bmatrix}$

13. $\begin{bmatrix} 15 & 10 & 100 \\ 12 & 15 & 87 \end{bmatrix}$

14. $\begin{bmatrix} 5 & 7 & 15 \\ 12 & 2 & 7 \\ 6 & 15 & 0 \end{bmatrix}$

15. $\begin{bmatrix} 3 & 2 \\ -4 & -1 \\ 4 & -1 \\ 4 & 1 \end{bmatrix}$

8.2 Exam questions

Note: Mark allocations are available with the fully worked solutions online.
1. D
2. B
3. C

8.3 Addition, subtraction and scalar operations with matrices

8.3 Exercise

1. a. $\begin{bmatrix} 4 & 4 \\ 8 & -2 \end{bmatrix}$
   b. $\begin{bmatrix} 4 & 4 \\ 8 & -2 \end{bmatrix}$
   c. This cannot be done.
   d. $\begin{bmatrix} -4 & -10 \\ 2 & 6 \end{bmatrix}$
   e. $\begin{bmatrix} -12 & 0 & 0 \\ -6 & 0 & -3 \end{bmatrix}$

2. a. This cannot be done.
   b. $\begin{bmatrix} 23 & 37 & -25 \\ 19 & 22 & -22 \end{bmatrix}$
   c. $\begin{bmatrix} 7 & 3 & -17 \\ 17 & -2 & -4 \end{bmatrix}$
   d. $\begin{bmatrix} 24 & 3 & -9 \\ 30 & 5 & -2 \end{bmatrix}$
   e. $\begin{bmatrix} -17 & 0 & -8 \\ -13 & -7 & -2 \end{bmatrix}$

3. a. $\begin{bmatrix} 4 & 5 \\ 2 & 0 \end{bmatrix}$ **b.** $\begin{bmatrix} 2 & 10 \\ -6 & 4 \end{bmatrix}$

   c. Not possible   **d.** $\begin{bmatrix} 2 & -5 \\ 8 & -4 \end{bmatrix}$

   e. $\begin{bmatrix} 3 & -3 & 2 \\ 0 & 1 & -4 \end{bmatrix}$

4. D

5. $\begin{bmatrix} 26.00 & 90.00 & 81.50 & 159.95 \end{bmatrix}$

6. a. $\begin{bmatrix} -18 & 24 \\ 18 & -12 \end{bmatrix}$ **b.** $\begin{bmatrix} 12 & -16 \\ -12 & 8 \end{bmatrix}$

   c. $\begin{bmatrix} -3 & 4 \\ 3 & -2 \end{bmatrix}$ **d.** $\begin{bmatrix} -2.4 & 3.2 \\ 2.4 & -1.6 \end{bmatrix}$

7. a. $\begin{bmatrix} -14 & -6 \\ 19 & -6 \end{bmatrix}$ **b.** $\begin{bmatrix} 12 & -12 \\ 3 & 3 \end{bmatrix}$

8. a. $\begin{bmatrix} 12 & 15 \\ 3 & 6 \end{bmatrix}$ **b.** $\begin{bmatrix} 0.4 & 0.8 \\ 0.2 & 0.6 \end{bmatrix}$ **c.** $\begin{bmatrix} 20 & 25 \\ 5 & 10 \end{bmatrix}$

   d. $\begin{bmatrix} 5 & 7 \\ 1.5 & 3.5 \end{bmatrix}$ **e.** $\begin{bmatrix} 16 & 26 \\ 6 & 16 \end{bmatrix}$ **f.** $\begin{bmatrix} 4 & 3.5 \\ 0.5 & 0 \end{bmatrix}$

9. a. $\begin{bmatrix} 4 & 12 & 0 & 8 & 8 \\ 8 & 8 & 16 & 4 & 12 \\ 4 & 0 & -8 & 4 & -4 \\ 8 & 4 & 12 & 4 & 0 \\ 4 & 4 & 20 & 4 & 4 \end{bmatrix}$

   b. $\begin{bmatrix} -0.4 & -0.6 & 0 & -0.1 & 0.2 \\ -0.2 & -0.3 & 0.4 & -0.1 & -0.3 \\ -0.3 & 0 & -0.2 & -0.1 & -0.1 \\ -0.2 & -0.2 & 0.3 & -0.1 & -1 \\ -0.5 & -0.1 & -0.8 & -0.1 & 0 \end{bmatrix}$

   c. $\begin{bmatrix} -9 & -9 & 0 & 3 & 12 \\ 0 & -3 & 24 & 0 & 0 \\ -6 & 0 & -12 & 0 & -6 \\ 0 & -3 & 18 & 0 & -30 \\ -12 & 0 & -9 & 0 & 3 \end{bmatrix}$

   d. $\begin{bmatrix} 14 & 30 & 0 & 14 & 8 \\ 16 & 18 & 16 & 8 & 24 \\ 12 & 0 & -8 & 8 & -4 \\ 16 & 10 & 12 & 8 & 20 \\ 16 & 8 & 46 & 8 & 6 \end{bmatrix}$

   e. $\begin{bmatrix} -5.5 & -7.5 & 0 & -0.5 & 4 \\ -2 & -3.5 & 8 & -1 & -3 \\ -4 & 0 & -4 & -1 & -2 \\ -2 & -2.5 & 6 & -1 & -15 \\ -7 & -1 & -9.5 & -1 & 0.5 \end{bmatrix}$

10. a. $E = \begin{bmatrix} 6 & 10 & -9 \\ 12 & -2 & -20 \end{bmatrix}$ **b.** $D = \begin{bmatrix} 6 & 0 \\ 5 & 2 \end{bmatrix}$

   c. $C = \begin{bmatrix} \dfrac{13}{2} & \dfrac{13}{2} \\ 5 & 11 \end{bmatrix}$

11. $a = 9, b = -5, c = -8, d = -4$

12. a. $E = \begin{bmatrix} 4 & 7 & -1 \\ 1 & 0 & 2 \end{bmatrix}$ **b.** $D = \begin{bmatrix} 5 & -3 \\ 4 & 3 \end{bmatrix}$

   c. $C = \begin{bmatrix} 6 & 13.5 \\ 3.5 & 11 \end{bmatrix}$

13. b, e

14. a. $a_{21} = 4, b_{21} = 2$ and $c_{21} = 3$

   b. Store A $= \begin{bmatrix} 14 & 13 \\ 9 & 12 \end{bmatrix}$

15. a. Semester 1 $= \begin{bmatrix} 72 & 84 \\ 76 & 68 \\ 81 & 82 \end{bmatrix}$; semester 2 $= \begin{bmatrix} 78 & 74 \\ 76 & 77 \\ 89 & 85 \end{bmatrix}$

   b. $C = \dfrac{1}{2}(A + B)$

   c. $\begin{bmatrix} 75 & 79 \\ 76 & 72.5 \\ 85 & 83.5 \end{bmatrix}$

16. a. Centre A and centre C   **b.** $\begin{bmatrix} 34 & 35 \\ 25 & 35 \\ 14 & 9 \end{bmatrix}$

   c. $\begin{bmatrix} 40\,000 & 26\,000 \\ 50\,000 & 36\,000 \\ 80\,000 & 56\,000 \end{bmatrix}$ **d.** $\begin{bmatrix} 36\,000 & 23\,400 \\ 45\,000 & 32\,400 \\ 72\,000 & 50\,400 \end{bmatrix}$

17. a. $B = \begin{bmatrix} 0.25 & 0.40 & 0.20 & 0.15 \end{bmatrix}$

   b. $A = 800B$

   c. $\begin{bmatrix} 200 & 320 & 160 & 120 \end{bmatrix}$

18. a. $\begin{bmatrix} 154 & 214 & 1138 \\ 207 & 180 & 1422 \end{bmatrix}$ **b.** $\begin{bmatrix} 14 & 19\frac{5}{11} & 103\frac{5}{11} \\ 18\frac{9}{11} & 16\frac{4}{11} & 129\frac{3}{11} \end{bmatrix}$

19. a. New stock level $(N) = 40\%$ of current stock levels (matrix S)
$$= 0.4 \times S$$

   b. $\begin{bmatrix} 9 & 60 & 45 & 2 & 436 \\ 14 & 84 & 58 & 2 & 320 \\ 18 & 92 & 52 & 3 & 480 \\ 13 & 280 & 92 & 2 & 160 \\ 31 & 168 & 64 & 3 & 340 \end{bmatrix}$

20. a. True   **b.** Not true   **c.** Not true   **d.** True

8.3 Exam questions

Note: Mark allocations are available with the fully worked solutions online.

1. C

2. E

3. D

8.4 Multiplying matrices

8.4 Exercise

1. a. $\begin{bmatrix} 14 & 14 \\ 12 & -18 \end{bmatrix}$ **b.** $\begin{bmatrix} -4 & 14 \\ 30 & 0 \end{bmatrix}$ **c.** $\begin{bmatrix} 7 & 0 \\ 0 & 7 \end{bmatrix}$

2. a. A is a 2×3 matrix.

     B is a 3×2 matrix.

     C is a 2×2 matrix.

   b. i. AB does exist.   **ii.** AC does not exist.

     iii. BA does exist.   **iv.** CA does exist.

     v. BC does exist.

3. a. $A = 2 \times 2$ $B = 2 \times 3$ $C = 2 \times 2$

   b. i, iii, iv and **vi**

c. $AB = 2 \times 3$ $AC = 2 \times 2$ $CA = 2 \times 2$ $CB = 2 \times 3$

d. $AB = \begin{bmatrix} 16 & 0 & 16 \\ 3 & 9 & 12 \end{bmatrix}$, $AC = \begin{bmatrix} 12 & 16 \\ 9 & 21 \end{bmatrix}$,

$CA = \begin{bmatrix} 6 & -6 \\ -9 & 27 \end{bmatrix}$, $CB = \begin{bmatrix} -6 & -2 & -8 \\ 21 & -1 & 20 \end{bmatrix}$

4. a. i. $\begin{bmatrix} 14 & 7 \\ 7 & 7 \end{bmatrix}$ **ii.** $\begin{bmatrix} 13 & 11 \\ 5 & 8 \end{bmatrix}$

    b. i. $\begin{bmatrix} a & b \\ c & d \end{bmatrix}$ **ii.** $\begin{bmatrix} a & b \\ c & d \end{bmatrix}$

    c. N in part **b** is the identity matrix.

5. a. Column permutation of R

    b. They all change.

    c. Row 1 is now row 3.

6. a. Row permutation of R

    b. Row 1 moves to row 3.

    c. $P = \begin{bmatrix} 0 & 0 & 1 \\ 1 & 0 & 0 \\ 0 & 1 & 0 \end{bmatrix}$

7. A

8. a. D   **b.** A   **c.** D   **d.** E   **e.** B

9. a. $\begin{bmatrix} 24 & 26 \\ -9 & -3 \end{bmatrix}$ **b.** $\begin{bmatrix} 32 & 34 \\ 6 & 12 \end{bmatrix}$ **c.** $\begin{bmatrix} 16 & 20 \\ 10 & -5 \end{bmatrix}$

    d. $\begin{bmatrix} 2 & 44 \\ -3 & 9 \end{bmatrix}$ **e.** $\begin{bmatrix} 25 & 19 \\ 8 & 17 \end{bmatrix}$

10. C

11. a. $\begin{bmatrix} 550 & 160 \\ 750 & 220 \\ 990 & 350 \end{bmatrix}$

    b. $\begin{bmatrix} 1.08 & 0 \\ 0 & 0.92 \end{bmatrix}$

    c. The washing machines are priced at $594, $810 and $1069, and the dryers are priced at $147, $202 and $322.

12. a. $\begin{bmatrix} 1.50 & 2.50 \\ 2.00 & 3.00 \\ 2.75 & 3.25 \end{bmatrix}$

    b. $\begin{bmatrix} 1.15 & 0 \\ 0 & 0.88 \end{bmatrix}$

    c. New price for lettuce is $1.73, $2.30 and $3.16. New price for potatoes (per kg) is $2.20, $2.64 and $2.86.

13. a. $\begin{bmatrix} 2.50 & 0.90 \\ 3.50 & 1.90 \\ 4.00 & 2.50 \end{bmatrix}$ **b.** $\begin{bmatrix} 0.85 & 0 \\ 0 & 1.15 \end{bmatrix}$ **c.** $\begin{bmatrix} 2.13 & 1.04 \\ 2.98 & 2.19 \\ 3.40 & 2.88 \end{bmatrix}$

14. a. $\begin{bmatrix} 25.00 & 9.90 \\ 35.00 & 19.90 \\ 95.00 & 75.00 \\ 140.00 & 128.00 \end{bmatrix}$ **b.** $\begin{bmatrix} 0.80 & 0 \\ 0 & 0.80 \end{bmatrix}$

    c. $\begin{bmatrix} 20.00 & 7.92 \\ 28.00 & 15.92 \\ 76.00 & 60.00 \\ 112.00 & 102.40 \end{bmatrix}$

15. a. $\begin{bmatrix} 25.00 & 26.00 \\ 62.50 & 45.00 \\ 55.00 & 30.00 \\ 12.50 & 22.00 \end{bmatrix}$

b. $\begin{bmatrix} 51.00 \\ 107.50 \\ 85.00 \\ 34.50 \end{bmatrix}$

    c. i. Year 10/11   **ii.** Year 10/11

16. a. A row permutation of D

    b. Row 1 is moved to row 2.

    c. Column 3 is moved to column 2.

    d. D is not a permutation matrix since it can only have a single 1 in each column and row.

    e. $P = \begin{bmatrix} 0 & 0 & 1 \\ 1 & 0 & 0 \\ 0 & 1 & 0 \end{bmatrix}$

17. a. Total sales at each store $= \begin{bmatrix} 7700 & 1600 \\ 4950 & 2240 \\ 5500 & 2560 \\ 3850 & 1920 \end{bmatrix}$

    b. Store A = $9300, store B = $7190, store C = $8060 and store D = $5770

    c. i. Store A has the highest sales for iPads at $7700.

    ii. Store A has the highest total sales at $9300.

18. a. $\begin{bmatrix} 12 & 10 & 10 & 12 \\ 15 & 25 & 15 & 25 \\ 10 & 10 & 10 & 10 \\ 8 & 20 & 5 & 18 \end{bmatrix}$

    b. $\begin{bmatrix} 15 & 0 & 0 & 0 \\ 0 & 20 & 0 & 0 \\ 0 & 0 & 30 & 0 \\ 0 & 0 & 0 & 32 \end{bmatrix}$

    c. $\begin{bmatrix} 180 & 200 & 300 & 384 \\ 225 & 500 & 450 & 800 \\ 150 & 200 & 300 & 320 \\ 120 & 400 & 150 & 576 \end{bmatrix}$

    d. $\begin{bmatrix} 1064 \\ 1975 \\ 970 \\ 1246 \end{bmatrix}$

19. a. February $= \begin{bmatrix} 600 & 500 & 0 \\ 480 & 750 & 840 \\ 240 & 1000 & 0 \end{bmatrix}$

    March $= \begin{bmatrix} 1200 & 1000 & 420 \\ 840 & 1500 & 1260 \\ 1200 & 1750 & 0 \end{bmatrix}$

    April $= \begin{bmatrix} 1440 & 750 & 1680 \\ 600 & 1500 & 2100 \\ 1560 & 250 & 420 \end{bmatrix}$

    b. $\begin{bmatrix} 3240 & 2250 & 2100 \\ 1920 & 3750 & 4200 \\ 3000 & 3000 & 420 \end{bmatrix}$

    c. $\begin{bmatrix} 27 & 9 & 5 \\ 16 & 15 & 10 \\ 25 & 12 & 1 \end{bmatrix}$

    d. $\begin{bmatrix} 7590 \\ 9870 \\ 6420 \end{bmatrix}$ Store A earned $7590. Store B earned $9870. Store C earned $6420.

20. a. i. $\begin{bmatrix} 2 & 1 \\ 1 & 1 \end{bmatrix}$ **ii.** $\begin{bmatrix} 3 & 2 \\ 2 & 1 \end{bmatrix}$ **iii.** $\begin{bmatrix} 5 & 3 \\ 3 & 2 \end{bmatrix}$ **iv.** $\begin{bmatrix} 8 & 5 \\ 5 & 3 \end{bmatrix}$

b. The numbers produced in the solution matrix are Fibonacci numbers. If we let F_n represent the nth Fibonacci number, then the situation can be represented as follows.

For a matrix $A = \begin{bmatrix} 1 & 1 \\ 1 & 0 \end{bmatrix}$, A^n can be written as

$$\begin{bmatrix} F_{n+1} & F_n \\ F_n & F_{n-1} \end{bmatrix}.$$

c. $A^8 = \begin{bmatrix} F_9 & F_8 \\ F_8 & F_7 \end{bmatrix}$

$\quad = \begin{bmatrix} 34 & 21 \\ 21 & 13 \end{bmatrix}$

d. $A^{30} = \begin{bmatrix} 1 & 1 \\ 1 & 0 \end{bmatrix}^{30}$

$\quad = \begin{bmatrix} 1\,346\,269 & 832\,040 \\ 832\,040 & 514\,229 \end{bmatrix}$

Given that $A_{30} = \begin{bmatrix} F_{31} & F_{30} \\ F_{30} & F_{29} \end{bmatrix}$, the 30th Fibonacci number, F_{30}, is 832 040.

8.4 Exam questions

Note: Mark allocations are available with the fully worked solutions online.

1. A

2. E

3. D

8.5 The inverse of a matrix and its determinant

8.5 Exercise

1. det $A = 6$
det $B = -42$

2. det $A = 48$
det $B = 0$

3. a. -11 **b.** $\dfrac{1}{3}$ **c.** 13 **d.** 0

4. A, B and C will have an inverse; D has no inverse because it is a singular matrix; that is, its determinant is equal to 0.

5. $A^{-1} = \dfrac{1}{4}\begin{bmatrix} 8 & -6 \\ -6 & 5 \end{bmatrix}$

6. a. $|C| = -10$

b. $C^{-1} = \dfrac{1}{-10}\begin{bmatrix} 4 & 1 \\ -10 & -5 \end{bmatrix}$ or $\begin{bmatrix} -0.4 & -0.1 \\ 1 & 0.5 \end{bmatrix}$

7. a. $-\dfrac{1}{11}\begin{bmatrix} 5 & -8 \\ -7 & 9 \end{bmatrix}$ **b.** $3\begin{bmatrix} \frac{1}{3} & \frac{1}{6} \\ -1 & \frac{1}{2} \end{bmatrix}$

c. $\dfrac{1}{13}\begin{bmatrix} 8 & -9 \\ -3 & 5 \end{bmatrix}$ **d.** No inverse exists.

8. $\dfrac{1}{2}\begin{bmatrix} 3 & -5 \\ -5 & 9 \end{bmatrix}$

9. $AB = BA = I$

10. A

11. E

12. a. -0.005 **b.** $\begin{bmatrix} -90 & 40 \\ 50 & -20 \end{bmatrix}$

13. a. i. -28 **ii.** $\begin{bmatrix} \frac{1}{14} & \frac{3}{28} \\ \frac{3}{14} & -\frac{5}{28} \end{bmatrix}$

b. i. 6 **ii.** $\begin{bmatrix} -1 & -\frac{2}{3} \\ -\frac{1}{2} & -\frac{1}{2} \end{bmatrix}$

c. i. -1 **ii.** $\begin{bmatrix} -4 & 3 \\ 7 & -5 \end{bmatrix}$

d. i. -0.1 **ii.** $\begin{bmatrix} -2.5 & 10 \\ 2 & -4 \end{bmatrix}$

e. i. 0
ii. Does not exist — singular matrix.

f. i. 0
ii. Does not exist — singular matrix.

g. i. 96 **ii.** $\begin{bmatrix} -\frac{5}{24} & \frac{1}{4} & \frac{1}{24} \\ \frac{1}{4} & -\frac{1}{2} & \frac{1}{4} \\ \frac{1}{24} & \frac{1}{4} & -\frac{5}{24} \end{bmatrix}$

h. i. -3072 **ii.** $\begin{bmatrix} \frac{1}{16} & \frac{1}{16} & -\frac{1}{16} & 0 \\ \frac{1}{12} & 0 & \frac{1}{12} & -\frac{1}{6} \\ 0 & \frac{1}{4} & 0 & -\frac{1}{4} \\ -\frac{5}{48} & -\frac{3}{16} & \frac{1}{48} & \frac{1}{3} \end{bmatrix}$

i. i. -72 **ii.** $\begin{bmatrix} -\frac{1}{3} & \frac{1}{6} & \frac{1}{6} & \frac{1}{6} \\ \frac{1}{3} & -\frac{2}{3} & \frac{1}{3} & \frac{1}{3} \\ \frac{1}{9} & \frac{1}{9} & -\frac{2}{9} & \frac{1}{9} \\ \frac{1}{12} & \frac{1}{12} & \frac{1}{12} & -\frac{1}{6} \end{bmatrix}$

j. i and **ii** Not possible — J is not a square matrix.

14. a. $X = \begin{bmatrix} 3 & -44 \\ -3 & 61 \end{bmatrix}$ **b.** $X = \begin{bmatrix} 44 \\ -59 \end{bmatrix}$

15. a. $X = \begin{bmatrix} 9 & -3 \\ -17 & 7 \end{bmatrix}$ **b.** $X = \begin{bmatrix} 58 \\ -102 \end{bmatrix}$

16. a. $X = \dfrac{1}{30}\begin{bmatrix} 21 & -33 \\ 15 & 15 \end{bmatrix} = \begin{bmatrix} 0.7 & -1.1 \\ 0.5 & 0.5 \end{bmatrix}$

b. $X = \dfrac{1}{30}\begin{bmatrix} 36 & -45 \\ 18 & 0 \end{bmatrix} = \begin{bmatrix} 1.2 & -1.5 \\ 0.6 & 0 \end{bmatrix}$

c. $X = \dfrac{1}{30}\begin{bmatrix} 15 & 25 \\ -18 & 30 \end{bmatrix} = \begin{bmatrix} 0.5 & 0.83 \\ -0.6 & 1 \end{bmatrix}$

d. $X = \begin{bmatrix} 20 \\ 10 \end{bmatrix}$

17. a. $X = \begin{bmatrix} 2 \\ -1 \end{bmatrix}$ b. $X = \begin{bmatrix} 1 \\ 1 \end{bmatrix}$

18. $X = \begin{bmatrix} 5 & 6 \\ -3 & -\dfrac{8}{3} \end{bmatrix}$

8.5 Exam questions

Note: Mark allocations are available with the fully worked solutions online.

1. E

2. A

3. A

8.6 Dominance and communication matrices

8.6 Exercise

1. a. i. A–D, A–D (2)
   ii. A–C–D, A–E–D (2)
   iii. A–B–E–D (1)

   b. $A = \begin{bmatrix} 0 & 1 & 1 & 2 & 1 \\ 0 & 0 & 0 & 0 & 1 \\ 0 & 0 & 0 & 1 & 0 \\ 0 & 0 & 0 & 0 & 0 \\ 0 & 0 & 0 & 1 & 0 \end{bmatrix}, A^2 = \begin{bmatrix} 0 & 0 & 0 & 2 & 1 \\ 0 & 0 & 0 & 1 & 0 \\ 0 & 0 & 0 & 0 & 0 \\ 0 & 0 & 0 & 0 & 0 \\ 0 & 0 & 0 & 0 & 0 \end{bmatrix}$

2. a. i. A–D (1)
   ii. A–B–D, A–C–D, A–C–D, A–E–D (4)
   iii. A–C–E–D (1)

   b. $A = \begin{bmatrix} 0 & 1 & 1 & 1 & 1 \\ 0 & 0 & 0 & 1 & 0 \\ 0 & 0 & 0 & 2 & 1 \\ 0 & 0 & 0 & 0 & 0 \\ 0 & 0 & 0 & 1 & 0 \end{bmatrix}, A^2 = \begin{bmatrix} 0 & 0 & 0 & 4 & 0 \\ 0 & 0 & 0 & 0 & 0 \\ 0 & 0 & 0 & 0 & 0 \\ 0 & 0 & 0 & 0 & 0 \\ 0 & 0 & 0 & 0 & 0 \end{bmatrix}$

3. a. i. A–D (1)
   ii. A–B–D, A–B–D, A–C–D, A–C–D (4)
   iii. None possible

   b. i. A–D (1)
   ii. A–B–D, A–E–D, A–C–D, A–C–D (4)
   iii. A – C – E – D, A – C – E – D (2)

4. a. $A = \begin{bmatrix} 0 & 1 & 2 & 1 \\ 0 & 0 & 0 & 2 \\ 0 & 0 & 0 & 1 \\ 0 & 0 & 0 & 0 \end{bmatrix}, A^2 = \begin{bmatrix} 0 & 0 & 0 & 4 \\ 0 & 0 & 0 & 0 \\ 0 & 0 & 0 & 0 \\ 0 & 0 & 0 & 0 \end{bmatrix}$

   b. $A = \begin{bmatrix} 0 & 1 & 2 & 1 & 1 \\ 0 & 0 & 0 & 1 & 0 \\ 0 & 0 & 0 & 1 & 1 \\ 0 & 0 & 0 & 0 & 0 \\ 0 & 0 & 0 & 1 & 0 \end{bmatrix}, A^2 = \begin{bmatrix} 0 & 0 & 0 & 4 & 2 \\ 0 & 0 & 0 & 0 & 0 \\ 0 & 0 & 0 & 1 & 0 \\ 0 & 0 & 0 & 0 & 0 \\ 0 & 0 & 0 & 0 & 0 \end{bmatrix}$

5. One-stage routes: L_1–L_2 (1)
   Two-stage routes: L_1–B–L_2, L_1–A–L_2, L_1–A–L_2 (3)
   Three-stage routes: L_1–A–B–L_2, L_1–A–B–L_2 (2)
   6 routes overall

6. a. A–B–D   b. D–C–A   c. D–C–B, D–C–A–B

7. a. i. O–G
   ii. O–R–G
   iii. O–C–R–G, O–C–O–G, O–S–O–G

b. i. None possible
   ii. R–C–S
   iii. R–G–O–S, R–C–O–S

c. i. None possible   ii. S–O–R
   iii. S–O–C–R

8. P

9. E

10. a. First: D, second: C, third: B and fourth: A

    b. $A + A^2 = \begin{bmatrix} 0 & 0 & 0 & 0 \\ 1 & 0 & 0 & 0 \\ 2 & 1 & 0 & 0 \\ 3 & 2 & 1 & 0 \end{bmatrix}$

    Sum of the first row is 0. A is placed fourth.
    Sum of the second row is 1. B is placed third.
    Sum of the third row is 3. C is placed second.
    Sum of the fourth row is 6. D is placed first (as it has the highest sum).

11. a. First: C, second: E, third: A, fourth: B and fifth: D

    b. $A + A^2 = \begin{bmatrix} 0 & 1 & 0 & 2 & 0 \\ 0 & 0 & 0 & 1 & 0 \\ 2 & 3 & 0 & 4 & 1 \\ 0 & 0 & 0 & 0 & 0 \\ 1 & 2 & 0 & 3 & 0 \end{bmatrix}$

    Sum of the first row is 3. A is placed third.
    Sum of the second row is 1. B is placed fourth.
    Sum of the third row is 10. C is placed first (as it has the highest sum).
    Sum of the fourth row is 0. D is placed fifth.
    Sum of the fifth row is 6. E is placed second.

12. a. C   b. Z

13. D

14.

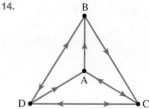

$$\text{Communication matrix} = \begin{array}{c} \\ A \\ B \\ C \\ D \end{array} \begin{array}{c} \begin{array}{cccc} A & B & C & D \end{array} \\ \begin{bmatrix} 0 & 1 & 1 & 0 \\ 0 & 0 & 1 & 1 \\ 1 & 0 & 0 & 1 \\ 1 & 1 & 1 & 0 \end{bmatrix} \end{array}$$

15. a. 'Talk to' (vertical) so C talks to A, B and D.
    b. 'Receive calls' (horizontal) so B receives calls from C and D.
    c. Since they cannot call themselves.
    d. 'Call' or 'talk to' (vertical) so D cannot call C.

16. $\text{Communication matrix} = \begin{array}{c} \\ P \\ Q \\ R \\ S \end{array} \begin{array}{c} \begin{array}{cccc} P & Q & R & S \end{array} \\ \begin{bmatrix} 0 & 0 & 1 & 0 \\ 1 & 0 & 1 & 1 \\ 1 & 1 & 0 & 1 \\ 1 & 1 & 0 & 0 \end{bmatrix} \end{array}$

17. a. 'Talk to' (vertical) so C talks to A and D.
    b. 'Receive calls' (horizontal) so B receives calls from D.

c. 'Call/talk to' (vertical) B can only call D.

d. D cannot call A.

e. D can receive calls from A, B and C.

18. a. F

b. A and C

c. F, E, D, B and A/C

8.6 Exam questions

Note: Mark allocations are available with the fully worked solutions online.

1. B

2. D

3. A

8.7 Transition matrices and Leslie matrices

8.7 Exercise

1. a. i. $\begin{bmatrix} 0.5 & 0.5 \\ 0.5 & 0.5 \end{bmatrix}$ ii. $\begin{bmatrix} 0.63 & 0.38 \\ 0.38 & 0.63 \end{bmatrix}$

   iii. $\begin{bmatrix} 0.56 & 0.44 \\ 0.44 & 0.56 \end{bmatrix}$ iv. $\begin{bmatrix} 0.51 & 0.49 \\ 0.49 & 0.51 \end{bmatrix}$

   b. $\dfrac{1}{2}\begin{bmatrix} 1 & 1 \\ 1 & 1 \end{bmatrix}$

2. a. i. $\begin{bmatrix} 1 & 0 \\ 0 & 1 \end{bmatrix}$ ii. $\begin{bmatrix} 1 & 0 \\ 0 & 1 \end{bmatrix}$

   iii. $\begin{bmatrix} 1 & 0 \\ 0 & 1 \end{bmatrix}$ iv. $\begin{bmatrix} 1 & 0 & 0 \\ 0 & 1 & 0 \\ 0 & 0 & 1 \end{bmatrix}$

   b. The identity matrix

3. a.
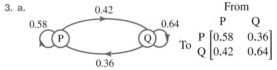
   To $\begin{array}{c} P \\ Q \end{array}\begin{bmatrix} 0.58 & 0.36 \\ 0.42 & 0.64 \end{bmatrix}$ (From P Q)

   b. $\begin{bmatrix} 0.69 & 0.27 \\ 0.31 & 0.73 \end{bmatrix}$

4. a. $\begin{array}{c} A \\ B \end{array}\begin{bmatrix} 0.68 & 0.23 \\ 0.32 & 0.77 \end{bmatrix}$ (A B)

   b. $\begin{array}{c} \text{North} \\ \text{South} \end{array}\begin{bmatrix} 0.80 & 0.15 \\ 0.20 & 0.85 \end{bmatrix}$ (North South)

5. a. $\begin{bmatrix} 0.9 & 0.08 \\ 0.1 & 0.92 \end{bmatrix}$

   b. 120 wagons at point A and 130 wagons at point B

   c. 43% of the wagons at point A and 57% of the wagons at point B

6. a. $\begin{bmatrix} 0.84 & 0.7 \\ 0.16 & 0.3 \end{bmatrix}$ b. $\begin{bmatrix} 0.65 & 0.20 & 0.03 \\ 0.15 & 0.68 & 0 \\ 0.20 & 0.12 & 0.97 \end{bmatrix}$

7. a. $T = \begin{bmatrix} 0.52 & 0.16 & 0.14 \\ 0.12 & 0.40 & 0.07 \\ 0.36 & 0.44 & 0.79 \end{bmatrix}$

   b. Store A will have 694 customers, store B will have 369 customers and store C will have 1937 customers.

c. Store A: 23.0%, store B: 12.2%, store C: 64.9%

d. Store A: $\dfrac{17}{74}$, store B: $\dfrac{9}{74}$, store C: $\dfrac{24}{37}$

8. a. 68.1% b. 32.0%

9. a. $\begin{bmatrix} 0.385 & 0.14 & 0 & 0.0775 \\ 0.28 & 0.66 & 0 & 0.01 \\ 0.18 & 0.19 & 1 & 0.005 \\ 0.155 & 0.01 & 0 & 0.9075 \end{bmatrix}$

   b. T has no steady state because the matrix T^2 contains zero elements.

   c. A

10. a. $\begin{bmatrix} 0.6 & 0.1 & 0 & 0.05 \\ 0.2 & 0.8 & 0.05 & 0 \\ 0.1 & 0.1 & 0.9 & 0 \\ 0.1 & 0 & 0.05 & 0.95 \end{bmatrix}$

    b. i. Texcal, 4128 customers

    ii. Oilmart, 4116 customers

    iii. CP, 5600 customers

11. a. There are 100 Jersey cows expected to be milked on day 4.

    b. Nine more cows were milked on day 4 compared to day 1.

12. $S_2 = \begin{bmatrix} 0.65 & 0.23 \\ 0.35 & 0.77 \end{bmatrix}\begin{bmatrix} 48 \\ 41 \end{bmatrix} + \begin{bmatrix} 5 \\ -2 \end{bmatrix}$

13. a. 61%

    b. 82%

    c. 16 additional people go to supermarket A and 7 additional people go to supermarket B.

14. a. Week 2: Sweet Cola: 1139; Tangy Lemon: 1042
    Week 3: Sweet Cola: 1296; Tangy Lemon: 1066
    Week 4: Sweet Cola: 1433; Tangy Lemon: 1110

    b. 543

15. a. 21 students b. 8 students

16. a. 86 sheep

    b. Difference $= 220 - 212 = 8$ sheep

17. 2022: city A: 59 210; city B: 69 945
    2023: city A: 63 045; city B: 68 265
    2024: city A: 66 554; city B: 66 911

18. a. 0.078

    b. Dry: $\dfrac{11}{12}$, wet: $\dfrac{1}{12}$

    c. They should insure the event as there is a very good chance that the day will be dry.

19. B

20. a. $\begin{bmatrix} 55 \\ 32 \\ 25 \end{bmatrix}$

    b. $\begin{bmatrix} 0.6 & 2.5 & 0.5 \\ 0.5 & 0 & 0 \\ 0 & 0.6 & 0 \end{bmatrix}$

    c. 570

21. a. 264 shoppers b. 56 shoppers

22. Loss of 5229 (6000 − 771) Victorians to Queensland

8.7 Exam questions

Note: Mark allocations are available with the fully worked solutions online.

1. D
2. B
3. C

8.8 Review

8.8 Exercise

Multiple choice

1. C
2. C
3. A
4. D
5. A
6. E
7. E
8. D
9. C
10. B
11. D
12. A
13. C
14. D
15. D
16. C
17. B

Short answer

18. a. A is a 3×2 rectangular matrix.
    B is a 3×3 square matrix.
    b. a_{32} is 2; b_{32} is 0.
    c. a_{21}, b_{31}

19. a. $\begin{bmatrix} 4 & 5 \\ 8 & 4 \end{bmatrix}$ b. Not possible   c. $\begin{bmatrix} 2 & -10 & 0 \\ 6 & -6 & 6 \end{bmatrix}$

20. a. $\begin{bmatrix} 1.12 & 0 \\ 0 & 0.95 \end{bmatrix}$ b. $\begin{bmatrix} 278.88 & 27.55 \\ 761.60 & 46.55 \end{bmatrix}$

21. 4×2

22. a. Store A: 283; store B: 117
    b. Store A: 72.5%; store B: 27.5%

23. A: order (3×3); lower triangular matrix; $2, 2 = 4$;
    7 is in x_{31}.
    B: order (3×1); column matrix; $2, 2$ doesn't exist;
    7 is in x_{11}.
    C: order (2×2); identity matrix; $2, 2 = 1$;
    7 is not in the matrix.
    D: order (3×3); symmetrical matrix; $2, 2 = 0$; 7 is in x_{33}.

24. a. PR is a row permutation of R.
    b. Row 1 moves to row 2.
    c. $P = \begin{bmatrix} 0 & 1 & 0 \\ 1 & 0 & 0 \\ 0 & 0 & 1 \end{bmatrix}$

25. a. C can 'talk to' (vertical) B and D.
    b. B can 'receive calls' (horizontal) from C and D.

c. D can talk to all three; therefore, D can talk to the most people.

d. C cannot call A or itself; therefore, A and C.

Extended response

26. a. January: $A = \begin{bmatrix} 450 & 1200 & 1500 \\ 310 & 1000 & 1200 \\ 250 & 750 & 600 \end{bmatrix}$,

    February: $B = \begin{bmatrix} 320 & 600 & 900 \\ 158 & 580 & 920 \\ 130 & 345 & 700 \end{bmatrix}$,

    March: $C = \begin{bmatrix} 540 & 1300 & 1400 \\ 212 & 1080 & 1569 \\ 278 & 850 & 900 \end{bmatrix}$

    b. $A + B + C = \begin{bmatrix} 1310 & 3100 & 3800 \\ 680 & 2660 & 3689 \\ 658 & 1945 & 2200 \end{bmatrix}$

    c. $\begin{bmatrix} 0.50 \\ 1.50 \\ 4.50 \end{bmatrix}$
    Store A: \$22 405; store B: \$20 890.50;
    store C: \$13 146.50

    d. \$56 442

    e. $T = \begin{bmatrix} 0.80 & 0.05 & 0.10 \\ 0.15 & 0.70 & 0.30 \\ 0.05 & 0.25 & 0.60 \end{bmatrix}$

    f. Store A: 36; store B: 49; store C: 35

    g. Store A: 31; store B: 52; store C: 37

27. a. $\begin{bmatrix} 0 \\ 0 \\ 100 \end{bmatrix}$

    b. $\begin{bmatrix} 0 & 0 & 200 \\ 0.02 & 0 & 0 \\ 0 & 0.05 & 0 \end{bmatrix}$

    c.

Age	Eggs	Nymphs	Adults
0	0	0	100
1	20 000	0	0
2	0	400	0
3	0	0	20
4	4000	0	0
5	0	80	0
6	0	0	4
7	800	0	0
8	0	16	0

    d. The number of eggs, nymphs and adults falls and rises over the 8 years.

28. a. 160 customers   b. 3 customers

8.8 Exam questions

Note: Mark allocations are available with the fully worked solutions online.

1. a. Brie and Dex
   b. Elena $\rightarrow$ Dex $\rightarrow$ Brie $\rightarrow$ Chai
   c. Alex$\rightarrow$Brie $\rightarrow$ Dex and Alex $\rightarrow$ Elena $\rightarrow$ Dex

2. a. 1×3

b. 2800 shoppers

c. i. $Q = \begin{bmatrix} 1104 & 1296 & 1056 \\ 621 & 729 & 594 \\ 575 & 675 & 550 \end{bmatrix}$

ii. 594 shoppers were in the clothing area at Westmall.

d. Total amount $= [135 \ 143 \ 131] \times \begin{bmatrix} 21.30 \\ 34.00 \\ 14.70 \end{bmatrix} = [9663.20]$

e. $K = \begin{bmatrix} 1.05 & 0 & 0 \\ 0 & 0.85 & 0 \\ 0 & 0 & 0.99 \end{bmatrix}$

3. E

4. a. If the determinant of a matrix is zero, there is no inverse. But the determinant is 1; therefore, there is an inverse.

b. $\begin{bmatrix} -7 & 9 \\ -4 & 5 \end{bmatrix}$

c. Two sandwich bars

5. a. 3×2

b. Add up column 2 : $50 + 20 + 40 = 110$

c. $L = \begin{bmatrix} 50 \\ 20 \\ 40 \end{bmatrix}$

d. $R^T = \begin{bmatrix} 3 & 6 & 22 & 19 \\ 1 & 10 & 7 & 2 \\ 1 & 3 & 10 & 26 \end{bmatrix}$

e. The number of cars that parked for two hours in area C.

9 Undirected graphs, networks and trees

LEARNING SEQUENCE

Fully worked solutions for this topic are available online.

9.1 Overview

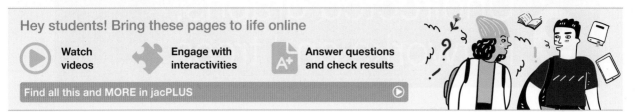
9.1.1 Introduction

The study of networks and decision mathematics is a branch of graph theory. The graphs referred to in this area of study are not graphs such as bar graphs or graphs associated with the Cartesian plane; these graphs refer to geometric representations of situations where they are defined as sets of vertices and sets of edges.

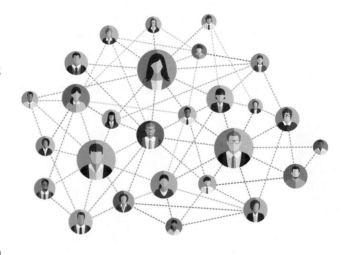

Undirected graphs are the simplest form of network. These graphs show two-way connections between vertices and are not concerned with the direction of the path between vertices. The social connections between a group of friends could be illustrated using an undirected graph, as a friendship is a reciprocal relationship. The network of airports in Australia can be represented as an undirected graph, because if a plane flew into an airport from a city, it would fly out of that airport and (most often) back to the first city.

Early studies of graph theory were done in response to difficult problems. The mathematician Francis Guthrie first posed the Four Colour Theorem in 1852. In 1976, mathematicians at the University of Illinois were able to harness their knowledge of graph theory and the power of computer algorithms to test all possible combinations and prove the theorem.

KEY CONCEPTS

This topic covers the following key concepts from the VCE Mathematics Study Design:
- the concepts, conventions and terminology of graphs including planar graphs and Euler's rule, and directed (digraphs) and networks
- use of matrices to represent graphs, digraphs and networks and their application
- the concepts, conventions and notations of walks, trails, paths, cycles and circuits
- Eulerian trails and Eulerian circuits: the conditions for a graph to have an Eulerian trail or an Eulerian circuit, properties and applications
- Hamiltonian paths and cycles: properties and applications
- trees and spanning trees
- minimum spanning trees in a weighted connected graph and their determination by inspection or by Prim's algorithm
- use of minimal spanning trees to solve minimal connector problems
- determination of the shortest path between two specified vertices in a graph, digraph or network by inspection
- Dijkstra's algorithm and its use to determine the shortest path between a given vertex and each of the other vertices in a weighted graph or network.

Source: VCE Mathematics Study Design (2023–2027) extracts © VCAA; reproduced by permission.

9.2 Basic concepts of a network

9.2.1 Definition of a network

What do the telephone system, the Australian Army, your family tree and the internet have in common? The answer is that they can all be considered **networks**.

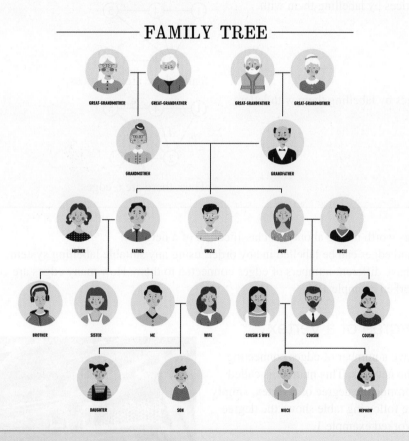

In the case of the telephone system, the objects are telephones (and exchanges, satellites, ...). In the case of the Australian Army, the objects are units (platoons, companies, regiments, divisions, ...). In the case of the internet, the objects are computers, while your family tree is made up of parents, siblings, grandparents, cousins, aunts, ...

A network

A network is a collection of objects connected to each other in some specific way.

The mathematical term for these objects is *vertex*. Consider the network represented in the diagram. This is perhaps the simplest possible network. It consists of two vertices (circles labelled 1 and 2) and one *connection* between them. This connection is called an **edge**.

In the case of the telephone system, the edges are the cables connecting homes and exchanges; in the Australian Army they are the officers of various ranks; in the family tree, the links between the generations and between spouses can be considered as edges.

The first distinguishing features of a network are the total number of vertices and the total number of edges.

WORKED EXAMPLE 1 Identifying the number of vertices and edges

Count the number of vertices and edges in the network shown.

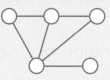

THINK	WRITE/DRAW
1. Count the vertices by labelling them with numbers.	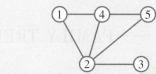 Thus, there are 5 vertices.
2. Count the edges by labelling them with letters.	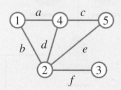 Thus, there are 6 edges.

There are two things worth noting about this classification of a network:
1. The vertices and edges can be labelled in any order, using any suitable labelling system.
2. Vertices may have different numbers of edges connected to them. How many edges are connected to vertex 2 in Worked example 1?

9.2.2 The degree of a vertex

Each vertex may have a number of edges connecting it with the rest of the network. This number is called the **degree**. To determine the degree of a vertex, simply count its edges. The following table shows the degree of each vertex in Worked example 1.

Vertex	1	2	3	4	5
Degree	2	4	1	3	2

A vertex with degree 0 is *not connected* to any other vertex, and is called an **isolated vertex**.

An edge which connects a vertex to itself is called a **loop** and contributes 2 towards the degree.

If two (or more) edges connect the *same pair* of vertices, they are called **parallel edges** (or *multiple edges*) and all count towards the degree. Otherwise, if there is only *one* connection between two vertices, the connection is called a *simple*, or *single*, connection.

WORKED EXAMPLE 2 Determining the degree of a vertex

Determine the degree of each vertex in the figure shown.

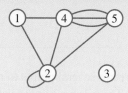

THINK	WRITE
1. Node 1 has 2 simple edges.	Degree of node $1 = 2$
2. Node 2 has 3 simple edges and 1 loop.	Degree of node $2 = 3 + 2$ $= 5$
3. Node 3 has no edges — it's an isolated node.	Degree of node $3 = 0$
4. Node 4 has 2 simple edges and 3 parallel edges.	Degree of node $4 = 2 + 3$ $= 5$
5. Node 5 has 1 simple edge and 3 parallel edges.	Degree of node $5 = 1 + 3$ $= 4$

9.2.3 Representations of networks

So far we have seen the graphical representation of a network as a two-dimensional collection of vertices and edges. Hence, networks are sometimes called *graphs*. Other ways to represent the network without losing any of its essential features are by:
1. labelling vertices and labelling edges according to their vertices
2. matrix representation.

To label vertices, simply list them. If there are three vertices labelled A, B and C, write $V = \{A, B, C\}$. To label edges according to their vertices, identify the vertices that the edge connects. If an edge connects vertex 1 with vertex 3, we represent the edge as (1, 3). If there is a loop at vertex 4, its edge is (4, 4). If there are 2 parallel edges between vertices 2 and 4, we write (2, 4), (2, 4).

WORKED EXAMPLE 3 Labelling vertices and edges

Label the vertices and edges for the figure shown, as in Worked example 2.

THINK	WRITE
1. Label the vertices.	$V = \{1, 2, 3, 4, 5\}$
2. Examine each edge, in turn.	
Vertex 1–vertex 4	$(1, 4)$
Vertex 1–vertex 2	$(1, 2)$
Vertex 2–vertex 2 (loop)	$(2, 2)$
Vertex 2–vertex 4	$(2, 4)$
Vertex 2–vertex 5	$(2, 5)$
Vertex 4–vertex 5 (3 parallel edges)	$(4, 5), (4, 5), (4, 5)$

▶

3. Combine vertices and edges into a list.	$V = \{1, 2, 3, 4, 5\}$ $E = \{(1,4), (1,2), (2,2), (2,4),$ $(2,5), (4,5), (4,5), (4,5)\}$

There are several points to note about this representation:
1. There is no 3 in the list of edges (E). This implies it is an isolated vertex.
2. The number of pairs in $E = \{(1,4), (1,2), ...\}$ is 8, which must be the same as the number of edges.
3. The number of times a vertex appears anywhere inside E equals the degree of the vertex. For example, the digit 4 appears 5 times, so the degree of vertex 4 is 5.
4. From this representation of V and E we can construct (or reconstruct) the original graph.

WORKED EXAMPLE 4 Constructing a network

Construct a graph (network) from the following list of vertices and edges.
$V = \{A, B, C, D, E\}$
$E = \{(A,B), (A,C), (A,D), (B,C), (B,D), (C,E), (D,E), (E,E)\}$

THINK	DISPLAY/WRITE
1. Start with a single vertex, say vertex A, and list the vertices to which it is connected.	Vertex A is connected to B, C and D.
2. Construct a graph showing these connections.	
3. Take the next vertex, say B, and list the vertices to which it is connected.	Vertex B is connected to A (already done), C and D (twice: parallel edge).
4. Add the edges from Step 3.	
5. Repeat Steps 3 and 4 for vertex C.	Vertex C is connected to A (already done), B (already done) and E. 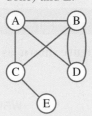
6. Repeat Steps 3 and 4 for vertex D and, finally, add the loop (E, E). As a check, count the edges in the list E (9) and compare this number with the number of edges in your final graph.	Vertex D is connected to A (already done), B (already done) and E. Vertex E is connected to C (already done), D (already done) and E (loop). 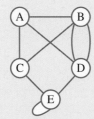

There may be other geometric configurations that can be drawn from the same vertex and edge lists, but they would be **isomorphic** (or equivalent) to this one.

9.2.4 Matrix representation of networks

A method of representing a network in concise form is through the use of a *matrix*. Recall that a matrix is a rectangular collection, or 'grid' of numbers. To represent the network, write the names of the vertices above the columns of the matrix and to the left side of the rows of the matrix. The number of edges connecting vertices is placed at the intersection of the corresponding row and column. This is best shown with an example.

tlvd-3938

WORKED EXAMPLE 5 Representing a network as a matrix

Represent the network shown (from Worked example 4) as a matrix.

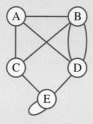

THINK

1. Set up a blank matrix, putting the vertex names across the top and down the side. Thus there are 25 possible entries inside the matrix.

2. Consider vertex A. It is connected to each of the vertices B, C and D once, so put 1 in the corresponding columns of row 1 and in the corresponding row of column 1.

3. Consider vertex B. It is connected to C once and D twice. Put 1 and 2 in the corresponding columns of row 2 and in the corresponding rows of column 2.

WRITE

$$
\begin{array}{c c c c c c}
 & A & B & C & D & E \\
A & & & & & \\
B & & & & & \\
C & & & & & \\
D & & & & & \\
E & & & & &
\end{array}
$$

$$
\begin{array}{c c c c c c}
 & A & B & C & D & E \\
A & & 1 & 1 & 1 & \\
B & 1 & & & & \\
C & 1 & & & & \\
D & 1 & & & & \\
E & & & & &
\end{array}
$$

$$
\begin{array}{c c c c c c}
 & A & B & C & D & E \\
A & & 1 & 1 & 1 & \\
B & 1 & & 1 & 2 & \\
C & 1 & 1 & & & \\
D & 1 & 2 & & & \\
E & & & & &
\end{array}
$$

4. Repeat for vertices C, D and E. Vertex C is connected to vertex E once, so put 1 in the corresponding column of row 3 and in the corresponding row of column 3 (shown in red).

Vertex D is connected to vertex E once, so put 1 in the corresponding column of row 4 and in the corresponding row of column 4 (shown in black).

Vertex E is connected to itself once (loop), so put 1 in the corresponding column 5, row 5. *Note:* Only one entry is needed for loops (shown in green). A value of 1 in the leading diagonal denotes a loop in the network, connecting a vertex to itself. This is important to understand when calculating the degree.

$$
\begin{array}{c} \\ A \\ B \\ C \\ D \\ E \end{array}
\begin{array}{c} \begin{matrix} A & B & C & D & E \end{matrix} \\ \begin{bmatrix} & & 1 & 1 & \\ 1 & & & 1 & 2 \\ 1 & 1 & & & 1 \\ 1 & 2 & & & 1 \\ & & 1 & 1 & 1 \end{bmatrix} \end{array}
$$

5. Complete the matrix by placing 0 in all unoccupied places.

6. Check your result by comparing the entries in the matrix with the original network representation. This is best done on a vertex-by-vertex basis.

$$
\begin{array}{c} \\ A \\ B \\ C \\ D \\ E \end{array}
\begin{array}{c} \begin{matrix} A & B & C & D & E \end{matrix} \\ \begin{bmatrix} 0 & 1 & 1 & 1 & 0 \\ 1 & 0 & 1 & 2 & 0 \\ 1 & 1 & 0 & 0 & 1 \\ 1 & 2 & 0 & 0 & 1 \\ 0 & 0 & 1 & 1 & 1 \end{bmatrix} \end{array}
$$

In matrix representation, the following rules apply:

$$
\begin{array}{c} \\ A \\ B \\ C \\ D \\ E \end{array}
\begin{array}{c} \begin{matrix} A & \quad & B & \quad & C & \quad & D & \quad & E \end{matrix} \\ \begin{bmatrix} 0 & & 1 & & 1 & & 1 & & 0 \\ 1 & + & 0 & + & 1 & + & 2 & + & 0 \\ 1 & & 1 & & 0 & & 0 & & 1 \\ 1 & & 2 & & 0 & & 0 & & 1 \\ 0 & & 0 & & 1 & & 1 & & 1 \end{bmatrix} \end{array}
\begin{array}{l} \\ = 4 \\ \\ \\ 3 + 1 = 4 \end{array}
$$

$$\underline{4} \qquad \qquad \underline{3 + 1 = 4}$$

1. The sum of a row (or a column) gives a degree of that vertex, except where a loop is present. Where a loop is present (denoted by 1 in the leading diagonal), add 1 to the sum of the row or column.

2. If an entire row and its corresponding entire column have only 0s, then that vertex is isolated.

3. The matrix is diagonally symmetric.

9.2 Exercise

1. **WE1** Count the number of vertices and edges in the following networks.

a.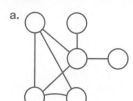

b.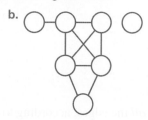

2. Count the number of vertices and edges in the following networks.

a.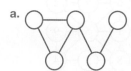

b.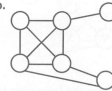

c.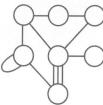

d.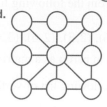

3. **MC** The number of vertices and edges in the figure is:

A. vertices = 7, edges = 7.
B. vertices = 7, edges = 10.
C. vertices = 7, edges = 11.
D. vertices = 7, edges = 11.
E. vertices = 11, edges = 7.

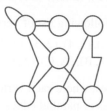

4. **WE2** Determine the degree of the labelled vertices in each diagram.

a.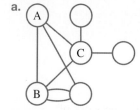

b.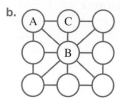

5. **MC** The degree of vertex A in the figure is:

A. 3
B. 4
C. 6
D. 5
E. 7

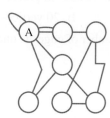

6. Determine the degree of the labelled vertices in each diagram.

a.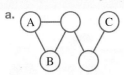

b.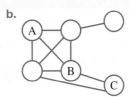

c.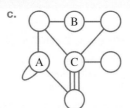

d.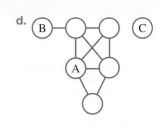

7. **WE3** List the vertices and label *all* the edges, according to their vertices, in each of these diagrams.

a.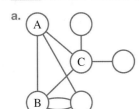

b.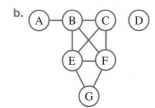

8. List the vertices and label *all* the edges, according to their vertices, in each of these diagrams.

a.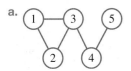

b.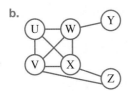

c.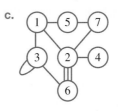

d.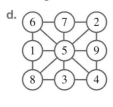

9. **WE4** a. Construct a network from the following list of vertices and edges.
   $V = \{1, 2, 3, 4, 5, 6\}$
   $E = \{(1, 2), (1, 4), (1, 6), (2, 3), (2, 6), (3, 4), (4, 6)\}$

   WE5 b. Represent the network in part **a** as a matrix.

10. Construct a network from the following list of vertices and edges.
    $V = \{A, B, C, D, E, F, G\}$
    $E = \{(A, B), (A, C), (A, F), (B, F), (D, E), (D, F), (E, F), (F, G)\}$

11. Construct a network from the following list of vertices and edges.
    a. $V = \{1, 2, 3, 4\}$ $E = \{(1, 2), (1, 4), (2, 3), (3, 4)\}$
    b. $V = \{A, B, C, D, E\}$ $E = \{(A, B), (A, C), (A, C), (B, B), (B, C), (B, D), (C, D)\}$

12. Consider a network of 4 vertices, where each vertex is connected to each of the other 3 vertices with a single edge (no loops, isolated vertices or parallel edges).

    a. List the vertices and edges.
    b. Construct a diagram of the network.
    c. List the degree of each vertex.

13. Repeat question **12** for:

    i. a network of 5 vertices
    ii. a network of 8 vertices.

14. Using the results from questions **12** and **13**, predict the number of edges for a similar network of:

    a. 10 vertices   b. 20 vertices   c. 100 vertices.

    Note that the increase in the number of edges is one of the problems that had to be overcome in the design of computer networks.

15. Construct a network representing the following family tree. Use a single node to represent each married couple.

Allan and Betty had 3 children: Charles, Doris and Earl.

Charles married Frances and had 2 children, George and Harriet.

Doris married Ian and had 1 child, John.

Earl married Karen and had 3 children, Louise, Mary and Neil.

16. Copy and complete the matrix representation of the network shown. The first few entries are shown.

$$
\begin{array}{c}
\\
1\\
2\\
3\\
4\\
5
\end{array}
\begin{array}{cccccc}
1 & 2 & 3 & 4 & 5\\
\left[\begin{matrix} 0 & 1 & 1 & 0 & 0\\ 1 & & & & \\ 1 & & & & \\ 0 & & & & \\ 0 & & & & \end{matrix}\right]
\end{array}
$$

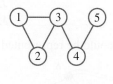

17. Represent the following networks by matrices.

a.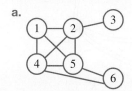

b.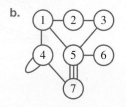

c.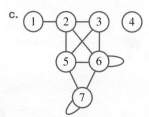

18. Construct networks from the following matrix representations.

Note that the number of rows = number of columns = number of vertices. Watch out for loops.

a. $\begin{bmatrix} 0 & 1 & 1 & 0\\ 1 & 0 & 1 & 0\\ 1 & 1 & 0 & 1\\ 0 & 0 & 1 & 0 \end{bmatrix}$

b. $\begin{bmatrix} 0 & 1 & 0 & 0 & 0\\ 1 & 0 & 1 & 1 & 0\\ 0 & 1 & 0 & 2 & 2\\ 0 & 1 & 2 & 1 & 1\\ 0 & 0 & 2 & 1 & 0 \end{bmatrix}$

c. $\begin{bmatrix} 0 & 1 & 1 & 0 & 0 & 0\\ 1 & 0 & 0 & 0 & 0 & 1\\ 1 & 0 & 0 & 2 & 0 & 2\\ 0 & 0 & 2 & 0 & 0 & 1\\ 0 & 0 & 0 & 0 & 0 & 0\\ 0 & 1 & 2 & 1 & 0 & 0 \end{bmatrix}$

9.2 Exam questions

Question 1 (1 mark)

Source: VCE 2021, Further Mathematics Exam 1, Section B, Module 2, Q1; © VCAA.

MC Consider the graph below.

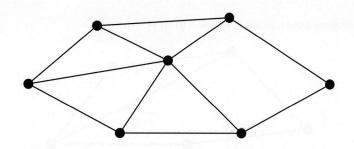

The number of vertices with a degree of 3 is

A. 1   **B.** 2   **C.** 3   **D.** 4   **E.** 5

Question 2 (1 mark)

Source: VCE 2020, Further Mathematics Exam 1, Section B, Module 2, Q8; © VCAA.

MC The adjacency matrix below shows the number of pathway connections between four landmarks: J, K, L and M.

$$\begin{array}{c} & \begin{array}{cccc} J & K & L & M \end{array} \\ \begin{array}{c} J \\ K \\ L \\ M \end{array} & \left[\begin{array}{cccc} 1 & 3 & 0 & 2 \\ 3 & 0 & 1 & 2 \\ 0 & 1 & 0 & 2 \\ 2 & 2 & 2 & 0 \end{array}\right] \end{array}$$

A network of pathways that could be represented by the adjacency matrix is

A.

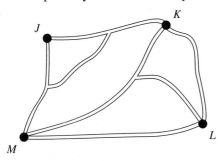

B.

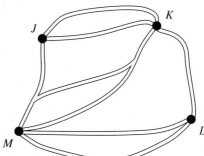

C.

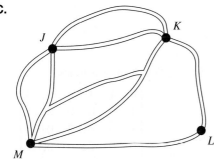

D.

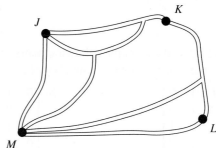

E.

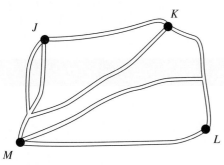

Question 3 (1 mark)

Source: VCE 2019, Further Mathematics Exam 1, Section B, Module 2, Q1; © VCAA.

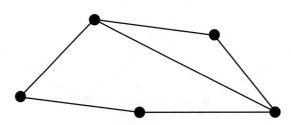

MC In the graph shown above, the sum of the degrees of the vertices is

A. 5   **B.** 6   **C.** 10   **D.** 11   **E.** 12

More exam questions are available online.

9.3 Planar graphs and Euler's formula

9.3.1 Types of graphs

In graph theory, graphs are made up of vertices, with edges connecting the vertices.

Degenerate graph

A graph with *no* edges is called a **degenerate graph** (or *null* graph).

Complete graph

A graph where all vertices are connected directly to all other vertices without parallel edges or loops is called a **complete graph**.

The figure on the left is degenerate; the one on the right is complete. How many edges would there be in a complete graph of 6 vertices?

A degenerate graph

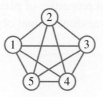

A complete graph

Formula for the number of edges

For a complete graph, if E = number of edges and V = number of vertices, then $E = \dfrac{V(V-1)}{2}$.

WORKED EXAMPLE 6 Calculating the number of vertices in a complete graph

a. **For this complete graph, calculate the number of edges.**
b. **For a complete graph with 12 vertices, calculate the number of edges.**

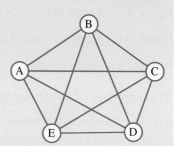

THINK

WRITE

a. 1. This is a complete graph, so write the formula for the number of edges.

$E = \dfrac{V(V-1)}{2}$

---

2. Substitute $V = 5$ into the formula.

$$E = \frac{5(5-1)}{2}$$
$$= 10$$

3. Write the answer.

There are 10 edges.

b. 1. This is a complete graph, so write the formula for the number of edges.

$$E = \frac{V(V-1)}{2}$$

2. Substitute $V = 12$ into the formula.

$$E = \frac{12(12-1)}{2}$$
$$= 66$$

3. Write the answer.

There are 66 edges.

9.3.2 Planar graphs

Planar graphs are a special kind of network or graphs in which there are no intersecting edges. The edges and vertices in a planar graph divide the graph into a number of faces, as shown in the following diagram.

When counting the faces, remember to include the infinite face — the region surrounding the graph. The additional properties of planar graphs will allow us to map two-dimensional and even three-dimensional objects into graphs.

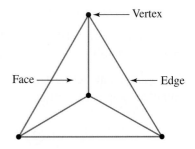

Planar graphs

If a graph has no edges that cross, then it is a planar graph.

Consider the following graphs.

a. 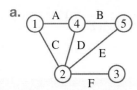 **b.**   **c.**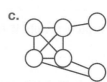

Figure **a** is a planar graph because none of the paths {A, B, C, D, E, F} cross each other.

Figure **b** is apparently not a planar graph because the path (A, D) crosses the path (B, C).

Is figure **c** a planar graph?

The regions of a planar graph

Consider the simplified version of the graph in Figure **a**, as shown. Note that the large circular vertices have been replaced by small black circles. Otherwise, this network is identical to that in Figure **a**.

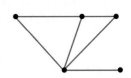

Now, observe how this planar graph can be divided into 3 *regions*: region I, region II and region III.

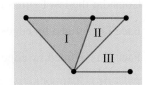

Note also that one of the regions (III) will always be *infinite*, because it continues beyond the bounds of the diagram. All the other regions have a *finite* area. These regions are also called **faces**, for a reason which will soon become apparent.

The reason one region becomes infinite is demonstrated by the fact that, when you look at three-dimensional objects, you can't see all the faces at the same time, no matter from which angle you look.

Converting non-planar graphs

Although it may *appear* that a graph is not planar, modifying the graph may make it clearly planar.

There is no specific method, but by trial and error it may be possible to remove all the crossing paths. (It may also help to imagine the nodes as nails in a board and the edges as flexible rubber bands.) Alternatively, it may be possible to move the vertices so that the connecting edges don't cross. If there are no crossings left, the graph is planar.

tlvd-3939

WORKED EXAMPLE 7 Converting a graph to a planar graph

Convert the graph shown to a planar graph. Indicate the faces (regions) of the planar graph.

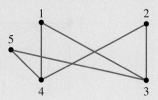

THINK	DISPLAY/WRITE
1. Confirm that the graph is non-planar. $E = \{(1,3), (1,4), (2,3), (2,4), (3,5), (4,5)\}$	Edge $(1,3)$ crosses $(2,4)$. Edge $(3,5)$ crosses $(1,4)$ and $(2,4)$.
2. Two crossings could be eliminated by swapping vertex 2 and vertex 3. Redraw the modified graph. Check that all the edges are connected to the same vertices.	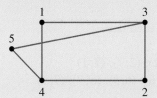 $E = \{(1,3), (1,4), (2,3), (2,4), (3,5), (4,5)\}$
3. Placing node 5 inside the rectangle is one way of eliminating all crossings. Note that this planar graph is only one of several possible answers.	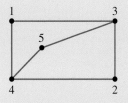
4. Define the faces (regions).	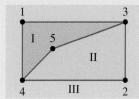

The *degree* of each face is the number of edges defining that region.
Consider the last figure in Worked example 7.

Face I is defined by edges $(1,3), (1,4), (4,5)$ and $(5,3)$, so its degree $= 4$.

Face II is defined by edges $(3,5), (5,4), (4,2)$ and $(2,3)$, so its degree $= 4$.

Face III is defined by edges $(1,3), (1,4), (4,2)$ and $(2,3)$, so its degree $= 4$.

In almost all cases, each region will have a degree of *at least* 3.
Why? Can you think of exceptions?

WORKED EXAMPLE 8 Determining the degree of each face

Determine the degree of each face of the graph shown in the figure.

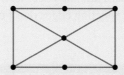

THINK	DISPLAY/WRITE
Define the edges and faces of the graph.	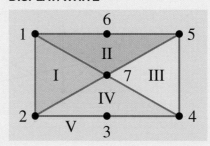

$$E = \{(1,2), (2,3), (3,4), (4,5), (5,6),$$
$$(6,1), (1,7), (5,7), (4,7), (2,7)\}$$

Count the edges for each face.
For example:
Face I — edges $(1,2), (2,7), (7,1)$

Face II — edges $(1,6), (6,5), (5,7), (7,1)$

Face I — degree $= 3$
Face II — degree $= 4$
Face III — degree $= 3$
Face IV — degree $= 4$
Face V — degree $= 6$

9.3.3 Converting three-dimensional solids to planar graphs

Another application of planar graphs is the conversion of the graph representing a three-dimensional solid (with flat faces) to a planar graph.

WORKED EXAMPLE 9 Converting a 3D solid to a planar graph

The figure shows a cube with vertices
$V = \{A, B, C, D, E, F, G, H\}$.
Convert this to a planar graph.

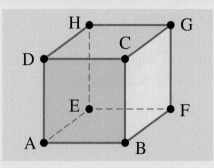

THINK

1. List the edges (12 in all).

2. Imagine the three-dimensional cube 'collapsing' to a two-dimensional graph. Try collapsing the face A–B–C–D into the face E–F–G–H.

3. Check the edges to see that they are the same as in Step 1. Note also the edges (A, E), (B, F), (C, G) and (D, H), which link the 'collapsed' faces.

DISPLAY/WRITE

$E = \{(A, B), (A, D), (A, E), (B, C),$
$(B, F), (C, D), (C, G), (D, H), (E, F),$
$(E, H), (F, G), (G, H)\}$

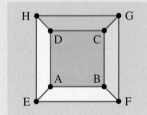

There are some other interesting features of this planar graph:
1. The planar graph is, in a sense, a two-dimensional 'projection' of the original cube.
2. The original 'base' of the cube (A–B–F–E) has become the *infinite* region of the planar graph.

9.3.4 Euler's formula

By now it may be clear that there is a mathematical relationship between the vertices, edges and faces of planar graphs.

In fact, it is the same relationship, known as **Euler's formula**, that you may have learned when studying solid geometry:

> **Euler's formula**
>
> **vertices = edges − faces + 2**
> $$V = E - F + 2$$

WORKED EXAMPLE 10 Verifying Euler's formula

Verify Euler's formula for the 'cube' of the last figure in Worked example 8.

THINK	DISPLAY/WRITE
1. List the vertices.	$V = \{A, B, C, D, E, F, G, H\}$
2. Count the vertices.	$V = 8$
3. List the edges.	$E = \{(A, B), (A, D), (A, E), (B, C), (B, F), (C, D), (C, G),$ $(D, H), (E, F), (E, H), (F, G), (G, H)\}$
4. Count the edges.	$E = 12$
5. Define the faces (regions).	

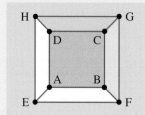

There are 6 faces in all: $\{I, II, III, IV, V, VI\}$
$F = 6$

6. Confirm Euler's formula by substitution.	$V = E - F + 2$ $8 = 12 - 6 + 2$ $8 = 8$ Therefore, Euler's formula is verified.

Note that the cube is a form of prism (an object with a uniform cross-section), and all prisms can be converted to planar graphs using the above technique of one face 'collapsing' into another.

 Resources

 Interactivities Planar graphs (int-6467)
Euler's formula (int-6468)

9.3 Exercise

Students, these questions are even better in jacPLUS

 Receive immediate feedback and access sample responses

Access additional questions

Track your results and progress

Find all this and MORE in jacPLUS

1. **WE6** **a.** For the following complete graph, calculate the number of edges.

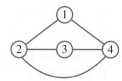

b. For a complete graph with 8 vertices, calculate the number of edges.

2. **WE7** Convert the following graph to a planar graph.

3. **a.** Redraw the following network diagram so that it is a planar graph.

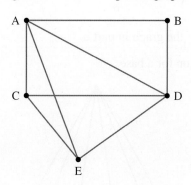

b. **WE8** Determine the degree of each vertex of the graph in part **a.**

4. Modify the following graphs so that their representations are planar.

a.

b.

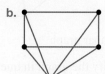

c.

d.

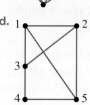

5. **MC** The graph represented in the figure is apparently not planar because:

A. edge (A, C) crosses edge (B, E).
B. edge (A, D) crosses edge (E, F).
C. edges (A, E), (F, E), (C, E) and (B, E) intersect.
D. vertex EE has a degree of 4.
E. none of the above.

6. Determine the degree of each vertex of the graph shown.

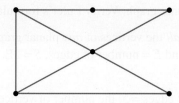

7. **MC** A complete graph with 7 vertices would have:

A. 7 edges.
B. 14 edges.
C. 21 edges.
D. 28 edges.
E. 42 edges.

8. a. **WE9** Convert the three-dimensional triangular prism shown to a planar graph.

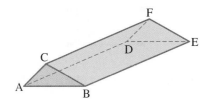

b. **WE10** Verify Euler's formula for the graph in part **a.**

9. Consider a 'pyramid' with an octagon for a base.

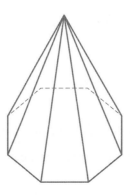

Convert the representation to a planar graph.

10. a. By moving vertex F only, modify the graph in question **5** so that it is clearly planar.
b. State the number of faces in your planar graph.
c. Determine the degree of each vertex.

11. a. By moving vertex 5 only, modify the graph so that it is clearly planar.
b. State the number of faces in your planar graph.
(*Hint:* You may have to use curved edges to connect all the vertices.)

12. a. **MC** The degree of vertex E in the figure is:

A. 1
B. 2
C. 3
D. 4
E. 5

b. Verify Euler's formula for the figure in part **a.**

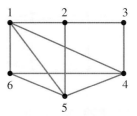

13. Convert a triangular pyramid to a planar graph. Verify Euler's formula for your graph.

14. Show that the sum of the degrees of *all* the vertices of *any* planar graph is always an even number. Also show that if S = sum of the degrees and E = number of edges, $S = 2E$. (This is known as the handshaking lemma.)

15. **MC** In a planar graph, the number of edges = 5, the number of vertices = 4, therefore the number of faces is:

A. 1
B. 3
C. 9
D. 11
E. unable to be determined from the given information.

16. Convert the rectangular pyramid shown into a planar graph.

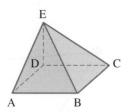

17. The diagram shown is a crude floor plan for a small house with 6 rooms, labelled A, B, ... , F.

Convert this plan to a planar graph where rooms are considered as vertices.

(*Hint:* What should the edges be?)

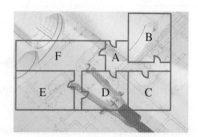

18. Eight people in a room shake hands with each other once.

a. Determine the number of handshakes.

b. Represent the handshakes as a complete graph.

c. Represent the handshakes as a matrix.

9.3 Exam questions

Question 1 (1 mark)

Source: VCE 2021, Further Mathematics Exam 1, Section B, Module 2, Q5; © VCAA.

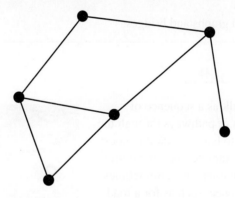

MC Consider the following five statements about the graph above:
- The graph is planar.
- The graph contains a cycle.
- The graph contains a bridge.
- The graph contains an Eulerian trail.
- The graph contains a Hamiltonian path.

How many of these statements are true?

   A. 1   **B.** 2   **C.** 3   **D.** 4   **E.** 5

Question 2 (1 mark)

Source: VCE 2020, Further Mathematics Exam 1, Section B, Module 2, Q1; © VCAA.

MC A connected planar graph has seven vertices and nine edges.

The number of faces that this graph will have is

   A. 1   **B.** 2   **C.** 3   **D.** 4   **E.** 5

Question 3 (1 mark)

Source: VCE 2019, Further Mathematics Exam 1, Section B, Module 2, Q2; © VCAA.

MC Consider the graph below.

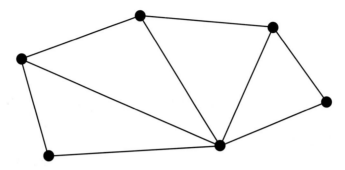

The minimum number of extra edges that are required so that an Eulerian circuit is possible in this graph is

   A. 0   **B.** 1   **C.** 2   **D.** 3   **E.** 4

More exam questions are available online.

9.4 Walks, trails, paths, cycles and circuits

LEARNING INTENTION

At the end of this subtopic you should be able to:
- solve exploring and travelling problems, using walks, trails, paths, Eulerian trails and circuits, and Hamiltonian paths and cycles
- apply properties of connected graphs and bridges.

9.4.1 Walks, trails and paths

In planar graphs we can define a **walk** as a sequence of edges and look at various sequences or pathways through the network. Sometimes you may wish to have a walk that goes through *all* nodes only once — for example, for a travelling salesperson who wishes to visit each town once. Sometimes you may wish to use all edges only once, such as for a road repair gang repairing all the roads in a shire.

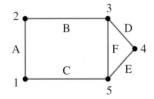

There are different ways of naming a walk. For example, consider travelling from node 1 to node 3 in the figure. A walk could be specified via node 2, namely A–B, or by specifying the vertices, 1–2–3. Alternatively one could take the walk C–E–D, or C–F. Each of these routes is a walk.

Connected graphs

If there is a walk between all possible pairs of vertices, then it is a connected graph.

For example, in the figure on the left, there is no walk between vertices 1 and 2, nor between vertices 3 and 4, so it is not a connected graph. However, if we add a single edge between vertices 1 and 2, as in the figure on the right, the entire graph becomes connected.

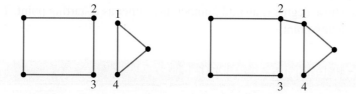

Bridges

A bridge is an edge in a connected graph whose deletion will cause the graph to no longer be connected.

In the above example on the right, the edge between vertices 1 and 2 is a bridge. When this edge is removed, the graph is no longer connected.

9.4.2 Euler trails

Consider a walk where no routes are repeated.

Trails

A trail is a walk in which no edges are repeated. A trail can pass through the same vertex more than once.

The trail in the figure above is 1–3–4–3.

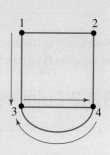

Consider a trail where every edge is used only once, as in our road repair gang example.

Euler trails

An Euler trail is a trail that uses every edge exactly once.

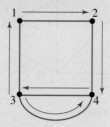

In the figure above, an Euler trail would be 3–1–2–4–3–4.

1. For an Euler trail to exist, all vertices must be of an even degree *or* there must be exactly two vertices of odd degree.
2. If the degrees of *all* the vertices are even numbers, start with *any* vertex. In this case, the starting vertex and the ending vertex are the same.
3. If there are two vertices whose degree is an odd number, use either as a starting point. The other vertex of odd degree must be the ending point.

WORKED EXAMPLE 11 Identifying an Euler trail

Identify an Euler trail in the figure shown.

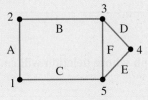

THINK

1. Determine a starting vertex. Since there are vertices (3 and 5) whose degree is an odd number, use one of these to start.

2. Attempt to find a path which uses each edge exactly once. You are allowed to visit a vertex more than once; it is only edges that are restricted to one use.

WRITE

Use vertex 3 as the start.

B–A–C–F–D–E or B–A–C–E–D–F or D–E–F–B–A–C are all Euler trails. There are several other possible Euler trails.

Note that the starting vertex and ending vertex are not the same, but we started and ended on the vertices with odd degree (vertices 3 and 5).

9.4.3 Euler circuits

A circuit in mathematical terms has much the same meaning as any circuit.

Circuits

A circuit is a trail beginning and ending at the same vertex.

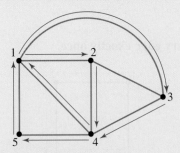

In the figure above, 3–4–5–1–2–4–1–3 is a circuit.

With our road repair gang example, it would be desirable that the Euler trail started and finished at the same point. This kind of Euler trail is called an Euler circuit.

Euler circuits

An Euler circuit is an Euler trail where the starting and ending vertices are the same.

An Euler circuit in the figure above could be 6–4–1–3–2–1–5–6.

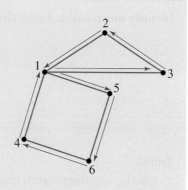

It is important to note that an Euler circuit *cannot* exist for planar graphs that have *any* vertices whose degree is odd. In such graphs there is no Euler circuit. Therefore, the planar graph of Worked example 11 does not contain an Euler circuit because vertices 3 and 5 were of odd degree.

WORKED EXAMPLE 12 Identifying an Euler circuit

Identify an Euler circuit for the planar graph shown.

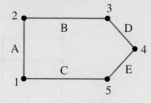

THINK	WRITE
1. Confirm that all vertices have even degree.	Vertex 1 degree = 2 Vertex 2 degree = 2 Vertex 3 degree = 2 Vertex 4 degree = 2 Vertex 5 degree = 2
2. Pick any vertex to start and determine a trail that uses each edge and ends at the same vertex.	Start with vertex 1. Then the Euler circuit could be: A–B–D–E–C or C–E–D–B–A

9.4.4 An Euler circuit algorithm

For some networks, it may be difficult to determine an Euler circuit, even after determining that all vertices have even degree. Here is an algorithm that 'guarantees' an Euler circuit.

Consider a network where all vertices are of even degree. Let $V = \{1, 2, 3, ...\}$ be the list of vertices.

Step 1. Choose a starting vertex from the list V. Call this vertex A.

Step 2. From vertex A, find the *smallest possible* path which returns to vertex A. This is a '*subcircuit*' of the original network. Let S be the list of vertices in this subcircuit.

Step 3. For each vertex in S, choose a single vertex in turn as the starting vertex of a different subcircuit. It should also be as small as possible, and not use any previously used edge.

Step 4. For each of these new subcircuits (if there are any), add any new vertices to the list in S.

Step 5. Repeat Steps 3 and 4 until there are no more new vertices, edges or subcircuits left — that is, until the lists S and V are the same.

Step 6. Join the subcircuits at their intersection points.

tlvd-3940

WORKED EXAMPLE 13 Applying the Euler circuit algorithm

Identify one possible Euler circuit for the network shown using the Euler circuit algorithm.

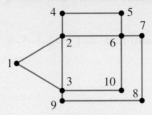

THINK	**DISPLAY/WRITE**
1. Choose a starting vertex, and find its smallest subcircuit. The subcircuit is marked in pink.	Vertex 1 is selected. The path 1–2–3–1 is the smallest possible. 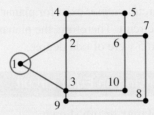
2. Create the list S from the first subcircuit. Find new subcircuits, not using any edges already used (apply Step 3 of the algorithm). The new subcircuit is marked in green. Note that this subcircuit does not use any previously used edges.	$S = \{1, 2, 3\}$ From vertex 2, there is a subcircuit 2–4–5–6–2. 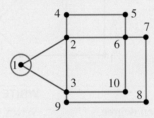
3. Add to the list S (apply Step 4 of the algorithm).	$S = \{1, 2, 3, 4, 5, 6\}$
4. Find the new subcircuits (reapply Step 3). The new subcircuit is marked in orange.	From vertex 3, there is a subcircuit 3–9–8–7–6–10–3. 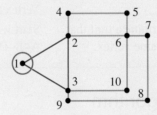
5. Add to list S (reapply Step 4). Check that all vertices are in the list (Step 5).	$S = \{1, 2, 3, 4, 5, 6, 7, 8, 9, 10\}$ $= V$, so stop.
6. Apply Step 6. Form the Euler trail, starting with the first subcircuit, and proceeding through all the other subcircuits at their intersections. Note that the second subcircuit is in the 1st set of square brackets [ ] and the next subcircuit is in the 2nd set of square brackets [ ].	The red circuit and the blue circuit are connected at vertex **2**. The red circuit and the green circuit are connected at vertex **3**. Euler circuit = 1–[**2**–4–5–6–**2**]–[**3**–9–8–7–6–10–**3**] – 1
7. List the Euler circuit.	1–2–4–5–6–2–3–9–8–7–6–10–3–1

9.4.5 Paths and cycles

Consider a walk in which no vertices are repeated, and a path that begins and ends at the same vertex.

Paths and cycles

A path is a walk in which no vertices are repeated (except possibly the start and finish). The path in the figure could be 1–2–3–4–5.

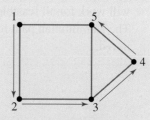

A cycle is a path beginning and ending at the same vertex.

The cycle in the figure could be 1–2–3–4–5–1.

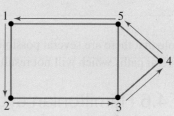

In Euler trails and circuits each edge was used *exactly once*, while vertices could be re-used. Now, consider the case where it is desirable to use each vertex *exactly once*, as in our travelling salesperson problem mentioned at the start of this section.

Hamiltonian paths

A Hamiltonian path uses every vertex exactly once.

The Hamiltonian path in the figure could be 1–2–3–4–5.

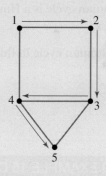

It is important to note that not all edges need to be used. Furthermore, there can be only up to 2 vertices with degree 1 (dead ends). In this case, these would be the start and/or the finishing vertices.

WORKED EXAMPLE 14 Determining a Hamiltonian path

Determine a Hamiltonian path in the planar graph shown.

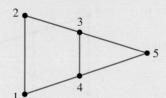

<table>
<tr><td>

THINK

1. Choose a starting node. If there is a vertex with degree = 1, then use it to start.

2. Attempt to visit each vertex. This will work for all feasible planar graphs. The Hamiltonian path found is shown in red.

</td><td>

WRITE

Since there are no vertices with degree 1, choose any node to start. Choose vertex 1.

The path connecting nodes 1–2–3–4–5 was chosen as one of the Hamiltonian paths.

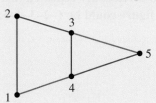

</td></tr>
</table>

Note that there are several possible Hamiltonian paths for the planar graph in Worked example 14 and there are several paths which will not result in a Hamiltonian path. Can you find such a path?

9.4.6 Hamiltonian cycles

When determining a Hamiltonian path, sometimes it is desirable to start and finish with the same vertex. For example, our travelling salesperson may live in one of the towns (vertices) she visits and would like to start and finish at her home town after visiting all the other towns once. This is similar to the concept of an Euler circuit.

Hamiltonian cycles

A Hamiltonian cycle **is a Hamiltonian path which starts and finishes at the same vertex.**

The Hamiltonian cycle in this figure could be **1–2–3–5–4–1.**

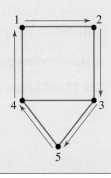

WORKED EXAMPLE 15 Determining a Hamiltonian cycle

Determine a Hamiltonian cycle in the planar graph shown. (This is the same graph used in Worked example 14.)

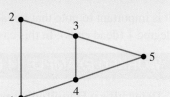

<table>
<tr><td>

THINK

1. Choose a starting (and finishing) vertex.

2. Attempt to visit each vertex and return to vertex 1.
The Hamiltonian cycle found is shown in pink.

</td><td>

WRITE

Choose vertex 1.

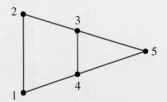

The path connecting nodes 1-2-3-5-4-1 was chosen as one of the Hamiltonian cycles.

</td></tr>
</table>

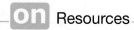
9.4 Exercise

1. **WE11** Starting at vertex 1, identify an Euler trail in the figure shown.

2. Choosing the other vertex of degree 3 in the figure used for question **1**, identify another Euler trail.

3. Starting at vertex 1, identify an Euler trail in the figure shown.

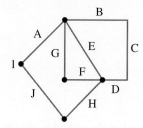

4. Starting with a vertex of degree 4 from the figure in question **3**, identify another Euler trail.

5. **MC** Select which edge in the following connected graph is a bridge.

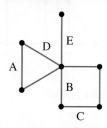

A. A   **B.** B   **C.** C   **D.** D

6. **WE12** Starting at vertex 1, identify an Euler circuit in the figure shown.

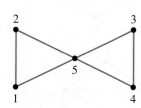

7. a. MC Select the additional edge that should be added to the planar graph shown so that it could be possible to define an Euler circuit.

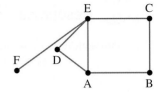

    A. CA   **B.** FA   **C.** FD

    D. EB   **E.** None of the above

  b. State which edge in the original planar graph is a bridge.

8. MC Select which of the networks shown have Euler circuits.

    A. Both   **B.** Neither   **C.** Figure **a** only

    D. Figure **b** only   **E.** None of the above

  a.

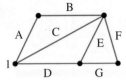

  b.

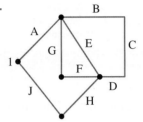

9. Using the graph shown, determine:

  a. an Euler trail

  b. an Euler circuit starting at vertex 1 (circled in pink).

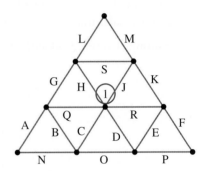

10. WE13 Use the Euler circuit algorithm to identify an Euler circuit in the graph shown.

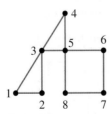

11. MC Select which of the following has an Euler circuit.

  A.

  B.

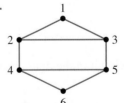

  C.

  D.

  E.

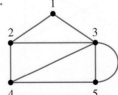

12. **WE14** Starting at vertex 2, determine a Hamiltonian path for the graph shown.

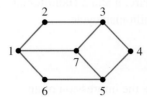

13. **MC** The path shown in pink in the figure is:
    A. an Euler trail.
    B. an Euler circuit.
    C. a Hamiltonian path.
    D. a Hamiltonian cycle.
    E. none of the above.

14. **WE15** Starting at vertex 2, determine a Hamiltonian cycle for the graph shown.

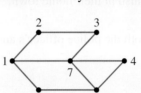

15. **MC** The network shown has:
    A. an Euler circuit and a Hamiltonian path.
    B. a Hamiltonian path and a cycle.
    C. an Euler trail and a Hamiltonian cycle.
    D. an Euler trail only.
    E. none of the above.

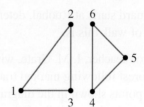

16. Starting at vertex 7, determine a Hamiltonian path for the graph shown.

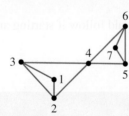

17. **MC** Select which of the following paths is a Hamiltonian cycle for the figure shown.
    A. 2–3–4–5–6–1–2
    B. 2–3–4–5–2
    C. 2–5–3–4–5–6–1–2
    D. 2–5–6–4–3–2
    E. 2–1–6–4–5–2

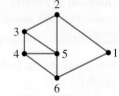

18. For the network shown:
    a. determine an Euler trail
    b. determine an Euler circuit
    c. determine a Hamiltonian path
    d. determine a Hamiltonian cycle.

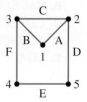

19. **a.** Using the network shown, determine what two edges should be added to the network so that it has both an Euler circuit and a Hamiltonian cycle.

b. Determine the Euler circuit and Hamiltonian cycle.

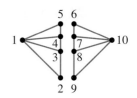

20. The police commissioner wishes to give the impression of an increased police presence on the roads. The roads that need to be covered are depicted in the network diagram. A speed camera is set up once during the day on each of the roads.

a. Determine a walk that a police officer could follow in order not to travel more than once on any road.

b. State what type of walk this is.

The police officer knows that there is a dirt track linking the towns of Larebil and Yrtnuoc. They wish to meet the commissioner's instructions but also wish to start and finish in their home town, Larebil.

c. Determine a walk that would meet both the police officer's and the commissioner's requirements.

d. State what type of walk this is.

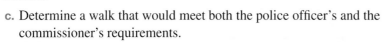

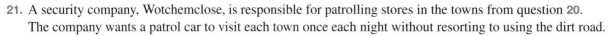

21. A security company, Wotchemclose, is responsible for patrolling stores in the towns from question **20**. The company wants a patrol car to visit each town once each night without resorting to using the dirt road.

a. If the security guard starts at Ruobal, determine a walk that will meet the company's requirements.

b. State what type of walk this is.

22. A physical education teacher, I. M. Grate, wishes to plan an orienteering course through a forest following marked tracks. They have placed checkpoints at the points shown in the diagram. The object of any orienteer is to visit each of the checkpoints once to collect a mark.

a. Suggest a walk that an efficient orienteer could follow if the course starts at C and finishes at B.

b. Determine a walk that the orienteer should follow if starting and finishing at C.

c. State what type of walks these are.

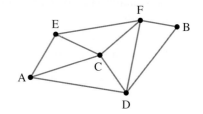

9.4 Exam questions

▶ **Question 1 (1 mark)**

Source: VCE 2020, Further Mathematics Exam 1, Section B, Module 2, Q2; © VCAA.

MC Consider the graph below.

Which one of the following is not a Hamiltonian cycle for this graph?

A. *ABCDFEGA*

B. *BAGEFDCB*

C. *CDFEGABC*

D. *DCBAGFED*

E. *EGABCDFE*

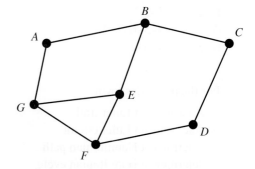

Question 2 (1 mark)

Source: VCE 2017, Further Mathematics Exam 1, Section B, Module 2, Q6; © VCAA.

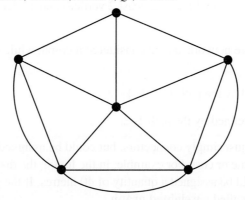

MC An Eulerian trail for the graph above will be possible if only one edge is removed.

In how many different ways could this be done?

A. 1   **B.** 2   **C.** 3   **D.** 4   **E.** 5

Question 3 (1 mark)

Source: VCE 2014, Further Mathematics Exam 1, Section B, Module 5, Q3; © VCAA.

MC The diagram below shows the network of roads that Stephanie can use to travel between home and school.

The numbers on the roads show the time, in minutes, that it takes her to ride a bicycle along each road. Using this network of roads, the shortest time that it will take Stephanie to ride her bicycle from home to school is

A. 12 minutes

B. 13 minutes

C. 14 minutes

D. 15 minutes

E. 16 minutes

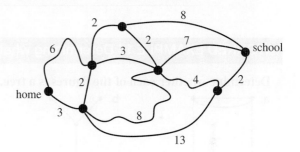

More exam questions are available online.

9.5 Trees and their applications

> **LEARNING INTENTION**
>
> At the end of this subtopic you should be able to:
> - apply properties of graphs and subgraphs
> - apply properties of trees
> - determine the shortest path between two selected vertices of a graph
> - apply an algorithm to calculate a shortest path
> - determine the minimum spanning tress by applying Prim's algorithm
> - determine the shortest path by applying Dijkstra's algorithm.

9.5.1 Graphs and subgraphs

There are many applications where only part of the network is required as a solution to a problem. This section will look at such problems involving **subgraphs** and **trees**. To begin with, we need a few more definitions.

Until now we have used the term *network* to refer to a collection of vertices and edges. A network can also be called a graph. In practice, a graph should have at least 2 vertices and 1 edge. All or part of a graph can be considered as a *subgraph*.

For example, in the figure, the entire network can be considered a graph, while the path in pink can be considered a subgraph.

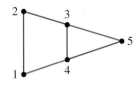

Another subgraph could be defined by the path 1–2–3–4–1.

A 'minimum' subgraph could be defined by the path 1–2.

Often the edges in a graph are not just simply connectors, but could be assigned some quantity, such as distance, time or cost. For example, in the figure, the distance between vertex 1 and vertex 2 could be assigned a quantity of 40 metres. If the graph contains such quantities, then it is called a **weighted graph**.

Trees

A *tree* is a connected subgraph which cannot contain any:
1. **loops**
2. **parallel (or multiple) edges**
3. **cycles.**

WORKED EXAMPLE 16 Determining whether a graph is a tree

Determine whether each of the figures is a tree, and if not, explain why not.

a. b. c. d.

THINK

1. Examine each figure in turn, looking for loops, parallel edges or cycles.

2. Examine Figure **a**.

3. Examine Figure **b**.

4. Examine Figure **c**.

5. Examine Figure **d**.

WRITE

Figure **a** has parallel edges (at the top) so it is not a tree.

Figure **b** has no loops, parallel edges or cycles so it is a tree.

Figure **c** has a cycle (at the top) so it is not a tree.

Figure **d** has a loop (at the bottom) so it is not a tree.

The advantage of trees within a network is that the tree could determine an 'efficient' connection between vertices in the sense that there is a minimum distance, cost or time.

9.5.2 Shortest paths

Sometimes it may be useful to determine the **shortest path** between two selected vertices of a graph. For example, when going shopping, a person may leave home and travel east via the playground or north via the parking lot, and still end up at the same shop. In one case, the distance travelled may be the minimum.

WORKED EXAMPLE 17 Determining the shortest path between two selected vertices

Determine the shortest path between nodes A and F in the figure shown. Nodes are labelled A, ... , G and distances (in metres) between them are labelled in blue.

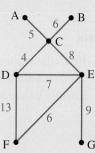

THINK	WRITE
1. From the starting node, by inspection, determine the possible trees between A and F.	(1) A–C–E–D–F (2) A–C–D–F (3) A–C–D–E–F (4) A–C–E–F
2. For each tree, calculate the total distance travelled.	(1) $5 + 8 + 7 + 13 = 33$ (2) $5 + 4 + 13 = 22$ (3) $5 + 4 + 7 + 6 = 22$ (4) $5 + 8 + 6 = 19$
3. Choose the path with the shortest distance.	Path A–C–E–F is the shortest path, with a distance of 19 m.

When choosing the possible paths in Step 1, there is no point in finding paths that are *not* trees. There will *always* be a tree which covers the same vertices over a smaller distance. Non-tree paths will include cycles and loops, which only add to the total distance.

9.5.3 A shortest path algorithm

Sometimes it can be difficult to list all the paths between the starting and ending vertex. Here is an algorithm that 'guarantees' the shortest path — assuming that the starting vertex is already chosen.

Step 1. From the starting vertex, find the *shortest* path to all other *directly connected* vertices. Include all such vertices, including the starting one in the list $S = \{A, B, ... \}$.

Step 2. Choose a vertex (V) directly connected to the vertices in S and find the shortest path to the starting vertex. Generally, there is one possible path for each degree of V, although some obvious paths can be eliminated immediately.

Step 3. Add the new vertex, V, to the list S.

Step 4. Repeat Steps 2 and 3 until all vertices are in S. Determine the shortest path to the vertex you want.

WORKED EXAMPLE 18 Determining the shortest path

Determine the shortest path from vertex 1 to vertex 9. Vertices are labelled in black, distances in blue. (*Note:* Lines are not to scale.)

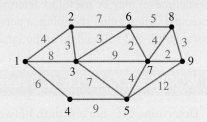

THINK

1. From the starting vertex (1), find the shortest path to each of the vertices directly connected to it.

2. Determine the set of vertices, S.

3. Apply Step 2 of the algorithm for a vertex connected directly to one in S.

4. Apply Step 3 of the algorithm and add to the set of vertices, S.

5. Reapply Step 2 of the algorithm for another vertex directly connected to one in S.

6. Reapply Step 3 of the algorithm and add to the set of vertices, S.

7. Reapply Step 2 for another vertex directly connected to one in S.

8. Reapply Step 3 and add to the set of vertices, S.

9. Reapply Step 2 for another vertex directly connected to one in S.

10. Reapply Step 3 and add to the set of vertices, S.

WRITE

From	To	Via	Distance	
1	2	—	4	shortest path to 2
1	3	—	8	
1	3	2	$4+3=7$	shortest path to 3
1	4	—	6	shortest path to 4

$S = \{1, 2, 3, 4\}$

Select vertex 5.

From	To	Via	Distance	
1	5	3	$7+7=14$	shortest path to 5
1	5	4	$6+9=15$	

$S = \{1, 2, 3, 4, 5\}$

Select vertex 6.

From	To	Via	Distance	
1	6	2	$4+7=11$	
1	6	3	$7+3=10$	shortest path to 6
1	6	7	$7+9+2=18$	

$S = \{1, 2, 3, 4, 5, 6\}$

Select vertex 7.

From	To	Via	Distance	
1	7	6	$10+2=12$	shortest path to 7
1	7	3	$7+9=16$	
1	7	5	$14+4=18$	

$S = \{1, 2, 3, 4, 5, 6, 7\}$

Select vertex 8.

From	To	Via	Distance	
1	8	6	$10+5=15$	shortest path to 8
1	8	7	$12+4=16$	

$S = \{1, 2, 3, 4, 5, 6, 7, 8\}$

11. Reapply Step 2 for another vertex directly connected to one in S.	Select vertex 9.

Select vertex 9.

From	To	Via	Distance
1	9	8	$15 + 3 = 18$
1	9	7	$12 + 2 = 14$ shortest path to 9
1	9	5	$14 + 12 = 26$

12. Reapply Step 3 and add to the set of vertices, S.

$S = \{1, 2, 3, 4, 5, 6, 7, 8, 9\}$. Stop, because all vertices are in the list.

13. Apply Step 4 and determine the shortest path.

1 to 9 via 7   (see Step 11)
1 to 7 via 6   (see Step 7)
1 to 6 via 3   (see Step 5)
1 to 3 via 2   (see Step 1)
1 to 2   (see Step 1)
Path $= 1–2–3–6–7–9$
Distance $= 4 + 3 + 3 + 2 + 2 = 14$

9.5.4 Spanning trees and Prim's algorithm

In the network shown, the vertices represent school buildings and the edges represent footpaths. The numbers represent the distance, in metres, between the buildings. The school council has decided to cover some of the footpaths so that the students can access any building during rainy weather without getting wet. Three possible trees that would accomplish this are shown in the following figures.

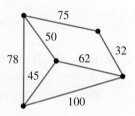

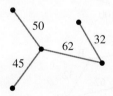
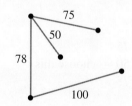
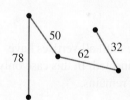

Note that each of these trees includes all the vertices of the original network. These trees are called **spanning trees** because of this property. In practice, the school council would like to make the total distance of covered footpaths as small as possible, in order to minimise cost. In this case, they would have the **minimum spanning tree**. Can you determine which of the figures is the minimum spanning tree?

Minimum spanning trees and Prim's algorithm

One method of determining the minimum spanning tree is called **Prim's algorithm**. The steps are as follows:

Step 1. Choose the edge that has the smallest value in the network. If 2 or more edges are the smallest, choose any of these.

Step 2. Inspect the 2 vertices included so far and select the smallest edge leading from either vertex. Again, if there is a 'tie', arbitrarily choose any one.

Step 3. Inspect all vertices included so far and select the smallest edge leading from any included vertex. If there is a 'tie', choose one randomly.

Step 4. Repeat Step 3 until all vertices in the graph are included in the tree.

Determine the minimum spanning tree for the network representing footpaths in a school campus.

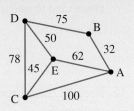

THINK	WRITE
1. Find the edge with the smallest distance. This can be done by listing all the edges and choosing the smallest.	A–B = 32, by inspection

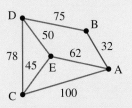

2. Inspect A and B and find the shortest edge connecting one of these to a third vertex.	A–E = 62 — choose this A–C = 100 B–D = 75

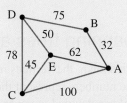

3. Inspect A, B and E and find the shortest edge connecting one of these to another vertex.	A–C = 100 E–C = 45 — choose this E–D = 50 B–D = 75

4. Continue until all vertices have been connected. In this case only vertex D remains.	B–D = 75 E–D = 50 — choose this C–D = 78

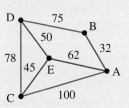

5. Since all vertices have been connected, this is the minimum spanning tree. Calculate the total distance of the minimum spanning tree.	Total distance = 32 + 62 + 45 + 50 = 189 m

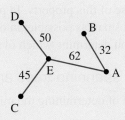

9.5.5 Maximum spanning tree

In some cases you may be required to calculate the **maximum spanning tree** instead of the minimum spanning tree. In this case, Prim's algorithm works by finding the largest edges at each stage instead of the smallest edges.

WORKED EXAMPLE 20 Determining the maximum spanning tree

The figure shown represents a telephone network connecting 6 towns,
A, B, … , F. The numbered edges represent the 'capacity' of the telephone
connection between the towns connected; that is, the maximum number of
calls that can be made at the same time along that edge.
A telephone engineer wishes to determine the maximum capacity of the
system in terms of a tree connecting all the towns so that calls can be routed
along that tree.

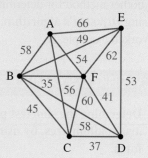

THINK	WRITE
1. Because this is a maximum spanning tree, find the edge with the largest capacity.	This is edge A−E (66). 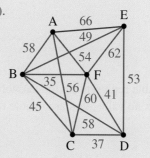
2. Inspect edges from vertices A and E, and find the edge with the largest capacity.	This is edge E−F (62). 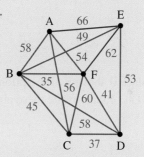
3. Inspect edges from vertices A, E and F, and find the edge with the largest capacity.	This is edge F−C (60). 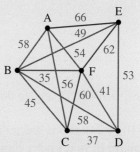
4. Repeat until all towns are connected.	A−B (58) and B−D (58) are the edges selected. 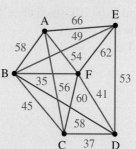
5. Determine maximum capacity.	Maximum capacity $= 66 + 62 + 60 + 58 + 58$ $= 304$ telephone calls at the same time.

9.5.6 Dijkstra's algorithm

Another method for determining the shortest path between a given vertex and each of the other vertices is by using **Dijkstra's algorithm**. An effective way to use this algorithm is by using a tabular method, as shown in Worked example 21.

WORKED EXAMPLE 21 Determining the shortest path by using Dijkstra's algorithm

Determine the shortest path from A to E, where the distances are in kilometres, by using Dijkstra's algorithm in tabular form.

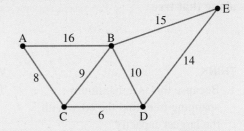

THINK

1. Construct a table representing the network. Where there is no value, you are unable to take that path.

WRITE

	A	B	C	D	E
A	X	16	8		
B	16	X	9	10	15
C	8	9	X	6	
D		10	6	X	14
E		15		14	X

2. Starting at E, place 0 above it. The options are B (15 km) and D (14 km). Write these above the letters.

	A	B	C	D	E
		15		14	0
A	X	16	8		
B	16	X	9	10	15
C	8	9	X	6	
D		10	6	X	14
E		15		14	X

3. Go to the shortest distance, D, and go down vertically.
   B: Add 14 and 10 (= 24); this is greater than the 15 already above B, so it's not a choice.
   C: Add 14 and 6 (= 20); since there's nothing above C, write this above C.
   E: We started at E so we don't want to go back there.

	A	B	C	D	E
		15	20	14	0
A	X	16	8		
B	16	X	9	10	15
C	8	9	X	6	
D		10	6	X	14
E		15		14	X

4. Now E and D are done, the next lowest is B; again, work vertically.

   A: Add 15 and 16 (= 31); since there's nothing above A, write this there.

   C: Add 15 and 9 (= 34); this is greater than 20, so leave it as 20.

   D: Already been to D.

   E: Already been to E.

	31	15	20	14	0
	A	**B**	**C**	**D**	**E**
A	X	16	8		
B	16	X	9	10	15
C	8	9	X	6	
D		10	6	X	14
E		15		14	X

5. Now that E, D and B are done, the next lowest is C; again work vertically.

   A: Add 20 and 8 (= 28); since this is lower than 31 that is currently there, replace it with 28.

   B: Already been to B.

   D: Already been to D.

	28 31	15	20	14	0
	A	**B**	**C**	**D**	**E**
A	X	16	8		
B	16	X	9	10	15
C	8	9	X	6	
D		10	6	X	14
E		15		14	X

6. The shortest distance from A to E is then the number above E, in this case 28.

The shortest distance from A to E is 28 km.

 Resources

9.5 Exercise

Students, these questions are even better in jacPLUS

| | **Receive immediate feedback and access sample responses** | | **Access additional questions** | **Track your results and progress** |

Find all this and MORE in jacPLUS

1. **WE16** Determine *all* the trees connecting vertices A and B, without going through vertex F.

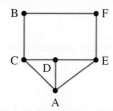

2. **MC** Consider the following paths:

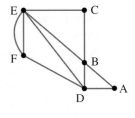

   i. A–B–C–E–F–D–A   **ii.** A–D–F–E–C–B

   iii. A–B–D–A   **iv.** A–B–D–F–E–C

   Select which (if any) are trees.

   A. All are trees.   **B.** i and iv are trees.   **C.** ii and iv are trees.

   D. iv is a tree.   **E.** None are trees.

3. Consider the figure shown. State which of the paths marked in pink on the following figures are trees.

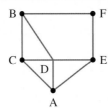

   a.    **b.**   **c.**    **d.**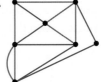

4. Determine *all* the trees that connect vertices A and B without going through vertex F.

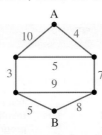

5. **WE17** Determine the shortest path from A to B, where the distances (in blue) are in kilometres.

6. The network shown represents the time (blue numerals in minutes) that it takes to walk along pathways connecting 8 features in a botanical garden. Vertices 1 and 8 are entrances. Calculate the minimum time to walk between the entrances, along pathways only.

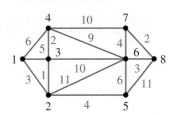

7. **a.** **WE18** Referring to the network shown, where distances are in km, calculate:

    i. the shortest path from B to F

    ii. the shortest path from A to C.

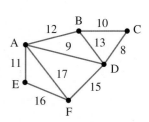

  b. Referring to the network in part **a**, where distances are in km, calculate the shortest path from E to C.

  c. **MC** The total length of the minimum spanning tree in part **a** is:

    A. 20   **B.** 37   **C.** 40   **D.** 51   **E.** 66

8. **WE19** Calculate the minimum spanning tree for the network shown.

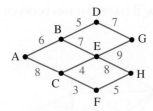

9. **MC** **WE20** In the figure shown, the calculations using Prim's algorithm for the minimum spanning tree would be:

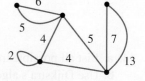

   A. $2+4+4+5+7$
   B. $4+4+5+7$
   C. $4+4+6+7$
   D. $4+5+7+4$
   E. 20

10. Draw the minimum spanning tree in each of the following graphs and calculate the total length.

   a.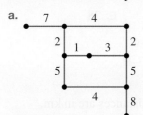
   b.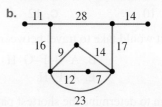

11. **WE21** Determine the shortest path from A to E, where the distances are in kilometres, by using Dijkstra's algorithm in tabular form.

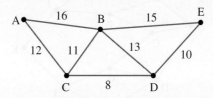

12. Determine the shortest path from A to E, where the distances are in kilometres, by using Dijkstra's algorithm in tabular form.

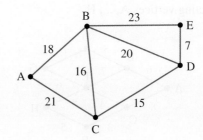

13. **MC** Select the true statement.

   A. A Hamiltonian cycle is a tree.
   B. A tree can contain multiple edges or loops.
   C. A Hamiltonian path is not a tree.
   D. A tree can visit the same vertex more than once.
   E. A tree can have one edge.

Questions 14–16 *refer to the network shown in the figure.*

Vertices are labelled A, B, ..., H and the *time* it takes to travel between them, in minutes, is given by the numbers in blue.

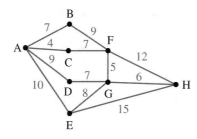

14. a. List all possible trees connecting A and H that pass through B.
    b. Use Dijkstra's algorithm to determine the shortest time for A–H.

15. **MC** The total number of possible trees connecting A and H is given by:

    A. 8   **B.** 10   **C.** 12   **D.** 15   **E.** 28

16. **MC** The shortest time it would take to travel between A and H is given by the tree:

    A. A–B–F–H   **B.** A–B–F–G–H   **C.** A–C–F–H
    D. A–C–F–G–H   **E.** A–E–H

17. Use Dijkstra's algorithm to determine the shortest path from A to K, where the distances are in km.

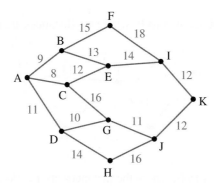

18. The figure shows a network connecting vertices A, ..., H.

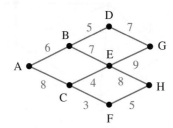

    a. Determine the number of different trees connecting A to G.
    b. Calculate the shortest path connecting A to G.
    c. Determine the number of different trees connecting D to F.
    d. Calculate the shortest path connecting D to F.

19. Using Prim's algorithm, determine the length of the minimum spanning tree in each of the graphs shown.

a.

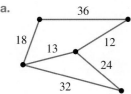

b.

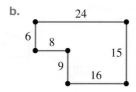

c.

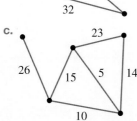

d.

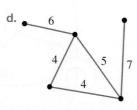

20. Draw the minimum spanning tree in each of the following graphs and calculate the total length.

a.

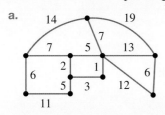

b.

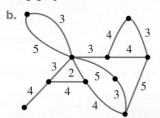

21. Flyemsafe Airlines wishes to service six cities. The directors have decided that it is too costly to have direct flights between all the cities. The airline needs to minimise the number of open routes while maximising the total number of passengers that it can carry.

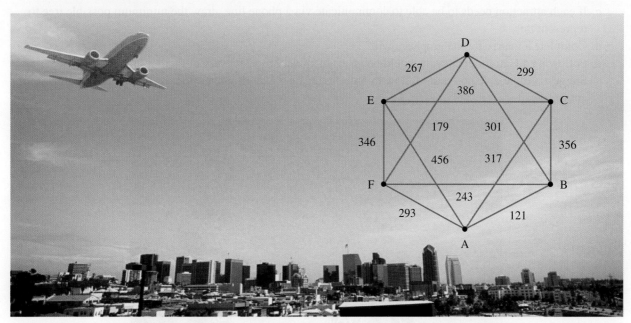

The network diagram shown has edges representing routes and vertices representing cities. The numbers on the edges are projected capacities.
Calculate:

a. the maximum spanning tree that will meet the airline's requirements

b. the total carrying capacity of this tree.

22. A fairground has 5 main attractions which are joined by paths to the entrance/exit gate. The numbers show the distance along the paths in metres.

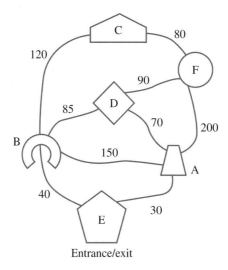

a. Draw an undirected graph to represent the fairground and then write down:

  i. the number of edges
  ii. the number of vertices
  iii. the degree of each vertex.

b. Calculate the minimum distance a person would have to walk to visit every attraction, beginning and ending at the entrance/exit.

c. If each attraction needs to be able to communicate via a phone line, draw the minimum possible tree to represent this.

d. Complete a matrix for the graph shown.

e. Following a Hamiltonian cycle would be an efficient way to visit every attraction in the fairground. Suggest a route a visitor could follow in order to create a Hamiltonian cycle, beginning and ending at the entrance/exit.

9.5 Exam questions

Question 1 (1 mark)

*Source: VCE 2020, Further Mathematics Exam 1, Section B, Module 2, Q3; © VCAA.

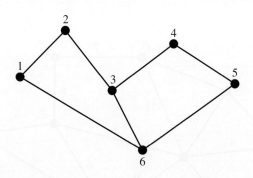

MC Which one of the following is not a spanning tree for the network above?

A.

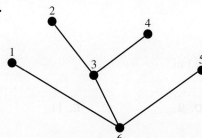

B.

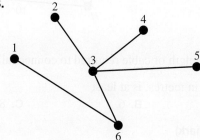

C.

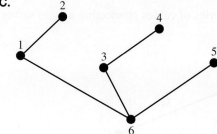

D.

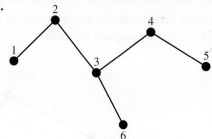

E.

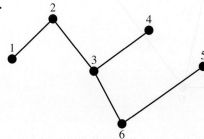

Question 2 (1 mark)

Source: VCE 2020, Further Mathematics Exam 1, Section B, Module 2, Q5; © VCAA.

MC The network below shows the distances, in metres, between camp sites at a camping ground that has electricity.

The vertices *A* to *I* represent the camp sites.

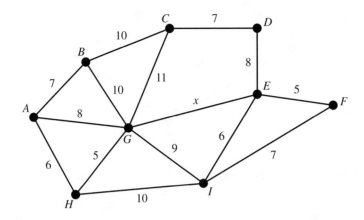

The minimum length of cable required to connect all the camp sites is 53 m.

The value of *x*, in metres, is at least

A. 5   **B.** 6   **C.** 8   **D.** 9   **E.** 11

Question 3 (1 mark)

Source: VCE 2019, Further Mathematics Exam 1, Section B, Module 2, Q5; © VCAA.

MC The following diagram shows the distances, in metres, along a series of cables connecting a main server to seven points, *A* to *G*, in a computer network.

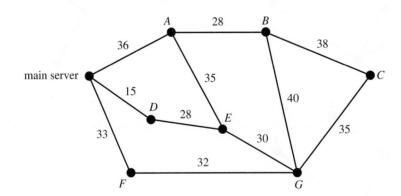

The minimum length of cable, in metres, required to ensure that each of the seven points is connected to the main server directly or via another point is

A. 175   **B.** 203   **C.** 208   **D.** 221   **E.** 236

More exam questions are available online.

9.6 Review

9.6.1 Summary

doc-38040

9.6 Exercise

Multiple choice

Questions 1 and 2 refer to the network shown.

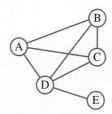

1. **MC** The sum of the degrees of all the vertices in the network is:

   A. 5   B. 10   C. 12   D. 13   E. 14

2. **MC** The list of edges can be written as:

   A. {A, B, C, D, E}
   B. {(A, A), (B, B), (C, C), (D, D), (E, E)}
   C. {(A, B), (A, C), (A, D), (B, C), (B, D), (C, D), (D, E)}
   D. {(A, B), (A, C), (A, D), (B, C), (C, D), (D, E)}
   E. {(A, B, C, D), (B, A, C, D), (C, A, B, D), (D, A, B, C, E), (E, D)}

3. **MC** From the following matrix representation of a network with vertices A, B, C, D, E, the degree of vertex A is:

$$
\begin{array}{c}
\ \ \ A\ \ B\ \ C\ \ D\ \ E \\
\begin{array}{c} A \\ B \\ C \\ D \\ E \end{array}
\left[
\begin{array}{ccccc}
1 & 2 & 1 & 0 & 1 \\
2 & 0 & 1 & 0 & 1 \\
1 & 1 & 0 & 0 & 1 \\
0 & 0 & 0 & 0 & 0 \\
1 & 0 & 1 & 0 & 0
\end{array}
\right]
\end{array}
$$

   A. 1
   B. 2
   C. 5
   D. 6
   E. 10

4. **MC** Using the matrix in question **3**, the isolated vertex is:

    A. A   **B.** B   **C.** C   **D.** D   **E.** E

5. **MC** A certain planar graph with 6 vertices can be divided into 4 regions (or faces). The number of edges it has is:

    A. 2   **B.** 6   **C.** 8   **D.** 10   **E.** 4

6. **MC** Consider the networks **i**, **ii**, **iii** and **iv**. Select the connected graphs.

i.

ii.

iii.

iv.

    A. i only   **B.** i, ii and iii   **C.** i and iii   **D.** ii and iv   **E.** All are connected.

7. **MC** Select the Hamiltonian path in the graph shown.

    A. C–G–A–H–G–F–E–D–B–A–C–B–E
    B. C–G–H–A–B–D–E–B–C
    C. C–G–H–A–B–D–E–F
    D. A–B–C–D–E–F–G–H
    E. There is no Hamiltonian path in this graph.

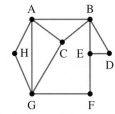

8. **MC** For the graph in question **7**, an Euler trail:

    A. does not exist because there are 2 vertices whose degree is 3.
    B. exists because there are exactly 2 vertices whose degree is 3.
    C. exists because there is an even number of vertices.
    D. must not start on a vertex with odd degree.
    E. requires the starting vertex and the ending vertex to be the same.

9. Consider the graph shown to answer the following questions.

    a. **MC** The shortest path from H to E is:

      A. 24   **B.** 25   **C.** 26   **D.** 27   **E.** 28
    b. **MC** The length of the minimum spanning tree for the graph is:

      A. 39   **B.** 40   **C.** 43   **D.** 49   **E.** 50

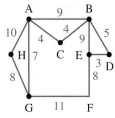

10. **MC** A connected planar graph has 12 edges. This graph could have:

    A. 5 vertices and 6 faces.
    B. 5 vertices and 8 faces.
    C. 6 vertices and 8 faces.
    D. 6 vertices and 9 faces.
    E. 7 vertices and 9 faces.

11. For the network in question **9**:

    a. label the vertices and edges in terms of the vertices

    b. write a matrix representation of the network.

12. Draw a diagram of a network represented by the following:

$V = \{A, B, C, D, E, F\}$ and $E = \{(A, B), (A, C), (B, C), (B, D), (B, E), (C, F), (D, E), (E, F)\}$

Calculate the sum of the degrees of all the vertices.

13. a. Convert the network in the figure to a planar graph.

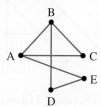

    b. Confirm Euler's formula for the result.

14. Consider the following figure.

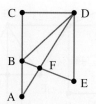

    a. Starting at vertex A, determine an Euler circuit for the given figure.

    b. Starting at vertex A, determine two different Hamiltonian cycles for the given figure.

15. In a town there are five friends: Paul, Shiva, Kevin, Asha and Carlo. The friends' houses are linked by the number of footpaths given in the matrix.

$$\begin{bmatrix} 0 & 2 & 1 & 0 & 1 \\ 2 & 0 & 0 & 1 & 1 \\ 1 & 0 & 0 & 1 & 1 \\ 0 & 1 & 1 & 0 & 1 \\ 1 & 1 & 1 & 1 & 0 \end{bmatrix}$$

Due to an oversight, the names of all the friends were not listed with the rows and columns; however, the following information is known:

- Shiva and Paul have two footpaths between their houses.
- There is only one path between Kevin and each of his friends.
- Paul and Carlo have one path linking their houses to each other.
- The second column in the matrix representation above represents Paul's edges.

a. Redraw the matrix with the names in their correct places above the columns and beside the rows.
b. Using the answer to part **a**, draw the network of footpaths and houses.
c. Redraw the network and add a path to it that will enable an Euler circuit to be travelled by the group of friends.
d. Explain why the friends would want an extra footpath in their network of footpaths.
e. Using the new network diagram, identify an Euler circuit.

16. Determine a Hamiltonian path for the network shown.

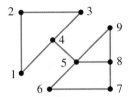

17. The figure shown represents rail connections between 11 towns. A telegraph system is to be set up connecting all the towns. Calculate the shortest total distance for this system. Note that messages can be relayed from one station to another via a third station.

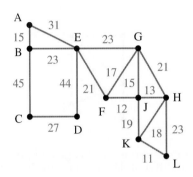

Extended response

18. In a state forest there are six camp sites: Angel Valley, Booming Falls, Crater Face, Dewey Seat, Echidna Spike and Fearful Drop. The camp sites have walking tracks linking them to various other camp sites. The matrix representing the walks and camp sites is given as shown.

$$
\begin{array}{c c}
 & \begin{array}{cccccc} A & B & C & D & E & F \end{array} \\
\begin{array}{c} A \\ B \\ C \\ D \\ E \\ F \end{array} &
\left[\begin{array}{cccccc}
1 & 0 & 1 & 3 & 1 & 0 \\
0 & 0 & 1 & 1 & 0 & 0 \\
1 & 1 & 0 & 1 & 0 & 1 \\
3 & 1 & 1 & 0 & 0 & 1 \\
1 & 0 & 0 & 0 & 0 & 1 \\
0 & 0 & 1 & 1 & 1 & 0
\end{array} \right]
\end{array}
$$

a. State the number of loops in the matrix.
b. State the number of multiple edges in the matrix.
c. State the degree of vertex A.
d. State the degree of vertex D.
e. Represent the matrix as a planar graph.
f. Draw any tree from the graph.

19. The plan shown represents a botanical garden in the town of Lovely Banks. There are 6 features (oak tree, kiosk, flower beds, pine tree, gazebo and lake) and entrances. Paths connecting various features are drawn as blue lines and can be taken as straight lines connecting the dots for each feature.

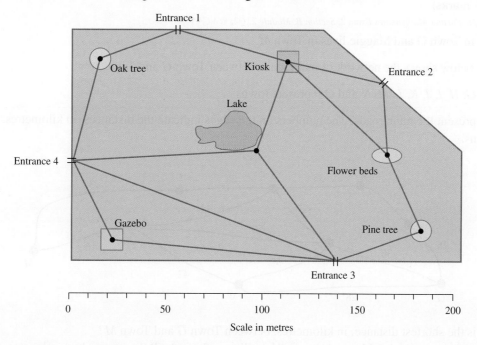

Entrance 1

Oak tree

Kiosk

Entrance 2

Lake

Entrance 4

Flower beds

Gazebo

Pine tree

Entrance 3

Scale in metres

a. Draw this 'network' of 10 vertices as a planar graph.

b. A visitor arriving at Entrance 2 wishes to travel along all paths exactly once. This is not possible, but can become so by adding a new path between two features. Add this new path and determine the visitor's path. Explain what kind of path this is.

c. Measure the distance along the paths using a ruler and the scale at the bottom of the plan. Quote distances to the nearest 5 metres. (Do not include the path from part **b.**)

d. Calculate the shortest path that connects all the vertices (including entrances). State what kind of path this is. Make a drawing of this path. (Do not include the path from part **b.**)

e. Calculate the shortest path between Entrance 4 and Entrance 2. (Do not include the path from part **b.**)

20. An airfreight company operates out of Melbourne (M) to Geelong (G), Ballarat (B), Castlemaine (C), Seymour (S), Mansfield (F) and Warragul (W). A planar graph of the company's flight network is outlined, with distances shown in kilometres.

Note that some flight paths, such as Melbourne to Warragul, are prohibited so they do not appear on the network.

a. Verify Euler's formula for this network.

b. Represent the network as a matrix.

c. The company wishes to fly to all six of these satellite towns and return to Melbourne. Mathematically, determine what type of path this is. Determine the two possible flight paths for this.

d. Draw a spanning tree for this network.

e. Now determine the minimum spanning tree and the total distance represented. Comment on the usefulness of this for the company.

f. Use Dijkstra's algorithm to determine the shortest distance between Geelong and Mansfield.

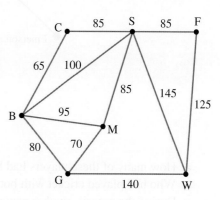

Question 1 (2 marks)

Source: VCE 2021, Further Mathematics Exam 2, Section B, Module 2, Q2; © VCAA.

George lives in Town G and Maggie lives in Town *M*.

The diagram below shows the network of main roads between Town *G* and Town *M*.

The vertices *G, H, I, J, K, L, M, N* and *O* represent towns.

The edges represent the main roads. The numbers on the edges indicate the distances, in kilometres, between adjacent towns.

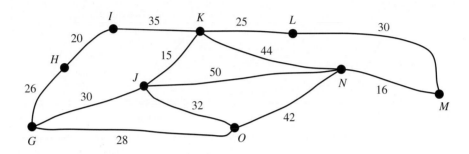

  a. What is the shortest distance, in kilometres, between Town *G* and Town *M*?
  b. George plans to travel to Maggie's house. He will pass through all the towns shown above.
     George plans to take the shortest route possible.
     Which town will George pass through twice?

Question 2 (3 marks)

Source: VCE 2020, Further Mathematics Exam 2, Section B, Module 2, Q1; © VCAA.

The Sunny Coast Cricket Club has five new players join its team: Alex, Bo, Cameron, Dale and Emerson.

The graph below shows the players who have played cricket together before joining the team.

For example, the edge between Alex and Bo shows that they have previously played cricket together.

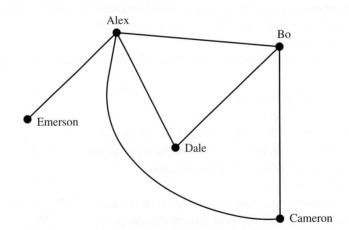

  a. How many of these players had Emerson played cricket with before joining the team?
  b. Who had played cricket with both Alex and Bo before joining the team?
  c. During the season, another new player, Finn, joined the team.
     Finn had not played cricket with any of these players before.
     Represent this information on the **graph above**.
     (Answer on the graph above.)

Question 3 (4 marks)

Source: VCE 2020, Further Mathematics Exam 2, Section B, Module 2, Q3; © VCAA.

A local fitness park has 10 exercise stations: *M* to *V*.

The edges on the graph below represent the tracks between the exercise stations.

The number on each edge represents the length, in kilometres, of each track.

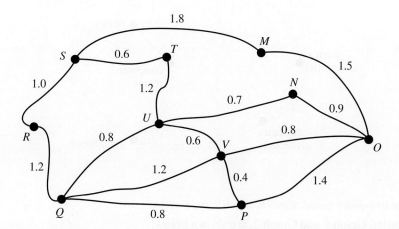

The Sunny Coast cricket coach designs three different training programs, **all starting at exercise station *S***.

Training program number	Training details
1	The team must run to exercise station *O*.
2	The team must run along all tracks just once.
3	The team must visit each exercise station and return to exercise station *S*.

a. What is the shortest distance, in kilometres, covered in training program 1?

b. **i.** What mathematical term is used to describe training program 2?

   ii. At which exercise station would training program 2 finish?

c. To complete training program 3 in the minimum distance, one track will need to be repeated.
   Complete the following sentence by filling in the boxes provided.
   This track is between exercise station _____ and exercise station _____.

Question 4 (3 marks)

Source: VCE 2019, Further Mathematics Exam 2, Section B, Module 2, Q1; © VCAA.

Fencedale High School has six buildings. The network below shows these buildings represented by vertices. The edges of the network represent the paths between the buildings.

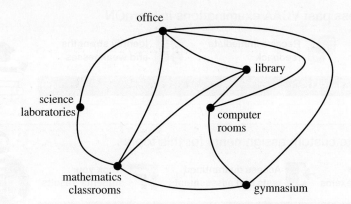

a. Which building in the school can be reached directly from all other buildings?

b. A school tour is to start and finish at the office, visiting each building only once.

    i. What is the mathematical term for this route?

    ii. Draw in a possible route for this school tour on the diagram below.

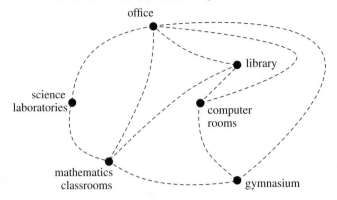

Question 5 (1 mark)

Source: VCE 2019, Further Mathematics Exam 1, Section B, Module 2, Q4; © VCAA.

MC Two graphs, labelled Graph 1 and Graph 2, are shown below.

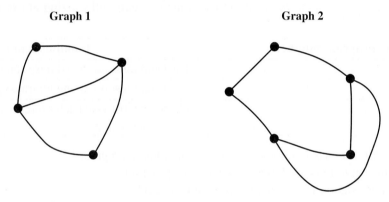

Which one of the following statements is **not** true?

    A. Graph 1 and Graph 2 are isomorphic.

    B. Graph 1 has five edges and Graph 2 has six edges.

    C. Both Graph 1 and Graph 2 are connected graphs.

    D. Both Graph 1 and Graph 2 have three faces each.

    E. Neither Graph 1 nor Graph 2 are complete graphs.

More exam questions are available online.

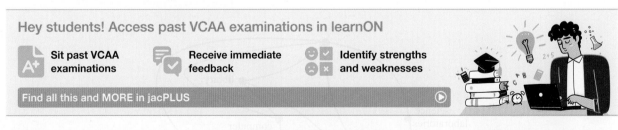

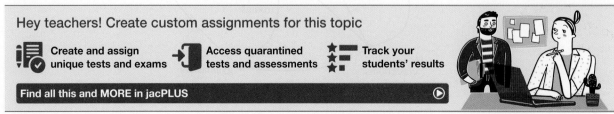

Answers

Topic 9 Undirected graphs, networks and trees

9.2 Basic concepts of a network

9.2 Exercise

1. a. Vertices $= 6$, edges $= 8$
    b. Vertices $= 7$, edges $= 9$
2. a. Vertices $= 5$, edges $= 5$
    b. Vertices $= 6$, edges $= 9$
    c. Vertices $= 7$, edges $= 11$
    d. Vertices $= 9$, edges $= 16$
3. C
4. a. $\deg(A) = 3$, $\deg(B) = 4$, $\deg(C) = 4$
    b. $\deg(A) = 3$, $\deg(B) = 8$, $\deg(C) = 3$
5. D
6. a. $\deg(A) = 2$, $\deg(B) = 2$, $\deg(C) = 1$
    b. $\deg(A) = 3$, $\deg(B) = 4$, $\deg(C) = 2$
    c. $\deg(A) = 3$, $\deg(B) = 2$, $\deg(C) = 6$
    d. $\deg(A) = 4$, $\deg(B) = 1$, $\deg(C) = 0$
7. a. $V = \{A, B, C, D, E, F\}$
       $E = \{(A, B), (A, D), (A, E), (B, D),$
       $\qquad\qquad (B, E), (B, E), (C, D), (D, F)\}$
    b. $V = \{A, B, C, D, E, F, G\}$
       $E = \{(A, B), (B, C), (B, E), (B, F),$
       $\qquad\qquad (C, E), (C, F), (E, F), (E, G), (F, G)\}$
8. a. $E = \{(1, 2), (1, 3), (2, 3), (3, 4), (4, 5)\}$
    b. $V = \{U, V, W, X, Y, Z\}$
       $E = \{(U, V), (U, W), (U, X), (V, W),$
       $\qquad\qquad (V, X), (V, Z), (W, X), (W, Y), (X, Z)\}$
    c. $V = \{1, 2, 3, 4, 5, 6, 7\}$
       $E = \{1, 2), (1, 3), (1, 5), (2, 4), (2, 6),$
       $\qquad\qquad (2, 6), (2, 6), (2, 7), (3, 3), (3, 6), (5, 7)\}$
    d. $V = \{1, 2, 3, 4, 5, 6, 7, 8, 9\}$
       $E = \{(1, 5), (1, 6), (1, 8), (2, 5), (2, 7), (2, 9),$
       $\qquad\qquad (3, 4), (3, 5), (3, 8), (4, 5), (4, 9), (5, 6),$
       $\qquad\qquad (5, 7), (5, 8), (5, 9), (6, 7)\}$

9. a.
    b.
	1	2	3	4	5	6
1	0	1	0	1	0	1
2	1	0	1	0	0	1
3	0	1	0	1	0	0
4	1	0	1	0	0	1
5	0	0	0	0	0	0
6	1	1	0	1	0	0

10.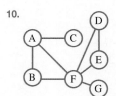

11. a.   b.

12. a. $V = \{1, 2, 3, 4\}$
       $E = \{(1, 2), (1, 3), (1, 4), (2, 3), (2, 4), (3, 4)\}$
    b.
    c. Degree of each vertex $= 3$

13. i. a. $V = \{1, 2, 3, 4, 5\}$
         $E = \{1, 2), (1, 3), (1, 4), (1, 5), (2, 3),$
         $\qquad\qquad (2, 4), (2, 5), (3, 4), (3, 5), (4, 5)\}$
       b.
       c. Degree of each vertex $= 4$
    ii. a. $V = \{1, 2, 3, 4, 5, 6, 7, 8\}$
        $E = \left\{ \begin{array}{l} (1, 2), (1, 3), (1, 4), (1, 5), (1, 6), \\ (1, 7), (1, 8), (2, 3), (2, 4), (2, 5), \\ (2, 6), (2, 7), (2, 8), (3, 4), (3, 5), (3, 6), \\ (3, 7), (3, 8), (4, 5), (4, 6), (4, 7), \\ (4, 8), (5, 6), (5, 7), (5, 8), (6, 7), \\ (6, 8), (7, 8) \end{array} \right\}$
        b.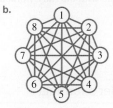
        c. Degree of each vertex $= 7$

14. a. 45   b. 190   c. 4950

15.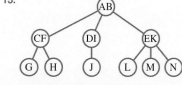

16.
	1	2	3	4	5
1	0	1	1	0	0
2	1	0	1	0	0
3	1	1	0	1	0
4	0	0	1	0	1
5	0	0	0	1	0

17. a. $\begin{bmatrix} 0 & 1 & 0 & 1 & 1 & 0 \\ 1 & 0 & 1 & 1 & 1 & 0 \\ 0 & 1 & 0 & 0 & 0 & 0 \\ 1 & 1 & 0 & 0 & 1 & 1 \\ 1 & 1 & 0 & 1 & 0 & 1 \\ 0 & 0 & 0 & 1 & 1 & 0 \end{bmatrix}$

b. $\begin{bmatrix} 0 & 1 & 0 & 1 & 1 & 0 & 0 \\ 1 & 0 & 1 & 0 & 0 & 0 & 0 \\ 0 & 1 & 0 & 0 & 1 & 0 & 0 \\ 1 & 0 & 0 & 1 & 0 & 0 & 1 \\ 1 & 0 & 1 & 0 & 0 & 1 & 3 \\ 0 & 0 & 0 & 0 & 1 & 0 & 0 \\ 0 & 0 & 0 & 1 & 3 & 0 & 0 \end{bmatrix}$

c. $\begin{bmatrix} 0 & 1 & 0 & 0 & 0 & 0 & 0 \\ 1 & 0 & 1 & 0 & 1 & 1 & 0 \\ 0 & 1 & 0 & 0 & 1 & 1 & 0 \\ 0 & 0 & 0 & 0 & 0 & 0 & 0 \\ 0 & 1 & 1 & 0 & 0 & 1 & 1 \\ 0 & 1 & 1 & 0 & 1 & 1 & 1 \\ 0 & 0 & 0 & 0 & 1 & 1 & 1 \end{bmatrix}$

18. a.

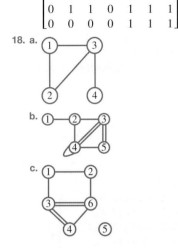

b.

c.

9.2 Exam questions

Note: Mark allocations are available with the fully worked solutions online.

1. E

2. E

3. E

9.3 Planar graphs and Euler's formula

9.3 Exercise

1. a. 6 edges   **b.** 28 edges

2.

3. a.

(diagram)

b. Face 1: degree $= 3$
Face 2: degree $= 3$
Face 3: degree $= 3$
Face 4: degree $= 3$
Face 5: degree $= 4$

4. Answers may vary. Below are suggested solutions.

a.

(diagram)

b.

c. Already planar

d.

5. B

6. Face 1: degree $= 4$
Face 2: degree $= 3$
Face 3: degree $= 3$
Face 4: degree $= 6$

7. C

8. a.

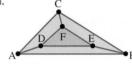

b. $V = 9$, $E = 16$, $F = 9$, so $9 = 6 - 9 + 2$

9.

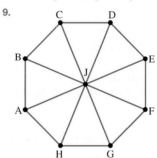

10. a.

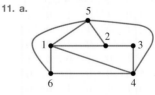

b. 5

c. Face I: degree $= 3$
Face II: degree $= 3$
Face III: degree $= 4$
Face IV: degree $= 4$
Face V: degree $= 4$

11. a.

(diagram)

b. 6

12. a. D

b. $V = 8$, $E = 13$, $F = 7$, so $8 = 13 - 7 + 2$

13. $V = 4$, $E = 6$, $F = 4$, so $4 = 6 - 4 + 2$

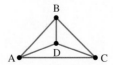

14. Since the degree of a single node is determined by the number of edges (E) 'leaving' it, and each such edge must be the 'entering' edge of another node, each edge is counted twice in the sum of degrees (S). Thus the sum must be an even number. And since each edge is counted twice, $S = 2E$.

15. B

16.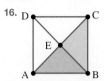

17.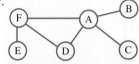

The edges should represent the doorways between rooms. (*Note:* The hall space between rooms E and D belongs to room F. Similarly, the hall space between rooms B and C belongs to room A.)

18. a. 28

b.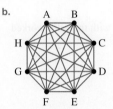

c.
$$
\begin{array}{c|cccccccc}
 & A & B & C & D & E & F & G & H \\
\hline
A & 0 & 1 & 1 & 1 & 1 & 1 & 1 & 1 \\
B & 1 & 0 & 1 & 1 & 1 & 1 & 1 & 1 \\
C & 1 & 1 & 0 & 1 & 1 & 1 & 1 & 1 \\
D & 1 & 1 & 1 & 0 & 1 & 1 & 1 & 1 \\
E & 1 & 1 & 1 & 1 & 0 & 1 & 1 & 1 \\
F & 1 & 1 & 1 & 1 & 1 & 0 & 1 & 1 \\
G & 1 & 1 & 1 & 1 & 1 & 1 & 0 & 1 \\
H & 1 & 1 & 1 & 1 & 1 & 1 & 1 & 0 \\
\end{array}
$$

9.3 Exam questions

Note: Mark allocations are available with the fully worked solutions online.

1. D
2. D
3. C

9.4 Walks, trails, paths, cycles and circuits

9.4 Exercise

1. A–B–C–D–E–F–G
2. E–F–G–D–C–B–A
3. A–B–C–D–E–G–F–H–J
4. A–J–H–E–G–F–D–C–B
5. E
6. 1–2–5–3–4–5–1
7. a. B   b. FE
8. D
9. i. A–G–L–M–S–H–J–K–F–P–E–R–D–O–C–Q–B–N
   ii. H–S–M–L–G–A–N–B–Q–C–O–P–F–E–D–R–K–J

10. 1–3–4–5–6–7–8–5–3–2–1
11. D
12. 2–1–6–5–4–3–7
13. C
14. 2–1–6–5–4–7–3–2
15. E
16. 7–6–5–4–3–2–1
17. A
18. a. 2–1–3–4–5–2–3
    b. Cannot be done (odd vertices).
    c. 1–2–3–4–5
    d. 1–3–4–5–2–1
19. a. Join 4 to 7 and 3 to 8.
    b. Euler circuit = 3–2–1–3–4–1–5–4–7–6–10–7–8–10–9–8–3;
       Hamiltonian cycle = 3–2–1–5–4–7–6–10–9–8–3
20. a. Y–R–N–Y–S–L–N–L
    b. An Euler trail
    c. L–S–Y–R–N–Y–L–N–L
    d. An Euler circuit
21. a. R–Y–S–L–N
    b. A Hamiltonian path
22. a. C–D–A–E–F–B
    b. C–A–E–F–B–D–C
    c. Hamiltonian path, Hamiltonian cycle

9.4 Exam questions

Note: Mark allocations are available with the fully worked solutions online.

1. D
2. E
3. C

9.5 Trees and their applications

9.5 Exercise

1. A–C–B; A–D–C–B; A–E–D–C–B
2. C
3. b, d
4. A–C–B; A–D–B; A–C–D–B; A–D–C–B;
   A–E–D–B; A–E–D–C–B
5. 17 km
6. 16 minutes
7. a. i. 28 km   ii. 17 km
   b. 28 km
   c. B
8.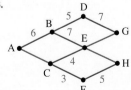
9. B

10. a. 32

b. 86

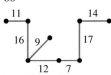

(1 of 2 possible answers)

11. 30 km

12. 41 km

13. E

14. a. A–B–F–H; A–B–F–G–H; A–B–F–G–E–H

b. 22

15. C

16. D

17. 44 km

18. a. 5   **b.** 18   **c.** 7   **d.** 19

19. a. $12 + 13 + 18 + 24 = 67$

b. $6 + 8 + 9 + 16 + 15 = 54$

c. $5 + 10 + 14 + 26 = 55$

d. $4 + 4 + 6 + 7 = 21$

20. a. 49

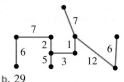

b. 29

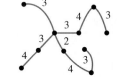

21. a.

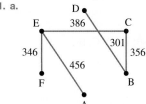

b. 1845 passengers

22. a.

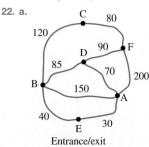

Entrance/exit

i. 9

ii. 6

iii. A–4, B–4, C–2, D–3, E–2, F–3

b. 430 m

c.

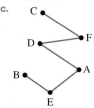

d.

	A	B	C	D	E	F
A	0	1	0	1	1	1
B	1	0	1	1	1	0
C	1	1	0	0	0	1
D	1	0	1	0	0	0
E	1	1	0	0	0	0
F	1	0	1	1	0	0

e. E–A–D–F–C–B–E

9.5 Exam questions

Note: Mark allocations are available with the fully worked solutions online.

1. B

2. D

3. B

9.6 Review

9.6 Exercise

Multiple choice

1. E

2. C

3. D

4. D

5. C

6. C

7. C

8. B

9. a. C   **b.** A

10. C

Short answer

11. a. $V = \{A, B, C, D, E, F, G, H\}$

$E = \{(A, B), (A, C), (A, G), (A, H), (B, C),$
$(B, D), (B, E), (D, E), (E, F), (F, G), (G, H)\}$

b.

	A	B	C	D	E	F	G	H
A	0	1	1	0	0	0	1	1
B	1	0	1	1	1	0	0	0
C	1	1	0	0	0	0	0	0
D	0	1	0	0	1	0	0	0
E	0	1	0	1	0	1	0	0
F	0	0	0	0	1	0	1	0
G	1	0	0	0	0	1	0	1
H	1	0	0	0	0	0	1	0

12. Diagram may be somewhat different from figure.
Sum of degrees = 16

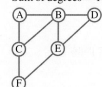

13. a.

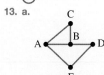

b. $V = 5, E = 6, F = 3$, thus $5 = 6 - 3 + 2$

14. a. A–B–C–D–B–F–D–E–F–A

b. A–B–C–D–E–F–A or A–F–E–D–C–B–A

15. a.

	Shiva	Paul	Matt	Carlo	Kevin
Shiva	0	2	1	0	1
Paul	2	0	0	1	1
Matt	1	0	0	1	1
Carlo	0	1	1	0	1
Kevin	1	1	1	1	0

b.

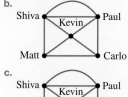

c.

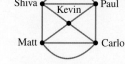

d. So they could start and finish at the same house

e. M–C–P–S–M–K–P–S–K–C–M

16. 1–2–3–4–5–6–7–8–9

17. Distance = 199 km

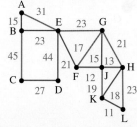

Extended response

18. a. 1

b. 3

c. 7

d. 6

e.

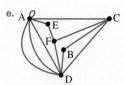

f. Sample responses can be found in the worked solutions in the online resources.

19. a.

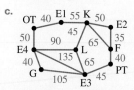

b. Connect flowerbed to lake. An Euler circuit:
E2–K–L–F–K–E1–OT–E4–L–E3–E4–G–E3–PT–F–E2
Others are possible.

c.

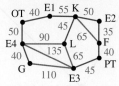

d. Minimum spanning tree; distance = 400 m

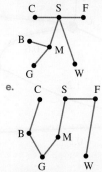

e. E2–K–L–E4, distance = 185 m

20. a. $V = 7$, $E = 11$, $F = 6$, so $7 = 11 - 6 + 2$

b.
$$\begin{bmatrix} 0 & 1 & 1 & 0 & 1 & 0 & 0 \\ 1 & 0 & 1 & 0 & 0 & 0 & 1 \\ 1 & 1 & 0 & 1 & 1 & 0 & 0 \\ 0 & 0 & 1 & 0 & 1 & 0 & 0 \\ 1 & 0 & 1 & 1 & 0 & 1 & 1 \\ 0 & 0 & 0 & 0 & 1 & 0 & 1 \\ 0 & 1 & 0 & 0 & 1 & 1 & 0 \end{bmatrix}$$

c. A Hamiltonian cycle: M–G–W–F–S–C–B–M and
M–B–C–S–F–W–G–M

d. Sample responses can be found in the worked solutions in the online resources.

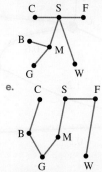

Total distance = 510 km
It's of little use because they need to start and finish in Melbourne and flight paths would backtrack.

f. 240 km

9.6 Exam questions

Note: Mark allocations are available with the fully worked solutions online.

1. a. 86 km

b. Town K

2. a. One other player

b. Cameron and Dale

c. See the figure at the bottom of the page.*

3. a. 3.2 km

b. i. Eulerian trail

ii. Eulerian trails start and finish at vertices with an odd degree. The training program starts at S, with a degree of 3, and will finish at P, also with a degree of 3.

c. This track is between exercise station S and exercise station T. See the figure at the bottom of the page.**

4. a. The office

b. i. Hamiltonian cycle

ii. See the figure at the bottom of the page.***

5. A

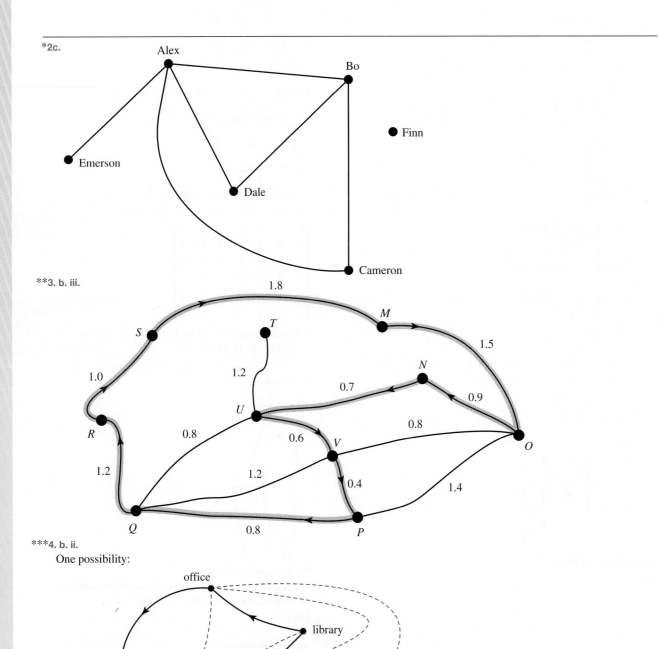

*2c.

**3. b. iii.

***4. b. ii.

One possibility:

10 Directed graphs and network flow

Fully worked solutions for this topic are available online.

10.1 Overview

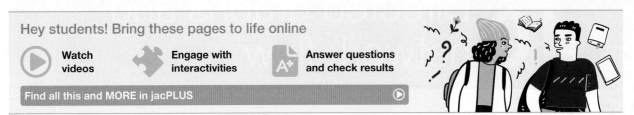

10.1.1 Introduction

Directed graphs show all the existing connections between vertices (or nodes) and also the direction in which movement can take place from one vertex to another. The world wide web is a directed network, where websites are the vertices and the communication between websites are the edges. Not all websites communicate with each other and some websites can send information to another website, but the receiving website cannot send information back. The first paper written on this branch of mathematics was published by Leonard Euler in 1736, having been asked to solve the famous problem of the seven bridges of Konigsberg. Konigsberg (now Kaliningrad in Russia) had seven bridges that crossed the Pregel River and linked the mainland to two islands. Euler was asked to provide a walk around the city that crossed each of the bridges only once. He solved the problem using directed graphs.

Graph theory is used to solve problems in computer science, business and organisational structures, transportation, including trains and the trucking industry, logistics, chemistry, physics, electric circuits, biology, neuro pathways and housing development infrastructure, such as water and electricity supplies. In the modern world, a company such as Netflix or Google uses algorithms based on the mathematics of networks to collect data on customer activity and hence ensure its recommendations are highly personalised.

KEY CONCEPTS

This topic covers the following key concepts from the VCE Mathematics Study Design:
- use of networks to model flow problems: capacity, sinks and sources
- solution of small-scale network flow problems by inspection and the use of the 'maximum-flow minimum-cut' theorem to aid the solution of larger-scale problems
- use of a bipartite graph and its tabular or matrix form to represent a matching problem
- determination of the optimum assignment(s) of people or machines to tasks by inspection or by use of the Hungarian algorithm for larger-scale problems
- construction of an activity network from a precedence table (or equivalent) including the use of dummy activities where necessary
- use of forward and backward scanning to determine the earliest starting times (EST) and latest starting times (LST) for each activity
- use of earliest starting times and latest starting times to identify the critical path in the network and determine the float times for non-critical activities
- use of crashing to reduce the completion time of the project or task being modelled.

Source: VCE Mathematics Study Design (2023–2027) extracts © VCAA; reproduced by permission.

10.2 Precedence tables and activity networks

10.2.1 Activity charts and networks

In any process, ranging from our daily schedule to major construction operations, tasks need to be completed within a certain period of time.

Consider Frieda's morning schedule, where she needs to eat her breakfast, download her email and read her email. The first task takes 6 minutes, the second task takes 1 minute, and the last task takes 2 minutes. Frieda needs to complete all these tasks in 7 minutes. How might she accomplish this?

Clearly, she needs to be able to do some tasks simultaneously. Although this seems like a simple problem, let us look at what might happen each minute.

Time	Activity	Activity
1st minute		Download email
2nd minute	Eat breakfast	
3rd minute	Eat breakfast	
4th minute	Eat breakfast	
5th minute	Eat breakfast	Read email
6th minute	Eat breakfast	Read email
7th minute	Eat breakfast	

More complex activities require more planning and analysis. A network diagram can be used to represent the 'flow' of activities.

In the figure shown, the **edges** of the network represent the three activities of downloading (B), reading (C) and eating (A). The left **vertex** (or **node**) represents the start of all activity, the right vertex the end of all activity, and the middle vertex indicates that activity B must occur before activity C can begin. In other words, activity B is the **immediate predecessor** of activity C.

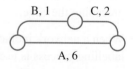

Another way of representing this information is in an activity chart.

Activity letter	Activity	Predecessor	Time (min)
A	Eat breakfast	—	6
B	Download email	—	1
C	Read email	B	2

This chart also shows that activity B (downloading) is the immediate predecessor of activity C (reading), and that activities A and B have no predecessors.

An alternative network diagram is also shown.

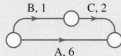

The activities can be undertaken only in a certain sequence, so arrowheads are placed on the edges. Because of the implied direction, these networks are called **directed graphs** or *directed networks*. (The edges in a directed graph represent a *one-way path* between the vertices, as opposed to undirected graphs, where the edges represent a two-way path between the vertices.)

We can use the network diagram to help Frieda reduce the total time spent on the tasks. If the tasks were spread out in a straight line, so that no two tasks were completed at the same time, then they would take her 9 minutes. The diagram shows that some of Frieda's tasks can be carried out simultaneously. Let us investigate the time savings that can be made.

To determine the time saving, first determine the **earliest start time** for each activity.

Forward scanning

By **forward scanning** through a network we can calculate the earliest start times for each activity and the earliest completion time for the whole project.

Earliest start time

The *earliest start time* (EST) is the earliest that any activity can be started after all prior activities have been completed.

The EST is determined by looking at all the previous activities, starting with the immediate predecessors and working back to the start of the project. An activity can start no earlier than the *completion* of such predecessors. Obviously, the EST for the first activity is 0.

The EST can be recorded on a network diagram by using triangles and boxes, as shown in the diagram. The activity is represented by the edge between the vertices. The duration (T_X) of the activity is represented as the number above the edge. The earliest start time of activity X (EST_X) is recorded in the triangle preceding the edge.

When a network includes two or more activities, the same labelling process is used.

The purpose of the boxes beneath the triangles will be explained in a later section.

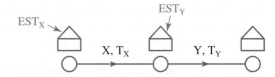

▶ tlvd-3943

WORKED EXAMPLE 1 Calculating the earliest completion time using forward scanning

Use forward scanning to determine the earliest completion time for Frieda's initial three tasks.

THINK	WRITE/DRAW
1. Begin with the network diagram.	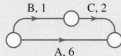

2. Add boxes and triangles near each of the vertices.

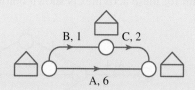

3. The earliest start time (EST) for each vertex is entered in the appropriate triangle. Vertices with no immediate predecessors are given the value of zero.

As activities B and A have no immediate predecessor, their earliest start time is zero.

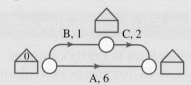

4. Move to another vertex and enter the earliest start time (EST) in its triangle. In the case of activity C, it must wait one minute while its immediate predecessor, B, is completed.

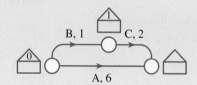

5. The last vertex's earliest start time is entered. When more than one edge joins at a vertex, then the earliest start time is the largest value of the paths to this vertex. This is because all tasks along these paths must be completed before the job is finished. There are two paths converging at the final vertex. The top path takes 3 minutes and the bottom path takes 6 minutes to complete. The larger value is entered in the triangle.

Path B–C = 1 + 2
        = 3 minutes
Path A = 6 minutes

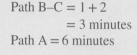

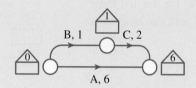

6. The earliest completion time is the value in the triangle next to the end vertex.

All tasks can be completed in 6 minutes.

It is important for anybody planning many tasks to know which tasks can be delayed and which tasks must be completed immediately. In Worked example 1, the eating must be commenced immediately if the 6-minute time is to be attained, whereas downloading the email could be delayed three minutes and there would still be enough time for it to be read while eating.

Let us now extend Frieda's activity chart to a more complex set of activities for her morning routine.

Activity letter	Activity	Predecessor	Time (min)
A	Prepare breakfast	—	4
B	Cook breakfast	A	2
C	Eat breakfast	B, E, G	6
D	Have shower	A	4
E	Get dressed	D	4
F	Brush teeth	C, H	2
G	Download email	A	1
H	Read email	B, E, G	2
Total time			25

The network diagram for these activities is shown below.

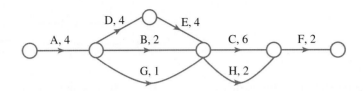

WORKED EXAMPLE 2 Calculating the earliest completion time

Using all the activities listed in Frieda's morning routine, calculate the earliest completion time and hence identify those tasks that may be delayed without extending the completion times.

THINK	WRITE/DRAW
1. Add the boxes and triangles to the directed network diagram.	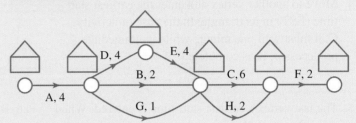
2. Begin forward scanning. The earliest start time (EST) for the first three vertices in the path can be entered immediately.	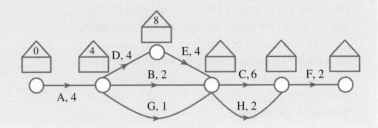

3. Calculate the time values for the paths to the fourth vertex. Enter the largest value (or longest time) into the appropriate triangle.	$A–D–E = 4+4+4$
	    $= 12$ minutes
	$A–B = 4+2$
	    $= 6$ minutes
	$A–G = 4+1$
	    $= 5$ minutes
	
	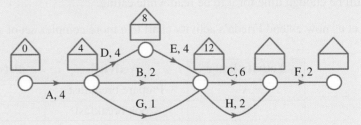

4. Repeat step 3 for the next vertex. Note that calculations begin by using the time from the previous vertex (12 minutes).

$$A–E–C = 12 + 6$$
$$= 18 \text{ minutes}$$
$$A–E–H = 12 + 2$$
$$= 14 \text{ minutes}$$

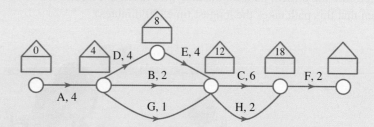

5. There is only one path to the last activity (F). Add its time requirement to that of the previous vertex (18 minutes).

$$A–C–F = 18 + 2$$
$$= 20 \text{ minutes}$$
Earliest completion time is 20 minutes.

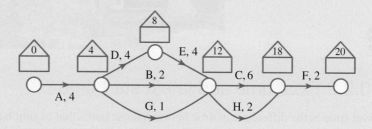

6. The time in the last triangle indicates the earliest completion time.

Earliest completion time = 20 minutes

7. Identify sections of the network where there was a choice of paths. There are two such sections in the network. Examine the first one (the 4th vertex).

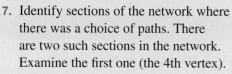

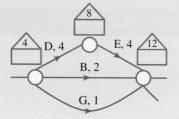

8. List and total the time for each path through this section of the network. Activities on the path with the largest value cannot be delayed.

$$D–E = 4 + 4$$
$$= 8 \text{ minutes}$$
$$B = 2 \text{ minutes}$$
$$G = 1 \text{ minute}$$
Activities B and G can be delayed.

9. Repeat step 8 for the next section identified in step 7.

$$C = 6 \text{ minutes}$$
$$H = 2 \text{ minutes}$$
Activity H can be delayed.

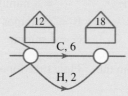

Critical paths

The path through the network which follows those activities that cannot be delayed without causing the entire project to be delayed is called the **critical path**.

Therefore the critical path for the activities listed in Frieda's morning routine would be A–D–E–C–F. It is easily seen that this path takes the longest time (20 minutes).

10.2.2 Float time and latest start time

Float time is the difference in time between those paths that cannot be delayed and those that can. When planning projects, paths with float time are often delayed if there is a cost saving; otherwise, they are done as soon as possible if this is more appropriate. The **latest start time** for such activities is defined as the latest time they may begin without delaying the project.

WORKED EXAMPLE 3 Calculating float times

Work out the float time for activities B and G in Worked example 2, and hence identify the latest starting time for these activities.

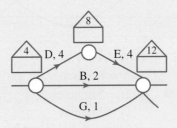

THINK	WRITE
1. List the alternative paths for the section containing activities B and G and the times for these alternatives.	$\begin{aligned} D{-}E &= 4+4 \\ &= 8 \text{ minutes} \end{aligned}$ $B = 2 \text{ minutes}$ $G = 1 \text{ minute}$
2. Subtract the smaller times separately from the maximum time.	$\begin{aligned} \text{Float time for activity } B &= 8-2 \\ &= 6 \text{ minutes} \end{aligned}$ $\begin{aligned} \text{Float time for activity } G &= 8-1 \\ &= 7 \text{ minutes} \end{aligned}$

3. Look up the earliest completion time for the activity on the critical path and subtract the activity times.

D–E is on the critical path.
Earliest completion time = 12 minutes
Latest start time for activity B = 12 − 2
$$= 10 \text{ minutes}$$
Latest start time for activity G = 12 − 1
$$= 11 \text{ minutes}$$

The float times indicate the amount of time for which these activities can be delayed without delaying the completion of all tasks. Furthermore, activity B could begin up to 6 minutes (4 + 6) after the start of the critical activity (D), while G could begin up to 7 minutes (4 + 7) after the same critical activity (D). There will be a more formal treatment of float time in the next section.

 **Resources**

Interactivity Critical path analysis (int-6290)

10.2 Exercise

Students, these questions are even better in jacPLUS

Receive immediate feedback and access sample responses

Access additional questions

Track your results and progress

Find all this and MORE in jacPLUS

1. **WE1, 2, 3** Consider the following figure.

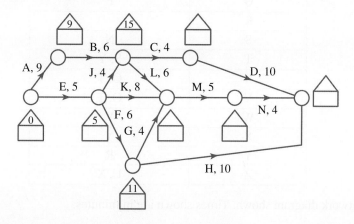

a. Complete a forward scan for the critical path network shown.
b. Determine the earliest completion time.
c. Calculate the maximum time that path J can be delayed without increasing the earliest completion time.

2. The project plan for a new computer software program is shown in the figure. Time is measured in days.

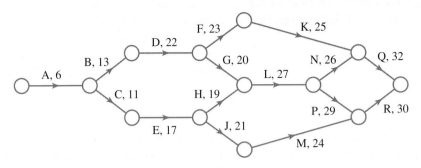

a. Determine the earliest completion time.
b. Calculate the maximum time that path K can be delayed without increasing the earliest completion time.

3. Prepare a network diagram from the activity chart.

Activity	Immediate predecessor
D	—
E	D
F	D
G	E, F

4. Prepare a network diagram from the activity chart.

Activity	Immediate predecessor
N	—
O	N
P	O, T
Q	P
R	—
S	N
T	S, Y
U	O, T
V	O, T
W	V
X	Y
Y	R
Z	U, X

5. **MC** Consider the network diagram shown. Times shown are in minutes.

a. Select the true statement.

  A. Activity A is an immediate predecessor of F.
  B. Activity D is an immediate predecessor of F.
  C. Activity F must be done before activity D.
  D. Activity F must be done before activity E.
  E. Activity D is an immediate predecessor of E.

b. The minimum time taken to complete all activities is:

  A. 19 minutes.   **B.** 21 minutes.   **C.** 23 minutes.   **D.** 28 minutes.   **E.** 49 minutes.

c. Determine the critical path for the network shown.
d. Produce an activity chart.

6. Refer to the diagram shown.

   a. Use forward scanning to determine the earliest completion time.

   b. Identify tasks that may be delayed without increasing the earliest completion time.

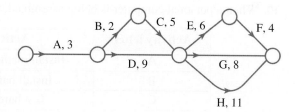

7. Refer to the network diagram shown.

   a. **MC** The number required in the triangle above the vertex after activities B and F is:

   A. 0   **B.** 4   **C.** 5   **D.** 8   **E.** 13

   b. **MC** The number required in the triangle above the vertex after activity E is:

   A. 5   **B.** 9   **C.** 10
   D. 18   **E.** none of these.

   c. **MC** The earliest completion time for all tasks is:

   A. 27   **B.** 24   **C.** 21   **D.** 18   **E.** 15

   d. Produce an activity chart.

   e. For the network diagram shown:

      i. calculate the critical path

      ii. determine which activities have float time and hence calculate their float times

      iii. determine the latest start time for all non-critical activities.

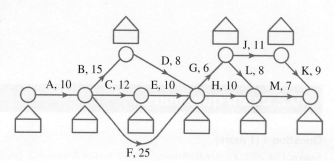

8. a. Calculate the earliest start time for each vertex in the network shown.

   b. Hence, calculate the earliest completion time for the project.

   c. Produce an activity chart.

   d. For the network diagram shown:

      i. calculate the critical path

      ii. determine which activities have float time.

9. Prepare a network diagram from each of the activity charts.

a.

Activity	Immediate predecessor
A	–
B	–
C	A

b.

Activity	Immediate predecessor
A	—
B	A
C	A
D	C
E	B
F	B
G	F
H	D, E, G
I	J, H
J	D, E, G

10. When a personal computer is being assembled, the following processes must be performed.

Activity letter	Activity	Predecessor	Time (min)
A	Install memory board	—	2
B	Install hard drive	A	20
C	Test hard drive	B, E	4
D	Install I/O ports	A	5
E	Install DVD drive	D	3
F	Test DVD drive	E	5
G	Install operating system	C, F	10
H	Test assembled computer	G	12
Total time			61

a. Construct a network diagram.
b. Determine the minimum time in which all tasks could be completed.

10.2 Exam questions

Question 1 (1 mark)

Source: VCE 2020, Further Mathematics Exam 1, Section B, Module 2, Q6; © VCAA.

MC The activity network below shows the sequence of activities required to complete a project.

The number next to each activity in the network is the time it takes to complete that activity, in days.

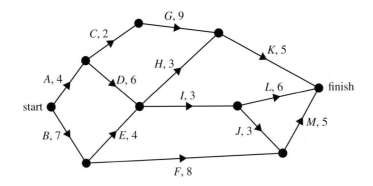

The minimum completion time for this project, in days, is

A. 18   **B.** 19   **C.** 20   **D.** 21   **E.** 22

Question 2 (1 mark)

Source: VCE 2017, Further Mathematics Exam 1, Section B, Module 2, Q4; © VCAA.

MC The directed graph below shows the sequence of activities required to complete a project.

The time to complete each activity, in hours, is also shown.

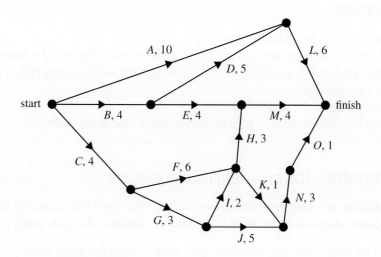

The earliest starting time, in hours, for activity N is

A. 3 **B.** 10 **C.** 11

D. 12 **E.** 13

Question 3 (1 mark)

Source: VCE 2016, Further Maths, Exam1, Sec B, Mod 2, Q6.

MC The directed graph below shows the sequence of activities required to complete a project.

All times are in hours.

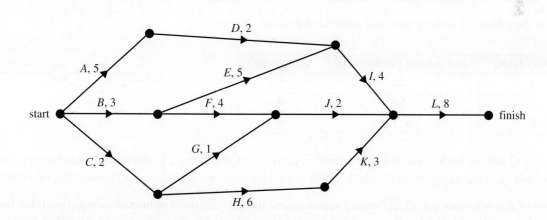

The number of activities that have exactly two immediate predecessors is

A. 0 **B.** 1 **C.** 2

D. 3 **E.** 4

More exam questions are available online.

10.3 Critical path analysis with backward scanning and crashing

LEARNING INTENTION

At the end of this subtopic you should be able to:
- define and apply two rules to successfully complete a critical path analysis of a network
- apply forward and backward scanning to determine the earliest starting times (EST) and latest starting times (LST) for each activity
- introduce a dummy edge to the network diagram
- use crashing to reduce the completion time of the project or task being modelled.

10.3.1 Requirements for critical path analysis

With more complex projects requiring the coordination of many activities, it is necessary to record more information on the network diagrams and to display the information more formally using charts.

In the previous section the float times and the critical path were worked out using somewhat informal methods. In this section a more formal method will be shown to enable float times to be calculated and the critical path to be determined. This method involves **backward scanning**.

Along with the informal rules and techniques already developed in the previous section, we need to define two more rules that must be followed in order to successfully complete a critical path analysis of a network.

> **Rule 1:**
>
> **Two vertices can be connected directly by a maximum of one edge.**

Consider the following activity table and associated network diagram.

Activity letter	Immediate predecessor	Time (min)
A	—	5
B	—	4
C	A, B	6

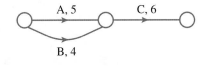

For activity C to have both A and B as predecessors, activities A and B must be drawn as parallel edges. Clearly this does not meet the requirement of rule 1, which allows for only one edge (activity) connecting two vertices.

Violation of this rule does not affect forward scanning and the calculation of minimum completion time, but will cause problems when identifying the critical path using the method of backward scanning described later in this section. A method for dealing with parallel edges will be suggested below.

> **Rule 2:**
>
> **An activity must be represented by exactly one edge.**

Consider the two network diagrams shown here.

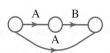

This example indicates two separate flows along the same edge. If A were a water pipe, how could you keep the two flows separate?

This example suggests that A can happen at the same time as B while still being its immediate predecessor.

Backward scanning

To complete a critical path analysis, a procedure called backward scanning must be performed.

Backward scanning starts at the end vertex and moves backward through the network, subtracting the time of each edge from the earliest start time of each succeeding vertex.

When two or more paths are followed back to the same vertex, the smallest such difference is recorded. The results of each backward scanning step yield the latest start time for each activity. Latest start time is the latest time an activity can start without delaying the project.

Earliest finish time (EFT) for an activity is equal to the earliest start time (EST) of the activity plus its duration (T).

The following diagram illustrates the activity X.

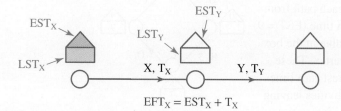

The earliest finish time for activity X (EFT_X) will be the earliest start time for activity X (EST_X) plus the duration of activity X (T_X).

$$EFT_X = EST_X + T_X$$

The following diagram illustrates two activities X and Y, where activity Y directly follows activity X.

As previously established, EST_X is represented by the triangle preceding activity X (in pink on the diagram). The latest start time for activity X (LST_X) is represented by the blue box preceding the activity. The same applies for activity Y and all following activities.

Note that EFT cannot be read from the triangles or boxes; it must be calculated.

Float time, also called 'slack', is the maximum time that an activity can be delayed without delaying a subsequent activity on the critical path and thus affecting the earliest completion time.

From the above, it can be seen that there is a relationship between float time and the other quantities.

Namely: float time for activity X = latest start time for activity Y (LST_Y) − earliest start time for activity

X (EST_X) − duration of activity X (T_X)

That is: float time$_X$ = LST_Y − EST_X − T_X, where activity Y directly follows activity X.

The technique of backward scanning is best explained with an example.

tlvd-3944

WORKED EXAMPLE 4 Using forward and backward scanning to display the critical path

The network diagram shown has been constructed for a project manager. Use forward and backward scanning to clearly display the critical path and to list any float times.

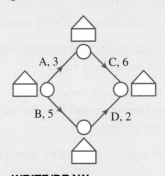

THINK

1. Forward scan through the network and record the earliest start time (EST) for each activity in the appropriate triangle.

WRITE/DRAW

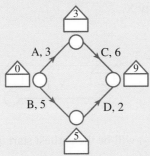

2. a. Begin backward scanning. Start at the end vertex and trace backwards along all paths from this vertex.

   b. Subtract the times of the activities along each path from the earliest finish time (EST = 9) and record the value in the box at the previous vertex. These values are the latest start times (LST) for the activities leaving this vertex.

Along path C: 9 − 6 = 3
Along path D: 9 − 2 = 7

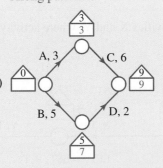

Latest start time for activity C (LST_C) = 3
Latest start time for activity D (LST_D) = 7

3. Repeat the process backwards through the diagram. Where two (or more) paths come together (activities A and B), record the *smaller* value in the box.

Along path A: $3 - 3 = 0$
Along path B: $7 - 5 = 2$
Smaller value $= 0$

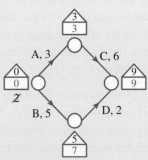

4. The critical path can now be clearly identified. It is the path that has the same numbers in both the triangles and boxes at any vertex. Remember to include *all* such vertices in the critical path.

Critical path is A–C (as shown in blue).

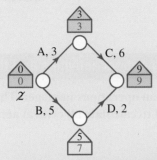

5. a. Float times are calculated now. Construct a table with the headings shown.
Note that all activities that were on the critical path have float times of zero. Therefore, A and C should (and do) have a float time of zero.

Activity X	Activity time (T_X)	Earliest start time of this activity (EST_X)	Latest start time of the following activity (LST_Y)	Float time
A	3	0	3	0
B	5	0	7	2
C	6	3	9	0
D	2	5	9	2

b. Record the times from the triangles in the earliest start time (EST_X) column, the times in the boxes in the latest start time (LST_Y) column as well as the activity times (T_X). Calculate float times using the equation:
$float_X = LST_Y - EST_X - T_X$
In this example, the float times are also the difference between the corresponding times in the boxes and triangles. This is not the rule in the general case.

For activity D: float $= 9 - 5 - 2$
$\qquad = 2$
For activity C: float $= 9 - 3 - 6$
$\qquad = 0$
For activity B: float $= 7 - 0 - 5$
$\qquad = 2$
For activity A: float $= 3 - 0 - 3$
$\qquad = 0$

10.3.2 Earliest completion times

Worked example 4 is fairly simple as the critical path could easily be determined by direct inspection. There is only one path that is not on the critical path, so the calculation of float time is also simple.

In the real world, the problems are more complicated and require the use of the formal method. Float times are important for the efficient management of any project. They enable the manager to determine what delays can be tolerated in the project.

For example, the manager in charge of a building site is able to tell subcontractors that they have a time window in which they must work. The subcontractors can then arrange their schedules to incorporate this time window.

tlvd-3945

WORKED EXAMPLE 5 Calculating critical activities and the earliest completion time

The following chart has been given to an operations manager. The activities have been simplified to letter names. The manager is required to calculate all critical activities and the earliest completion time for the project by:

a. creating a network diagram
b. completing a forward scan and determining the earliest completion time
c. completing a backward scan and identifying the critical path
d. calculating the float times for each activity.

Activity letter	Immediate predecessor	Time (days)
A	—	3
B	—	4
C	B	6
D	A	15
E	A	4
F	E	7
G	E	2
H	C, G	4
I	F, H	2
J	D, I	3
K	D, I	4
M	J	2
N	K	3

THINK

a. Construct the network diagram from the table.

WRITE/DRAW

a.

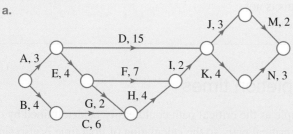

b. 1. Draw boxes and triangles at each vertex. Forward scan through the network. Start at zero for the first vertex and then add the times taken for a path (T_X) and write it in the triangle at the next vertex (EST_Y). Adding the times of the paths (T_X) to the times in the triangle at the previous vertex (EST_X) gives the next value to be entered (EST_Y). When two paths converge at a vertex, the larger time value is entered as all immediate predecessors need to be completed before the next activity can begin.

b. $A = 3$ days
$A\text{–}E = 3 + 4$
$\qquad = 7$ days
$B = 4$ days
The blue triangle (EST_H) may be reached by following two paths:
$A\text{–}E\text{–}G = 3 + 4 + 2$
$\qquad = 9$ days
$B\text{–}C = 4 + 6$
$\qquad = 10$ days, so the larger of the times, 10, is entered in the triangle.

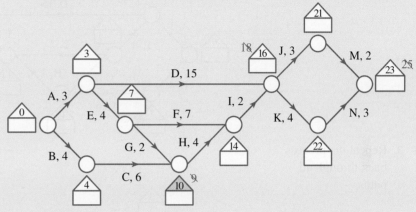

2. The earliest completion time can be read from the last triangle.

Earliest completion time $= 25$ days

c. 1. Starting at the end vertex, begin the backward scan. Enter the earliest completion time in the last box. Subtract the times of any paths ending at this vertex (M and N) from the value in the last box and enter the result in the appropriate boxes to calculate the latest start time (LST) of that activity.

c. Blue box value (LST_M) $= 25 - 2$
$\qquad\qquad\qquad\qquad = 23$ days

Pink box value (LST_N) $= 25 - 3$
$\qquad\qquad\qquad\qquad = 22$ days

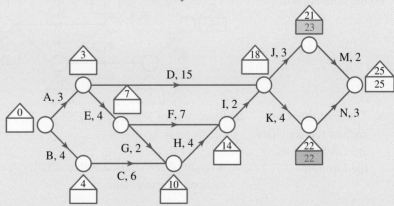

2. Repeat the process backwards through the network diagram. Where the paths converge, the smaller value is entered in the box.

Backtracking to the green box (LST_J) via activity J = 23 − 3
$$= 20$$
Backtracking to the green box (LST_K) via activity K = 23 − 4
$$= 18$$
Enter the smaller of the values (18) in the green box.

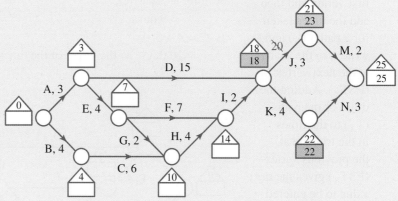

3. Repeat the process back through all paths.

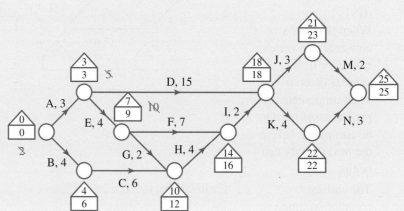

4. Vertices which have the earliest start time (triangles) equal to the latest start time (boxes) are identified as being on the critical path.

A–D–K–N is the critical path.

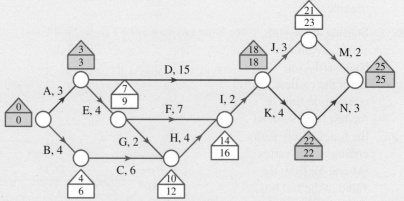

d. 1. a. The float times for each of the activities are calculated using the formula.

d. Float time of $X = LST_Y − EST_X − T_X$
Float (A) = 3 − 0 − 3
$$= 0$$
A result of zero indicates that activity A is on the critical path, with no float time available.

b. The section showing activity A from part **c** step 4 clearly shows the latest start time of the following activity (LST_Y) (pink), the earliest start time (EST_X) (blue) and the activity time (T_X) (green).

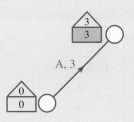

A, 3

2. a. The best way to keep organised and to calculate float times is to set up a table.

b. Add columns for earliest start times (EST), latest start times (LST) and float times to the original table.

Repeat step 1 for all activities.

Activity letter	Immediate predecessor	T_X	EST_X	LST_Y	Float time$_X$ = LST_Y − EST_X − T_X
A	—	3	0	3	$3 - 0 - 3 = 0$
B	—	4	0	6	$6 - 0 - 4 = 2$
C	B	6	4	12	$12 - 4 - 6 = 2$
D	A	15	3	18	$18 - 3 - 15 = 0$
E	A	4	3	9	$9 - 3 - 4 = 2$
F	E	7	7	16	$16 - 7 - 7 = 2$
G	E	2	7	12	$12 - 7 - 2 = 3$
H	C, G	4	10	16	$16 - 10 - 4 = 2$
I	F, H	2	14	18	$18 - 14 - 2 = 2$
J	D, I	3	18	23	$23 - 18 - 3 = 2$
K	D, I	4	18	22	$22 - 18 - 4 = 0$
M	J	2	21	25	$25 - 21 - 2 = 2$
N	K	3	22	25	$25 - 22 - 3 = 0$

Note that all activities that were on the critical path have float times of zero. It is important to note that if even a single activity is 'floated' by having its start delayed, then the entire network diagram should be redrawn and float times recalculated.

If the manager employed extra workers for a critical activity, its duration time could be reduced, hence reducing the completion time for the project. The reduction in the duration time of an activity is called **crashing**. Crashing may result in a different critical path. This will be explored further in a later section.

10.3.3 Dummy activities

A **dummy activity** is an edge that must be added to avoid a network with two or more activities having the same name or occurring in parallel.

Earlier in this section we set up two rules:

Rule 1. Two vertices can be connected directly by a maximum of one edge.

Rule 2. An activity must be represented by exactly one edge.

A table and a drawing were presented in which there were parallel edges (breaking rule 1). A method to overcome this problem, explained by example, will be given.

Introduce a dummy edge to the network diagram shown.

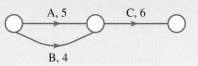

Activity letter	Immediate predecessor	Time (min)
A	—	5
B	—	4
C	A, B	6

THINK

1. Construct edges A and B.
   C must follow both A and B. This is clearly a problem causing parallel edges (violating rule 1 for networks).

2. Construct C after activity A. Introduce a dummy activity (B′) label and allocate a time of zero.
   Note: Use a dotted line to show dummy activity.
   Therefore, not only is A the immediate predecessor of C, but B (via B′ with time = 0) is also effectively the immediate predecessor of C.

WRITE/DRAW

The introduction of the dummy activity with a time value of zero enables scanning to take place along both edges, A and B. Additionally, the critical path can be shown more clearly.

From the table shown construct a network diagram.

Activity letter	Immediate predecessor	Time (days)
A	—	5
B	—	4
C	A, B	4
D	B	6

THINK

1. Construct edges A and B.

2. Construct edge D following B, as D has only one immediate predecessor.

WRITE/DRAW

3. C must follow both A and B (clearly a problem). Construct C after activity A. Introduce a dummy activity (B′) from B to C and allocate it a time of zero.

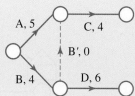

Worked example 7 provides a method of not only avoiding parallel edges but also avoiding A being shown as the immediate predecessor of D, which (from the table) it clearly is not.

10.3.4 Crashing

As discussed earlier, crashing is a method of speeding up the completion time of a project by shortening the critical path. Follow the same method as in previous sections to calculate the new critical path and minimum completion time.

WORKED EXAMPLE 8 Applying crashing to reduce the completion time

Take the critical path found in the network in Worked example 5 reproduced here.
To shorten the overall completion time of the project, activity A is to be shortened to 2 days and activity D is to be shortened to 12 days. Determine the new critical path and the new minimum completion time for the project.

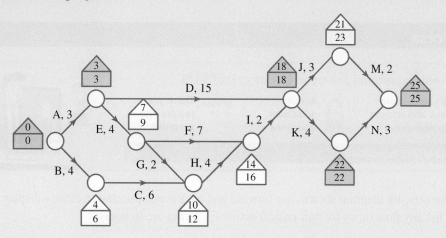

THINK

1. Redraw the network with the new completion times.

WRITE/DRAW

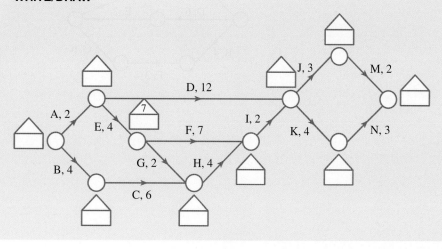

2. Recalculate the earliest
start times (EST)
and latest start times
(LST) by completing
forward and backward
scanning.

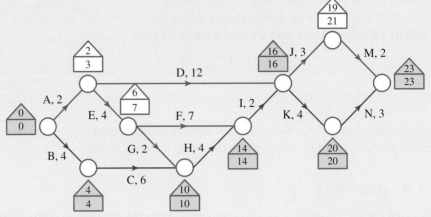

3. Write the answer.

The new critical path is B–C–H–I–K–N. The minimum finishing time is
23 days.

10.3 Exercise

1. **WE4** For the network diagram shown, use forward and backward scanning to clearly display the critical
path and to list any float times for non-critical activities. Times are in hours.

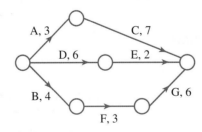

2. Perform a backward scan on the network shown.

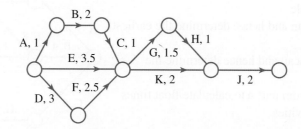

Determine:
  a. the critical path
  b. float times for non-critical activities.

3. **WE5** The manufacturing of bicycles can be considered as a 7-step process:
  A — Collect all the parts — 12 minutes
  B — Paint frame — 35 minutes   (requires A to be completed first)
  C — Assemble brakes — 16 minutes   (requires A to be completed first)
  D — Assemble gears — 20 minutes   (requires B to be completed first)
  E — Install brakes — 12 minutes   (requires C to be completed first)
  F — Install seat — 5 minutes   (requires C to be completed first)
  G — Final assembly — 18 minutes   (requires D and E to be completed first)

  a. Construct an activity chart.
  b. Construct a network diagram.
  c. Determine the earliest completion time using forward and backward scanning.
  d. Determine the critical path.

4. **MC** In the bicycle manufacturing system described in question **3**, activities with float time are:
  A. A, B, C, D, E, F, G.   **B.** A, B, D, G.   **C.** C, E, F.
  D. C only.   **E.** none.

5. For the network diagram shown, use forward and backward scanning to clearly display the critical path and to list any float times. Times are in minutes.

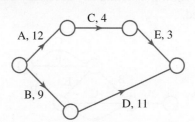

6. Complete the figure by forward and backward scanning and hence:
  a. determine the earliest completion time
  b. indicate the critical path.
  Times are in days.
  c. Imagine now that activity E can be completed in 9 days.
    Explain how this affects the answers to parts **a** and **b**.
  d. **MC** The float time for activity D in the figure is:
    A. 1 day.   **B.** 2 days.   **C.** 3 days.
    D. 4 days.   **E.** 7 days.

7. From the network diagram shown:

a. produce an activity table
b. complete a forward scan and hence determine the earliest completion time
c. complete a backward scan and hence determine the critical path
d. use the activity table from part a to calculate float times for all non-critical activities.

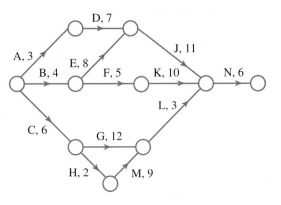

8. From the following activity table:

Activity	Immediate predecessor	Activity time (h)
A	—	5
B	—	3
C	A	4
D	A	7
E	B	4
F	C	9
G	D, E	5
H	G	3
J	G	6
K	F, H	4

a. construct a network diagram
b. determine the earliest completion time
c. by forward and backward scanning, determine the critical path
d. determine the float time for all non-critical activities.

9. **WE6** Redraw the network diagram shown, inserting any necessary dummy activities so that rule 1 for critical path problems is not violated.

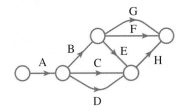

10. **WE7** From the following activity table, construct a network diagram and indicate the location and direction of the dummy activity.

Activity	Immediate predecessor	Activity time (h)
A	—	3
B	—	5
C	A	7
D	B	7
E	B, C	1
F	D, E	2

11. Consider the following activity table.

Activity	Immediate predecessor	Activity time (min)
A	—	7
B	—	6
C	A, B	8
D	C	12
E	C	7
F	D, E	9

a. By creating one (or more) dummy activities, construct a proper critical path network diagram.
b. **WE8** From your network diagram in part a, determine:
   i. the earliest completion time
   ii. the critical path
   iii. float times for non-critical activities.

12. Convert the following activity table into a network diagram.

Activity	Immediate predecessor	Activity time (h)
A	—	1000
B	—	600
C	—	800
D	A, B	1100
E	B	400
F	B	100
G	C	600
H	D, E	1600

13. From the following activity table:

Activity	Immediate predecessor	Activity time (h)
A	—	11
B	—	9
C	A	2
D	A	5
E	B	12
F	C	3
G	D	3
H	E	4
J	E, F, G	7

a. construct a network diagram, adding any dummy activities that may be required
b. determine the earliest completion time
c. determine the critical path by forward and backward scanning
d. determine the float time for all non-critical activities.

14. Redraw the network diagram shown, inserting any necessary dummy activities so that rule 1 for critical path problems is not violated.

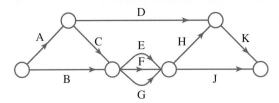

15. From the network diagram shown (activity times are recorded in days):

   a. forward scan to determine the earliest completion time
   b. backward scan to determine the critical path
   c. determine the float time for activity X
   d. imagine that crashing results in J being completed in 5 days. Explain how this affects the earliest completion time and the critical path.

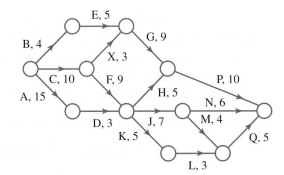

16. Given the network diagram shown, determine the:

   a. four missing values in the boxes
   b. critical path
   c. float time for activity F.

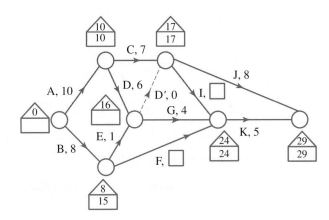

17. Given the network diagram shown, determine the:

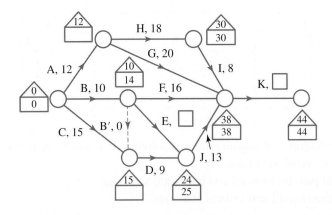

   a. four missing values in the boxes
   b. critical path
   c. float time for activity F.

18. The network diagram shows the activities, along with the times (in hours), needed to complete a particular project.
Also drawn below is the corresponding activity chart showing earliest start times rather than activity times. There is an activity X, which is yet to be drawn on the network diagram.

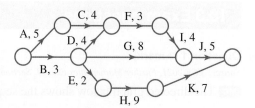

Activity	Immediate predecessor(s)	EST
A	—	0
B	—	0
C		5
D	B	3
E	B	3
F		9
G	B	
H	E	5
I	F	12
J	G, I, X	17
K	H	
X	C, D	8

a. Use the information in the network diagram to complete the table above by filling in the shaded boxes.
b. Draw and label activity X on the network diagram, including its direction and time.
c. State the critical path.
d. Determine the latest start time for activity H.

19. A school is building a new library. The separate stages required for construction and the number of weeks taken to complete them are shown on the directed network.

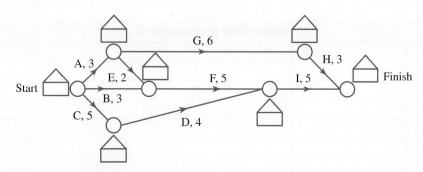

a. Calculate the earliest start time for stage F.
b. Calculate the minimum time it will take to build the library.
c. Determine the critical path.
d. Calculate the slack time for stage H.
e. If the overall time of the project was to be reduced, determine which stages could be shortened.
f. If stage A was reduced to 2 weeks and stage F was also reduced to 2 weeks, determine the new critical path and the new minimum time for completion of the library.

Question 1 (1 mark)

Source: VCE 2021, Further Mathematics Exam 1, Section B, Module 2, Q6; © VCAA.

MC The directed graph below shows the sequence of activities required to complete a project.

The time taken to complete each activity, in hours, is also shown.

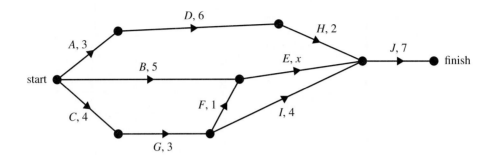

The minimum completion time for this project is 18 hours.

The time taken to complete activity E is labelled x.

The maximum value of x is

A. 2	**B.** 3	**C.** 4	**D.** 5	**E.** 6

Question 2 (1 mark)

Source: VCE 2019, Further Mathematics Exam 1, Section B, Module 2, Q7; © VCAA.

MC A project involves nine activities, A to I.

The immediate predecessor(s) of each activity is shown in the table below.

Activity	Immediate predecessor(s)
A	–
B	A
C	A
D	B
E	B, C
F	D
G	D
H	E, F
I	G, H

A directed network for this project will require a dummy activity.

The dummy activity will be drawn from the end of

A. activity B to the start of activity C.

B. activity B to the start of activity E.

C. activity D to the start of activity E.

D. activity E to the start of activity H.

E. activity E to the start of activity F.

▶ **Question 3 (1 mark)**

Source: VCE 2019, Further Mathematics Exam 1, Section B, Module 2, Q8; © VCAA.

MC The directed network below shows the sequence of activities, *A* to *S*, that is required to complete a manufacturing process.

The time taken to complete each activity, in hours, is also shown.

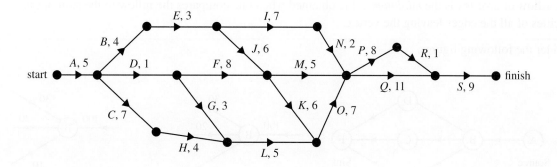

The number of activities that have a float time of 10 hours is

A. 0　　　　　**B.** 2　　　　　**C.** 2　　　　　**D.** 3　　　　　**E.** 4

More exam questions are available online.

10.4 Flow problems

> **LEARNING INTENTION**
>
> At the end of this subtopic you should be able to:
> * solve problems on flow capacities and maximum flow
> * determine the minimum cut–maximum flow.

10.4.1 Flow capacities and maximum flow

There are many examples of networks throughout the modern world. The road and rail systems that link all parts of our country, such as the one in the image, and the airline flight paths that not only link us to other places within the country but also to places overseas, are examples of networks that move people and products.

Other networks such as the information superhighway (the internet), telephone lines and the postal system allow the transfer of information.

All these networks have common attributes. The attribute dealt with in this section is

network flow. There are networks that allow for the flow to be in both directions along a path and those that allow for flow in one direction only. The *direction* of flow needs to be displayed clearly on any diagram: simple arrows on the path suffice. The *quantity* of flow is just as important.

The network's starting vertex is called the **source**. This is where all flows commence. The flow goes through the network to the end vertex, which is called the **sink**.

The **flow capacity** (capacity) of an edge is the amount of flow that an edge can allow through if it is not connected to any other edges.

The **inflow** of a vertex is the total of the flows of all edges leading into the vertex.

The **outflow** of a vertex is the *minimum* value obtained when one compares the inflow to the sum of the capacities of all the edges leaving the vertex.

Consider the following figures.

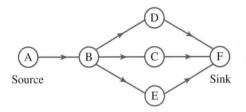

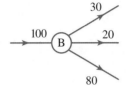

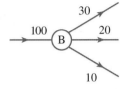

All flow commences at A. It is therefore the source. All flow converges on F, indicating it is the sink.

B has an inflow of 100 but now the capacity of the edges leaving B is 130 (80 + 20 + 30). The outflow is the smaller of these two inflow/outflow values, which is 100.

B has a maximum inflow of 100. The flow capacity of the edges leaving B is 30 + 20 + 10 = 60. The outflow is the smaller of these two inflow/outflow values, which is 60.

WORKED EXAMPLE 9 Creating a network diagram to indicate the direction and quantity of the flow

Convert the information presented in the following table to a network diagram, clearly indicating the direction and quantity of the flow.

From	To	Quantity (litres per minute)	Demand (E)
Rockybank Reservoir (R)	**Marginal Dam (M)**	**1000**	—
Marginal Dam (M)	**Freerange (F)**	**200**	**200**
Marginal Dam (M)	**Waterlogged (W)**	**200**	**200**
Marginal Dam (M)	**Dervishville (D)**	**300**	**300**

THINK

Construct and label the required number of vertices.
Note: The vertices are labelled with the names of the source of the flow and the corresponding quantities are recorded on the edges. Link the vertices with edges and record flow direction and quantity on these.

WRITE/DRAW

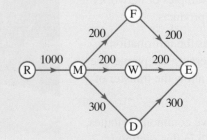

Note: In Worked example 9, the location Demand (E) does not exist. It is preferable for a network diagram to have both a single source and a single sink, so Demand (E) was included to simplify the diagram. The reason for this will become clear in the following worked examples.

Worked example 9 is a simple case of a network in which the direction and quantity of flow are evident. Such a network diagram allows for analysis of the flow in the network; it allows us to see if various edges in the network are capable of handling the required flow.

> ## Flow capacity
>
> The *flow capacity of the network* is the total flow possible through the entire network.

tlvd-4981

WORKED EXAMPLE 10 Determining flow capacity

Use the information from Worked example 9 to determine, by inspection:
a. the flow capacity of the network
b. whether the flow through the network is sufficient to meet the demand of all towns.

THINK	WRITE/DRAW
a. 1. Examine the flow into and out of the Marginal Dam vertex. Record the smaller of the two at the vertex. This is the maximum flow through this point in the network.	a. Even though it is possible for the reservoir to send 1000 L/min (in theory), the maximum flow that the dam can pass on is 700 L/min (the minimum of the inflow and the sum of the capacities of the edges leaving the dam). 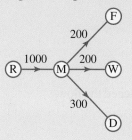
2. In this case, the maximum flow through Marginal Dam is also the maximum flow of the entire network.	Maximum flow is 700 L/min.
b. 1. Determine that the maximum flow through Marginal Dam meets the total flow demanded by the towns.	b. 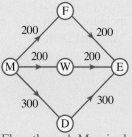 Flow through Marginal = 700 L/min Flow demanded = 200 + 300 + 200 $\qquad\qquad = 700$ L/min
2. If the requirements of step 1 are able to be met, then we can determine that the flow into each town is equal to the flow demanded by them.	By inspection of the table in Worked example 9, all town inflows equal town demands (capacity of edges leaving the town vertices).

10.4.2 Excess flow capacity

Consider what would happen to the system if Rockybank Reservoir continually discharged 1000 L/min into Marginal Dam while its output remained at 700 L/min.

Such flow networks enable future planning. Future demand may change, the population may grow or a new industry that requires more water may come to one of the towns. Worked example 11 will examine such a case.

Excess flow capacity is the surplus of the capacity of an edge minus the flow into the edge.

WORKED EXAMPLE 11 Calculating changing flow capacity

A new dairy factory (Creamydale (C)) is to be set up on the outskirts of Dervishville. The factory will require 250 L/min of water.
a. Determine whether the original flow to Dervishville is sufficient.
b. If the answer to part a is no, determine whether there is sufficient flow capacity into Marginal Dam to allow for a new pipeline to be constructed directly to the factory from Marginal Dam, to meet the factory's demand.
c. Determine the maximum flow through the network if the new pipeline is constructed.

THINK	WRITE/DRAW
a. 1. Add the demand of the new factory to Dervishville's original flow requirements. If this value exceeds the flow into Dervishville, then the new demand cannot be met.	a. 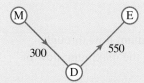
2. The new requirements exceed the flow.	The present network is not capable of meeting the new demand.
b. 1. Reconstruct the network including a new edge for the factory after Marginal Dam.	b.

2. Repeat step 1 from Worked example 10 to determine the outflow of vertex M.

Marginal Dam inflow $= 1000$
Marginal Dam outflow
$= 200 + 200 + 300 + 250$
$= 950$

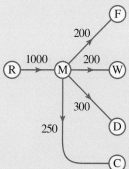

3. Determine if the flow is sufficient.

There is excess flow capacity of 300 into Marginal Dam that is greater than the 250 demanded by the new factory. The existing flow capacity to Marginal Dam is therefore sufficient.

c. This answer can be gained from part **b** step 2 above.

c. The maximum flow through the new network is 950.

The maximum flow through most simple networks can be determined using these methods, but more complex networks require different methods.

10.4.3 Minimum cut–maximum flow method

To determine the maximum flow, the network first needs to be divided or cut into two parts. A **cut** in a network diagram is a line drawn through a number of edges which stops *all* flow from the source to the sink. The value of the cut is the total flow of the edges that are cut.

The *minimum cut* is the cut with the minimum value.

The *maximum flow* through a network is equal to the value of the minimum cut.

tlvd-4982

WORKED EXAMPLE 12 Calculating the minimum cut and the maximum flow through the network

For the network diagram shown:
a. **make three cuts**
b. **calculate the value of each cut**
c. **determine the value of all possible cuts to give the value of the minimum cut and hence the maximum flow through the network.**

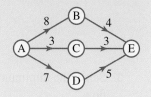

THINK

a. 1. Isolate the source, A, by cutting any edges leaving it.
Cut 1 = {AB, AC, AD}

WRITE/DRAW

a.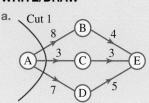

2. Isolate the sink, E, by cutting any edges leading into it.
Cut 2 = {BE, CE, DE}

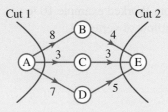

3. Place a cut through each of the three paths going from A to E.
Cut 3 = {AD, AC, BE}

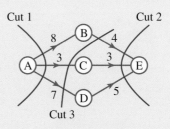

b. Add the values of the edges crossed by each of the cuts in part **a**.

b. Value of cut 1 = 8 + 3 + 7
$$= 18$$
Value of cut 2 = 4 + 3 + 5
$$= 12$$
Value of cut 3 = 7 + 3 + 4
$$= 14$$

c. 1. Identify the cuts in the diagram.
Cut 1 = {AB, AC, AD}
Cut 2 = {BE, CE, DE}
Cut 3 = {AD, AC, BE}
Cut 4 = {AD, CE, BE}
Cut 5 = {AB, CE, DE}
Cut 6 = {AB, AC, DE}

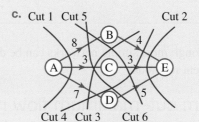

2. Calculate the value of all cuts.

c. Value of cut 1 = 8 + 3 + 7
$$= 18$$
Value of cut 2 = 4 + 3 + 5
$$= 12$$
Value of cut 3 = 7 + 3 + 4
$$= 14$$
Value of cut 4 = 7 + 3 + 4
$$= 14$$
Value of cut 5 = 8 + 3 + 5
$$= 16$$
Value of cut 6 = 8 + 3 + 5
$$= 16$$

3. Select the minimum value as the maximum flow.

The value of the minimum cut is 12, so the maximum flow through the network has a value of 12.

As can be seen in Worked example 12, ensuring that all cuts have been made is a complicated procedure. The diagram becomes cluttered.

There are two cuts missing from the diagram. Can you find them?

Performing cuts at the source and sink first enables an upper limit for the value of the minimum cut to be set. In part **b**, it was clear that the value of the minimum cut in the diagram must be less than or equal to 12. In this case it was 12.

In some networks it is possible to produce a cut in which an edge actually heads back *inside* the cut rather than being directed out of the cut, as is required and as all edges in Worked example 12 do. If an edge does this, then its flow value is ignored in the calculation of the cut value. The inside of a cut is the side on which the source vertex lies. Sometimes the inside of a cut is shaded.

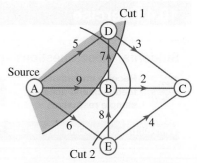

The shaded section represents the inside of cut 1.

WORKED EXAMPLE 13 Determining the values of the cuts

Determine the values of the cuts made on the network diagram.

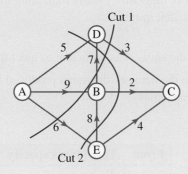

THINK

1. Determine the overall direction of the flow.

2. Determine the edges that are crossed by the cuts and identify any which head back into the cut.

3. Calculate the cut values.

WRITE

The flow is from A to C, so the inside of the cut is on the left-hand side where the source vertex, A, is.

Cut 1 = {AE, AB, BD, DC}
BD heads back inside the cut, so ignore its flow value.
Cut 2 = {AD, BD, BC, BE, AE}
BE heads back inside the cut, so ignore its flow value.
Note that for this cut, BD heads outside the cut.

$$Cut\,1 = 6 + 9 + 3$$
$$= 18$$
$$Cut\,2 = 5 + 7 + 2 + 6$$
$$= 20$$

 Resources

Interactivities Network flow (int-6287)
            Network flow cuts (int-6288)

10.4 Exercise

1. **WE9, 10** Consider the following flow table.

From	To	Flow capacity
R	S	250
S	T	200
T	U	100
T	V	100
U	V	50

   a. Convert the flow tables into network diagrams, clearly indicating the direction and quantity of the flow.
   b. From the network diagram shown, determine:

   i. the flow capacity
   ii. whether the flow rate through the network is sufficient to meet the demand.

   c. Introduce new edges, as shown, to the network diagrams produced in part a.
   d. Calculate the new network flow capacities.

2. Consider the following flow table.

From	To	Flow capacity
E	F	8
E	G	8
G	H	5
G	J	3
F	H	2
F	J	6
J	K	8
H	K	8

   a. Convert the flow tables into network diagrams, clearly indicating the direction and quantity of the flow.
   b. From the network diagram shown, determine:

   i. the flow capacity
   ii. whether the flow rate through the network is sufficient to meet the demand.

   c. Introduce new edges, as shown, to the network diagrams produced in part a.
   d. Calculate the new network flow capacities.

3. **WE11** **a.** Convert the following flow tables into network diagrams, clearly indicating the direction and quantity of the flow.

i.

From	To	Flow capacity
A	B	100
A	C	200
B	C	50
C	D	250
D	E	300

ii.

From	To	Flow capacity
M	N	20
M	Q	20
N	O	15
N	R	5
Q	R	10
O	P	12
R	P	12

b. For each of the network diagrams shown, determine the flow capacity and whether the flow through the network is sufficient to meet the demand.

c. Introduce new edges, as shown, to each of the network diagrams produced in part **a**.

d. Calculate the new network flow capacities.

4. For vertex B in the network shown, state:

a. the inflow at B

b. the edge capacities flowing out of B

c. the outflow from B.

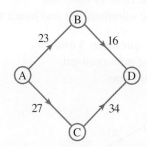

5. For vertex B in the network shown, state:

a. the inflow at B

b. the edge capacities flowing out of B

c. the outflow from B.

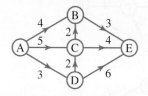

6. Answer the following questions.

i. Convert the following flow diagrams to flow tables (as in question **3**).

a.

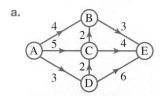

b.

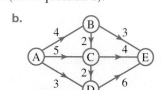

c.

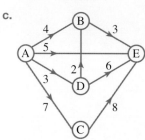

d.

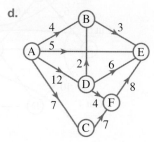

ii. Calculate the capacity of each of the networks in part **i**.

7. **WE12** Consider the network flow diagram shown, representing the flow of water between a reservoir (A) and a town (H). The other vertices (B, ... , G) represent pumping stations.

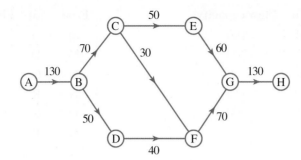

Using the minimum cut method, determine the maximum flow through the network.

8. Consider the network flow diagram shown.

   a. Determine the four cuts shown in the network that are invalid.
   b. Determine the value of each cut.
   c. Determine the minimum cut and hence the maximum flow.

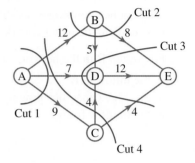

For the networks in questions 9 and 10:
a. draw in the line of minimum cut
b. state the maximum flow.

9. **WE13**

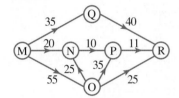

10.

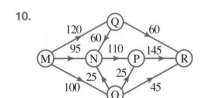

11. Consider the following network diagram.

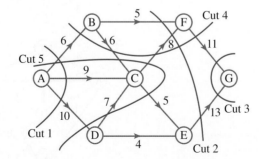

   a. Determine if any of the five cuts in the network diagram are invalid.
   b. Determine the value of each cut.
   c. Determine the minimum cut and hence the maximum flow.

12. For each of the following networks:

a.

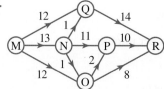

b.

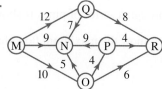

   i. draw in the line of minimum cut
   ii. state the maximum flow.

13. A network diagram of streets and freeways flowing into a city at peak hour is shown.

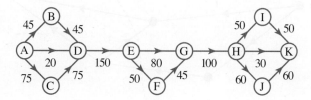

a. Copy the diagram and clearly label the freeways.
b. Explain what would happen to the traffic at:

   i. vertex E
   ii. vertex H.

c. Add a new edge to the diagram so that bottlenecks would not occur at peak hour.

10.4 Exam questions

Question 1 (1 mark)

Source: VCE 2020, Further Mathematics Exam 1, Section B, Module 2, Q9; © VCAA.

MC The flow of liquid through a series of pipelines, in litres per minute, is shown in the directed network below.

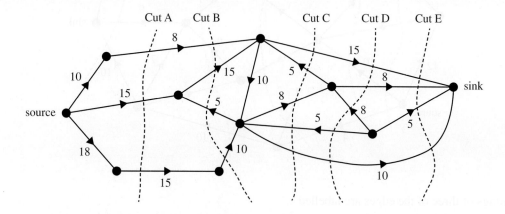

Five cuts labelled A to E are shown on the network.

The number of these cuts with a capacity equal to the maximum flow of liquid from the source to the sink, in litres per minute, is

A. 1 B. 2 C. 3 D. 4 E. 5

Question 2 (1 mark)

Source: VCE 2019, Further Mathematics Exam 1, Section B, Module 2, Q3; © VCAA.

MC The flow of water through a series of pipes is shown in the network below.

The numbers on the edges show the maximum flow through each pipe in litres per minute.

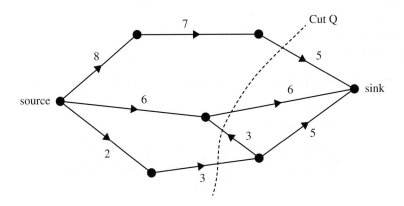

The capacity of Cut Q, in litres per minute, is
  A. 11   **B.** 13   **C.** 14   **D.** 16   **E.** 17

Question 3 (1 mark)

Source: VCE 2017, Further Mathematics Exam 1, Section B, Module 2, Q8; © VCAA.

MC The flow of oil through a series of pipelines, in litres per minute, is shown in the network below.

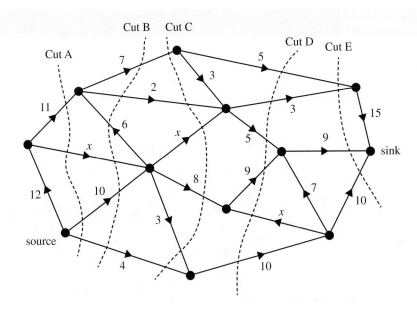

The weightings of three of the edges are labelled x.

Five cuts labelled A–E are shown on the network.

The maximum flow of oil from the source to the sink, in litres per minute, is given by the capacity of
  A. Cut A if $x = 1$ **B.** Cut B if $x = 2$ **C.** Cut C if $x = 2$
  D. Cut D if $x = 3$ **E.** Cut E if $x = 3$

More exam questions are available online.

10.5 Bipartite graphs and allocation problems

LEARNING INTENTION

At the end of this subtopic you should be able to:
- define the concept of bipartite graphs
- represent and interpret information as bipartite graphs
- determine the optimum assignment(s) of people or machines to tasks
- apply the Hungarian algorithm for larger-scale problems.

10.5.1 Bipartite graphs

With network flow problems, the task was to determine the maximum flow through a network. Another possible problem could be to determine the *exact* flows through each path in a network, so that as much of the capacity of each edge as possible is used. This is a class of situation called assignment (or allocation) problems.

For example, we could be generating electricity at three power plants for distribution to five towns, so that each town gets the required amount of electricity, regardless of the plant it came from. To start this technique, we need to define the concept of **bipartite graphs**.

In the previous section on network flow, the minimum cut method was introduced to help solve the flow problem. Consider the cut through the network as dividing the network into two parts: 'flow in and flow out', or 'supply and demand'. Imagine separating the graph along the cut into its two parts. This graph is known as a bipartite graph. A bipartite graph is one where the vertices can be separated into two types of vertex — supply and demand.

Consider a typical network flow problem with a cut defined in the figure on the left. By separating the graph, as in the figure on the right, we have divided the graph into two parts. The supply vertices are M and W, while the demand vertices are F, D and E.

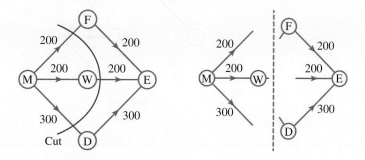

Representing information as bipartite graphs

In many practical situations not only flows, but goods, money and even people can be represented by bipartite graphs.

The Shiny Shoe Company makes shoes at two factories: one at Alphaville, producing 200 pairs per day, and the other at Beta City, producing 300 pairs per day.

It ships them to three distributors: Fartown, who want 50 pairs per day, Giver River, who want 250 pairs per day and Hamlet, who want 200 pairs per day.

Represent this information as a bipartite graph.

THINK

1. Determine all the supply vertices, starting with a single 'supply' vertex.

2. Represent the supply side of the graph.

3. Determine all the demand vertices, ending with a single 'demand' vertex.

4. Add the demand side of the graph.

WRITE/DRAW

Supply vertices are S, A and B.

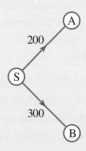

Demand vertices are F, G, H and D.

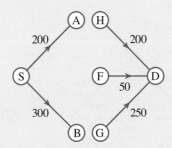

The next section concerns the problem of allocating the 'flow' of shoes between the suppliers and the distributors. Sometimes the supply and demand vertices have quite distinct types.

For example, the supply vertices might be students, while the demand vertices could be the subjects that they study or the sports they play. The key to bipartite graphs is that there is a separation between the two sides and that there is some sort of flow from supply to demand. The separation does not always have to be literal, as in the next example.

tlvd-4983

WORKED EXAMPLE 15 Interpreting a bipartite graph

The following table lists four students and the four subjects offered
in the Science program in Year 12.
a. Represent this information as a bipartite graph.
b. Determine whether the following statements are True or False.
   i. Alice studies more Science subjects than Carla.
   ii. Carla and Betty together study all the Science subjects.
   iii. Alice and Betty together study more subjects than Carla.

Student	Subjects taken
Alice	Biology, Chemistry
Betty	Chemistry, Physics
Carla	Biology, Chemistry, Psychology
Diane	Psychology

THINK

a. 1. Determine the supply vertices and represent
the supply side of the graph. In this case
they are (arbitrarily) assigned to the
students.

2. Determine the demand vertices. Represent
the demand side and show the connections
as arrows joining supply vertices to demand
vertices.
Note there is no quantity of flow between
supply and demand.

WRITE/DRAW

a. Supply vertices are Alice, Betty, Carla and
Diane.

(Supply)

Alice •

Betty •

Carla •

Diane •

Demand vertices are Biology, Chemistry,
Physics and Psychology.

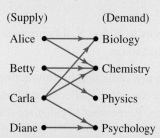

b. i. 1. Determine the number of subjects studied
by Alice and Carla.

2. Determine the truth of the statement.

ii. 1. Determine the number of subjects studied
by Betty and Carla.

2. Determine the truth of the statement.

b. i. Alice studies two subjects (Biology and
Chemistry).
Carla studies three subjects (Biology,
Chemistry, Psychology).
Clearly, the statement is false.

ii. Betty studies Chemistry and Physics.
Carla studies Biology, Chemistry and
Psychology.
Clearly, the statement is true.

Note that in part **b iii**, the fact that Alice and Betty both studied Chemistry does not mean that two (distinct) subjects were studied.

10.5.2 The allocation problem

Consider a situation where there are a number of jobs to be done and the same number of people to do them. However, each person can do each job in a different amount of time (or at a different cost). How can the jobs be allocated one per person, so that time (or cost) is minimised? This is known as **optimal allocation**. Sometimes the optimal allocation may be obvious, but in most cases it is not, and may even require some trial and error work.

In the general case, the jobs and people can be put in an allocation matrix, as in the following worked example.

	Job X	Job Y	Job Z
Alan	10	4	9
Bob	8	11	10
Carl	9	8	7

So, Alan would take 10 hours to do job X, 4 hours to do job Y and 9 hours to do job Z. Similarly, Bob would take 8 hours to do job X, and so on.

The first step is to perform row reduction by subtracting the smallest value in each row from all the numbers in that row. This may produce the optimal allocation in one step. This is the simplest possible case: the optimal allocation is almost obvious. Remember to calculate the total time (or cost) and ensure it is indeed the minimum.

tlvd-4984

WORKED EXAMPLE 16 Determining the optimal allocation and hence the minimum time

A building site has three more jobs to be done by the three remaining workers, Alan, Bob and Carl. The times taken by each to do the three jobs are given in the following table. Determine the optimal allocation and hence state the minimum time.

	Job X	Job Y	Job Z
Alan	10	4	9
Bob	8	11	10
Carl	9	8	7

| | THINK | WRITE/DRAW |

THINK

1. Generate or write the matrix of people against job times.

2. Step 1a. Perform row reduction. Determine the smallest value in each row, and subtract it from all numbers in that row.

3. Subtract 4 from each number in row A. Subtract 8 from each number in row B. Subtract 7 from each number in row C.

4. Step 1b. Attempt an allocation. Cover all the zeroes with the smallest number of straight (horizontal/vertical, but not diagonal) lines. If the number of lines equals the number of jobs to be allocated, go to step 2. If not, further steps to be taken will be shown in another worked example.

5. Step 2a. Produce a bipartite graph. Show *all* possible allocations, where there are zeroes connecting people to jobs.

6. Step 2b. List the possible allocations and determine the smallest total.

WRITE/DRAW

1.

	X	Y	Z
A	10	4	9
B	8	11	10
C	9	8	7

2. The smallest number in row A is 4.
The smallest number in row B is 8.
The smallest number in row C is 7.

3.

	X	Y	Z
A	6	0	5
B	0	3	2
C	2	1	0

4.

	X	Y	Z
A	~~6~~	~~0~~	~~5~~
B	~~0~~	~~3~~	~~2~~
C	~~2~~	~~1~~	~~0~~

In this case there are three lines and three jobs.

5.

6. There is only 1 possible allocation: $A \rightarrow Y, B \rightarrow X, C \rightarrow Z$

Total time $= 4 + 8 + 7$

$\qquad\quad = 19$ hours

Hence, the minimum time is 19 hours.

10.5.3 The Hungarian algorithm

Unfortunately, there are two cases where the basic algorithm will not produce the optimal allocation:

1. When there are not enough zeros to draw the right number of straight lines in step 1b

2. When there are too many zeros so that one person may be allocated more than one job in step 2a

In either case, we need to use the **Hungarian algorithm**, which involves the following steps, replacing step 2 of the previous example.

Step 2. Perform column reduction (when there are not enough zeros). This is similar to row reduction and is performed when step 1 has not produced an optimal allocation. This may produce an optimal allocation using the method of steps 1b and 1c. If not, proceed to step 3.

Step 3. Modify the original matrix according to the Hungarian algorithm. This involves an addition and subtraction operation involving various quantities in the matrix from step 2 and will be detailed in the following worked example. This should produce an optimal allocation.

Step 4. Produce a bipartite graph and list all possible allocations.

WORKED EXAMPLE 17 Applying the Hungarian algorithm to minimise the total cost

Amy, Beth, Cate and Dana offer quotes on how much each of them will charge to complete four different jobs, P, Q, R and S. The table summarises these charges (in dollars).

	P	Q	R	S
Amy	17	24	42	21
Beth	25	18	19	20
Cate	29	14	31	22
Dana	11	20	17	14

Use the Hungarian algorithm to minimise the total cost to complete the four jobs, one job per person.

THINK

1. Step 1a. Perform row reduction.
Row A: Subtract 17
Row B: Subtract 18
Row C: Subtract 14
Row D: Subtract 11

2. Step 1b. Attempt an allocation.
Draw the minimum number of straight lines to cover all the zeros.

3. Step 2a. Perform column reduction.
Column P: Subtract 0 (do nothing)
Column Q: Subtract 0 (do nothing)
Column R: Subtract 1
Column S: Subtract 2

4. Step 2b. Attempt an allocation.
Draw the minimum number of straight lines to cross out all the zeros.

5. Step 3. Perform the Hungarian algorithm.
Step 3a. Determine the smallest uncovered number from step 2b.
Step 3b. Add this number to all covered numbers in the matrix from step 2b. At the intersections of straight lines, add this number twice (circled at right).

WRITE/DRAW

	P	Q	R	S
A	0	7	25	4
B	7	0	1	2
C	15	0	17	8
D	0	9	6	3

	P	Q	R	S
A	0	7	25	4
B	7	0	1	2
C	15	0	17	8
D	0	9	6	3

Only 2 lines — cannot continue allocation.

	P	Q	R	S
A	0	7	24	2
B	7	0	0	0
C	15	0	16	6
D	0	9	5	1

	P	Q	R	S
A	0	7	24	2
B	7	0	0	0
C	15	0	16	6
D	0	9	5	1

Only 3 lines are formed. As 4 lines are required, allocation cannot continue.

The smallest uncovered number is 1.

	P	Q	R	S
A	1	8	24	2
B	0	2	1	1
C	16	1	16	6
D	1	10	5	1

6. Step 3c. Subtract the overall smallest number from all the numbers in the matrix.

The smallest overall number is 1.

	P	Q	R	S
A	0	7	23	1
B	8	1	0	0
C	15	0	15	5
D	0	9	4	0

Step 3d. Attempt an allocation. Repeat step 3 until a possible allocation is found.

In this case there are 4 lines and 4 jobs, so go to step 4.

	P	Q	R	S
A	0	7	23	1
B	8	1	0	0
C	15	0	15	5
D	0	9	4	0

Step 4a. Produce a bipartite graph. Show all possible allocations, where there are zeros connecting people to jobs.

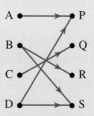

7. Step 4b. List possible allocations and determine the smallest total.

- By inspection, Cate must be allocated Q (she's the only one at Q).
- Beth must be allocated R (she's the only one at R).
- Therefore Dana must be allocated S (she's the only one left at S).
- Therefore Amy must be allocated P.

Amy → P, Beth → R, Cate → Q, Dana → S
Minimum cost = 17 + 19 + 14 + 14
= $64
(numbers from the original matrix in the question)

Sometimes the optimal allocation is carried out to maximise a quantity such as a score. In this process, all elements in the matrix are subtracted from the largest one first, and in doing so, modifying the situation to a minimisation problem. From then on, the procedure is exactly the same as that set out in Worked example 17.

 Resources

 Interactivity The allocation problem and the Hungarian algorithm (int-6289)

10.5 Exercise

1. **WE14** An electricity company produces 4000 kWh, 5000 kWh and 6000 kWh at its three hydroelectric plants. This electricity is supplied to four towns as follows: 20% to Town A, 25% to Town B, 15% to Town C and the rest to Town D. Represent this information with a bipartite graph.

2. An oil company supplies petrol to three towns, A, B and C. Within Town A there are two sub-depots where petrol is stored temporarily before being delivered to service stations. The other two towns have only a single depot each.

   Town A gets 30 000 and 40 000 litres per week for each of its sub-depots, while Town B gets 10 000 litres and Town C gets 25 000 litres. However, the demand for petrol is as follows: Town A: 60 000, Town B: 15 000 and Town C: 30 000 litres.

   a. Represent this information in a bipartite graph.
   b. Suggest a delivery system so that all towns get the required amount of petrol each week.

3. **WE15** Five diners (Albert, Brian, Chris, David and Earl) go to the pub for dinner and place the orders as shown in the table.

Diner	Dishes
Albert	Soup, fish
Brian	Fish
Chris	Soup, beef, desert
David	Beef, dessert
Earl	Fish, dessert

   a. Represent this information as a bipartite graph.
   b. **MC** Select the true statement.

      A. Albert and Brian between them have more kinds of dishes than Chris and David.
      B. Chris and David between them have tried all the dishes.
      C. David and Earl between them have more kinds of dishes than Brian and Chris.
      D. Brian and Chris between them have more kinds of dishes than David and Earl.
      E. All the above statements are false.

4. A publisher produces 1000 copies per month of the latest bestseller by Thomas Wolf at two factories. Thomas Wolf is co-author of *Midnight Assassin: A Murder in America's Heartland*, and author of *The Called Shot: Babe Ruth, the Chicago Cubs, and the Unforgettable Major League Baseball Season of 1932*.

   The first factory produces 400 copies.

   The books are then distributed to two states (Queensland and Victoria), with Queensland requiring 350 copies and Victoria getting the rest.

   Represent this information as a bipartite graph.

5. **MC** Five students have various hobbies as indicated by the bipartite graph. From this graph it can be said that:

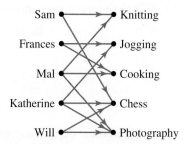

A. Sam and Mal have all the hobbies between them.
B. Mal and Frances, in total, have more hobbies than Sam and Will.
C. Mal and Sam each have the same number of hobbies.
D. Katherine and Sam, in total, have fewer hobbies than Frances and Mal.
E. Will had fewer hobbies than all other students.

For questions 6 and 7, perform row reduction on the matrices, which represent times (in hours), and attempt an optimal allocation for the minimum time. State the minimum time.

6. **WE16**
$$\begin{bmatrix} 6 & 3 & 7 \\ 2 & 4 & 5 \\ 3 & 5 & 2 \end{bmatrix}$$

7.
$$\begin{bmatrix} 4 & 3 & 7 & 3 \\ 9 & 4 & 6 & 5 \\ 5 & 6 & 7 & 8 \\ 4 & 8 & 3 & 5 \end{bmatrix}$$

For questions 8 and 9, perform an optimal allocation on the matrices by:
1. row reduction

2. column reduction

3. the Hungarian algorithm.

State the minimum total allocation and at which stage each matrix was solved.

8. **WE17**
$$\begin{bmatrix} 6 & 9 & 9 & 4 \\ 10 & 9 & 9 & 7 \\ 4 & 9 & 6 & 3 \\ 5 & 8 & 8 & 6 \end{bmatrix}$$

9.
$$\begin{bmatrix} 5 & 23 & 14 & 9 \\ 11 & 29 & 6 & 14 \\ 21 & 17 & 14 & 13 \\ 20 & 27 & 22 & 8 \end{bmatrix}$$

10. a. Perform row reduction on the matrix, which represents times (in hours), and attempt an optimal allocation for the minimum time. State the minimum time.

$$\begin{bmatrix} 16 & 14 & 20 & 13 \\ 15 & 16 & 17 & 16 \\ 19 & 13 & 13 & 18 \\ 22 & 26 & 20 & 24 \end{bmatrix}$$

b. Draw the bipartite graph from part a.

11. A mother wishes to buy presents (a game, a doll and a toy truck) for her three children from three different stores. From the three different stores, the game costs ($30, $45, $40), the doll costs ($50, $50, $60) while the truck costs ($35, $30, $30).
Show that the optimal allocation yields a total of $110.

12. `MC` Given the matrix shown, the total value of the optimal allocation is:

A. 9   B. 11   C. 15   D. 16   E. 19

$$\begin{bmatrix} 7 & 3 & 7 \\ 3 & 3 & 5 \\ 6 & 5 & 5 \end{bmatrix}$$

13. A government department wishes to purchase five different cars for its fleet. For reasons of fairness it must not purchase more than one car from any dealer. It receives quotes from the dealers according to the following table (with cost in thousands of dollars).

	D1	D2	D3	D4	D5
Car 1	20	15	17	16	18
Car 2	17	15	19	17	16
Car 3	18	19	16	19	16
Car 4	19	19	17	21	17
Car 5	24	19	17	17	17

a. Perform row and column reduction.
b. Draw the bipartite graph after reduction.
c. i. Determine the optimal allocation.
   ii. State the minimum cost.

14. Perform an optimal allocation on the following matrices by row reduction, column reduction and the Hungarian algorithm.
State the minimum total allocation and at which stage each matrix was solved.

a. $\begin{bmatrix} 10 & 15 & 12 & 17 \\ 17 & 21 & 19 & 14 \\ 16 & 22 & 17 & 19 \\ 23 & 26 & 29 & 27 \end{bmatrix}$ b. $\begin{bmatrix} 12 & 10 & 11 & 13 & 11 \\ 11 & 11 & 13 & 12 & 12 \\ 12 & 16 & 13 & 16 & 12 \\ 9 & 10 & 9 & 11 & 9 \\ 14 & 11 & 11 & 11 & 11 \end{bmatrix}$

15. A company wishes to hire four computer programmers (A, B, C and D) to develop four software packages. Because of time constraints each programmer can accept at most one job. The quotes (in weeks) given by the four programmers are shown in the following table.

	Job 1	Job 2	Job 3	Job 4
A	30	40	50	60
B	70	30	40	70
C	60	50	60	30
D	20	80	50	70

a. Perform row (and if necessary, column) reduction.
b. Perform the Hungarian algorithm (if necessary).
c. Display possible allocations using a bipartite graph.
d. Determine:

   i. the optimal allocation
   ii. the total time required to complete the four tasks.

16. Four workers (Tina, Ursula, Vicky, Wendy) need to be optimally allocated four tasks (one per worker). The times (in minutes) that each worker takes are shown in the table.

	Job 1	Job 2	Job 3	Job 4
T	100	50	35	55
U	60	45	70	55
V	40	70	50	30
W	70	50	70	70

Perform an optimal allocation, stating:

a. the tasks allocated to each worker

b. the total time required to complete all tasks.

17. David Lloyd George High School will be competing in a Mathematics competition. There are four categories (Algebra, Calculus, Functions and Geometry) and four possible competitors (Ken, Louise, Mark and Nancy) — one per category.

To determine the best competitor for each category, a test is given and the results (as percentages) for each student recorded.

	Algebra	Calculus	Functions	Geometry
Ken	60	78	67	37
Louise	45	80	70	90
Mark	60	35	70	86
Nancy	42	66	54	72

a. Determine the row-reduced matrix.

b. Determine the column-reduced matrix.

c. Determine the optimal allocation using a bipartite graph.

d. Determine the average score for the team.

18. In a Meals-on-Wheels program in a remote region of Victoria, there are four elderly people (P1, P2, P3, P4) who require meals to be delivered.

There are four volunteers (V1, V2, V3, V4) who live at varying distances (specified in kilometres) away.

	P1	P2	P3	P4
V1	13	17	14	23
V2	8	12	17	9
V3	9	17	14	11
V4	21	16	13	14

a. Determine the optimum allocation.

b. State the total distance travelled given that each volunteer delivers exactly one meal.

10.5 Exam questions

Question 1 (1 mark)

Source: VCE 2020, Further Mathematics Exam 1, Section B, Module 2, Q7; © VCAA.

MC Four friends go to an ice-cream shop.

Akiro chooses chocolate and strawberry ice cream.

Doris chooses chocolate and vanilla ice cream.

Gohar chooses vanilla ice cream.

Imani chooses vanilla and lemon ice cream.

This information could be presented as a graph.

Consider the following four statements:
- The graph would be connected.
- The graph would be bipartite.
- The graph would be planar.
- The graph would be a tree.

How many of these four statements are true?

A. 0
B. 1
C. 2
D. 3
E. 4

Question 2 (1 mark)

Source: VCE 2020, Further Mathematics Exam 2, Section B, Module 2, Q2; © VCAA.

A cricket team has 11 players who are each assigned to a batting position.

Three of the new players, Alex, Bo and Cameron, can bat in position 1, 2 or 3.

The table below shows the average scores, in runs, for each player for the batting positions 1, 2 and 3.

		Batting position		
		1	**2**	**3**
Player	**Alex**	22	24	24
	Bo	25	25	21
	Cameron	24	25	19

Each player will be assigned to one batting position.

To which position should each player be assigned to **maximize** the team's score? Write your answer in the table below.

Player	Batting position
Alex	
Bo	
Cameron	

Jacaranda Maths Quest 12 General Mathematics VCE Units 3 & 4 Seventh Edition

Source: VCE 2016, Further Mathematics Exam 1, Section B, Module 2, Q8; © VCAA.

MC Five children, Alan, Brianna, Chamath, Deidre and Ewen, are each to be assigned a different job by their teacher. The table below shows the time, in minutes, that each child would take to complete each of the five jobs.

	Alan	Brianna	Chamath	Deidre	Ewen
Job 1	5	8	5	8	7
Job 2	5	7	6	7	4
Job 3	9	5	7	5	9
Job 4	7	7	9	8	5
Job 5	4	4	4	4	3

The teacher wants to allocate the jobs so as to minimise the total time taken to complete the five jobs.

In doing so, she finds that two allocations are possible.

If each child starts their allocated job at the same time, then the first child to finish could be either
- **A.** Alan or Brianna.
- **B.** Brianna or Deidre.
- **C.** Chamath or Deidre.
- **D.** Chamath or Ewen.
- **E.** Deidre or Ewen.

More exam questions are available online.

10.6 Review

10.6.1 Summary

doc-38041

Hey students! Now that it's time to revise this topic, go online to:

Access the topic summary

Review your results

Watch teacher-led videos

Practise VCAA exam questions

Find all this and MORE in jacPLUS

10.6 Exercise

Multiple choice

1. Use the following table to answer the questions.

Activity	Time	Immediate predecessor
A	12	—
B	11	A
C	13	—
D	24	A, C
E	11	D, B
F	21	E

a. **MC** The activities that come before activity E are:

A. D and B.   **B.** A, D and B.   **C.** A, B, C and D.   **D.** A, C and D.   **E.** F.

b. **MC** The correct diagram for the table is:

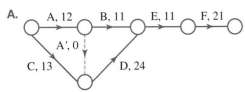

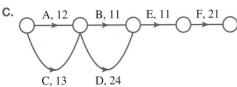

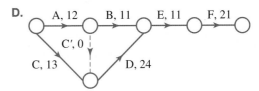

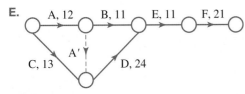

c. **MC** The earliest start time for activity E is:

A. 23   **B.** 24   **C.** 37   **D.** 36   **E.** none of these.

d. **MC** The earliest completion time for the network is:

A. 56   **B.** 55   **C.** 68   **D.** 69   **E.** 70

2. **MC** The times missing from the table for activities B and C respectively are:

Activity	Activity time	Earliest start time	Earliest finish time	Float time
A	3	0	3	0
B		0	7	2
C	6	3		0
D	2	5	9	2

A. 5 and 9
B. 5 and 8
C. 0 and 9
D. 9 and 5
E. 9 and 0

3. **MC** The source and sink, respectively, for the network flow diagram are:

A. V and U
B. U and V
C. S and V
D. V and S
E. R and S

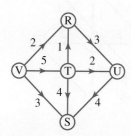

4. **MC** The outflow from vertex T is:

A. 35
B. 27
C. 31
D. 24
E. more than 1 possible outflow.

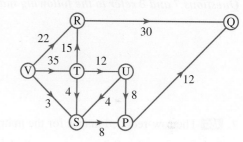

5. **MC** The diagram that shows the minimum cut is:

A.

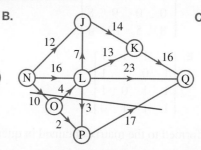

B.

C.

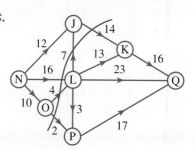

D.

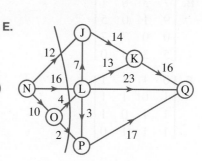

E.

6. **MC** Select the option that does *not* show valid cuts.

A. a and c **B.** d and e **C.** a and d

D. c only **E.** They are all cuts.

a.
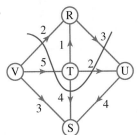

b.
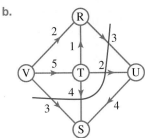

c.
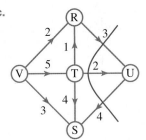

d.
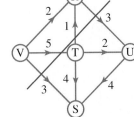

e.
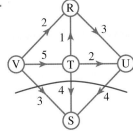

Questions 7 and 8 refer to the following matrix.

$$\begin{bmatrix} 15 & 11 & 16 & 6 \\ 13 & 5 & 2 & 12 \\ 4 & 2 & 11 & 14 \\ 12 & 7 & 12 & 10 \end{bmatrix}$$

7. **MC** The row-reduced matrix for the matrix shown above is:

A. $\begin{bmatrix} 9 & 5 & 10 & 0 \\ 11 & 3 & 2 & 10 \\ 2 & 0 & 9 & 12 \\ 5 & 0 & 5 & 5 \end{bmatrix}$

B. $\begin{bmatrix} 11 & 6 & 14 & 0 \\ 9 & 3 & 0 & 6 \\ 0 & 0 & 9 & 8 \\ 8 & 5 & 10 & 4 \end{bmatrix}$

C. $\begin{bmatrix} 9 & 5 & 10 & 0 \\ 11 & 3 & 0 & 10 \\ 2 & 0 & 9 & 12 \\ 5 & 0 & 5 & 3 \end{bmatrix}$

D. $\begin{bmatrix} 9 & 5 & 10 & 6 \\ 11 & 3 & 2 & 10 \\ 2 & 2 & 9 & 12 \\ 7 & 5 & 5 & 5 \end{bmatrix}$

E. $\begin{bmatrix} 0 & 1 & 1 & 1 \\ 1 & 0 & 1 & 1 \\ 1 & 1 & 0 & 1 \\ 1 & 1 & 1 & 0 \end{bmatrix}$

8. **MC** After column reduction is performed to the matrix obtained in question 7, the resultant matrix is:

A. $\begin{bmatrix} 7 & 5 & 10 & 0 \\ 9 & 3 & 0 & 10 \\ 0 & 0 & 9 & 12 \\ 3 & 0 & 5 & 3 \end{bmatrix}$

B. $\begin{bmatrix} 7 & 3 & 8 & 1 \\ 9 & 1 & 0 & 5 \\ 0 & 0 & 7 & 7 \\ 5 & 3 & 3 & 0 \end{bmatrix}$

C. $\begin{bmatrix} 7 & 5 & 8 & 0 \\ 9 & 3 & 0 & 10 \\ 0 & 0 & 7 & 12 \\ 3 & 0 & 3 & 5 \end{bmatrix}$

D. $\begin{bmatrix} 11 & 6 & 14 & 0 \\ 9 & 3 & 0 & 6 \\ 0 & 0 & 9 & 8 \\ 8 & 5 & 10 & 4 \end{bmatrix}$

E. $\begin{bmatrix} 0 & 1 & 1 & 1 \\ 1 & 0 & 1 & 1 \\ 1 & 1 & 0 & 1 \\ 1 & 1 & 1 & 0 \end{bmatrix}$

9. **MC** For the graph of flights between cities, select the incorrect statement.

A. There are more flights leaving Melbourne and Brisbane than there are leaving Sydney and Perth.
B. Brisbane has the fewest departing flights.
C. All destinations have the same number of arriving flights.
D. Melbourne has twice as many departing flights as Brisbane.
E. Sydney has twice as many departing flights as Brisbane.

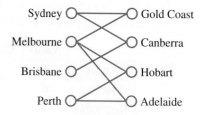

Short answer

10. Use the following flow table to produce a network flow diagram.

From	To	Flow quantity
A	B	13
A	C	6
B	C	10
B	D	4
C	D	3
C	E	14
D	F	10
E	F	15

11. a. Show the minimum cut on your diagram from question **10**.
    b. Hence, state the maximum flow.

12. Use the following table to answer the questions.

From	To	Flow quantity
A	B	13
A	C	6
A	G	16
B	C	10
B	D	4
B	G	2
C	D	3
C	E	14
D	F	10
E	F	15
G	D	3
G	H	10
H	F	13

a. Draw the network flow diagram for the table.
b. i. Calculate the maximum flow.
   ii. Show the minimum cut.

13. A gymnastics team of four — Annie, Beth, Consuela and Dianelle — has been selected. The team has been rated on four pieces of apparatus as follows:

	Vault	Uneven	Beam	Floor
Annie	5	7	5	9
Beth	6	10	10	7
Consuela	7	5	3	8
Dianelle	7	8	8	9

   a. Convert the table into a matrix for a minimisation problem.

   b. Perform a row reduction on the matrix.

   c. Perform a column reduction on the matrix produced in part **b.**

   d. **i.** Apply the Hungarian algorithm.

       ii. Hence, state the optimum team make-up.

Extended response

14. The following precedence table has been provided for the movement of stock within a store during a refit. The junior manager has been given the task of planning the operation so that the total time taken by the project is at a minimum and that staff are used most efficiently. Time is in hours.

Task	Time	Immediate predecessor
A	3	—
B	5	—
C	4	B
D	8	A, B
E	1	D, G
F	1	E
G	2	C
H	3	D
J	4	H
K	3	F, J

   a. Produce a network diagram from the information in the table.

   b. Determine, by forward scanning, the earliest completion time for the refit.

   c. Perform a backward scan and clearly show the critical path.

   d. Copy the table, adding columns for earliest start, earliest finish and float times. Complete the table.

   e. If the overall time of the project was to be reduced, determine which stages could be shortened. Explain your answer.

   f. If task D was reduced to 5 hours, determine the new critical path and the new minimum time for completion of the project.

15. The following network diagram was produced by a manager of a construction project. Unfortunately, they are not very competent and have forgotten to write the times for the activities on the diagram.
They have also lost most of the information from the original table that was used to produce the diagram. All the information that's left is in the table below on the next page.
They have decided that if they are to lose their job then you, as their junior, will go too, unless you can fill in the information missing from the diagram and the table.

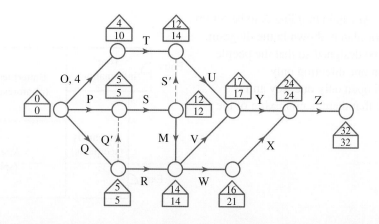

Activity	Immediate predecessor	Activity time	Earliest start time	Latest finish time	Float time
O	—				
P	—				1
Q	—				
R	Q				6
S	P, Q				
T	O				6
U	S, T				2
M	S				
V	R, M				
W	R, M				5
X	W				5
Y	U, V				
Z	X, Y				

a. Copy the diagram and indicate the (one) critical path.
   Copy the table.
b. First complete the missing float time column in the table.
c. Complete the earliest start time column in the table.
d. Complete the latest finish time column in the table.
e. Now, complete the time column for each activity and put these times into your copy of the diagram.

16. The network diagram represents the flow of salmon up the main tributaries of a river system during their spawning run, which leads from the estuary, E, to the lake, L.
    The figures represent the number of fish that pass along each section of the system every hour.

    a. Select the source and sink vertices.
    b. Determine the inflow and outflow for all other vertices.
    c. Determine if a pattern exists between the inflows and outflows. If so, give a possible reason for the pattern.
    d. Determine the flow capacity of the river network.
    e. Determine the maximum flow for the network.

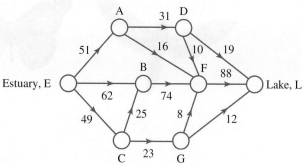

17. A Lepidoptera and Arachnid building is to be set up at the zoo. The floor plan is shown in the diagram. The building is to be designed so that the people can flow through in one direction only. Each doorway will open only one way and is designed to ensure that there is no mixing of the exhibits.

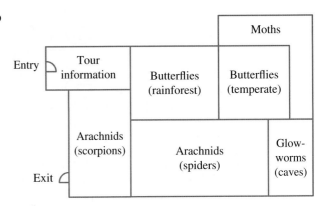

a. Draw the doors leading from one section to the next, clearly indicating in which direction they open. (The entry and exit doors have been completed for you.)
An analysis of the times spent in similar exhibits in zoos in other cities has provided the following table.

Section from	Section to	Arrival rate
Entry	Tour information	12
Tour information	Rainforest butterflies	13
Rainforest butterflies	Temperate butterflies	12
Temperate butterflies	Moths	2
Temperate butterflies	Glow-worms	4
Temperate butterflies	Arachnids	2
Moths	Glow-worms	4
Glow-worms	Spiders	6
Spiders	Scorpions	5
Scorpions	Exit	12

b. If the doors at the entry can be represented by an edge (A) with a capacity of 12, convert the information given in the chart and plan to a network flow diagram using the letters A to J.
c. Analyse the inflows, capacities and outflows and then describe what would happen to the number of people in the rainforest butterflies room.
d. i. Draw the minimum cut.
   ii. State at what rate people should be admitted so that they can flow smoothly through the building.

18. Imagine it is the year 2031, and the Australian Democrats have won the federal election with 71 members in the House of Representatives and 29 senators. The party has decided to do things differently when it comes to selecting its top four officials—namely prime minister, treasurer, foreign minister and speaker of the House of Representatives.

It will allow each of the 100 members of the caucus to 'elect' them according to the following rules:

i. Anyone can nominate for *all* positions.
ii. Members cast one vote for *each* position, but can pick a person for two or more positions.
iii. The four winners will be selected so that the *total* number of votes received for the four positions is a maximum.

There are five candidates in all — Alice Anderson, Boris Bologna, Cristina Colokis, Daniel Davis and Emily Eastwood — and the results of the poll are as shown in the table.

	Prime minister	Treasurer	Foreign minister	Speaker
A. Anderson	48	35	31	22
B. Bologna	1	10	5	4
C. Colokis	8	35	31	22
D. Davis	2	9	15	39
E. Eastwood	41	11	17	8

a. Since there are five people and only four positions, determine what should be done to the matrix so that an allocation algorithm can be applied.
b. Since the problem requires a maximum total, determine what should be done next so that the allocation algorithm can be applied.
c. Perform row reduction on your modified matrix.
d. Perform column reduction (if required) on your modified matrix.
e. Perform the Hungarian algorithm (if required) on your modified matrix.
f. Determine the optimal allocation.
g. Discuss the implications of the result in terms of candidates who got the most votes for a position (for example, D. Davis for speaker) actually getting the position.
h. If your answer to part g was negative, discuss why this might have happened.

10.6 Exam questions

Question 1 (1 mark)

Source: VCE 2021, Further Mathematics Exam 1, Section B, Module 2, Q8; © VCAA.

MC A network of roads connecting towns in an alpine region is shown below.

The distances between neighbouring towns, represented by the vertices, are given in kilometres.

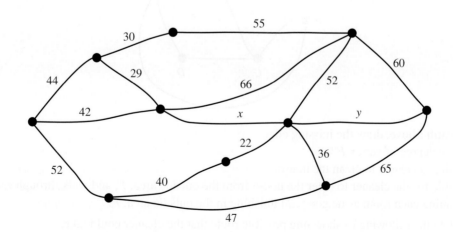

The region receives a large snowfall, leaving all roads between the towns closed to traffic.

To ensure each town is accessible by car from every other town, some roads will be cleared.

The minimum total length of road, in kilometres, that needs to be cleared is

A. 361 if $x = 50$ and $y = 55$

B. 361 if $x = 50$ and $y = 60$

C. 366 if $x = 55$ and $y = 55$

D. 366 if $x = 55$ and $y = 60$

E. 371 if $x = 55$ and $y = 65$

▶ **Question 2 (4 marks)**

Source: VCE 2021, Further Mathematics Exam 2, Section B, Module 2, Q1; © VCAA.

Maggie's house has five rooms, A, B, C, D and E, and eight doors.

The floor plan of these rooms and doors is shown below. The outside area, F, is shown shaded on the floor plan.

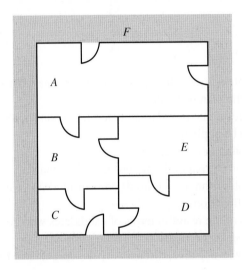

The floor plan is represented by the graph below.

On this graph, vertices represent the rooms and the outside area. Edges represent direct access to the rooms through the doors.

One edge is missing from the graph.

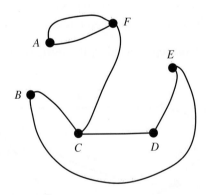

a. On the **graph above**, draw the missing edge. **(1 mark)**

b. What is the degree of vertex E? **(1 mark)**

c. Maggie hires a cleaner to clean the house.

It is possible for the cleaner to enter the house from the outside area, F, and walk through each room only once, cleaning each room as he goes and finishing in the outside area, F.

    i. Complete the following to show **one** possible route that the cleaner could take. **(1 mark)**

    ii. What is the mathematical term for such a journey? **(1 mark)**

Question 3 (2 marks)

Source: VCE 2020, Further Mathematics Exam 2, Section B, Module 2, Q4; © VCAA.

Training program 1 has the cricket team starting from exercise station S and running to exercise station O.

For safety reasons, the cricket coach has placed a restriction on the maximum number of people who can use the tracks in the fitness park.

The directed graph below shows the capacity of the tracks, in number of people per minute.

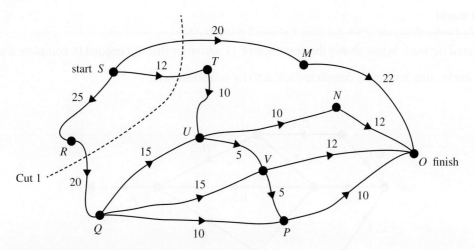

a. How many different routes from S to O are possible? **(1 mark)**

b. When considering the possible flow of people through this network, many different cuts can be made. Determine the capacity of Cut 1, shown above. **(1 mark)**

Question 4 (4 marks)

Source: VCE 2020, Further Mathematics Exam 2, Section B, Module 2, Q5; © VCAA.

The Sunny Coast cricket clubroom is undergoing a major works project.

This project involves nine activities: A to I.

The table below shows the earliest start time (EST) and duration, in months, for each activity.

The immediate predecessor(s) is also shown.

The duration for activity C is missing.

Activity	EST	Duration	Immediate predecessor(s)
A	0	2	–
B	0	5	–
C	5		A, B
D	7	7	C
E	7	9	C
F	5	3	B
G	14	4	D
H	8	9	F
I	18	2	E, G, H

The information in the table above can be used to complete a directed network.

This network will require a dummy activity.

a. Complete the following sentence by filling in the boxes provided.
This dummy activity could be drawn as a directed edge from the end of activity _____ to the start of
activity _____.
(1 mark)
b. What is the duration, in months, of activity C?
(1 mark)
c. Name the four activities that have a float time.
(1 mark)
d. The project is to be crashed by reducing the completion time of one activity only.
What is the minimum time, in months, that the project can be completed in?
(1 mark)

Question 5 (1 mark)

Source: VCE 2018, Further Mathematics Exam 1, Section B, Module 2, Q5; © VCAA.

MC The directed network below shows the sequence of 11 activities that are needed to complete a project.

The time, in weeks, that it takes to complete each activity is also shown.

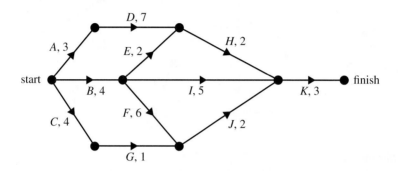

How many of these activities could be delayed without affecting the minimum completion time of the project?
- **A.** 3
- **B.** 4
- **C.** 5
- **D.** 6
- **E.** 7

More exam questions are available online.

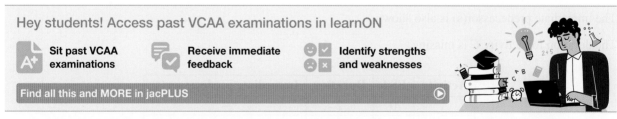

Hey students! Access past VCAA examinations in learnON

Sit past VCAA examinations

Receive immediate feedback

Identify strengths and weaknesses

Find all this and MORE in jacPLUS

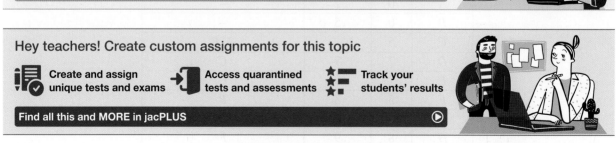

Hey teachers! Create custom assignments for this topic

Create and assign unique tests and exams

Access quarantined tests and assessments

Track your students' results

Find all this and MORE in jacPLUS

Answers

Topic 10 Directed graphs and network flow

10.2 Precedence tables and activity networks

10.2 Exercise

1 a.

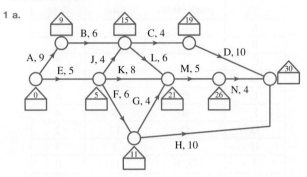

b. 30 minutes

c. 6

2. a. 147 days

b. 26 days

3.

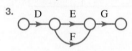

4.

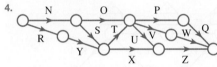

5. a. B

b. D

c. A–C–F

d.

Activity	Immediate predecessor	Activity time
A	—	7
B	—	9
C	A	12
D	B	8
E	B	4
F	C, D	9

6. a. 23 minutes   **b.** B, C, E, F, G

7. a. E

b. D

c. A

d.

Activity	Immediate predecessor	Activity time
A	—	3
B	—	4
C	—	5
D	A	6
E	B, F	5
F	C	8
G	D	18
H	E	8
J	E	6

e. i. A–D–G

ii. Float (H) = 1, Float (J) = 3, Float (E) = 1, Float (B) = 10, Float (C) = 1, Float (F) = 1

iii. Activity B can be delayed 10 minutes, activity C can be delayed 1 minute, activity E can be delayed 1 minute, activity F can be delayed 1 minute, activity H can be delayed 1 minute, activity J can be delayed 3 minutes.

8. a.

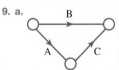

b. 61 minutes

c.

Activity	Immediate predecessor	Activity time
A	—	10
B	A	15
C	A	12
D	B	8
E	C	10
F	A	25
G	D, E, F	6
H	D, E, F	10
J	G	11
K	J	9
L	G	8
M	H, L	7

d. i. A–F–G–J–K   **ii.** M, L, H, C, E, B, D

9. a.

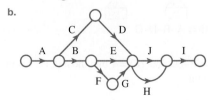

b.

10. a.

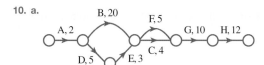

b. 49 minutes

10.2 Exam questions

Note: Mark allocations are available with the fully worked solutions online.
1. E

2. D

3. C

10.3 Critical path analysis with backward scanning and crashing

10.3 Exercise

1. Critical path = B–F–G; float (C) = 3 h, float (E) = 5 h, float (A) = 3 h, float (D) = 5 h

2. a. The critical path is D–F–G–H–J.

 b. Float (K) = 8 − 5.5 − 2 = 0.5 hours
 Float (C) = 5.5 − 3 − 1 = 1.5 hours
 Float (B) = 4.5 − 1 − 2 = 1.5 hours
 Float (A) = 2.5 − 0 − 1 = 1.5 hours
 Float (E) = 5.5 − 0 − 3.5 = 2 hours

3. a.

Activity	Activity	Immediate predecessor	Activity time
A	Collect parts	—	12
B	Paint frame	A	35
C	Assemble brakes	A	16
D	Assemble gears	B	20
E	Install brakes	C	12
F	Install seat	C	5
G	Final assembly	D, E	18

 b.

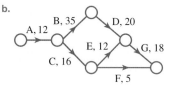

 c. 85 minutes

 d.

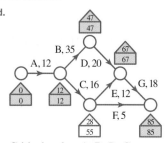

 Critical path = A–B–D–G

4. C

5. Critical path = B–D; float (E) = 1 min, float (C) = 1 min, float (A) = 1 min

6. a. 31 days

 b. Critical path = A–C–E–G

 c. The completion time is reduced to 28 days. The critical path is not affected.

 d. D

7. a, d

Activity	Immediate predecessor	Activity time	Float time
A	—	3	2
B	—	4	0
C	—	6	2
D	A	7	2
E	B	8	0
F	B	5	4
G	C	12	2
H	C	2	3
J	D, E	11	0
K	F	10	4
L	G, M	3	2
M	H	9	3
N	J, K, L	6	0

 b. 29

 c. B–E–J–N

8. a.

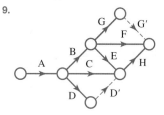

 b. 24 hours

 c. A–D–G–H–K

 d. Float (B) = 5, float (E) = 5, float (C) = 2, float (F) = 2, float (J) = 1

9.

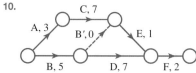

10.

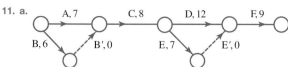

11. a.

A, 7   C, 8   D, 12   F, 9
B, 6   B′, 0   E, 7   E′, 0

 b. i. 36 minutes

  ii. A–C–D–F

  iii. Float (E) = 27 − 15 − 7 = 5 minutes and float (B) = 7 − 0 − 6 = 1 minutes

12.

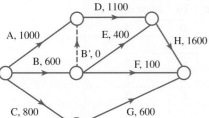

13. a.

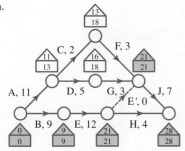

b. 28 h

c. B–E–E'–J

d. Float (A) = 2, float (C) = 5, float (F) = 5,
float (D) = 2, float (G) = 2, float (H) = 3

14.

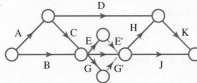

15. a. 35 days

b. C–F–J–M–Q

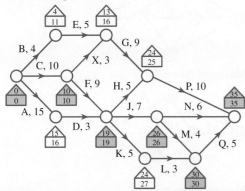

c. 3 days

d. When J is reduced to 5 days, the earliest completion
time is reduced to 34 days. The critical path becomes
C–F–H–P.

16. a.

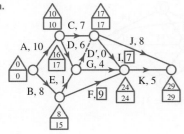

b. A–C–I–K

c. Float (F) = 7

17. a.

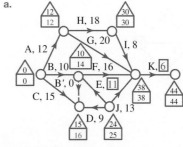

b. A–H–I–K

c. Float (F) = 12

18. a. C: A; F: C, D; EST G = 3; EST K = 14

b.

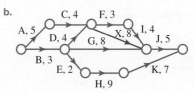

c. A–C–X–J

d. Latest start time for H = 6.

19. a. 5 weeks   **b.** 15 weeks

  c. A–E–F–I   **d.** 3 weeks

  e. A–E–F–I   **f.** C–D–I, 14 weeks

10.3 Exam questions

Note: Mark allocations are available with the fully worked
solutions online.

1. B

2. B

3. D

10.4 Flow problems

10.4 Exercise

1. a.

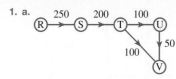

b.i. 150   **ii.** Yes

c.

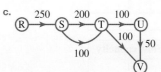

d. 150

2. a.

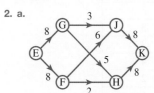

b.i. 15   **ii.** No

c.

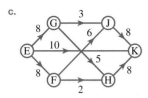

d. 25

3. a. i.

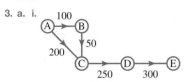

ii.

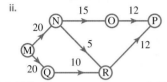

b. i. 250; no ****ii.** 24; yes

c. i.

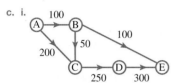

ii.

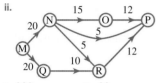

d. i. 300 ****ii.** 29

4. a. 23 **b.** 16 **c.** 16

5. a. 6 **b.** 3 **c.** 3

6. i. a.

From	To	Flow capacity
A	B	4
A	C	5
A	D	3
B	E	3
C	B	2
C	E	4
D	C	2
D	E	6

b.

From	To	Flow capacity
A	B	4
A	C	5
A	D	3
B	E	3
B	C	2
C	E	4
C	D	2
D	E	6

c.

From	To	Flow capacity
A	B	4
A	C	7
A	D	3
A	E	5
B	E	3
C	E	8
D	B	2
D	E	6

d.

From	To	Flow capacity
A	B	4
A	C	7
A	D	12
A	E	5
C	F	7
D	B	2
D	E	6
D	F	4
F	E	8

ii. a. 10 **b.** 10 **c.** 18 **d.** 22

7. Minimum cut = maximum flow
= 110

8. a. Cut 2

b. Cut 1 = 28; cut 3 = 28; cut 4 = 27

c. Maximum flow = 24

9. a.

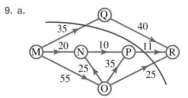

b. Maximum flow = 71

10. a.

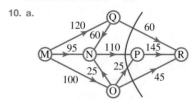

b. Maximum flow = 240

11. a. Cut 4

b. Cut 1 = 25; cut 2 = 26; cut 3 = 24; cut 5 = 42

c. Maximum flow = 20

12. a. i.

ii. Maximum flow = 31

b. i.

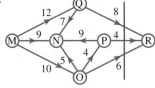

ii. Maximum flow = 18

13. a.

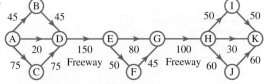

b. i. There would be a traffic jam.

ii. The traffic should flow smoothly as the inflow is less than the capacity of flows leading from H.

iii.

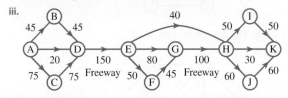

10.4 Exam questions

Note: Mark allocations are available with the fully worked solutions online.

1. C

2. C

3. B

10.5 Bipartite graphs and allocation problems

10.5 Exercise

1.

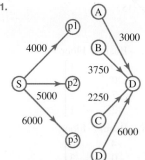

2. a.

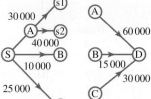

b. Send 30 000 from S1 to A, 30 000 from S2 to A, 10 000 from S2 to B, 5000 from B to B, 5000 from B to C and 25 000 from C to C. (This may not be the cheapest method.)

3. a.

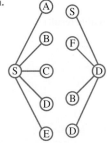

b. D

4.

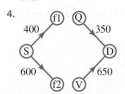

5. B

6. 7

7. 15

8. 24 — Hungarian algorithm

9. 36 — column reduction

10. a. 61

b.

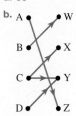

11. Sample responses can be found in the worked solutions in the online resources.

12. B

13. a.
$$\begin{bmatrix} 3 & 0 & 2 & 1 & 3 \\ 0 & 0 & 4 & 2 & 1 \\ 0 & 3 & 0 & 3 & 0 \\ 0 & 2 & 0 & 4 & 0 \\ 5 & 2 & 0 & 0 & 0 \end{bmatrix}$$

b.

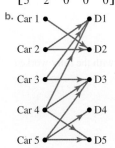

c. i. C1 → D2, C2 → D1, C5 → D4, C3 → D3, C4 → D5 or
C1 → D2, C2 → D1, C5 → D4, C4 → D3

ii. Total = $82 000

14. a. 67 — column reduction

b. 53 — row reduction

15. a.
$$\begin{bmatrix} 0 & 10 & 10 & 30 \\ 40 & 0 & 0 & 40 \\ 30 & 20 & 20 & 0 \\ 0 & 60 & 20 & 50 \end{bmatrix}$$

b.
$$\begin{bmatrix} 0 & 0 & 0 & 20 \\ 50 & 0 & 0 & 40 \\ 40 & 20 & 20 & 0 \\ 0 & 50 & 10 & 40 \end{bmatrix} \text{ (one possible result)}$$

c.

d. i. A → J2, B → J3, C → J4, D → J1 or A → J3, B → J2, C → J4, D → J1

   ii. Total = 130 minutes

16. a. T → J3, U → J1, V → J4, W → J2

   b. 175 minutes

17. a.

	A	C	F	G
K	18	0	11	41
L	45	10	20	0
M	26	51	16	0
N	30	6	18	0

b.

	A	C	F	G
K	0	0	0	41
L	27	10	9	0
M	8	51	5	0
N	12	6	7	0

c.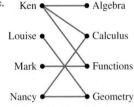

Ken → Algebra, Louise → Geometry,
Mark → Functions, Nancy → Calculus

d. 71.5%

18. a. V1 → P2, V2 → P4, V3 → P1, V4 → P3 or V1 → P3, V2 → P4, V3 → P1, V4 → P2

   b. Total = 48 km

10.5 Exam questions

Note: Mark allocations are available with the fully worked solutions online.

1. C

2.

Player	Batting position
Alex	3
Bo	1
Cameron	2

3. B

10.6 Review

10.6 Exercise

Multiple choice

1. a. C   **b.** A   **c.** C   **d.** D

2. A

3. D

4. E

5. E

6. A

7. C

8. A

9. D

Short answer

10.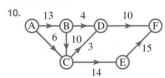

11. a.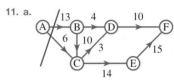

   b. The maximum flow is 19.

12. a.

(diagram)

   b. i. The maximum flow is 33.

   ii.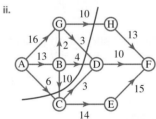

13. a.

	Vault	Uneven	Beam	Floor
Annie	5	3	5	1
Beth	4	0	0	3
Consuela	3	5	7	2
Dianelle	3	2	2	1

b.

	Vault	Uneven	Beam	Floor
Annie	4	2	4	0
Beth	4	0	0	3
Consuela	1	3	5	0
Dianelle	2	1	1	0

c.

	Vault	Uneven	Beam	Floor
Annie	3	2	4	0
Beth	3	0	0	3
Consuela	0	3	5	0
Dianelle	1	1	1	0

d. i.

	Vault	Uneven	Beam	Floor
Annie	2	1	3	0
Beth	3	0	0	4
Consuela	0	3	5	1
Dianelle	0	0	0	0

   ii. A → F, B → U, C → V, D → B or A → F, B → B, C → V, D → U

Extended response

14. a.

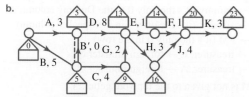

b.

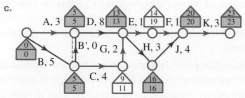

c.

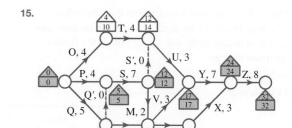

d. See the table at the bottom of the page.*

e. B, D, H, J, K

f. New critical path: B–C–G–H–J–K; new minimum completion time = 21 h

15.

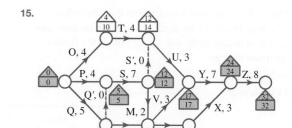

a. Critical path = Q–Q′–S–M–V–Y–Z

b, c, d, e. See the table at the bottom of the page.**

16. a. Source = estuary, E; sink = lake, L

b.

Vertex	Inflow	Outflow
A	51	47
B	87	74
C	49	48
D	31	29
F	108	88
G	23	20

*14 d.

Activity	T	EST	LST	Float time	Immediate predecessor
A	3	0	5	2	—
B	5	0	5	0	—
C	4	5	11	2	B
D	8	5	13	0	A, B
E	1	13	19	5	D, G
F	1	14	20	5	E
G	2	9	13	2	C
H	3	13	16	0	D, G
J	4	16	20	0	H
K	3	20	23	0	F, J

**15. b, c, d, e.

Activity	Immediate predecessor	T	EST	LST	Float time
O	—	4	0	10	6
P	—	4	0	5	1
Q	—	5	0	5	0
R	Q	3	5	14	6
S	P, Q	7	5	12	0
T	O	4	4	14	6
U	S, T	3	12	17	2
M	S	2	12	14	0
V	R, M	3	14	17	0
W	R, M	2	14	21	5
X	W	3	16	24	5
Y	U, V	7	17	24	0
Z	X, Y	8	24	32	0

c. The outflow for every vertex is less than the inflow. This may be due to the death of some fish on their way upstream or due to the fish slowing down the further upstream they go.

d. 162 fish/hour

e. 119 fish/hour

17. a.

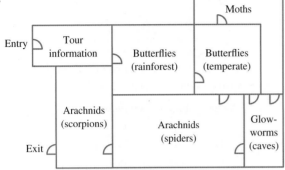

b.

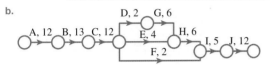

c. The room would quickly become crowded.

d. i.

D, 2   G, 6

A, 12   B, 13   C, 12   E, 4   H, 6   I, 5   J, 12

F, 2

ii. Five people per minute

18. a. Add a 'dummy' position, giving each candidate 0 votes for it.

b. Subtract each element from the largest in the matrix (48). The resultant matrix is:

	PM	Treasurer	FM	Speaker	Dummy
A	0	13	17	26	48
B	47	38	43	44	48
C	40	13	17	26	48
D	46	39	33	9	48
E	7	37	31	40	48

c.

	PM	Treasurer	FM	Speaker	Dummy
A	0	13	17	26	48
B	9	0	5	6	10
C	27	0	4	13	35
D	37	30	24	0	39
E	0	30	24	33	41

d.

	PM	Treasurer	FM	Speaker	Dummy
A	0	13	13	26	38
B	9	0	1	6	0
C	27	0	0	13	25
D	37	30	20	0	29
E	0	30	20	33	31

e.

	PM	Treasurer	FM	Speaker	Dummy
A	0	0	0	13	25
B	22	0	1	6	0
C	40	0	0	13	25
D	50	30	20	0	29
E	0	17	7	20	18

f. B → dummy (it's the only 0 in the Dummy column)
D → speaker (it's the only 0 in the Speaker column)
E → PM (it's the only 0 in the row)
A → treasurer, C → foreign minister or
C → treasurer, A → foreign minister

g. PM is not given to the top vote getter.
Because of the tie between A and C, either the treasurer or the foreign minister is not given to the top vote getter. D definitely deserved speaker and B definitely deserved dummy.

h. Although apparently democratic, it 'rewards' candidates who are relatively unpopular over the range of positions (particularly candidate E) who only get a lot of votes for 1 position, and 'punishes' those (particularly candidate A) who do get lots of votes across the positions. One could argue, in this case, that E got PM as she could do the 'least damage' in that job!

10.6 Exam questions

Note: Mark allocations are available with the fully worked solutions online.

1. B

2. a.

A — F
B
E
C — D

b. 2

c. i. $C − D − E − E − A$ or the other way
$A − B − E − D − C$

ii. Hamiltonian cycle

3. a. Ten different routes

b. 52

4. a. This dummy activity could be drawn as a directed edge from the end of activity B to the start of activity C.

b. Two months

c. A, F, H and E

d. Seventeen months

5. B

GLOSSARY

adjacency matrix a matrix that displays all of the possible one-stage pathways for a network. The sum of each row is equal to the outdegree of each vertex, and the sum of each column is equal to the indegree of each vertex

amortisation the process of paying off a loan by regular payments over a period of time

annuity an investment that has regular and constant payments over a period of time

annuity investment an investment that has regular deposits made over a period of time, resulting in the growth of the principal amount

association table a table that summarises how time periods are converted to numerical values, so that a regression line can be calculated for time series

associative a property whereby changing the placement of brackets in an expression does not alter the result; for example, $(A + B) + C = A + (B + C)$ shows that addition is associative

back-to-back stem plot a display used to compare two different sets of data. Back-to-back stem plots share the same stem, with one data set appearing on the left of the stem and the other data set appearing on the right of the stem

backward scanning a process of determining the critical path in a network. Backward scanning starts at the end node and moves backward through a network subtracting the time of each edge from the earliest start time of each succeeding node

bar chart a display of the categories of data on one axis (usually the horizontal axis) and the frequency of the data on the other axis (usually the vertical axis)

binary matrix a matrix that consists only of elements that are either 0 or 1

bipartite graph a graph where the nodes can be separated into two types of node — supply and demand

boxplot a display of numerical data that shows the five-number summary statistics of a data set

bridge an edge in a connected graph whose deletion will no longer cause the graph to be connected

categorical describes data that fit into non-numerical categories

causation the measure of how much the change in one variable is caused by another variable

circuit a trail beginning and ending at the same vertex

coefficient of determination this is the square of the product–moment correlation coefficient (r^2), and is a measure of how much a change in the response variable is influenced by a change in the explanatory variable

coincidence a correlation between two variables that does not represent a causal relationship

column matrix a matrix that has only one column

column permutation a matrix that has the same elements in each column as another matrix, but with the columns in a different order

communication matrix a matrix created from a communication network, in which a 1 represents a vertex that is able to communicate with another vertex

communication network a network that has the ability of the arrows in it to travel both ways, for example a set of people where they can have a one-way or two-way communication link

commutative a property whereby changing the order of terms in an expression does not alter the result; for example, $A + B = B + A$ shows that addition is commutative

complete graph a graph where there is an edge connecting each vertex to all other vertices in the graph, without parallel edges or loops

compounding period the time period in which interest is added to an investment or loan

conditional distribution a sub-population (sample) of a distribution, which is found in the middle of the table

confounding variable an outside influence that changes the relationship between a response variable and an explanatory variable

contingency table *see* two-way table

continuous describes numerical data that can take any value that lies within an interval

correlation a measure of the strength of the linear relationship between two variables

crashing the reduction in the duration time of an activity, which causes the critical path to be shortened

critical path the path through the network that follows those activities that cannot be delayed without causing the entire project to be delayed

cut in a network diagram, a line drawn through a number of edges that stops all flow from the source to the sink

cycle a path beginning and ending at the same vertex

cyclic time series data that show fluctuations upwards and downwards, but not according to season

degenerate graph a graph with no edges, also known as a null graph

degree the number of edges that are directly connected to a vertex

deseasonalising a technique involving calculating seasonal indices, which modifies the original time series in order to remove the effect of seasonal variation and exposing any other fluctuations that may be 'hidden' by seasonal variation

determinant in the 2×2 matrix $A = \begin{bmatrix} a & c \\ b & d \end{bmatrix}$ the determinant (det A or $|A|$) is the difference between the product of the diagonal elements: $ad - bc$. If $A = 0$, the inverse matrix will not exist

diagonal matrix a square matrix that only has non-zero elements on the main diagonal

Dijkstra's algorithm a method for determining the shortest path between a given vertex and each of the other vertices, often done in a tabular format

directed graph a graph where it is only possible to move along the edges of the graph in one direction

discrete describes numerical data that are counted in exact values, with the values often being whole numbers

dominance matrix a matrix that is the result of adding the matrices that represent the one-stage pathways (the adjacency matrix, A) and two-stage pathways (A^2). The dominance matrix consists of all the possible one- and two-stage pathways in the network

dominant vertex the vertex that holds the most influence over all the other vertices. It belongs to the row that has the highest sum in the dominance matrix

dot plot a display of discrete numerical data or categorical data. A scaled horizontal axis is used and each data value is indicated by a dot above this scale, resulting in a set of vertical 'lines' of evenly-spaced dots

dummy activity an edge that must be added to avoid a network with two or more activities having the same name or occurring in parallel

earliest finish time equal to the earliest start time (EST) of the activity plus its duration (T)

earliest start time the earliest time that any activity can be started after all prior activities have been completed

edge a connection between vertices

effective annual interest rate a rate used to compare the annual interest between loans with different compounding periods, such as daily, weekly or monthly: $r = \left(1 + \dfrac{i}{n}\right)^n - 1$

effective life the length of time that an asset is useful to a business

elements the entries in a matrix

equilibrium state or steady state the state reached in a Markov system when there is no change from one state matrix to the next; the fixed matrix T^∞

Euler circuit a circuit in which every edge is used once

Euler trail a trail in which every edge is used once

Euler's formula the mathematical relationship between the vertices, edges and faces of planar graphs:
$V = E - F + 2$

Excess flow capacity the surplus of the capacity of an edge less the flow into the edge

explanatory variable the variable represented on the x-axis and used in a bivariate analysis to explain or predict a difference in the response variable; also known as the independent variable

extrapolation making a prediction from a line of best fit that appears outside the parameters of the original data set

faces the distinct areas created by the non-intersecting edges and vertices of planar graphs

first-order recurrence relation a relation that relates a term in a sequence to the previous term in the same sequence

flat rate method a method of deprecation in which an item's value decreases by a fixed amount each unit time interval, generally each year. The depreciating value may be a dollar amount or a percentage of the original cost price

float time the difference in time between those paths that cannot be delayed and those that can

flow capacity the flow capacity of an edge is the amount of flow that an edge can allow through if it is not connected to any other edges

forecasting the process of predicting occurrences in the future based on the behaviour of existing data

forward scanning the process of looking through a network to calculate the earliest start time for each activity and the earliest completion time for the whole project

frequency table a display that tabulates data according to the frequencies of predetermined groupings

Hamiltonian cycle a cycle that reaches all vertices of a network

Hamiltonian path a path that reaches all vertices of a network

histogram a display of continuous numerical data similar to a bar chart where the width of each column represents a range of data values, and the height represents their frequencies

Hungarian algorithm a method for finding the optimal allocation, where the row reduction and column reduction will not produce the correct answer

identity matrix a square matrix in which all of the elements on the diagonal line from the top left to bottom right are 1 and all of the other elements are 0

immediate predecessor something that must occur before another thing can happen

indegree the number of edges moving into a vertex

inflow the inflow of a node is the total of the flows of all edges leading into the node

initial state the original, first state that the next state's probability is conditional on

initial state matrix a column vector with an entry for each state of the system; also known as a distribution vector, and denoted by S_0

initial value the first term in a sequence, from which the rest of the terms in a recurrence relation are generated

intercept the point at which a graph crosses one of the axes on a Cartesian plane

interest-only loan a loan where the borrower makes only the minimum repayment equal to the interest charged on the loan

interpolation making a prediction from a line of best fit that appears within the parameters of the original data set

interquartile range the difference between the lower and upper quartiles in a data set; $IQR = Q_3 - Q_1$

irregular time series data that show fluctuations that occur at random

isolated vertex a vertex that is not connected to any other vertex, and hence with degree 0

isomorphic describes networks or shapes that are equivalent to each other, for example two networks with the same edges and vertices

latest start time the difference in time between those paths that cannot be delayed and those that can

Leslie model a model used in ecology to describe changes in populations

linear relationship a relation between up to two variables of degree 1 that produces a straight line

logarithmic (log) form the logarithm of a number is the power of 10 that creates this number, i.e. $\log(10^n) = n$. This is used to make a logarithmic scale, a scale that uses logarithms instead of normal numbers to mark intervals. In base 10 each unit increase or decrease represents a tenfold increase or decrease in the quantity being measured

loop an edge in a graph that joins a vertex to itself

lower triangular matrix a type of square matrix in which all the elements above the main diagonal are zero

marginal distributions the sums (totals) of the rows or the columns in a two-way table that are found in the margins of the table

Markov system a system that illustrates the changing nature of an event between different states, according to fixed probabilities; also known as a Markov chain

matrix (plural *matrices*) a rectangular array of rows and columns that is used to store and display information

maximum spanning tree the spanning tree that covers the greatest distance for a network

mean commonly referred to as the average, the mean is calculated by dividing the sum of the data values by the number of data values. It is represented by either $\bar{x}$ (sample) or μ (population)

median the middle value of an ordered data set if there are an odd number of values, or halfway between the two middle values if there are an even number of values

minimum spanning tree the spanning tree that covers the greatest distance for a network

mode the value that occurs most often in a data set; the score with the highest frequency

multiplicative inverse a square matrix that when multiplied with another, results in the identity matrix. In other words, if $AA^{-1} = A^{-1}A = I$, then A^{-1} is called the multiplicative inverse of A

negatively skewed describes a distribution with higher frequencies on the right-hand side of the display

network flow an attribute that describes whether a network allows for the flow to be in both directions along a path, or in one direction only

network a system of interconnected objects

node a point on a network that may be connected to other nodes via edges

nominal describes categorical data that has no natural order or ranking

nominal interest rate the compound interest rate advertised by a financial institution, usually expressed as a percentage per annum. The nominal interest rate disregards the compounding period.

non-linear relationship when the relation between variables forms a shape or curve rather than a straight line

normal distribution a particular kind of distribution in which the data is symmetrical and has a bell shape centred on the mean

numerical describes data that can be counted or measured

one-stage pathway a pathway between two vertices that includes one edge only

optimal allocation allocation that gives the most efficient result (for example, minimising time or cost)

order the number of rows, m, and columns, n, of a matrix expressed as $m \times n$

ordinal describes categorical data that can be arranged into categories that have a natural order or ranking

outdegree the number of edges moving away from a vertex

outflow the outflow of a node is the minimum value obtained when one compares the inflow to the sum of the capacities of all the edges leaving the node

parallel boxplot a display where two or more boxplots share the same scale to enable comparisons between data sets

parallel dot plot a display where two or more dot plots share the same scale to enable comparisons between data sets

parallel edges where two (or more) edges connect the same pair of vertices. All the edges count towards the degree of the vertex

path a walk in which no vertices are repeated, except possibly the start and finish

Pearson's product–moment correlation coefficient a measure of the strength of a linear trend that is designated with the letter r and associated with a numerical value between -1 and $+1$. Values close to positive or negative '1' indicate strong linear trends; values close to zero indicate weak or no linear trends

permutation matrix an $n \times n$ matrix that is a row or column permutation of the identity matrix. Permutation matrices re-order the rows or columns of another matrix via matrix multiplication

perpetuity an annuity where a permanently invested sum of money provides regular payments which continue indefinitely

population all the units or individuals of interest; the total set of observations that could be made

positively skewed describes a distribution with higher frequencies on the left-hand side of the display

Prim's algorithm a set of logical steps that can be used to identify the minimum spanning tree for a weighted connected graph

quartiles these divide a set of data into quarters. The lower quartile (Q_1) is the median of the lower half of an ordered data set. The upper quartile (Q_3) is the median of the upper half of an ordered data set. The middle quartile (Q_2) is the median of the whole data set

range the difference between the highest and lowest values in a data set

reachability the concept of how it is possible to go from one vertex in a directed network to another

rectangular matrix a matrix in which the elements are arranged in a rectangle (as opposed to a square)

recurrence relation a relation that relates a term in a sequence to a combination of previous terms in the same sequence

reducing balance depreciation a method of depreciation, sometimes called 'diminishing value' depreciation, where the value of an asset is reduced by a fixed percentage of its previous value. This is an application of compound interest and the depreciated future value can be calculated using: $V_n = V_0 \left(1 - \dfrac{r}{100}\right)^n$

re-seasonalise the process of removing the effect of deseasonalising in order to produce more accurate predictions

residual plot where the residuals are plotted against the original x-values. This indicates whether or not a linear relationship exists between the two variables under investigation

response variable the variable represented on the y-axis in a bivariate analysis, influenced by the explanatory variable

row matrix a matrix that has only one row

row permutation a matrix that has the same elements in each row as another matrix, but with the rows in a different order

sample a limited number of individuals selected from the population with the purpose of representing characteristics of the whole population

scalar a scalar quantity can be any real number (as opposed to a matrix or matrix elements), such as negative or positive numbers, fractions or decimals

scalar multiplication each element of the matrix is multiplied by the same number, called a 'scalar'. A scalar quantity can be any real number, such as negative or positive numbers, fractions or decimals

scatterplot a visual display of bivariate data

seasonal indices a measure of by what factor a particular season is above or below the average of all seasons for the cycle

seasonal time series when data fluctuate during the year, as seasons change. The term *seasonal* is not specific to seasons of the year; it could also relate to other constant periods of highs and lows. The key feature is that the seasons occur at the same time each cycle

segmented bar chart a single bar that is divided up into segments. Each segment represents a particular group of the data. For univariate data, the segmented bar chart represents all the data being studied, and if comparing two categorical variables, multiple columns are used

sequence a related set of objects or events that follow each other in a particular order

shortest path the path from one vertex to another that covers the shortest distance possible

significant figures the number of digits that would occur in a if the number was expressed in scientific notation as $a \times 10^b$ or as $a \times 10^{-b}$

simple interest interest calculation based on the original amount borrowed or invested; also known as 'flat rate', as it is a constant amount: $I = \dfrac{V_0 rn}{100}$

singular matrix a matrix whose determinant is equal to 0, and hence does not have an inverse

sink a network's end node(s)

skewed describes a non-symmetric data distribution, i.e. when more data values occur on one side of the peak than on the other

slope indicates the change in the response variable as the explanatory variable increases by 1 unit

source a network's starting node(s)

spanning trees a tree that is a subgraph that includes all of the vertices of the original graph

square matrix a matrix that has the same number of rows and columns

standard deviation a measure of the spread of continuous numerical data around the mean

state matrix a matrix that represents a particular state in a Markov system after a particular number of transitions from the initial state

steady state *see* equilibrium state

stem plot an arrangement used for numerical data where each value is grouped according to its numerical place value (the 'stem') and then displayed horizontally as a single digit (the 'leaf')

subgraph all or part of a graph

symmetric when a data set displays a single peak and the data trail off both sides of this peak in roughly the same fashion

symmetrical matrix a matrix in which all its elements are symmetric with respect to the main diagonal, i.e. where a matrix is equivalent to its transpose. A symmetrical matrix must be square

time series data sets where the horizontal variable is time and, generally, where time goes up in even increments such as hours, days, weeks or years. Time series are used to see how a quantity varies with time

trail a walk in which no edges are repeated

transition matrix a matrix that summarises the probabilities of a Markov system

transpose The transpose of a matrix is found by switching its rows and columns.

transpose matrix *see* transpose

tree a simple connected graph with no circuits

trend a long-term increase or decrease in a time series

triangular matrix a type of square matrix in which either all the elements above or all the elements below the main diagonal are zero

two-stage pathway a pathway between two vertices that includes exactly two edges

two-way (or contingency) table a table in which the data are organised about two different categorical variables. The rows group one category and the columns group another category

unit cost method a method of depreciating an asset according to its use — the more it is used, the faster it will depreciate: $V_n = V_0 - nd$

upper triangular matrix a type of square matrix in which all the elements below the main diagonal are zero

variance a measure of the spread of a data set from the mean: $s^2 = \dfrac{\sum f(x - \bar{x})^2}{(\sum f) - 1}$ (sample variance)

vertex a point on a network that may be connected to other nodes via edges

walk any route taken through a network, including those that repeat edges and vertices

weighted graph a graph with values attached to the edges

zero matrix a square matrix that consists entirely of '0' elements

z-score also called the standardised score, it indicates the position of a certain score in relation to the mean. A z-score of 0 indicates the score is equal to the mean, a negative z-score indicates a score below the mean, and a positive z-score indicates a score above the mean

INDEX